CCSP Self-Study
CCSP SNRS Exam Certification Guide

Greg Bastien
Sara Nasseh, CCIE No. 5824
Christian Abera Degu

Cisco Press

800 East 96th Street
Indianapolis, Indiana 46240 USA

CCSP Self-Study: CCSP SNRS Exam Certification Guide

Greg Bastien

Sara Nasseh, CCIE No. 5824

Christian Abera Degu

Copyright © 2006 Cisco Systems, Inc.

Published by:
Cisco Press
800 East 96th Street
Indianapolis, IN 46240 USA

All rights reserved. No part of this book may be reproduced or transmitted in any form or by any means, electronic or mechanical, including photocopying, recording, or by any information storage and retrieval system, without written permission from the publisher, except for the inclusion of brief quotations in a review.

Printed in the United States of America 1 2 3 4 5 6 7 8 9 0

Library of Congress Cataloging-in-Publication Number: 2005922370

ISBN: 1587201534

First Printing December 2005

Warning and Disclaimer

This book is designed to provide information about selected topics for the Cisco CCSP SNRS exam. Every effort has been made to make this book as complete and as accurate as possible, but no warranty or fitness is implied.

The information is provided on an "as is" basis. The authors, Cisco Press, and Cisco Systems, Inc., shall have neither liability nor responsibility to any person or entity with respect to any loss or damages arising from the information contained in this book or from the use of the discs or programs that may accompany it.

The opinions expressed in this book belong to the author and are not necessarily those of Cisco Systems, Inc.

The Cisco Press self-study book series is as described, intended for self-study. It has not been designed for use in a classroom environment. Only Cisco Learning Partners displaying the following logos are authorized providers of Cisco curriculum. If you are using this book within the classroom of a training company that does not carry one of these logos, then you are not preparing with a Cisco trained and authorized provider. For information on Cisco Learning Partners please visit:www.cisco.com/go/authorizedtraining. To provide Cisco with any information about what you may believe is unauthorized use of Cisco trademarks or copyrighted training material, please visit: http://www.cisco.com/logo/infringement.html.

Feedback Information

At Cisco Press, our goal is to create in-depth technical books of the highest quality and value. Each book is crafted with care and precision, undergoing rigorous development that involves the unique expertise of members from the professional technical community.

Readers' feedback is a natural continuation of this process. If you have any comments regarding how we could improve the quality of this book or otherwise alter it to better suit your needs, you can contact us through e-mail at feedback@ciscopress.com. Please make sure to include the book title and ISBN in your message.

We greatly appreciate your assistance.

Corporate and Government Sales

Cisco Press offers excellent discounts on this book when ordered in quantity for bulk purchases or special sales.

For more information please contact: **U.S. Corporate and Government Sales** 1-800-382-3419 corpsales@pearsontechgroup.com

For sales outside the U.S. please contact: **International Sales** international@pearsoned.com

Trademark Acknowledgments

All terms mentioned in this book that are known to be trademarks or service marks have been appropriately capitalized. Cisco Press or Cisco Systems, Inc., cannot attest to the accuracy of this information. Use of a term in this book should not be regarded as affecting the validity of any trademark or service mark.

Publisher	John Wait
Editor-In-Chief	John Kane
Cisco Representative	Anthony Wolfenden
Cisco Press Program Manager	Jeff Brady
Executive Editor	Brett Bartow
Production Manager	Patrick Kanouse
Development Editor	Dan Young
Senior Project Editor	San Dee Phillips
Copy Editor and Indexer	Keith Cline
Technical Editors	Brian Done, David Lazarte, and Edward Storey
Book and Cover Designer	Louisa Adair
Composition	Interactive Composition Corporation

CISCO SYSTEMS

Corporate Headquarters
Cisco Systems, Inc.
170 West Tasman Drive
San Jose, CA 95134-1706
USA
www.cisco.com
Tel: 408 526-4000
    800 553-NETS (6387)
Fax: 408 526-4100

European Headquarters
Cisco Systems International BV
Haarlerbergpark
Haarlerbergweg 13-19
1101 CH Amsterdam
The Netherlands
www-europe.cisco.com
Tel: 31 0 20 357 1000
Fax: 31 0 20 357 1100

Americas Headquarters
Cisco Systems, Inc.
170 West Tasman Drive
San Jose, CA 95134-1706
USA
www.cisco.com
Tel: 408 526-7660
Fax: 408 527-0883

Asia Pacific Headquarters
Cisco Systems, Inc.
Capital Tower
168 Robinson Road
#22-01 to #29-01
Singapore 068912
www.cisco.com
Tel: +65 6317 7777
Fax: +65 6317 7799

Cisco Systems has more than 200 offices in the following countries and regions. Addresses, phone numbers, and fax numbers are listed on the Cisco.com Web site at www.cisco.com/go/offices.

Argentina • Australia • Austria • Belgium • Brazil • Bulgaria • Canada • Chile • China PRC • Colombia • Costa Rica • Croatia • Czech Republic Denmark • Dubai, UAE • Finland • France • Germany • Greece • Hong Kong SAR • Hungary • India • Indonesia • Ireland • Israel • Italy Japan • Korea • Luxembourg • Malaysia • Mexico • The Netherlands • New Zealand • Norway • Peru • Philippines • Poland • Portugal Puerto Rico • Romania • Russia • Saudi Arabia • Scotland • Singapore • Slovakia • Slovenia • South Africa • Spain • Sweden Switzerland • Taiwan • Thailand • Turkey • Ukraine • United Kingdom • United States • Venezuela • Vietnam • Zimbabwe

Copyright © 2003 Cisco Systems, Inc. All rights reserved. CCIP, CCSP, the Cisco Arrow logo, the Cisco *Powered* Network mark, the Cisco Systems Verified logo, Cisco Unity, Follow Me Browsing, FormShare, iQ Net Readiness Scorecard, Networking Academy, and ScriptShare are trademarks of Cisco Systems, Inc.; Changing the Way We Work, Live, Play, and Learn, The Fastest Way to Increase Your Internet Quotient, and iQuick Study are service marks of Cisco Systems, Inc.; and Aironet, ASIST, BPX, Catalyst, CCDA, CCDP, CCIE, CCNA, CCNP, Cisco, the Cisco Certified Internetwork Expert logo, Cisco IOS, the Cisco IOS logo, Cisco Press, Cisco Systems, Cisco Systems Capital, the Cisco Systems logo, Empowering the Internet Generation, Enterprise/Solver, EtherChannel, EtherSwitch, Fast Step, GigaStack, Internet Quotient, IOS, IP/TV, iQ Expertise, the iQ logo, LightStream, MGX, MICA, the Networkers logo, Network Registrar, *Packet*, PIX, Post-Routing, Pre-Routing, RateMUX, Registrar, SlideCast, SMARTnet, StrataView Plus, Stratm, SwitchProbe, TeleRouter, TransPath, and VCO are registered trademarks of Cisco Systems, Inc. and/or its affiliates in the U.S. and certain other countries.

All other trademarks mentioned in this document or Web site are the property of their respective owners. The use of the word partner does not imply a partnership relationship between Cisco and any other company. (0303R)

Printed in the USA

About the Authors

Greg Bastien, CCNP, CCSP, CISSP, is the chief technical officer of Virtue Technologies, Inc., and directs the actions of the engineering staff that supports several federal agencies. He holds a position as adjunct professor at Strayer University, teaching networking and network security courses. He completed his undergraduate and graduate degrees at Embry-Riddle Aeronautical University while on active duty as a helicopter flight instructor in the U.S. Army. Mr. Bastien lives with his wife and two sons in Potomac, Maryland.

Sara Nasseh, CCIE No. 5824, CISSP, is currently a senior consultant for Intercom Consulting and Federal Systems, working as a network architect and consultant for various federal agencies. She completed her Bachelor of Science degree in computer information systems at the University of Virginia College. She obtained her CCIE in April 2000 and has more than 11 years of experience in the data communications and network systems arena.

Christian Abera Degu, RHCE, CCSP, CCNP, CCDP, is a network architect with General Dynamics Network Systems supporting multiple civilian federal agencies. Mr. Degu completed his graduate degree in computer information systems. He resides in Alexandria, Virginia.

About the Technical Reviewers

Brian Done is a technical director for ManTech International Corporation. He has attained an M.B.A. degree with a major in information security (InfoSec), including numerous certifications such as the CCNP, CCDP, CCSP, NSA IAM, CHSP, CISM, ISSAP, ISSMP, and CISSP. In addition to his corporate duties, he is a principal InfoSec advisor providing support on diverse enterprise topics to the U.S. government. More information can be obtained at BrianDone.com or Leadership1st.org (his foundation).

David Lazarte is a CLD developer for Cisco security products. He was the developer for the SNRS v1.0 (SECUR 2.0) course and helped develop the exam. He is president and senior IT consultant for Amnetech Inc., (d/b/a American Network Technologies) and is also certified to teach career and technology education in the state of Texas. He is currently developing security courses for Cisco. He set up and taught the first local Cisco Networking Academy in the Houston Independent School District in Houston, Texas, in 1999. David taught and developed the curriculum for the Computer Maintenance Technology Magnet Program at HISD, including the local Cisco Networking Academy, and administered the local school network and databases. After leaving HISD, David provided Cisco design and support for large corporations and educational institutions, such as Houston Community College System (where he served as a member of the Technical Advisory Group for the Texas GigaPOP—Internet2), G.E. Power Systems, AIG, and Cardinal Health, providing network and security consulting, design, and implementations. He has more than 20 years of experience with computer systems and more than 15 years in networking. He has earned CCNP certification and a degree in electrical engineering technology from Del Mar College.

Edward Storey has more than 11 years of LAN/WAN/server infrastructure management, design, implementation, support, consulting, and sales engineering experience. He is a systems engineer for Cisco Systems, working with the Department of State. Ed holds both CCNP and CCIP certifications.

Dedications

This book is dedicated to the memory of my brother, Ali Reza Nasseh (November 27, 1970–February 22, 1998), for his courage, compassion, and infinite love of learning.

Acknowledgments

We sincerely appreciate the efforts of all those who helped to keep us focused throughout the process and our respective families (who put up with us during the process). We especially want to thank the editorial team of Brett Bartow, Dan Young, Andrew Cupp, San Dee Phillips, Patrick Kanouse, and Michelle Grandin. We also want to thank the technical reviewers for their hard work. Last but not least, we want to thank Alen and Roger for their assistance on the questions.

This Book Is Safari Enabled

The Safari® Enabled icon on the cover of your favorite technology book means the book is available through Safari Bookshelf. When you buy this book, you get free access to the online edition for 45 days.

Safari Bookshelf is an electronic reference library that lets you easily search thousands of technical books, find code samples, download chapters, and access technical information whenever and wherever you need it.

To gain 45-day Safari Enabled access to this book:

- Go to http://www.ciscopress.com/safarienabled
- Enter the ISBN of this book (shown on the back cover, above the bar code)
- Log in or Sign up (site membership is required to register your book)
- Enter the coupon code FMG3-KAKI-HF5J-NYII-LNFT

If you have difficulty registering on Safari Bookshelf or accessing the online edition, please e-mail customer-service@safaribooksonline.com.

Contents at a Glance

Foreword xxvi
Introduction xxvii

Part I	Overview of Network Security 3
Chapter 1	Network Security Essentials 5
Chapter 2	Defining and Detailing Attack Threats 25
Chapter 3	Defense in Depth 45

Part II	Managing Cisco Routers 59
Chapter 4	Basic Router Management 61
Chapter 5	Secure Router Administration 79

Part III	AAA 105
Chapter 6	Authentication 107
Chapter 7	Authentication, Authorization, and Accounting 125
Chapter 8	Configuring RADIUS and TACACS+ on Cisco IOS Software 149
Chapter 9	Cisco Secure Access Control Server 173
Chapter 10	Administration of Cisco Secure Access Control Server for Windows 199

Part IV	IOS Firewall Feature Set 215
Chapter 11	Securing Networks with Cisco Routers 217
Chapter 12	The Cisco IOS Firewall and Advanced Security Feature Set 241
Chapter 13	Cisco IOS Intrusion Prevention System 255
Chapter 14	Mitigating Layer 2 Attacks 279
Chapter 15	Context-Based Access Control 305
Chapter 16	Authentication Proxy and the Cisco IOS Firewall 329
Chapter 17	Identity-Based Networking Services 353
Chapter 18	Configuring 802.1x Port-Based Authentication 373

Part V   VPN   395

Chapter 19   Building a VPN Using IPsec   397

Chapter 20   Scaling a VPN Using IPsec with a Certificate Authority   433

Chapter 21   Troubleshooting the VPN Configuration on a Cisco Router   455

Chapter 22   Configuring Remote Access Using Easy VPN   479

Part VI   Enterprise Network Management   503

Chapter 23   Security Device Manager   505

Part VII   Scenarios   549

Chapter 24   Final Scenarios   551

Part VIII   Appendix   601

Appendix A   Answers to the "Do I Know This Already?" Quizzes and Q&A Sections   603

Index   642

Contents

Foreword xxvi

Introduction xxvii

Part I   Overview of Network Security   3

Chapter 1   Network Security Essentials   5

"Do I Know This Already?" Quiz   5

Foundation Topics   9

Defining Network Security   9

Balancing Business Needs with Network Security Requirements   9

Network Security Policies   9

Security Policy Goals   12

Security Guidelines   13

   Management Must Support the Policy   14

   The Policy Must Be Consistent   14

   The Policy Must Be Technically Feasible   15

   The Policy Should Not Be Written as a Technical Document   15

   The Policy Must Be Implemented Globally Throughout the Organization   15

   The Policy Must Clearly Define Roles and Responsibilities   16

   The Policy Must Be Flexible Enough to Respond to Changing Technologies and Organizational Goals   16

   The Policy Must Be Understandable   16

   The Policy Must Be Widely Distributed   17

   The Policy Must Specify Sanctions for Violations   17

   The Policy Must Include an Incident-Response Plan for Security Breaches   17

   Security Is an Ongoing Process   18

Network Security as a Process   18

Network Security as a Legal Issue   19

Foundation Summary   20

Network Security Policies   20

   Security Policy Goals   20

   Security Guidelines   21

Network Security as a Process   21

Q&A   23

Chapter 2   Defining and Detailing Attack Threats   25

"Do I Know This Already?" Quiz   25

Foundation Topics   29

Vulnerabilities   29

   Self-Imposed Network Vulnerabilities   29

      Lack of Effective Network Security Policy   30

      Network Configuration Weakness   31

      Technology Weakness   32

        Threats   34
        Intruder Motivations   34
            Lack of Understanding of Computers or Networks   35
            Intruding for Curiosity   35
            Intruding for Fun and Pride   35
            Intruding for Revenge   36
            Intruding for Profit   36
            Intruding for Political Purposes   36
        Types of Network Attacks   37
            Reconnaissance Attacks   37
            Access Attacks   38
            DoS Attacks   39
    Foundation Summary   40
        Vulnerabilities   40
            Self-Imposed Network Vulnerabilities   40
        Threats   41
        Intruder Motivations   41
        Types of Network Attacks   42
    Q&A   43

Chapter 3   **Defense in Depth**   45
    "Do I Know This Already?" Quiz   45
    Foundation Topics   49
        Overview of Defense in Depth   49
            Components Used for a Defense-in-Depth Strategy   50
            Physical Security   54
    Foundation Summary   55
    Q&A   57

Part II   **Managing Cisco Routers**   59

Chapter 4   **Basic Router Management**   61
    "Do I Know This Already?" Quiz   61
    Foundation Topics   65
        Router Configuration Modes   65
        Accessing the Cisco Router CLI   68
            Configuring CLI Access   70
        Cisco IOS Firewall Features   71
    Foundation Summary   73
        Router Configuration Modes   73
        Accessing the Cisco Router CLI   74
        Cisco IOS Firewall Features   74
    Q&A   77

Chapter 5   Secure Router Administration   79
　"Do I Know This Already?" Quiz   79
　Foundation Topics   84
　　Privilege Levels   84
　　Securing Console Access   84
　　Configuring the enable Password   85
　　　enable secret Command   86
　　service password-encryption Command   88
　　Configuring Multiple Privilege Levels   88
　　Warning Banners   90
　　Interactive Access   91
　　Securing vty Access   91
　　SSH Protocol   93
　　　Setting Up SSH on a Cisco IOS Router or Switch   94
　　　　Configuring a Router for SSHv2 Using a Host Name and Domain Name   94
　　　　Configuring a Router for SSHv2 Using RSA Key Pairs   95
　　　Secure Copy   96
　　Port Security for Ethernet Switches   97
　　　Configuring Port Security   98
　　AutoSecure   99
　Foundation Summary   102
　Q&A   103

Part III   AAA   105

Chapter 6   Authentication   107
　"Do I Know This Already?" Quiz   107
　Foundation Topics   111
　　Authentication   111
　　　Configuring Line Password Authentication   111
　　　Configuring Username Authentication   112
　　　Remote Security Servers   112
　　　　TACACS Overview   112
　　　　RADIUS Overview   114
　　　　Kerberos Overview   117
　　PAP, CHAP, and EAP Authentication   117
　　　PAP   118
　　　CHAP   118
　　　　MS-CHAP   119
　　　　MS-CHAP Version 2   119
　　　EAP   120
　Foundation Summary   122
　Q&A   123

Chapter 7	Authentication, Authorization, and Accounting 125
	"Do I Know This Already?" Quiz 125
	Foundation Topics 129
	AAA Overview 129
	Authentication 129
	Authorization 130
	Accounting 130
	Configuring AAA Services 130
	Configuring AAA Authentication 131
	Configuring Login Authentication Using AAA 132
	Enabling Password Protection at the Privileged Level 133
	Configuring PPP Authentication Using AAA 134
	Limiting the Number of Login Attempts 136
	Configuring AAA Authorization 136
	Configuring AAA Accounting 140
	Troubleshooting AAA 142
	Foundation Summary 146
	Q&A 147
Chapter 8	Configuring RADIUS and TACACS+ on Cisco IOS Software 149
	"Do I Know This Already?" Quiz 149
	Foundation Topics 153
	Configuring TACACS+ on Cisco IOS Software 153
	TACACS+ Authentication Example 155
	TACACS+ Authorization Example 156
	TACACS+ Accounting Example 157
	AAA TACACS+ Testing and Troubleshooting 157
	debug aaa authentication 157
	debug tacacs 158
	debug tacacs events 158
	Configuring RADIUS on Cisco IOS Software 160
	RADIUS Authentication Example 162
	RADIUS Authorization Example 163
	RADIUS Accounting Example 164
	RADIUS Configuration Testing and Troubleshooting 166
	Foundation Summary 169
	Q&A 171
Chapter 9	Cisco Secure Access Control Server 173
	"Do I Know This Already?" Quiz 173
	Foundation Topics 177
	Cisco Secure ACS for Windows 177
	Authentication 178
	Authorization 180
	Accounting 181

Administration   182
Replicating, Synchronizing, and Backing Up Databases   183
  Database Replication   183
  RDBMS Synchronization   186
  Database Backup   186
Cisco Secure ACS for Windows Architecture   187
  CSAdmin   188
  CSAuth   188
  CSDBSync   188
  CSLog   188
  CSMon   189
  CSTacacs and CSRadius   189
Authenticating Users   189
  Local Database   191
  Windows NT/2000 AD   192
  Generic LDAP User Database   193
  Token Server   193
Enabling User Changeable Passwords   194
Foundation Summary   195
Q&A   197

Chapter 10   Administration of Cisco Secure Access Control Server for Windows   199

"Do I Know This Already?" Quiz   199
Foundation Topics   203
Basic Deployment Factors for Cisco Secure ACS   203
  Hardware Requirements   203
  Operating System Requirements   204
  Browser Compatibility   204
  Performance Considerations   204
  AAA Clients   205
Installing Cisco Secure ACS for Microsoft Windows   205
  Cisco Secure ACS Deployment Sequence   208
Troubleshooting Cisco Secure ACS for Microsoft Windows   209
  Authentication Problems   210
  Troubleshooting Authorization Problems   210
  Administration Issues   210
Foundation Summary   212
Q&A   213

Part IV   IOS Firewall Feature Set   215

Chapter 11   Securing Networks with Cisco Routers   217

"Do I Know This Already?" Quiz   217
Foundation Topics   221
Defining ACLs   221

Determining When to Configure Access Lists 222
Types of IP ACLs 222
Standard IP ACLs 223
Extended IP ACLs 226
Reflexive ACLs 226
Time-Based ACLs 227
Certificate-Based ACLs 228
Configuring ACLs on a Router 228
Simple Network Management Protocol 230
Controlling Interactive Access Through a Browser 231
Disabling Directed Broadcasts 232
Routing Protocol Authentication 233
Defining Small Server Services 234
Disabling Finger Services 235
Disabling Network Time Protocol 235
Disabling Cisco Discovery Protocol 235
Foundation Summary 237
Q&A 239

Chapter 12 The Cisco IOS Firewall and Advanced Security Feature Set 241
"Do I Know This Already?" Quiz 241
Foundation Topics 245
Cisco IOS Firewall and Advanced Security Feature Set 245
Authentication Proxy 248
DoS Protection 248
Logging and Audit Trail 248
Port-to-Application Mapping 249
System-Defined Port Mapping 249
User-Defined Port Mapping 250
Host-Specific Port Mapping 251
URL Filtering 251
Foundation Summary 252
Q&A 253

Chapter 13 Cisco IOS Intrusion Prevention System 255
"Do I Know This Already?" Quiz 255
Foundation Topics 259
Cisco IOS IPS 259
Cisco IOS IPS Features 259
Cisco IOS IPS Functions 260
Cisco IOS IPS Restrictions 261
Memory Considerations 261
Unsupported Signatures 261
Unsupported CLI Features 262
Cisco IOS IPS Application 262

Cisco IOS IPS Configuration Tasks   262
Initializing the Cisco IOS IPS   263
Configuring the Notification Type   263
Configuring the Router Maximum Queue for Alarms   265
Defining the Protected Network   265
Working with Cisco IOS IPS Signatures and Rules   266
Loading IPS-Based Signatures   266
Creating and Applying IPS Rules   268
Verifying the Cisco IOS IPS Configuration   270
Cisco IOS IPS Deployment Strategies   273
Foundation Summary   275
Q&A   277

Chapter 14   Mitigating Layer 2 Attacks   279
"Do I Know This Already?" Quiz   279
Foundation Topics   283
Types of Attacks   283
CAM Table Overflow Attacks   283
Mitigating CAM Table Overflow Attacks   286
VLAN Hopping Attacks   288
Mitigating VLAN Hopping Attacks   289
STP Manipulation Attacks   289
Preventing STP Manipulation Attacks   290
MAC Address Spoofing—Man-in-the-Middle Attacks   291
Mitigating MAC Address Spoofing Attacks   292
Private VLAN Vulnerabilities   294
Defending Private VLANs   295
DHCP Starvation Attacks   296
Mitigating DHCP Starvation Attacks   296
IEEE 802.1x EAP Attacks   296
Mitigating IEEE 802.1x EAP Attacks   297
Factors Affecting Layer 2 Mitigation Techniques   298
Foundation Summary   301
Q&A   303

Chapter 15   Context-Based Access Control   305
"Do I Know This Already?" Quiz   305
Foundation Topics   309
Context-Based Access Control Features   309
Detecting and Protecting Against DoS Attacks   310
Generating Alerts and Audit Trails   310
How CBAC Works   311
UDP Sessions   313
ACL Entries   313
Handling Half-Open Sessions   314

CBAC Restrictions   314
Supported Protocols   314
   RTSP   315
   H.323   316
CPU and Performance Impact   317
   Throughput Performance   317
   Session Connection Improvements   317
   CPU Utilization Improvements   317
   Improvement Benefits   317
Configuring CBAC   318
   Selecting an Interface   318
   Configuring IP ACLs at the Interface   318
   Configuring Global Timeouts and Thresholds   319
   Port to Application Mapping   319
      PAM Configuration Task List   320
   Defining an Inspection Rule   320
      Configuring Generic TCP and UDP Inspection   322
      Configuring Java Inspection   322
   Applying the Inspection Rule to an Interface   323
Verifying and Debugging CBAC   323
   Debugging CBAC   323
      Generic debug Commands   324
      Transport-Level debug Commands   324
   Configuring CBAC Example   324
Foundation Summary   326
Q&A   327

Chapter 16   Authentication Proxy and the Cisco IOS Firewall   329
"Do I Know This Already?" Quiz   329
Foundation Topics   333
Understanding Authentication Proxy   333
   How Authentication Proxy Works   333
   What Authentication Proxy Looks Like   335
Authentication Proxy and the Cisco IOS Firewall   336
Configuring Authentication Proxy on the Cisco IOS Firewall   336
   Authentication Proxy Configuration Steps   337
      Step 1: Configure AAA   338
      Step 2: Configure the HTTP Server   339
      Step 3: Configure the Authentication Proxy   339
      Step 4: Verify the Authentication Proxy Configuration   339
   Authentication Proxy Configuration Examples   339
Using Authentication Proxy with TACACS+   344
   Step 1: Complete the Network Configuration   344
   Step 2: Complete the Interface Configuration   345
   Step 3: Complete the Group Setup   346

Using Authentication Proxy with RADIUS 347
Limitations of Authentication Proxy 349
Foundation Summary 350
Q&A 351

Chapter 17 **Identity-Based Networking Services** 353
"Do I Know This Already?" Quiz 353
Foundation Topics 357
IBNS Overview 357
IEEE 802.1x 357
802.1x Components 359
How 802.1x Works 359
Port State 362
Selecting EAP 362
EAP-MD5 363
Cisco Lightweight EAP 363
EAP Transport Layer Security 364
Protected EAP 365
EAP Flexible Authentication via Secure Tunneling 366
EAP Methods Comparison 368
Cisco Secure ACS 369
Foundation Summary 370
Q&A 371

Chapter 18 **Configuring 802.1x Port-Based Authentication** 373
"Do I Know This Already?" Quiz 373
Foundation Topics 377
802.1x Port-Based Authentication Configuration Tasks 378
802.1x Mandatory Configuration 379
Enabling 802.1x Authentication 379
Configuring the Switch-to-RADIUS Server Communication 381
802.1x Optional Configurations 382
Enabling Periodic Re-Authentication 382
Manually Re-Authenticating a Client Connected to a Port 383
Changing the Quiet Period 383
Changing the Switch-to-Client Retransmission Time 384
Setting the Switch-to-Client Frame-Retransmission Number 384
Enabling Multiple Hosts 385
Configuring a Guest VLAN 385
Resetting the 802.1X Configuration to the Default Values 386
Displaying 802.1x Statistics and Status 387
Foundation Summary 392
Q&A 393

Part V VPN 395

Chapter 19   Building a VPN Using IPsec 397

"Do I Know This Already?" Quiz 397

Foundation Topics 401

Configuring a Cisco Router for IPsec Using Preshared Keys 403

How IPsec Works 403

Step 1: Select the IKE and IPsec Parameters 404

Define the IKE (Phase 1) Policy 405

Define the IPsec Policies 408

Verify the Current Router Configuration 411

Verify Connectivity 411

Ensure Compatible Access Lists 412

Step 2: Configure IKE 412

Enable IKE 413

Create the IKE Policy 413

Configure Preshared Key 413

Verify the IKE Configuration 414

Step 3: Configure IPsec 414

Create the IPsec Transform Set 415

Configure IPsec SA Lifetimes 416

Create the Crypto ACLs 417

Create the Crypto Map 418

Apply the Crypto Map to the Correct Interface 419

Step 4: Test and Verify the IPsec Configuration 419

Configuring Manual IPsec 423

Configuring IPsec Using RSA-Encrypted Nonces 423

Configure the RSA Keys 424

Plan the Implementation Using RSA Keys 424

Configure the Router Host Name and Domain Name 424

Generate the RSA Keys 425

Verifying the RSA Keys 425

Enter Your Peer RSA Public Keys 426

Verify the Key Configuration 427

Manage the RSA Keys 427

Foundation Summary 428

Configure a Cisco Router for IPsec Using Preshared Keys 428

Verifying the IKE and IPsec Configuration 429

Explain the Issues Regarding Configuring IPsec Manually and Using RSA-Encrypted Nonces 430

Q&A 431

Chapter 20   Scaling a VPN Using IPsec with a Certificate Authority 433

"Do I Know This Already?" Quiz 433

Foundation Topics 437

Advanced IPsec VPNs Using Cisco Routers and CAs 437

Digital Signatures, Certificates, and Certificate Authorities 437
Overview of Cisco Router CA Support 438
SCEP 440
Configuring the Cisco Router for IPsec VPNs Using CA Support 440
   Step 1: Select the IKE and IPsec Parameters 441
   Step 2: Configure the Router CA Support 441
   Step 3: Configure IKE Using RSA Signatures 448
   Step 4: Configure IPsec 449
   Step 5: Test and Verify the Configuration 450

Foundation Summary 451
   Advanced IPsec VPNs Using Cisco Routers and CAs 451

Q&A 453

Chapter 21 Troubleshooting the VPN Configuration on a Cisco Router 455
"Do I Know This Already?" Quiz 455

Foundation Topics 459

show Commands 460
   show crypto ca certificates Command 460
   show crypto isakmp policy Command 461
   show crypto ipsec sa Command 462
   show crypto ipsec security-association lifetime Command 463
   show crypto ipsec transform-set Command 464
   show crypto isakmp key Command 464
   show crypto map Command (IPsec) 464
   show crypto key pubkey-chain rsa Command 465
   show crypto key mypubkey rsa Command 466

debug Commands 467
   debug crypto isakmp Command 467
   debug crypto key-exchange Command 467
   debug crypto engine Command 468
   debug crypto ipsec Command 468
   debug crypto pki messages Command 470
   debug crypto pki transactions Command 471

clear Commands 472
   clear crypto sa Command 473
   clear crypto isakmp Command 473
   clear crypto sa counters Command 474

Foundation Summary 475

Q&A 477

Chapter 22 Configuring Remote Access Using Easy VPN 479
"Do I Know This Already?" Quiz 479

Foundation Topics 483

Describe the Easy VPN Server 483
Describe the Easy VPN Remote 484

Easy VPN Server Functionality   486
  How Cisco Easy VPN Works?   488
  Configuring the Easy VPN Server   489
    Create IP Address Pool   490
    Prepare the Router for Easy VPN Server   491
    Configure the Group Policy Lookup   491
    Create the ISAKMP Policy for the Remote VPN Clients   492
    Define a Group Policy for a Mode Configuration Push   492
    Create the Transform Set   493
    Create the Dynamic Crypto Maps with RRI   494
    Apply the Mode Configuration to the Dynamic Crypto Map   494
    Apply the Dynamic Crypto Map to the Interface   495
    Enable IKE DPD   495
    Configure Xauth   496
  Easy VPN Modes of Operation   496
Foundation Summary   497
  Describe the Easy VPN Server   497
    Easy VPN Server Functionality   497
    Configuring the Easy VPN Server   497
    Easy VPN Modes of Operation   500
  Q&A   501

Part VI Enterprise Network Management   503

Chapter 23 Security Device Manager   505
"Do I Know This Already?" Quiz   505
Foundation Topics   509
Security Device Manager Overview   509
  Hardware Requirements   509
  Operating System Requirements   509
  Browser Compatibility   510
Installing SDM Software   510
SDM User Interface   511
SDM Wizards   513
  SDM LAN Wizard   513
  Using SDM to Configure a Firewall   515
    Selecting the Type of Firewall Configuration   515
    Configuring the Firewall Interfaces   516
    Applying the Firewall Configuration to the Router   518
    Advanced Firewall Configuration Options   519
  Using SDM to Configure a VPN   521
    Site-to-Site VPN   521
    Easy VPN Remote   525
    Easy VPN Server   528
    Dynamic Multipoint VPN   530

        Using SDM to Perform Security Audits   534
            Security Audit Wizard   534
            One-Step Lockdown   536
        Using the Factory Reset Wizard   536
    Using SDM Advanced Options   538
    Using SDM Monitor Mode   541
  Foundation Summary   545
  Q&A   547

Part VII Scenarios   549

Chapter 24   Final Scenarios   551
    Task 1—Configure Cisco Secure ACS for AAA on Miami Network Devices   556
    Task 2—Configure and Secure Miami Router   558
    Task 3—Configure 802.1x on Miami User Switches   561
    Task 4—Configure Miami User Switches and Router to Mitigate Layer 2 Attacks   561
    Task 5—Configure PEAP with Cisco Secure ACS   562
    Task 6—Prepare the Network for IPsec Using Preshared Keys   570
        Establish a Common Convention for Connectivity Between Locations   570
        Configure Initial Setup of the Router and Verify Connectivity   570
        Prepare for IKE and IPsec   571
        Define the Preshared Key   572
    Task 7—Configure IKE Using Preshared Keys   572
        Enable IKE   572
        Create the IKE Policy   572
        Configure the Preshared Key   573
        Verify the IKE Configuration   573
    Task 8—Configure IPsec Using Preshared Keys   573
        Configure Transform Sets and SA Parameters   574
        Configure IPsec SA Lifetimes   574
        Configure Crypto ACLs   574
        Configure Crypto Maps   574
        Apply the Crypto Map to the Interface   575
    Task 9—Configure IKE and IPsec on a Cisco Router   575
        Enable IKE   575
        Create an IKE Policy Using RSA Signatures   575
        Configure Transform Sets and SA Parameters   576
        Configure IPsec SA Lifetimes   576
        Configure Crypto ACLs   576
        Configure Crypto Maps   577
        Apply the Crypto Map to the Interface   577
    Task 10—Prepare the Network for IPsec Using Digital Certificates   577
        Configure Initial Setup of the Router and Verify Connectivity   577
        Prepare for IKE and IPsec   577
        Configure CA Support   578

Task 11—Test and Verify IPsec CA Configuration 579
    Display IKE Policies 579
    Display Transform Sets 580
    Display Configured crypto maps 580
    Display the Current State of IPsec SAs 580
    Clear Any Existing SAs 581
    Enable Debug Output for IPsec Events 581
    Enable Debug Output for ISAKMP Events 581
    Observe the IKE and IPsec Debug Outputs 581
    Verify IKE and IPsec SAs 583

Task 12—Configure Authentication Proxy on the Miami Router 584
    Configure AAA 584
    Configure the HTTP Server 585
    Configure Authentication Proxy 585
    Test and Verify the Authentication Proxy Configuration 585

Task 13—Configure CBAC on the Miami Router 585

Task 14—Configure Miami Router with IPS Using SDM 586

Task 15—Verify and Monitor Miami Router with IPS Using SDM 590

Task 16—Configure Easy VPN Server Using SDM 593

Task 17—Configure Easy VPN Remote Using SDM 597

Part VIII Appendix 601

Appendix A Answers to the "Do I Know This Already?" Quizzes and Q&A Sections 603

Index 642

Command Syntax Conventions

The conventions used to present command syntax in this book are the same conventions used in the IOS Command Reference. The Command Reference describes these conventions as follows:

- **Boldface** indicates commands and keywords that are entered literally as shown.
- *Italics* indicate arguments for which you supply actual values.
- Vertical bars (|) separate alternative, mutually exclusive elements.
- Square brackets [ ] indicate optional elements.
- Braces { } indicate a required choice.
- Braces within brackets [{ }] indicate a required choice within an optional element.

Foreword

CCSP SNRS Exam Certification Guide is an excellent self-study resource for the CCSP SNRS exam. Passing the exam validates the knowledge and ability to secure networks using Cisco routers and switches. It is one of several exams required to attain the CCSP certification.

Cisco Press Exam Certification Guide titles are designed to help educate, develop, and grow the community of Cisco networking professionals. The guides are filled with helpful features that allow you to master key concepts and assess your readiness for the certification exam. Developed in conjunction with the Cisco certifications team, Cisco Press books are the only self-study books authorized by Cisco Systems.

Most networking professionals use a variety of learning methods to gain necessary skills. Cisco Press self-study titles are a prime source of content for some individuals, and can also serve as an excellent supplement to other forms of learning. Training classes, whether delivered in a classroom or on the Internet, are a great way to quickly acquire new understanding. Hands-on practice is essential for anyone seeking to build, or hone, new skills. Authorized Cisco training classes, labs, and simulations are available exclusively from Cisco Learning Solutions Partners worldwide. Please visit www.cisco.com/go/training to learn more about Cisco Learning Solutions Partners.

I hope and expect that you'll find this guide to be an essential part of your exam preparation and a valuable addition to your personal library.

Don Field
Director, Certifications
Cisco Systems, Inc.
October 2005

Introduction

This book is designed to help you prepare for the Cisco SNRS certification exam. The SNRS exam is one in a series of exams required for the Cisco Certified Security Professional (CCSP) certification. This exam focuses on the application of security principles with regard to Cisco IOS routers, switches, and virtual private network (VPN) devices.

This book focuses on Cisco technology and software updates since the publication of *CCSP SECUR Exam Certification Guide*. These technologies include mitigating Layer 2 attacks, intrusion prevention systems, Security Device Manager, and 802.1x authentication and configuration, which are covered as new chapters. Other features, such as Secure Shell v2 and Cisco Secure Access Control Server version 3.3 are covered, too.

Who Should Read This Book

Network security is a complex business. It is important that you have extensive experience in and an in-depth understanding of computer networking before you can begin to apply security principles. The Cisco SNRS program was developed to introduce the security products associated with or integrated into Cisco IOS Software, explain how each product is applied, and explain how it can increase the security of your network. The SNRS program is for network administrators, network security administrators, network architects, and experienced networking professionals who are interested in applying security principles to their networks.

How to Use This Book

The book consists of 24 chapters. Each chapter tends to build upon the chapter that precedes it. The chapters that cover specific commands and configurations include case studies or practice configurations.

The chapters of the book cover the following topics:

- **Chapter 1, "Network Security Essentials"**—Chapter 1 is an overview of network security in general terms. This chapter defines the scope of network security and discusses the delicate "balancing act" required to ensure that you fulfill the business need without compromising the security of the organization. Network security is a continuous process that should be driven by a predefined organizational security policy.

- **Chapter 2, "Defining and Detailing Attack Threats"**—Chapter 2 discusses the potential network vulnerabilities and attacks that pose a threat to the network. This chapter provides you with a better understanding of the need for an effective network security policy.

- **Chapter 3, "Defense in Depth"**—Until recently, a network was considered secure if it had a strong perimeter defense. Network attacks are becoming much more dynamic and require a security posture that provides defense at many levels. Chapter 3 discusses the concepts that integrate all the security components into a single, effective security strategy.
- **Chapter 4, "Basic Router Management"**—This chapter details the administration of the Cisco IOS router and discusses the Cisco IOS Firewall feature set. This chapter focuses on the basic tasks that are required to manage an individual Cisco IOS router.
- **Chapter 5, "Secure Router Administration"**—This chapter explains how to secure the administrative access to the Cisco IOS router. It is important to secure this access to prevent unauthorized changes to the router.
- **Chapter 6, "Authentication"**— This chapter discusses the many different types of authentication and the advantages and disadvantages of each type.
- **Chapter 7, "Authentication, Authorization, and Accounting"**—AAA has become a key component of any security policy. AAA is used to verify which users are connecting to a specific resource, ensure that they are authorized to perform requested functions, and track which actions were performed, by whom, and at what time. Chapter 7 discusses the integration of AAA services into a Cisco IOS environment and how AAA can significantly impact the security posture of a network.
- **Chapter 8, "Configuring RADIUS and TACACS+ on Cisco IOS Software"**—TACACS+ and RADIUS are two key AAA technologies supported by Cisco IOS Software. Chapter 8 discusses the steps for configuring TACACS+ and RADIUS to communicate with Cisco IOS routers.
- **Chapter 9, "Cisco Secure Access Control Server"**—This chapter describes the features and architectural components of the Cisco Secure Access Control Server.
- **Chapter 10, "Administration of Cisco Secure Access Control Server for Windows"**—This chapter discusses the installation and configuration of the Cisco Secure Access Control Server on a Microsoft Windows 2000 Server.
- **Chapter 11, "Securing Networks with Cisco Routers"**—It is important to restrict access to your Cisco IOS router to ensure that only authorized administrators are performing configuration changes. There are many different ways to access the Cisco IOS router. Chapter 11 describes how to ensure that all nonessential services have been disabled to reduce any chances of accessing the router by exploiting open ports or running services.
- **Chapter 12, "The Cisco IOS Firewall and Advanced Security Feature Set"**—The Cisco IOS Firewall and Advanced Security feature set is a set of additional features available for Cisco IOS that provides security functionality on a router platform. These features are discussed in Chapter 12.

- **Chapter 13, "Cisco IOS Intrusion Prevention System"**—The Cisco IOS Intrusion Prevention System (IPS) feature set is the evolution of the Cisco IOS Intrusion Detection System (IDS). Cisco IPS products go beyond the IDS signature matching by incorporating features such as stateful pattern recognition, protocol analysis, traffic anomaly detection, and protocol anomaly detection. This chapter discusses the security features of the Cisco IOS IPS.

- **Chapter 14, "Mitigating Layer 2 Attacks"**—As the popularity of Ethernet switching and wireless local-area networks (WLANs) grow, the emphasis on Layer 2 security has become more important. This chapter discusses Layer 2 attacks, mitigations, and best practices and functionality.

- **Chapter 15, "Context-Based Access Control"**—CBAC is a Cisco IOS security feature that enables you to filter data based on an inspection of the data packet. This is incorporated as part of the Cisco IOS Security feature set and is used to greatly increase the security of the network perimeter.

- **Chapter 16, "Authentication Proxy and the Cisco IOS Firewall"**—Authentication proxy is a function that enables users to authenticate when accessing specific resources. The Cisco IOS Firewall is designed to interface with AAA servers using standard authentication protocols to perform this function. This functionality enables administrators to create a granular and dynamic per-user security policy.

- **Chapter 17, "Identity-Based Networking Services"**—Cisco Identity-Based Networking Services (IBNS) is a technology framework for delivering logical and physical network access authentication. IBNS combines several Cisco products that offer authentication, user policies, and access control to provide a comprehensive solution for increasing network access security. IBNS incorporates capabilities defined in the IEEE 802.1x standard. This chapter discusses IBNS and 802.1x features and functionality.

- **Chapter 18, "Configuring 802.1x Port-Based Authentication"**—This chapter describes how to configure 802.1x port-based authentication on a Catalyst switch to prevent unauthorized clients (supplicants) from gaining access to the network.

- **Chapter 19, "Building a VPN Using IPsec"**—Prior to the creation of VPN technology, the only way to secure communications between two locations was to purchase a "dedicated circuit." To secure communications across an enterprise would be tremendously expensive, and securing communications with remote users was simply cost-prohibitive. VPN technology enables you to secure communications that travel across the public infrastructure (that is, the Internet). VPN technology allows organizations to interconnect their different locations without having to purchase dedicated lines, greatly reducing the cost of the network infrastructure.

- **Chapter 20, "Scaling a VPN Using IPsec with a Certificate Authority"**—Cisco IOS devices are designed with a feature called CA interoperability support, which allows them to interact with a certificate authority (CA) when deploying IPsec. This functionality allows for a scalable and manageable enterprise VPN solution.

- **Chapter 21, "Troubleshooting the VPN Configuration on a Cisco Router"**—This chapter describes the numerous commands used to troubleshoot the configuration of VPNs using Cisco IOS and IPsec.
- **Chapter 22, "Configuring Remote Access Using Easy VPN"**—Cisco Easy VPN is a client/server application that allows for VPN security parameters to be "pushed out" to the remote locations that connect using a growing array of Cisco products.
- **Chapter 23, "Security Device Manager"**—The Cisco Security Device Manager (SDM) is a Java-based web management tool used for configuration and monitoring of Cisco IOS Software based routers. Cisco SDM is supported on a wide range of Cisco routers and Cisco IOS Software releases. This chapter provides general installation and configuration guidance for SDM.
- **Chapter 24, "Final Scenarios"**— This chapter provides a practical overview of topics discussed throughout the book. It consists of a scenario for an organization that requires your expertise with Cisco products to meet its constantly evolving business needs.
- **Appendix, "Answers to the "Do I Know This Already?" Quizzes and Q&A Sections"**

Each chapter follows the same format and incorporates the following tools to assist you by assessing your current knowledge and emphasizing specific areas of interest within the chapter:

- **Do I Already Know This Quiz?**—Each chapter begins with a quiz to help you assess your current knowledge of the subject. The quiz is divided into specific areas of emphasis that enable you to best determine where to focus your efforts when working through the chapter.
- **Foundation Topics**—The foundation topics are the core sections of each chapter. They focus on the specific protocols, concepts, or skills that you must master to successfully prepare for the examination.
- **Foundation Summary**—Near the end of each chapter, the foundation topics are summarized into important highlights from the chapter. In many cases, the foundation summaries are divided into charts, but in some cases, the important portions from each chapter are just restated to emphasize their importance within the subject matter. Remember that the foundation portions are in the book to assist you with your exam preparation. It is unlikely that you will be able to successfully complete the certification exam by just studying the foundation topics and foundation summaries, although they are a good tool for last-minute preparation just before taking the exam.
- **Q&A**—Each chapter ends with a series of review questions to test your understanding of the material covered. These questions are a great way to ensure that you not only understand the material, but that you also exercise your ability to recall facts.
- **CD-ROM-based practice exam**—This book includes a CD-ROM containing several interactive practice exams. It is recommended that you continue to test your knowledge and test-taking skills by using these exams. You will find that your test-taking skills will improve

by continued exposure to the test format. Remember that the potential range of exam questions is limitless. Therefore, your goal should not be to "know" every possible answer but to have a sufficient understanding of the subject matter so that you can figure out the correct answer with the information provided.

Figure I-1 depicts the best way to navigate through the book. If you think that you already have a sufficient understanding of the subject matter in a chapter, test yourself with the "Do I Know This Already?" quiz. Based on you score, you should determine whether to complete the entire chapter or to move on to the "Foundation Summary" and then on to the "Q&A" sections.

Figure I-1 *Completing the Chapter Material*

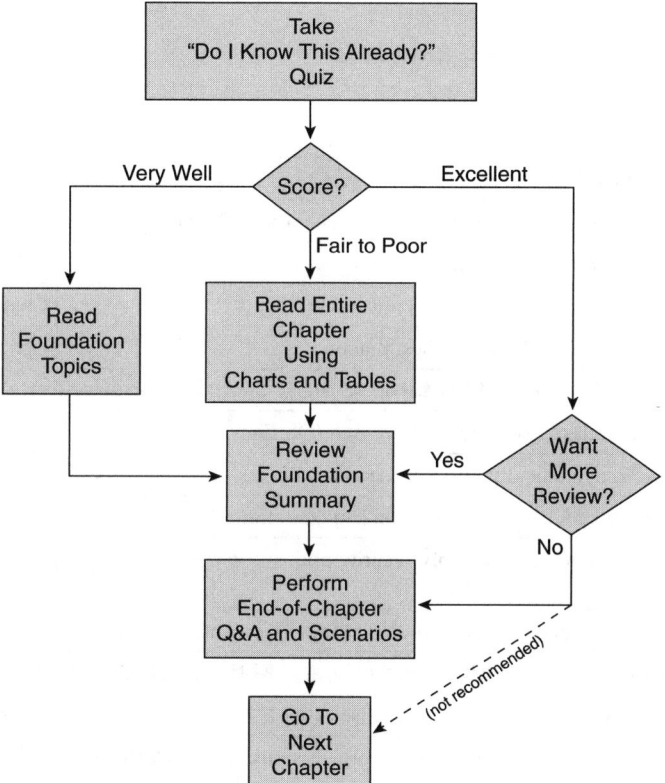

Certification Exam and This Preparation Guide

The questions for each certification exam are a closely guarded secret. The truth is that if you had the questions and could only pass the exam, you would be in for quite an embarrassment as soon as you arrived at your first job that required these skills. The point is to know the material, not just to successfully pass the exam. We do know which topics you must know to successfully complete this exam because they are published by Cisco. Coincidentally, these are the same topics required for you to be proficient when configuring Cisco IOS routers. It is also important to understand that this book is a "static" reference, whereas the exam topics are dynamic. Cisco can and does change the topics covered on certification exams often. This exam guide should not be your only reference when preparing for the certification exam. You can find a wealth of information available at Cisco.com that covers each topic in painful detail. The goal of this book is to prepare you as well as possible for the SNRS exam. Some of this is completed by breaking a 500-page (average) implementation guide into a 20-page chapter that is easier to digest. If you think that you need more detailed information on a specific topic, feel free to surf. Table I-1 lists each exam topic along with a reference to the chapter that covers the topic.

Table I-1 *SNRS Exam Topics and Chapter References*

Exam Topic	Chapter Where Topic Is Covered
Implement Layer 2 Security	
Utilize Cisco IOS and Cat OS commands to mitigate Layer 2 attacks	Chapter 14
Implement Cisco Identity-Based Networking Services	Chapter 17
Implement Cisco 802.1x port-based authentication	Chapter 18
Identify and describe Layer 2 security best practices	Chapter 14 and 24
Configure Cisco IOS Firewall Features to Meet Security Requirements	
Identify and describe the capabilities of the Cisco IOS Firewall feature set	Chapter 12
Configure CBAC to dynamically mitigate identified threats to the network	Chapter 15
Verify and troubleshoot CBAC configuration and operation	Chapter 15
Configure authentication proxy to apply security policies on a per-user basis	Chapter 16
Verify and troubleshoot authentication proxy configuration and operation	Chapter 16
Configure Cisco IOS Based IPS to Identify and Mitigate Threats to Network Resources	
Identify and describe the capabilities of the Cisco IOS IPS feature set	Chapter 13
Configure the IPS features to identify threats and dynamically block them from entering the network	Chapter 13
Verify and troubleshoot IDS operation	Chapter 13
Maintain and update the signatures	Chapter 13

continues

Table I-1 *SNRS Exam Topics and Chapter References (Continued)*

Exam Topic	Chapter Where Topic Is Covered
Configure Basic IPsec VPNs to Secure Site-to-Site and Remote Access to Network Resources	
Select the correct IPsec implementation based on specific stated requirements	Chapter 19
Configure IPsec encryption for site-to-site VPN using preshared keys	Chapter 19
Configure IPsec encryption for site-to-site VPN using certificate authority	Chapter 20 and 21
Verify and troubleshoot IPsec operation	Chapter 19 and 20
Configure Easy VPN Server	Chapter 22
Configure Easy VPN Remote using both hardware and software clients	Chapter 22
Troubleshoot Easy VPN	Chapter 22
Configure Authentication, Authorization, and Accounting to Provide Basic Secure Access Control for Networks	
Configure administrative access to the Cisco Secure ACS server	Chapter 10
Configure AAA clients on the Cisco Secure ACS (for routers)	Chapter 10
Configure users, groups, and access rights	Chapter 10
Configure router to enable AAA to use TACACS+	Chapter 8
Configure router to enable AAA to use a RADIUS server	Chapter 8
Verify and troubleshoot AAA operation	Chapter 8
Use Management Applications to Configure and Monitor Cisco IOS Security Features	
Initialize SDM communications on Cisco routers	Chapter 23
Perform a LAN interface configuration of a Cisco router using SDM	Chapter 23
Use SDM to define and establish a site-to-site VPN	Chapter 23

You will notice that not all the chapters map to a specific exam topic. This is because of the selection of evaluation topics for each version of the certification exam. Our goal is to provide the most comprehensive coverage to ensure that you are well prepared for the exam. In order to do this, we cover all the topics that have been addressed in different versions of this exam (past and present). Network security can (and should) be extremely complex and usually results in a series of interdependencies between systems operating in concert. This book will show you how one system (or function) relies on another, and each chapter of the book provides insight into topics in other chapters. Many of the chapters that do not specifically address exam topics provide a foundation that is necessary for a clear understanding of network security. Your short-term goal might be to pass this exam, but your overall goal is to become a qualified network security professional.

Note that because security vulnerabilities and preventive measures continue apace, Cisco Systems reserves the right to change the exam topics without notice. Although you can refer to the list of exam topics listed in Table I-1, always check the Cisco Systems website to verify the actual list of topics to ensure that you are prepared before taking an exam. You can view the current exam topics on any current Cisco certification exam by visiting its website at Cisco.com, clicking **Learning & Events > Career Certifications & Paths**. Note also that, if needed, Cisco Press might post additional preparatory content on the web page associated with this book at http://www.ciscopress.com/1587201534. It's a good idea to check the website a couple of weeks before taking your exam to be sure that you have up-to-date content.

Overview of the Cisco Certification Process

The network security market is currently in a position where the demand for qualified engineers vastly surpasses the supply. For this reason, many engineers consider migrating from routing/networking over to network security. Remember that "network security" is just "security" applied to "networks." This sounds like an obvious concept, but it is actually an important one if you are pursuing your security certification. You must be familiar with networking before you can begin to apply the security concepts. For example, the skills required to complete the CCNA will give you a solid foundation that you can expand into the network security field.

The requirements for and explanation of the CCSP certification are outlined at the Cisco Systems website. Go to Cisco.com and click **Learning & Events > Career Certifications & Paths**.

Taking the SNRS Certification Exam

As with any Cisco certification exam, it is best to be thoroughly prepared before taking the exam. There is no way to determine exactly what questions are on the exam, so the best way to prepare is to have a good working knowledge of all subjects covered on the exam. Schedule yourself for the exam and be sure to be rested and ready to focus when taking the exam.

The best place to find out the latest available Cisco training and certifications is http://www.cisco.com/en/US/learning/index.html.

Tracking CCSP Status

You can track your certification progress by checking https://www.certmanager.net/cisco/login.aspx?ReturnUrl=%2fCisco%2fDefault.aspx. You must create an account the first time you log on to the site.

How to Prepare for an Exam

The best way to prepare for any certification exam is to use a combination of the preparation resources, labs, and practice tests. This guide has integrated some practice questions and labs to help you better prepare. If possible, you want to get some hands-on time with the Cisco IOS routers.

There is no substitute for experience, and it is much easier to understand the commands and concepts when you can actually work with a Cisco IOS router. If you do not have access to a Cisco IOS router, you can choose from among a variety of simulation packages available for a reasonable price. Last, but certainly not least, Cisco.com provides a wealth of information about the Cisco IOS Software, all the products that operate using Cisco IOS Software, and the products that interact with Cisco routers. No single source can adequately prepare you for the SNRS exam unless you already have extensive experience with Cisco products and a background in networking or network security. At a minimum, you will want to use this book combined with the Technical Support & Documentation site resources (http://www.cisco.com/en/US/support/index.html) to prepare for this exam.

Assessing Exam Readiness

After completing a number of certification exams, we have found that you do not actually know whether you are adequately prepared for the exam until you have completed about 30 percent of the questions. At this point, if you are not prepared, it is too late. The best way to determine your readiness is to work through the "Do I Know This Already?" portions of the book, the review questions in the "Q&A" sections at the end of each chapter, and the case studies/scenarios. It is best to work your way through the entire book unless you can complete each subject without having to do any research or look up any answers.

Cisco Security Specialist in the Real World

Cisco has one of the most recognized names on the Internet. You cannot go into a data center or server room without seeing some Cisco equipment. Cisco-certified security specialists can bring quite a bit of knowledge to the table because of their deep understanding of the relationship between networking and network security. This is why the Cisco certification carries such clout. Cisco certifications demonstrate to potential employers and contract holders a certain professionalism and the dedication required to complete a goal. Face it, if these certifications were easy to acquire, everyone would have them.

Cisco IOS Software Commands

A firewall or router is not normally something to play with. That is to say that after you have it properly configured, you will tend to leave it alone until there is a problem or you need to make some other configuration change. This is the reason that the question mark (?) is probably the most widely used Cisco IOS Software command. Unless you have constant exposure to this equipment, it can be difficult to remember the numerous commands required to configure devices and troubleshoot problems. Most engineers remember enough to go in the right direction but will use the ? to help them use the correct syntax. This is life in the real world. Unfortunately, the question mark is *not* always available in the testing environment. Many questions on this exam require you to select the best command to perform a certain function. It is extremely important that you familiarize yourself with the different commands and their respective functions.

Rules of the Road

We have always found it confusing when different addresses are used in the examples throughout a technical publication. For this reason, we use the address space depicted in Figure I-2 when assigning network segments in this book. Note that the address space we have selected is all reserved space per RFC 1918. We understand that these addresses are not routable across the Internet and are not normally used on outside interfaces. Even with the millions of IP addresses available on the Internet, there is a slight chance that we could have chosen to use an address that the owner did not want published in this book.

Figure I-2 *Addressing for Examples*

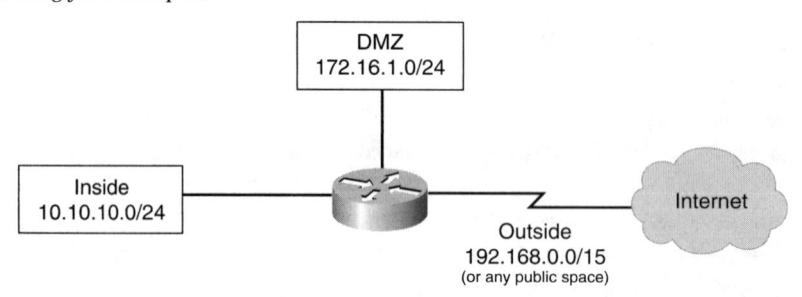

```
                       DMZ
                    172.16.1.0/24
                         |
    Inside               |              Internet
  10.10.10.0/24 ─────── Router ──────── (cloud)
                         |
                      Outside
                    192.168.0.0/15
                  (or any public space)
```

It is our hope that this will assist you in understanding the examples and the syntax of the many commands required to configure and administer Cisco IOS routers.

Exam Registration

The SNRS exam is a computer-based exam, with multiple-choice, fill-in-the-blank, list-in-order, and simulation-based questions. You can take the exam at any Pearson VUE (http://www.pearsonvue.com) or Prometric (http://www.prometric.com/Default.htm) testing center. Your testing center can tell you the exact length of the exam. Be aware that when you register for the exam, you might be told to allow a certain amount of time to take the exam that is longer than the testing time indicated by the testing software when you begin. This discrepancy is because VUE and Prometric want you to allow for some time to get settled and take the tutorial about the test engine.

Book Content Updates

Because Cisco Systems occasionally updates exam topics without notice, Cisco Press might post additional preparatory content on the web page associated with this book at http://www.ciscopress.com/1587201534. It is a good idea to check the website a couple of weeks before taking your exam, to review any updated content that might be posted online. We also recommend that you periodically check back to this page on the Cisco Press website to view any errata or supporting book files that may be available.

Part I: Overview of Network Security

Chapter 1 Network Security Essentials

Chapter 2 Defining and Detailing Attack Threats

Chapter 3 Defense in Depth

This chapter covers the following subjects:

- Defining Network Security
- Balancing Business Needs with Network Security Requirements
- Network Security Policies
- Network Security as a Process
- Network Security as a Legal Issue

CHAPTER 1

Network Security Essentials

Many believe that *network security* is actually two words that are mutually exclusive and that the only way to secure a computer is to segregate it completely. It is true that by restricting all physical and logical access to a computer system you can ensure complete security. Chances are that in most cases, complete security of a computer system by this method will render the "secure system" completely useless. Networking involves the sharing of assets and information and inherently involves a certain amount of risk. Computer networking is a complex and diverse undertaking, and securing computer networks adds several layers of complexity. Unlike Newton's third law of motion ("For every action, there is an equal and opposite reaction"), with computers and computer networks, for every action there might be no reaction at all, or there might be an unimaginable number of reactions that are neither equal nor opposite the action. Maintaining even a small computer network sometimes requires a delicate balancing act to ensure that all components (hardware, software, and so on) work and play well together, and that no single system or group of systems adversely affects the operations of the entire network. This "balancing act" only increases in difficulty as networks become bigger and more complex. Implementing security measures on a network can help ensure that systems perform as designed and, in many cases, provide the ability to logically separate systems that cause problems on the network until those systems can be dealt with. Of course, this is just an added benefit of network security. First and foremost, security is implemented to protect. Protect what? That is a good question . . . and the answer is probably different on every network. In this "information age," the commodity of our time is information in many forms. This "data" is considered property by most organizations and can be extremely valuable, or to a certain extent the release of such information can be costly. For this reason, network security has become a priority within most organizations.

"Do I Know This Already?" Quiz

The purpose of the "Do I Know This Already?" quiz is to help you determine whether you really need to read the entire chapter from beginning to end or just sections of the chapter. If you intend to read the chapter in its entirety, you do not necessarily need to answer these questions now.

The 10-question quiz, derived from the major sections in the "Foundation Topics" portion of this chapter, helps you determine how to spend your study time.

Chapter 1: Network Security Essentials

Table 1-1 outlines the major topics discussed in this chapter and the "Do I Know This Already?" quiz questions that correspond to those topics.

Table 1-1 *"Do I Know This Already?" Foundation Topics Section-to-Question Mapping*

Foundation Topics Section	Questions Covered in This Section
Defining Network Security	1–3
Balancing Business Needs with Network Security Requirements	4
Network Security Policies	5–7, 9
Network Security as a Process	8
Network Security as a Legal Issue	10

CAUTION The goal of self-assessment is to gauge your mastery of the topics in this chapter. If you do not know the answer to a question or are only partially sure of the answer, you should mark this question wrong for purposes of the self-assessment. Giving yourself credit for an answer you correctly guess skews your self-assessment results and might provide you with a false sense of security.

1. Network security has a direct affect on which of the following environment aspects?

    a. Implementation of products and services

    b. Organizational policies and directives

    c. Network architecture

    d. Process to control and monitor resources

    e. Scheduling of updates, patches, and routine maintenance

    f. All of the above

2. The definition of a corporate network security policy as it relates to users is most likely which of the following?

    a. A technical document that directs different aspects of a user's functions within the corporate environment

    b. A formal statement or document that specifies a set of rules that all users must follow while utilizing corporate resources

    c. A detailed action plan indicating what users should do in case of an emergency

    d. A document detailing the concept of operations as it pertains to the corporate network

    e. A technical document detailing corporate rules and regulations set forth by human resources

3. Which of the following are not components of a network security policy?

   a. Internet access
   b. E-mail access
   c. Server and workstation configuration
   d. VoIP fax and telephone access policy
   e. Disposure of trash
   f. All of the above

4. The network infrastructure policy ties together the following components, except for what?

   a. Network addressing schema
   b. Log consolidation and processing
   c. Problem resolution
   d. Quality of service
   e. Management and monitoring
   f. Naming convention

5. How can cost savings be realized through a standard set of operational policies and procedures?

   a. Savings through not having corrupted data
   b. Savings through not having downtime because of a DoS attack
   c. Savings through ensuring data integrity
   d. Savings through increased efficiency
   e. Savings through not having network "hiccups"
   f. All of the above

6. What high-level goals should a network security policy strive to achieve?

   a. Define requirements, identify components, detail contacts, and identify consequences
   b. Define requirements, guide configuration, define responsibilities, identify consequences, and define responses
   c. Collect requirements, provide guidance, define responsibilities, define standards, and define approach
   d. All of the above

7. A network security policy needs to follow standard guidelines, except for
   a. Technical people must support the policy.
   b. It must be consistent.
   c. It must be technically feasible.
   d. It must be implemented throughout the environment.
   e. It must define roles and responsibilities.
   f. It must be a nontechnical document.

8. The Cisco security wheel is used to depict what aspect of network security?
   a. Network monitoring
   b. The design process
   c. The network security life cycle
   d. Goal
   e. None of the above

9. What is the one biggest advantage of making the network security policy a nontechnical document?
   a. It makes writing and updating the document much easier.
   b. It simplifies the content for management.
   c. It provides a good framework for budget requests.
   d. It protects details on specific technologies.
   e. It limits use of technical jargon.
   f. It provides a framework for enforcement.

10. The protection of personal information maintained on networked computer systems is required by law.
    a. True
    b. False

The answers to the "Do I Know This Already?" quiz are found in the appendix. The suggested choices for your next step are as follows:

- **8 or less overall score**—Read the entire chapter. This includes the "Foundation Topics" and "Foundation Summary" sections and the "Q&A" section.

- **9 or 10 overall score**—If you want more review on these topics, skip to the "Foundation Summary" section and then go to the "Q&A" section. Otherwise, move on to Chapter 2, "Defining and Detailing Attack Threats."

Foundation Topics

Network security covers a broad range of topics that differ for nearly every organization but are all based on the same concepts and principles. This chapter defines network security as it applies to the Cisco SNRS exam and addresses the security policy, its goals and benefits, how it should be developed and by whom, how the policy should be implemented and maintained, and how to ensure that the policy remains effective as the organization continues to evolve.

Defining Network Security

Network security is the implementation of security devices, policies, and processes to prevent the unauthorized access to network resources or the alteration or destruction of resources or data. Security policies are defined later in this chapter and are the basis for the security implementation. The security devices and applications are tools that should be combined with standardized processes and procedures to ensure the enforcement of an effective network security policy.

Balancing Business Needs with Network Security Requirements

It is important to recognize that there is a tradeoff between a completely secure network and the needs of business. As mentioned in the chapter's opening, the only way to completely secure a computer is to physically disconnect it from the network and perhaps lock it in a secure area. Of course, this solution greatly reduces your ability to use whichever resources were to be shared on that system. The goal is to identify the risks and make an informed decision about how to address them.

Network Security Policies

Network security policies are created based on the security philosophy of the organization. The technical team uses the security policy as a guide to design and implement a security architecture that meets the needs of the organization while protecting the organization from potential threats. The corporate *security policy* is a formal statement that specifies a set of rules users must follow while accessing a corporate network.

The security policy is not a technical document; it is a business document that lays out the permitted and prohibited activities, and the efforts and responsibilities regarding network security. As defined in *Site Security Handbook* (RFC 2196), the security policy does not dictate how the business is operated. Rather, the business needs dictate the scope and depth of the security policy. Normally, a security policy is divided into several documents that address specific topics. These "usage policy statements" define the acceptable use of the network and the user's roles and responsibilities with respect to the network. The depth and scope of the network security policy documents depend on the size of the organization but should normally address the following topics:

- **Acceptable use of corporate assets policy**—This policy defines what is considered to be acceptable use of the corporate network and is normally the cornerstone of an organizational network security policy. It should address such items as e-mail use and Internet access. All employees should be briefed on the acceptable use policy on a regular basis to ensure that employees know what they are and are not authorized to do on the organization's computer network. Acceptable use policies should address the use of company e-mail and web browsing; define forbidden actions such as online pornography, gambling, and trading; and discuss possible ramifications if employees violate the policy.

- **Server and workstation configuration policy**—This policy defines which applications are to be configured on the network and should designate a specific build for each system. This policy is key to ensuring that all systems on the network adhere to a standard configuration, greatly reducing the time required for troubleshooting configuration problems. The definition of testing procedures for configuration/change management may be included in this policy or there may be a separate policy dedicated to individual testing procedures.

- **Patch management policy**—This defines how systems are upgraded and should outline how new patches are tested before being applied in the production environment. After a patch is approved for use in the production environment, it is added to the standard build to ensure that all new systems are configured with the approved system patch.

- **Network infrastructure policy**—This policy defines how the infrastructure is to be managed and who is responsible for its design, implementation, operations, and maintenance. This policy should address the following items:

    — Network addressing scheme

    — Naming convention

    — Configuration/change management

    — Quality of service

    — Management and monitoring of systems

    — Log consolidation and processing

    — Management of security event correlation and response

- **User account policy**—This policy defines which users should be given account permissions. A common user policy is called the *least-privilege policy*. This policy restricts users to the minimal user privileges that will allow users to complete their assigned tasks. This policy limits the number of users that have the ability to make configuration changes. This policy is similar to limiting the number of office personnel who have the combination to the office safe because if everyone knows the safe combination, there is not a need for the safe at all.

- **Other policies**—The number and scope of policies depends on the organization. Other policies can address such items as data handling, data backup, encryption, password requirements (length, type, and lifetime), and remote access.

Cisco recommends that three steps be implemented when establishing a network security policy:

- **Preparation**—When establishing a security policy, you should first create general-usage statements or a rough draft of the previously listed policy documents. Doing so will give you a good starting point. Next, perform a risk analysis to determine which risks you need to guard against and to define a level of acceptable risk. Risk levels are normally broken into three categories—high, medium, and low risk—but what is considered to be a risk must be defined for each organization. The final preparation item should be the designation of security team personnel and the definition of its duties.

- **Prevention**—This step defines how changes to your security posture are evaluated and implemented. Additionally, this step outlines how the security of the network should be managed and monitored. This management/monitoring should include the handling of log data, correlating and trending of log data, archiving of log data archive, and so on.

- **Response**—This step defines which actions are taken in the event of a problem on the network. It defines the individual responsibilities of members of the security team and addresses the following topics:
    - Reaction to an attempted security breach
    - How to isolate and handle a compromised network component
    - Evidence gathering and handling of log data
    - Working with law enforcement authorities
    - Network and system restoration
    - Policy review to ensure that any newly discovered vulnerabilities are compensated for

Creating the security policy is not normally a task for a single individual. A security team should include members from management, legal, human resources, and technical. The security policy must have the full support of management and must be enforceable based on applicable laws and regulations. Finally, the policy must be technically feasible.

Government regulation and good corporate governance have made the creation and enforcement of a network security policy commonplace within organizations that operate computer networks. Aside from the obvious benefits of maintaining standardized operational policies and procedures, there is one additional benefit of a security policy: cost savings. The cost savings can come in a variety of forms, including the following:

- **Savings by ensuring security of data**—Preventing unauthorized users from accessing the organization's data greatly reduces the chances of that data being corrupted. The cost of having to restore corrupted data can be tremendous.

- **Savings by preventing a denial-of-service (DoS) attack**—Although it is impossible to prevent a DoS attack, it is not too difficult to mitigate the attack by preventing access at multiple points on your network and at the Internet service provider (ISP). This type of defense requires significant coordination and cannot be implemented at the last minute.

- **Savings by preventing data manipulation**—Restricting access to only authorized users greatly reduces the risk of intentional data manipulation. Data manipulation is normally done to embarrass an organization and can have a lasting effect on its public image.

- **Savings by increasing efficiency**—An organization that standardizes its operations and clearly defines its practices will function more efficiently as a unit.

- **Savings by reducing "unknown" problems on the network**—"Unknown" problems on the network can result from the introduction of untested systems, configurations, or applications into the environment. By thoroughly testing and validating all practices, procedures, applications, and configurations prior to implementing them in a production environment, you greatly reduce the chances of creating these types of issues.

Security Policy Goals

In general, security policy goals can be summarized as five goals, including as a guide used by administrators in planning security efforts and responses; as a guide to the technical team that configures the equipment; as a guide for defining responsibilities and sanctions for users and administrators; as a guide for defining consequences for violating the policies; and as a guide for determining responses and escalations to recognized threats. Each specific goal is as follows:

- **Goal 1**—The first goal of the security policy is to guide the technical team by defining the requirements for the network and aid in the selection of equipment. The policy should define which functions a solution must perform, but not specify which solutions the technical team should implement. Because the security policy is not a technical document, a good policy does not dictate the exact equipment or configurations employed. For example, a good policy does not state that a Cisco PIX 515E Firewall will be used. Instead, the policy needs to define the minimum requirements for perimeter security, such as using a stateful inspection, proxy-based, or hybrid firewall.

- **Goal 2**—The second goal of the policy is to guide the technical team in configuring the equipment. For example, a security policy might state that the technical team should use its best effort to ensure that users cannot view websites that violate the acceptable use policy. However, the policy should identify forbidden content without listing which specific websites are acceptable and unacceptable.

- **Goal 3**—The third goal of the security policy is to define the responsibilities for users, administrators, and managers. Clearly defined responsibilities allow management and technical personnel to measure the performance of security efforts. When people know what is expected of them, they usually respond accordingly. Much of this would be addressed in the organization's acceptable use policy.

- **Goal 4**—The fourth goal of a security policy is to define consequences for violating the policies. If the security policy states that no programs will be downloaded from the Internet, for example, a stated penalty must apply to violations of that policy. This penalty allows users to understand that consequences apply to their actions.

- **Goal 5**—The fifth goal of a good policy is to define responses and escalations to recognized threats. Knowing how a threat is to be dealt with enables personnel to plan for the event. Failure to plan for a threat can result in confusion should that threat ever become a reality. Additionally, it is important to define escalation procedures for problems that are more difficult to pinpoint on the network. It is important that each member of the organization understand which steps to take in the event of a problem on the network.

Now that the general goals of the security policy have been discussed, it's time to consider some guidelines for a successful policy.

Security Guidelines

An effective security policy should follow standard guidelines. Many of these guidelines apply common sense, but it is best to define each guideline to ensure that all items are covered. At a minimum, it is important to address the following issues when developing a security policy:

- Management must support the policy.
- The policy must be consistent.
- The policy must be technically feasible.
- The policy should not be written as a technical document.
- The policy must be implemented globally throughout the organization.
- The policy must clearly define roles and responsibilities.
- The policy must be flexible enough to respond to changing technologies and organizational goals.

- The policy must be understandable.
- The policy must be widely distributed.
- The policy must specify sanctions for violations.
- The policy must include an incident-response plan for security breaches.
- Security is an ongoing process.

The next sections explore each of these guidelines.

Management Must Support the Policy

As in most business endeavors, unless management actively supports a policy, it will not be effective. A policy that restricts a business function or is considered to lack flexibility but addresses a critical need requires the full support of management; otherwise, it will not be followed. Security policies are designed to be restrictive. Every organization has employees who believe they should be able to do what is restricted by the policy. These same users will feel justified in violating the security policy if they believe that "management doesn't care." The security policy is designed to weigh the good of the organization before the needs of the individual.

The Policy Must Be Consistent

An effective policy must be consistent. Few things frustrate and confuse users more than inconsistent policies. Consistency of the policy refers to the application of the policy, not equal access for all employees within the organization. Access for each user should be determined by job function and need for such access. As mentioned in the section discussing security policies/user account policies, the consistent implementation of "least privilege" is a secure and highly recommended user policy.

However, policies must be consistent in both scope and goal. This means that policies should be applied to users based on their job function (for the organization) and not their position (within the organization). For example, it would not be acceptable to have a group of employees who do not use passwords because the manager of that group does not see the need. Likewise, managers should be required to maintain the same security standards as the average user. When one group of users needs additional resources, these resources should be allocated with the same care as any other access.

Inconsistent policies are difficult to implement and, by their nature, demonstrate a lack of consistency within the organization. Inconsistent or vague policies are also open to interpretation and might cause a debate as to which policy applies to a particular situation and can lead to an unacceptable risk for the organization.

The Policy Must Be Technically Feasible

This should be common sense. It is important to understand that the creation of a security policy is a management function. The security administrator should review the policy with management and ensure that they are advised as to its technical feasibility. The security administrator should recommend solutions that meet the business needs without compromising the security of the organization. The policy must also be feasible for the users; that is, the policy should not be so complicated that the user disregards it.

The Policy Should Not Be Written as a Technical Document

Writing the security policy in a nontechnical manner has a number of advantages. First, it is usually easier to understand than a technical document for the average employee. Second, because this document will be widely distributed, it is important that those receiving this document be able to understand it.

Making the policy a nontechnical document allows for the security concepts to be distributed without revealing the specific technologies used to accomplish the goals. For example, an administrator should be reluctant to distribute any document that specifies the make and model of the firewall to be used or that specifies exactly how systems will be monitored. Although the document is not normally for public dissemination, it is widely distributed within the organization, and many aspects of the policy can become public knowledge. Making public specific technical information related to the organization's security infrastructure may increase the vulnerability of that organization because a good cracker or hacker can use this information to form his attack strategy. Of course, some portions (or documents) will be more technical than others. Whereas acceptable use policies will be nontechnical, the server and workstation configuration document will be more technical consistent with the nature of the topic.

The implementation plan section of the security policy defines specific hardware and application information. Access to the implementation plan must be restricted to authorized personnel only. Normally, this access includes the security team members and anyone involved in the implementation of the security plan.

The Policy Must Be Implemented Globally Throughout the Organization

Because most organizations with more than one office location interconnect via private Frame Relay networks, virtual private networks (VPN), or other similar secure means, it is critical that all locations that have network connectivity have the same emphasis on security. Although a good security design limits and authenticates access between individual locations, a security hole at a single site presents a threat to the entire organization.

The Policy Must Clearly Define Roles and Responsibilities

Users need to be aware of what is allowed and prohibited. One can hardly blame a user for downloading music, for instance, if no policies prohibit such action.

Administrators must know their security goals and responsibilities if they are to be effective in accomplishing these goals. Management must also be aware of the role it is to play within the security efforts if these efforts are to be accomplished.

The Policy Must Be Flexible Enough to Respond to Changing Technologies and Organizational Goals

The security policy should be specific enough to define all requirements but not so inflexible that it does not account for changes in technology, growth within the organization, or infrastructure changes. The information technology (IT) field is ever evolving, and system and infrastructure upgrades and improvements occur constantly. The security policy is a living document and should constantly be reviewed and modified as necessary to ensure its relevance for the organization.

The Policy Must Be Understandable

Often, nontechnical employees have difficulty understanding technical terms and concepts. Therefore, it is imperative that the policy be clear and concise so that employees understand what is expected of them. Avoid using unnecessarily ambiguous or technology-specific terms so that nontechnical readers can understand their responsibilities. Many organizations present the policy to new employees during an orientation seminar and require them to acknowledge the training prior to receiving their network logon.

For example, a policy might address the issue of installing software in the following manner:

> Software Installation: The Information Systems department is responsible for ensuring that all computers operate efficiently and effectively. To this end, the Information Systems department configures all systems within the organization with approved software. Downloading any software to any computer is prohibited for all users except when specifically approved by the Information Systems department. Failure to abide by this policy is subject to disciplinary action as described in Section X.

The preceding statement starts by explaining the reasoning behind the policy. Following this, a statement defines who is responsible for installing applications. The reader is also directed to another part of the policy that lists the disciplinary actions that may be taken if this policy is disregarded.

Notice that the policy does not specifically state which applications are allowed or not allowed. This nonspecificity allows the Information Systems department to determine the appropriateness of applications without the need to rewrite the policy. The use of specific applications on the network is generally approved by a change control board (CCB), and approved applications are specified in guidelines, standard operating procedures (SOP), or other official-notification methods.

The Policy Must Be Widely Distributed

Many organizations publish their security policies on the Internet. You might find these published policies a valuable resource when starting to write your company's policies. Many large colleges and universities are a good source of sample policies.

Because the policy should not reveal any of your technical specifics, there is no reason for this policy not to be released to all employees. Distributing the policy to everyone within the organization and having each employee acknowledge receipt makes it easier to enforce the policy.

The Policy Must Specify Sanctions for Violations

Employees should be made aware of the policy and understand that violations of the policy have consequences. However, these sanctions should also be constructed in a way that does not limit the company to a specific action.

Because the legal rights of employees vary by city, state, and country, it is important to ensure that security team members from human resources and legal are involved when creating any policies that deal with disciplinary measures. The following policy is meant as an example only:

> Sanctions for Violating the Corporate Security Policy: Management may impose any disciplinary action it sees fit for any violations of the corporate security policies. These actions, although not limited to the following, may include verbal counseling, written letter of counseling, written letter of reprimand, or termination.

The Policy Must Include an Incident-Response Plan for Security Breaches

Any network has the potential of being compromised. Complex networks are more difficult to protect and can be more difficult to monitor. It is important to identify when your network is under attack and when the attack has resulted in a system or network breach. It is also important to develop an incident-response plan so that the security personnel know how to react to the compromise. Although the ultimate goal is to discover all breaches, some might go unnoticed. The policy must state the actions to take upon discovery of a breach. Most policies differentiate between breaches occurring from within the organization and those originating externally, such as from the Internet. The difference is because it is normally less difficult to identify the offending host if the attack originated from within the network and not from an internal resource that was exploited by an external source. Most organizations implement a stronger exterior-facing security perimeter, which greatly restricts the activity of a potential intruder. This design presents an additional risk because of a lack of internal controls, making it easier for an attacker on the internal network.

A sample policy section follows:

> Response to Internal Denial-of-Service (DoS) Attacks: Upon discovery of a DoS attack originating within the local-area network (LAN), the administrator will record and document the discovery for future forensics use. A secure machine should be utilized to track all packets originating from the source computer.
>
> The administrator will attempt to isolate the offending machine from the LAN. Next, the network segment where the attack is originating will be isolated.

Security Is an Ongoing Process

Security is a recurring process, not a single implementation effort. Figure 1-1 depicts what is commonly referred to as *the security wheel*. This figure, and many similar to it, appear throughout security documentation. It illustrates the recurring efforts necessary to effectively implement security within a corporation. These recurring efforts are referred to as the *security posture assessment* (SPA).

Figure 1-1 *The Security Wheel*

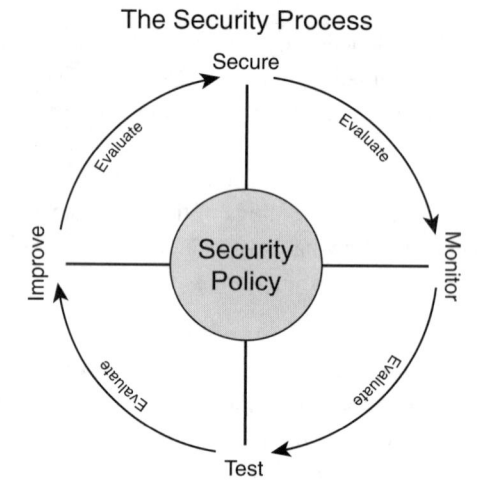

Network Security as a Process

The security wheel demonstrates the ongoing process to ensure that networks are secured and remain secure. The driving force behind the SPA is the security policy. The security policy states how often testing and monitoring must occur, which areas are tested, and how new security initiatives are implemented.

Four steps must be considered while implementing a security policy. Keep in mind that this is not a single process you complete after one round. This ongoing process must continue, and thus allow the network to evolve and constantly improve as new threats arise. Generally, each step continues as an ongoing process of its own, with each step relying on the other for input and improvement. The four steps of the security wheel are as follows:

- **Secure**—Actual implementation of a device or configuration fall under this step, including adding firewalls, intrusion detection systems, or AAA servers.

- **Monitor/respond**—After an implementation, the next step is to monitor that implementation. Monitoring enables the security administrators to understand better the challenges they face. Monitoring should occur at all times, with new monitoring implemented after changes to the network. Any issues discovered while monitoring the system need to be resolved upon discovery. (Return input to the "secure" step.)

- **Test**—Testing is critical to the SPA because without testing, there is no definitive feedback to determine the effectiveness of the implementations. Whereas the time between testing varies by company, all companies need to test to determine the effectiveness of their security efforts. Testing should always occur after significant changes are made to the network.

- **Manage/improve**—Management of all systems and of the process is important. It is also key to make improvements as necessary to systems, processes, policies, and so on. Whether these improvements involve new equipment or configuration changes, they result from previous testing and represent a prelude to increased efforts to secure the network.

Network Security as a Legal Issue

Consider the following scenario: An employee of Company X uses his computer (without authorization) to scan the Internet and eventually finds a server that belongs to Company Y that he is able to take control of using a documented exploit. The employee then uses that server to break into the database server at Insurance Company Z and steals the medical records of a celebrity containing sensitive and potentially damaging personal information. The stolen information is later distributed to the public. Who is responsible? Of course, the employee is ultimately responsible, but probably lacks the financial resources that make it worthwhile for the celebrity to seek legal recourse. However, companies X, Y, and Z could possibly become involved in legal action as a result of this theft.

Just as a person expects that a bank would take "reasonable steps" to ensure that her money is kept secure, organizations are expected to ensure that personal information is kept secure from public access. Many of the definitions for reasonable care are being created today, and these definitions constantly change in this fast-paced and fluid environment. The security policy mentioned earlier in this chapter is defined by RFC 2196, also known as the Internet Engineering Task Force's (IETF) *Site Security Handbook*.

Foundation Summary

The "Foundation Summary" section of each chapter lists the most important facts from the chapter. Although this section does not list every fact from the chapter that will be on your exam, a well-prepared candidate should (at a minimum) know all the details in each "Foundation Summary" section before taking the exam.

Network Security Policies

The following list outlines the key points for and advantages of having a security policy:

- Security policies are created based on the security philosophy of the organization.
- The technical team uses the security policy to design and implement the corporate security structure.
- The security policy is not a technical document.
- Read the *Site Security Handbook* (RFC 2196).
- The security policy should be developed by a security team, including members of management, legal, human resources, and technical staff.
- Having a good security policy saves in the following areas:
    - Savings by ensuring security of data
    - Savings by preventing DoS attacks
    - Savings by preventing data manipulation
    - Savings by increasing efficiency
    - Savings by reducing "unknown" problems on the network

Security Policy Goals

Many goals are achieved by implementing a strong yet balanced security policy. The following list outlines these goals:

- Guides the technical team in selecting their equipment
- Guides the technical team in configuring the equipment

- Defines the responsibilities for users, administrators, and management
- Defines the consequences for violating the policies
- Defines responses and escalations to recognized threats

Security Guidelines

The following list outlines the guidelines for developing and implementing a security policy:

- Management must support the policy.
- The policy must be consistent.
- The policy must be technically feasible.
- The policy should not be written as a technical document.
- The policy must be implemented globally throughout the organization.
- The policy must clearly define roles and responsibilities.
- The policy must be flexible enough to respond to changing technologies and organizational goals.
- The policy must be understandable.
- The policy must be widely distributed.
- The policy must specify sanctions for violations.
- The policy must contain an incident-response plan for security breaches.
- Security must be viewed as an ongoing process.

Network Security as a Process

The SPA is driven by the security policy. The security wheel demonstrates the four ongoing steps used to continuously improve the security of a network.

- **Secure**—Implement the equipment and processes and secure your organization's system configurations to reduce network exposure.
- **Monitor/respond**—Monitor the network to determine how changes have affected it and look for additional threats and respond to any newly discovered issues.

- **Test**—Test the current network and system configurations to determine whether any vulnerabilities exist.

- **Manage/improve**—Manage the process and make continuous improvements based on the results of your testing and vulnerabilities noted during the network monitoring or based on normal component upgrades and improvements.

Q&A

As mentioned in the section "How to Use This Book" in the Introduction, you have two choices for review questions: the Q&A questions here or the exam simulation questions on the CD-ROM. The questions that follow present a bigger challenge than the exam itself because they use an open-ended question format. By using this more difficult format, you can exercise your memory better and prove your conceptual and factual knowledge of this chapter. You can find the answers to these questions in the appendix.

1. Network security is the fine art of balancing security versus _____.

2. The policy that details information about software and hardware standards utilized within an organization and how they are installed is called _____.

3. A new vulnerability has come out, and the equipment that is being managed is vulnerable. What item will help to guide your actions to resolve this vulnerability?

4. What is the biggest benefit and business driver for network security from a manager's perspective?

5. One of the goals of a security policy is to define consequences for _____. This can be in the form of management or human resources discipline, demotion, or termination.

6. The security policy must have support from _____ because lack of such may impede various business functions or processes.

7. Security polices should define user access and control not by their position, but by what factor?

8. The security policy needs to be specific to define all the requirements, but it must also be _____ to respond to the industry.

9. So that an organization can respond quickly and efficiently to an exposed vulnerability or a suspected compromise of a network resource, the network security policy should also contain an _____.

10. Network security is considered a _____ because organizations are expected to ensure that personal information is kept secure from public access or leakage. Any leakage of this information may result in legal actions.

This chapter covers the following subjects:

- Vulnerabilities
- Threats
- Intruder Motivations
- Types of Network Attacks

CHAPTER 2

Defining and Detailing Attack Threats

This chapter discusses the potential network vulnerabilities and attacks that pose a threat to networks and explains the need for an effective network security policy.

"Do I Know This Already?" Quiz

The purpose of the "Do I Know This Already?" quiz is to help you decide whether you really need to read the entire chapter. If you already intend to read the entire chapter, you do not necessarily need to answer these questions now.

The 8-question quiz, derived from the major sections in the "Foundation Topics" portion of the chapter, helps you determine how to spend your limited study time.

Table 2-1 outlines the major topics discussed in this chapter and the "Do I Know This Already?" quiz questions that correspond to those topics.

Table 2-1 *"Do I Know This Already?" Foundation Topics Section-to-Question Mapping*

Foundation Topics Section	Questions Covered in This Section
Vulnerabilities	1–4
Threats	5, 6
Intruder Motivations	7
Types of Network Attacks	8

CAUTION The goal of self-assessment is to gauge your mastery of the topics in this chapter. If you do not know the answer to a question or are only partially sure of the answer, you should mark this question wrong for purposes of the self-assessment. Giving yourself credit for an answer you correctly guess skews your self-assessment results and might provide you with a false sense of security.

1. A good rule of thumb for "strong" passwords is that they follow which of the following guidelines?

    a. Should be upper- and lowercase, numbers, and special characters
    b. Should be complex and documented someplace
    c. Should be common words all strung together
    d. Should be documented so you can reference it
    e. Only a. and c.
    f. All of the above

2. The types of technology weaknesses are operating system, protocol, and _____ and _____.

    a. Communications, wiring
    b. Application, network equipment
    c. Network equipment, wiring
    d. Inexperienced staff, application
    e. None of the above

3. In general, which protocol is not considered "weak"?

    a. HTTP
    b. ICMP
    c. DNS
    d. RIPv1
    e. UDP
    f. All of the above

4. To help ensure that network equipment weaknesses are identified, a good rule of thumb is to _____.

    a. Test the unit in a production environment
    b. Trial the unit in a simulated load and protocol environment
    c. Trust the vendor and product engineers
    d. Read through all the documentation
    e. All of the above

5. Which of the following are considered intruders on a computer system?

    a. Crackers
    b. Hackers
    c. Phreakers
    d. Script kiddies
    e. a., b., and d
    f. All of the above

6. An intruder who enjoys the challenge of being able to bypass security measures is considered intruding for _____.

    a. Curiosity
    b. Fun and pride
    c. Revenge
    d. Profit
    e. Political purpose
    f. None of the above

7. What three major types of attacks are implemented by an intruder?

    a. DoS, reconnaissance, and access attacks
    b. Spoof, reconnaissance, and access attacks
    c. DoS, flood, and access attacks
    d. DoS, reconnaissance, and sweep
    e. Spoof, flood, and sweep
    f. None of the above

8. A virus, Trojan horse, or worm is considered which type of a function of an access attack?

    a. Interception
    b. Fabrication
    c. Modification
    d. Analysis
    e. All of the above

The answers to the "Do I Know This Already?" quiz are found in the appendix. The suggested choices for your next step are as follows:

- **6 or less overall score**—Read the entire chapter. This includes the "Foundation Topics" and "Foundation Summary" sections and the "Q&A" section.

- **7 or 8 overall score**—If you want more review on these topics, skip to the "Foundation Summary" section and then go to the "Q&A" section. Otherwise, move on to Chapter 3, "Defense in Depth."

Foundation Topics

Computer systems are a fundamental component of nearly every organization today. Large and small corporations, government agencies, and other organizations devote significant resources to maintaining their networks, and even the smallest organization is likely to use a computer to maintain its records and financial information. Because these systems can perform functions rapidly and accurately and because they make it easy to facilitate communication between organizations, computer networks continue to grow and become more interconnected. Any organization that wants to provide some public access to its network usually maintains a connection to the Internet. This access does not come without certain risks. This chapter defines some of the risks to networks and explains how an ineffective network security policy can further increase the chance of a network security breach. The key issues in securing and maintaining a computer network are confidentiality, integrity, and availability.

Vulnerabilities

To understand cyber-attacks, you must remember that a computer, no matter how advanced, is a machine that operates based on predetermined instruction sets. The operating systems and other software packages are compiled instruction sets that the computer uses to transform input into output. A computer cannot determine the difference between authorized and unauthorized input unless this information is written into the instruction sets. Any point in a software package that enables a user to alter the software or gain access to a system (that was not specifically designed into the software) is called a *vulnerability*. In many cases, crackers can exploit vulnerabilities and thus gain access to a network or computer. (See the section "Threats" later in this chapter.) It is possible to remotely connect to a computer on any of 65,535 available ports. As hardware and software technology continues to advance, the "dark side" continues to search for and discover new vulnerabilities. For this reason, most software manufacturers continue to produce patches for their products as vulnerabilities are discovered.

Self-Imposed Network Vulnerabilities

Most computer networks contain a combination of public and private data. A properly implemented security scheme protects all the data on the network while allowing some data to be accessed by outside entities, usually without the ability to change that data (for example, a corporate website). Other data, such as payroll information, should not be made available to the public and should be restricted only to specific users within the organization. Network security, when properly implemented, secures the corporate data, reduces the effectiveness of hacking attempts, and ensures that systems are used for their intended purposes. Networks designed to be freely available to the public need to be secured to ensure accuracy of the information and availability to the public.

Additionally, properly securing a network ensures that network resources are not used as an attack point against other networks.

Security attacks occur and damage networks because of the following three main reasons:

- Lack of effective network security policy
- Network configuration weaknesses
- Technology weaknesses

Lack of Effective Network Security Policy

Because a network security policy directs administrators regarding how communications should be enabled and implemented, this policy serves as the basis for all security efforts. Security policies have weaknesses for a number of reasons, including the following:

- **Politics**—Politics within an organization can cause a lack of consistency within the security policies or, worse, a lack of uniform application of the security policies. Many security policies make so many exceptions for management and business owners that they become meaningless.

- **Lack of a written security policy**—The lack of a written security policy is essentially the same as not having any policy. Publishing and widely distributing the security policy prevents confusion about it within the organization.

- **Lack of continuity**—When personnel change too frequently, people often take less care regarding the enforcement of security policies. When a system administrator leaves a position, for example, all the passwords used by that administrator should be changed. In an organization that changes administrators several times each year, there is a natural reluctance to change the passwords because users know they will be changed again soon because of administrator turnover.

- **Lack of disaster recovery planning**—A good disaster recovery plan must include contingencies for security breaches. Confusion that results from a disaster can hamper the success of forensics efforts because administrators might not be careful in their recovery efforts.

- **Lack of patch management within the security policy**—A good security policy allows for frequent hardware and software upgrades. A detailed procedure for implementing new hardware and software ensures that security does not become forgotten while implementing new equipment and software.

- **Lack of monitoring**—Failure to monitor logs and intrusion detection systems (IDS) exposes many organizations to attack without any knowledge that those attacks are occurring.

- **Lack of proper access controls**—Unauthorized network access is made easier when poorly designed access controls are implemented on the network. Improper password length, infrequent password changes, passwords written on sticky notes adhered to monitors, and freely shared passwords are security risks that potentially can lead to security breaches.

Network Configuration Weakness

As network devices become increasingly complex, the knowledge base required to configure systems correctly increases, too. This complexity represents more of an issue in smaller organizations in which a single administrator might be responsible for the LAN, WAN, servers, and workstations. In any organization, the most effective way to overcome network and system configuration issues is to establish and enforce a standardized baseline for all configurations. Configuration weaknesses normally fall into one of the following categories:

- **Misconfigured equipment**—A simple misconfiguration can cause severe security issues. Whether the error is caused through lack of knowledge of the system or a lack of attention to detail, the result might be an open vulnerability that leaves the system or network exposed to security threats and potential damage. Some areas of networking that are most susceptible to configuration errors are the firewall settings, access lists, Simple Network Management Protocol (SNMP) settings, and routing protocols.

- **Weak or exposed passwords**—Passwords that are too short, are easily guessed, or consist of common words make it easy for an intruder to gain access to company resources, networks, and data. A "strong" password should consist of at least eight characters and should include uppercase and lowercase letters, as well as numbers and special characters. Additionally, using the default password or administrator accounts is an especially poor practice. It is also important that users do not create a password that is too complex to remember. In such a scenario, users tend to write down their password on a stickie note, defeating the purpose of the password in the first place, and affix it to their monitor. One common method for creating and remembering passwords is the "vanity plate" method: Think of a word or phrase and convert it into the characters used on a vanity license plate, then change the case of a letter or two, and substitute one or more numbers for letters. Here is an example: In Virginia, for instance, a Honda owner is apparently not fond of mayonnaise. The Honda owner's license plate reads IH8 Mayo. You can drop in an underscore and an exclamation point and you get IH8_Mayo!. Not too fancy and easy to remember. Another password technique that poses a risk is the use of "common accounts" shared by many users. Common accounts prevent accurate accounting of which actions were taken by specific users and make it impossible to determine (to a legal standard) whether a specific user is responsible for a specific action.

- **Misconfigured Internet services**—Java applets, JavaScript, File Transfer Protocol (FTP) security settings, and Internet Protocol (IP) can all be configured in ways that are considered unsafe. Knowing exactly which services are required and which services are running ensures that Internet services do not create potential network security breaches.

- **Using default settings**—The default settings of many products are designed to assist in device configuration and production environment placement. One of the most common default settings is the default password, or the lack of a password by default. Other examples of default configuration settings include the following:
  - The default filters for the Cisco 3000 series VPN concentrators are insufficient protection for use in a production network.
  - By default, no access lists limit Telnet access on Cisco routers; if Telnet is enabled, you must ensure the access is limited to authorized source addresses only (from your management network).

These are just two examples of how default settings prove insufficient for production use.

Technology Weakness

All technologies have intrinsic weaknesses. These weaknesses can reside in the operating system, within the protocol, or within networking equipment. Each of these weaknesses is discussed further in the following sections:

- **Operating system weakness**—This weakness was discussed earlier in this chapter in the "Vulnerabilities" section. *Operating systems* are simple, coded instructions written for the computer. If an intruder can inject additional instructions into the system by exploiting a vulnerability within the operating system, the intruder might be able to affect how that system functions. Entire organizations are dedicated to discovering and exploiting operating system vulnerabilities. Some operating systems tend to present a greater challenge and therefore receive significantly more attention from hackers, crackers, and script kiddies. This is most likely why operating system developers dedicate tremendous resources to identifying and resolving possible vulnerabilities in a timely manner by releasing software patches. Many manufacturers have implemented an automated method to distribute patches and updates, such as Microsoft Windows Update, Linux up2date, and YUM (Yellow dog Updater Modified). It is now easier to immediately implement patches throughout the enterprise using automated techniques, thus ensuring the security posture of the network against the emerging threats that take advantage of new vulnerabilities.

- **Protocol weakness**—Some protocols suites, such as Transmission Control Protocol/Internet Protocol(TCP/IP), were designed without an emphasis on security aspects. Some of the security weaknesses of protocols are as follows:
  - As discussed in the previous section, some manufacturers receive significantly more attention from the hacker community. Microsoft has received more that its fair share of attention in this area. As a result, a significant number of vulnerabilities have been

identified within the Microsoft Windows operating system, Microsoft Windows products, and Microsoft networking. It is common for perimeter firewalls to block nearly all the ports utilized for Microsoft networking to help protect against this common vulnerability.

  — Network File System (NFS) is used by Novell and UNIX servers. There are no provisions for authentication or encryption. Additionally, because NFS uses a random selection of ports, it can be difficult for an administrator to limit access.

  — The TCP/IP suite consists of Internet Control Message Protocol (ICMP), User Datagram Protocol (UDP), and Transmission Control Protocol (TCP) and has several inherent weaknesses. For example, the header and footer on an IP packet can be intercepted and modified without leaving evidence of the change. ICMP packets are routinely used in denial-of-service (DoS) attacks, as discussed in the "Types of Network Attacks" section in this chapter.

  — As routing protocols were being developed, the emphasis was on functionality and communication and less on security. The lack of security in routing protocols resulted in a vulnerability that made it possible to inject incorrect information into the protocol and force a network disruption. There has been an increased emphasis on the security of routing protocols along with the general elevation in the priority of network security.

- **Application weaknesses**—Many applications are written without regard to security. The primary objective of application development is functionality. Service packs, upgrades, and patches are normally released by the application developer as vulnerabilities are identified. As technology continues to develop, however, security is becoming a greater priority and is now being written into newly created applications during design. Another factor that affects the relative strength of an application is the application's production timeline. The tremendous competition to get products to market sometimes forces software manufactures to introduce products that have not been thoroughly tested. In this case, the actual product testing might occur in a production environment.

- **Network equipment weakness**—Although all manufacturers strive to produce the best product possible, any system of sufficient complexity is prone to configuration errors or system design vulnerabilities. Additionally, all systems have their particular strengths and weaknesses. For example, one product might be efficient and secure when it processes a specific protocol or traffic for a specific application, but it might be weak or not support a different protocol or application. It is important to focus on exactly which type of network traffic you need to support and ensure that you implement the correct device in the correct location on the network. Additionally, always test your systems to ensure that they perform their functions as expected. Administrator who know the strengths and weaknesses of their equipment can overcome these shortcomings through the proper deployment and configuration of equipment.

Threats

Potential threats are divided into the following two categories, but their motivations tend to be more diverse and are detailed in the "Intruder Motivations" section of this chapter:

- **Structured threats**—A structured threat is an organized effort to breach a specific target. This can be the most dangerous threat because of its organized nature. Structured threats are preplanned attacks by a person or group seeking a specific target.

- **Unstructured threats**—Unstructured threats are by far the most common. These are usually a result of scanning the Internet by persons seeking targets of opportunity. Many different types of scanning files or "scripts" are available for download from the Internet and can be used to scan a host or network for vulnerabilities. Although the threat does not change, the attacker may change his methodology after vulnerabilities are identified.

Intruder Motivations

Several motivations might prompt someone to intrude on another's network. Although no text can list all the reasons that someone would decide to steal or corrupt data, some common themes become evident when looking at the motivations of previous intruders. To refine the discussion of intruder motivations, it is first necessary to define some terms. In the context of this chapter, *intruder* refers to someone who attempts to gain access to a network or computer system without authorization. Intruders can be further classified as *phreakers, crackers, hackers*, or *script kiddies*:

- **Phreaker**—Phreakers are individuals who have extensive knowledge of telephone networks and switching equipment. Their goal is to gain free access to telephone networks so that they can make local and long-distance calls.

- **Cracker**—Crackers use an advanced knowledge of networking and the Internet to compromise network security without proper authorization. Crackers are usually thought of as having a malicious intent.

- **Hacker**—Hackers investigate the integrity or security of a network or operating system. Usually relying on advanced programming techniques, hackers' motivations are not always malicious. *Ethical hackers* is a term that refers to security consultants; companies often hire ethical hackers to test current defenses (and thus perhaps expose weaknesses).

- **Script kiddie**—Script kiddies are novice hackers who rely heavily on publicly available scripts to test the security of a network and scan for vulnerabilities and weaknesses.

The dividing line between phreakers, crackers, hackers, and script kiddies is that phreakers, crackers, and hackers tend to be more skilled and normally develop their own tool sets; script kiddies, on the other hand, tend to be less skilled and use publicly available scripts. The motivation for someone to attempt to access, alter, or disrupt a network differs for each intruder. Some of the most common motivations are discussed in the following sections.

Lack of Understanding of Computers or Networks

Not all network or system intrusions are a directed effort. Sometimes, a user initiates a security breach through a lack of understanding. For example, an uneducated user with administrative rights on a Microsoft Windows 2000 or Microsoft Windows XP system can easily remove or change critical settings, resulting in an unusable system. Having too much trust combined with a lack of understanding can be equally dangerous. Often, network administrators open up their entire network to someone else when access to a single machine is all that is required. A poorly trained or inexperienced firewall administrator can easily open connectivity to a point that the firewall becomes ineffective. Another possibility is that a temporary firewall opening becomes a permanent opening because of a lack of procedures to ensure that temporary openings are closed after the need has passed. Although some security breaches occur without malicious intent, a good security policy can help prevent them.

Intruding for Curiosity

Sometimes, people are just curious about data contained in a system or network. One incident typical of this type is a 14-year-old boy who accessed a credit card company's computer systems. When asked why he broke into the system, he said that he wanted to see whether he could get in and he wanted to see what was there. Employees, for example, might attempt to access the payroll system to see whether their pay accords with coworkers. Alternatively, an employee might be curious about the financial status of the company or wonder whether employee personnel files contain anything interesting. Despite the focus of the curiosity, the common theme among those intruding because of curiosity is that the data suffers little or no damage (usually).

Intruding for Fun and Pride

Some intruders enjoy the challenge of bypassing security measures. Many times, the more sophisticated the security measures, the greater the challenge. Whether these intruders are crackers, hackers, or script kiddies, their motivation is fun, pride, or a combination of the two. When George Leigh Mallory was asked why he wanted to climb Mount Everest, his reply was "Because it is there." This seems to be the motivation for a number of potential intruders. On several bulletin boards and in several discussion groups, members list their latest conquests and the challenges posed to breaking into an organization's systems. The members of these groups applaud successful attempts and guide those who are unsuccessful. Additionally, many hacker websites publish tools, scripts, and code used to create viruses, worms, and Trojans. These websites are good places for a security administrator to monitor for information on the latest vulnerabilities and new techniques used for breaking into networks and systems.

Intruding for Revenge

Revenge can be powerful motivation. Disgruntled or former employees who have a good understanding of the network and know specific assets they want to target can cause substantial problems for an organization. Therefore, organizations should establish and enforce an "out-processing" procedure for employees to ensure that all the user accounts are removed or disabled and that any passwords for common accounts (which are not recommended in the first place) are changed. Whenever key personnel leave the organization, pay additional attention to network monitoring and intrusion detection to ensure that the former employee is not attempting to access the network using his knowledge of the network or perhaps any backdoor access that might have been put in place during his employment.

Intruding for Profit

Profit is a powerful motivator for breaking into computer networks and systems. Credit card information, unauthorized bank transfers, and manipulation of billing information can be extremely profitable if successful; however, not all intrusions for profit are based on money. In November 2002, a prominent news agency was accused of breaking into a Swedish company's computer system to steal data related to financial performance. The news agency was accused of obtaining this information to release it before the official announcement, thereby beating all the other news agencies to the story. At the time of this writing, there has been no determination whether this accusation has any merit. This example shows how profit might not always be directly related to transferring funds or obtaining credit card information. (Of course, the theft of corporate secrets could provide a competitor with a significant advantage in the marketplace.) Some intruders steal information such as credit card information and post it publicly to discredit the financial institution or to expose security vulnerabilities within that institution. Intruders have also taken information and then offered to return it for a price. As mentioned in Chapter 1, "Network Security Essentials," information is the currency of our time.

Intruding for Political Purposes

Because economies depend largely on electronic transactions, those economies are vulnerable to disruptions by an attacker. Cyber-warfare does exist and can pose a real threat to any economy. If disruption of an economy is desired, doing so through electronic means might become the ideal method for a number of reasons. These reasons include the ability to launch an attack from virtually any location, low equipment cost, low connectivity cost, and a lack of sufficient protection. In November 2002, a number of the primary Domain Name System (DNS) servers on the Internet were attacked through a distributed denial-of-service (DDoS) attack and were rendered inoperable for a number of hours. Although the motivation for this attack is unknown, a more sophisticated version could dramatically affect Internet traffic and disrupt many organizations that communicate over the Internet. Another more common political motivation is known as *hactivism*, which is the act of targeting an organization and disrupting its communications or defacing its website for political purposes.

Types of Network Attacks

Before learning about the characteristics of specific network attacks, you need to understand the different types of attacks. Attacks are defined by the goal of the attack rather than the motivation of the attacker. The three major types of network attacks, each with its own specific goal, are as follows:

- **Reconnaissance attacks**—A reconnaissance attack is designed not to inflict immediate damage to a system or network but only to map out the network and discover which address ranges are used, which systems are running, and which services are on those systems. One must "access" a system or network to some degree to perform reconnaissance, but normally the attacker does not cause any damage at that time.

- **Access attacks**—An access attack is designed to exploit vulnerability, as well as to gain access to a system on a network. When access is gained, the user can do the following:
  - Retrieve, alter, or destroy data
  - Add, remove, or change network resources, including user access
  - Install other exploits that can be used later to gain access to the system or network

- **DoS attacks**—A DoS attack is designed solely to cause an interruption to a system or network.

Reconnaissance Attacks

The *reconnaissance* attack term is misleading. The goal of this type of attack is actually to perform reconnaissance of a system or network, and the goal of the reconnaissance is to determine the makeup of the targeted system or network and to search for and map any vulnerabilities. A reconnaissance attack indicates potential for other more invasive attacks. Many reconnaissance attacks have been written into scripts that enable novice hackers or script kiddies to launch attacks on networks with a few mouse clicks. The following list identifies the more common reconnaissance attacks:

- **DNS queries**—A DNS query can provide a tremendous amount of information about an organization because the DNSs are designed to resolve IP address space to DNS names. DNS information is publicly available and simple to query. The two most informative DNS queries are the "DNS lookup" and the "whois query." The DNS lookup provides you with the specific IP address information for servers using a specific domain name. A whois query of the DNS system provides the unauthorized user with the following information:
  - Organization name
  - Organization ID (assigned by the American Registry of Internet Numbers)
  - Street address
  - Assigned public IP address space

- Public name server addresses
- Technical contact name, telephone number, and e-mail address

- **Ping sweep**—The output from a ping sweep can tell the unauthorized user the number of hosts active on a network.

- **Vertical scans**—Vertical scans scan the service ports of a single host and request different services at each port. This method enables the unauthorized user to determine which type of operating system is running and which services are running on the system.

- **Horizontal scans**—Horizontal scans scan an address range for a specific port or service. A common horizontal scan is the FTP sweep, which is the process of scanning a network segment searching for replies to connection attempts on port 21.

- **Block scans**—Block scans are a combination of the vertical and horizontal scans. In other words, they scan a network segment and attempt connections on multiple ports of each host on that segment.

Access Attacks

As the name implies, the goal of an access attack is to gain access to a system or a network. Having gained access, the user can perform many different functions. These functions fall into three distinct categories:

- **Interception**—If the unauthorized user can capture traffic going from the source to the destination, that user can store the data for later use. The data might be anything crossing the network segment connected to the sniffer (including confidential data such as personnel records, payroll, or research and development projects). If network management data is crossing the network, it is possible to acquire passwords for specific components and take control of that equipment. The methods used for intercepting traffic vary but usually require physical connectivity with the network. Upgrading from hub to switching technology greatly reduces the amount of traffic that can be captured by a network sniffer. The most effective way to protect your sensitive data is to save it in an encrypted format or to send it via an encrypted connection. The encryption prevents the intruder from being able to read the data. Figure 2-1 shows how interceptions can occur.

Figure 2-1 *Interceptions Can Occur if Data Is Sent in an Unencrypted Format*

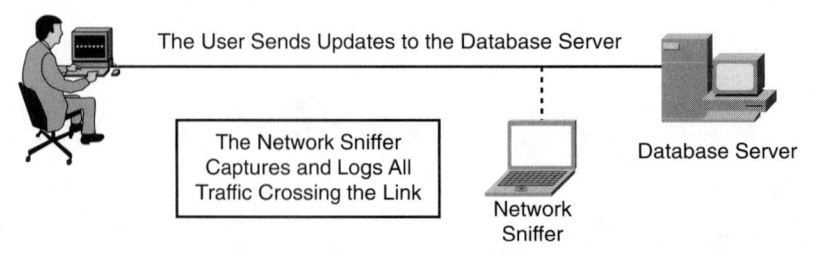

- **Modification**—Having access, the unauthorized user can now alter the resource. This not only includes altering file content, it also includes system configurations, unauthorized system access, and unauthorized privilege escalation. Unauthorized system access is completed by exploiting vulnerabilities in either the operating system or another application running on that system. *Unauthorized privilege escalation* refers to a user with a low level but authorized account attempting to gain higher-level or more privileged user account information to raise the unauthorized user's privilege level. This higher privilege level then enables the intruder to have greater control of the target system or network.

- **Fabrication**—Having access to the target system or network, the unauthorized user can create false objects and introduce them into the environment. This could include altering data or inserting packaged exploits such as a virus, worm, or Trojan horse that can continue to attack the network from within:

    — **Virus**—A computer virus can range from annoying to destructive. It consists of computer code that attaches itself to other software running on the computer. This way, each time the attached software opens, the virus reproduces and can continue to grow until it wreaks havoc on the infected system.

    — **Worm**—A worm is a virus that exploits vulnerabilities on networked systems to replicate itself. A worm scans a network looking for a system with a specific vulnerability. When it finds a host, it copies itself to that system and begins scanning from there, too.

    — **Trojan horse**—A Trojan horse is a program that usually claims to perform one function (such as a game) but does something completely different in addition the claimed function (such as corrupting the data on your hard disk). Many different types of Trojan horses get attached to systems, and the effects of these programs range from a minor irritation for the user to total destruction of the computer file system. Trojan horses are sometimes used to exploit systems by creating user accounts on systems that enable unauthorized users to gain access or upgrade their privilege level. Some Trojan horses capture data from the host system and send it back to a location where it can be accessed by the attacker. Other Trojan horses enable the attacker to take control of the system and enlist it in a DDoS attack, which is a common occurrence.

DoS Attacks

A DoS attack is designed to deny user access to systems or networks. These attacks usually target specific services and attempt to overwhelm them by making numerous requests concurrently. If a system is not protected and cannot react to a DoS attack, it can be easy to overwhelm that system by running scripts that generate multiple requests. It is possible to greatly increase the magnitude of a DoS attack by launching the attack from multiple systems against a single target. This practice is referred to as a *DDoS attack*. The use of Trojan horses in a DDoS was discussed in the previous section.

Foundation Summary

The "Foundation Summary" section of each chapter lists the most important facts from the chapter. Although this section does not list every fact from the chapter that will be on your SNRS exam, a well-prepared candidate should at a minimum know all the details in each "Foundation Summary" before going to take the exam.

Vulnerabilities

Vulnerability is anything that can be exploited to gain access to or gain control of a host or network.

Self-Imposed Network Vulnerabilities

An organization can create its own vulnerabilities by not ensuring that the following issues are resolved through process or procedure:

- The lack of an effective and consistent network security policy because of any of the following conditions:
    - **Politics**—Politics within an organization can cause a lack of consistency within the policies or a lack of uniform application of policies.
    - **Lack of a written security policy**—The lack of a written policy is essentially the same as not having a policy.
    - **Lack of continuity**—When personnel change too frequently, people often take less care to ensure that policies are enforced.
    - **Lack of disaster recovery planning**—The resultant confusion after a disaster often results in virtually all security efforts being dropped if the administrators are not careful in their recovery efforts.
    - **Lack of upgrade plans within the security policy**—A detailed procedure for implementing new hardware and software ensures that security does not become forgotten while implementing new equipment.
    - **Lack of monitoring**—Failure to monitor logs and intrusion detection systems appropriately exposes many organizations to constant attack without any knowledge that those attacks are occurring.
    - **Lack of proper access controls**—Improper password length, infrequent password changes, passwords written on notes attached to monitors, and freely shared passwords are all factors that can lead to security breaches.

- Configuration weakness within an organization can result in significant vulnerability exposure:
    - **Misconfigured equipment**—A simple misconfiguration can cause severe security issues.
    - **Weak or exposed passwords**—Passwords that are too short, are easily guessed, or consist of common words, especially when transmitted over the Internet, are cause for concern.
    - **Misconfigured Internet services**—Knowing exactly which services are required and which services are running ensures that Internet services do not create potential security breaches.
    - **Using default settings**—The default settings of many products are designed to assist in device configuration and production environment placement.
- All technologies have intrinsic weaknesses. These weaknesses might reside in the operating system, within the protocol, or within networking equipment:
    - All operating systems have weaknesses. You must take proper measures to make these systems as secure as possible.
    - Certain protocols can be exploited because of the way they were written and the functionality that was written into the protocol.
- Although all manufacturers strive to make the best product possible, any system of sufficient complexity is prone to human and mechanical errors. Additionally, all systems have their particular strengths and weaknesses. Knowing the nuances of your particular equipment is the best way to overcome technology weaknesses.

Threats

Computer networks face two different types of threats:

- **Structured threats**—Organized efforts to attack a specific target
- **Unstructured threats**—Neither organized nor targeting a specific host, network, or organization

Intruder Motivations

Four different names apply to potential intruders, categorized by their skill level and intent:

- **Phreaker**—Individuals who have extensive knowledge of telephone networks and switching equipment
- **Cracker**—More advanced and usually part of a structured threat

- **Hacker**—Can be involved in both structured and unstructured threats
- **Script kiddie**—Novice hacker using script files that perform most of the scanning and hacking functions

The motivations for intruders vary but generally fit into one of the following categories:

- Intruding by lack of understanding of computers or networks
- Intruding for curiosity
- Intruding for fun and pride
- Intruding for revenge
- Intruding for profit
- Intruding for political purposes

Types of Network Attacks

There are three major types of network attacks, each with its own specific goal:

- **Reconnaissance attacks**—An attack designed to gather information about a system or a network. The goal is to map the network, identify the systems and services, and to identify vulnerabilities that can be exploited later.
- **Access attacks**—An attack designed to exploit vulnerability and to gain access to a system on a network. After access has been gained, the user can do the following:
    — Retrieve, alter, or destroy data
    — Add, remove, or change network resources, including user access
    — Install other exploits that can be used later to gain access to the system or network
- **DoS attacks**—A DoS attack is designed solely to cause an interruption to a system or network.

Q&A

As mentioned in the section "How to Use This Book" in the Introduction, you have two choices for review questions: the Q&A questions here or the exam simulation questions on the CD-ROM. The questions that follow present a bigger challenge than the exam itself because they use an open-ended question format. By using this more difficult format, you can exercise your memory better and prove your conceptual and factual knowledge of this chapter. You can find the answers to these questions in the appendix.

1. An operating system application weakness that allows a user to alter software or gain access to a system or network is called a _____.

2. What are the three reasons that attacks can occur and damage networks?

3. If an organization has a high turnover rate for its system administrators, this is considered a security policy weakness because of _____.

4. Because smaller organizations have a limited IT staff, there is a greater chance that this organization would be susceptible to which kinds of configuration issues?

5. What is one of the most common default settings overlooked by most network and security administrators?

6. Describe an operating system weakness?

7. What is a structured threat and why does the Federal Bureau of Investigation (FBI) consider it important?

8. What are the five core reasons for intruding on a system or network?

9. The types of reconnaissance attacks are DNS queries, ping sweep, vertical scans, horizontal scans, and _____.

10. What is a DoS attack and why is it considered so destructive?

This chapter covers the following subject:

- Overview of Defense in Depth

CHAPTER 3

Defense in Depth

As technology continues to advance, network perimeters are becoming difficult to define. A secure perimeter can be difficult to maintain and no longer offers sufficient protection for a network. The most effective way to maintain a secure network is to implement layers of defense. This chapter looks at the combination of security devices, policies, and procedures that combine to make up a layered defense.

"Do I Know This Already?" Quiz

The purpose of the "Do I Know This Already?" quiz is to help you decide whether you really need to read the entire chapter. If you already intend to read the entire chapter, you do not necessarily need to answer these questions now.

The 10-question quiz, derived from the major sections in the "Foundation Topics" portion of the chapter, helps you determine how to spend your limited study time.

Table 3-1 outlines the major topics discussed in this chapter and the "Do I Know This Already?" quiz questions that correspond to those topics.

Table 3-1 *"Do I Know This Already?" Foundation Topics Section-to-Question Mapping*

Foundation Topics Section	Questions Covered in This Section
Overview of Defense in Depth	1–10

CAUTION The goal of self-assessment is to gauge your mastery of the topics in this chapter. If you do not know the answer to a question or are only partially sure of the answer, you should mark this question wrong for purposes of the self-assessment. Giving yourself credit for an answer you correctly guess skews your self-assessment results and might provide you with a false sense of security.

1. The security perimeter is considered _____.
   a. The connection to the Internet
   b. Extranet connections
   c. Remote access (VPN or dial-up)
   d. Security postures of other organizations
   e. All of the above

2. Which of the following is not a typical target of attack?
   a. Router
   b. Switch
   c. Networks and hosts
   d. PBX
   e. Applications
   f. Management components

3. A host can be compromised so that _____.
   a. An attacker can gain access to specific data
   b. Other attacks can be launched
   c. Access to the system is denied
   d. Data can be manipulated
   e. Data can be viewed
   f. All of the above

4. VPN technology is beneficial because it _____.
   a. Provides cost savings
   b. Connects secure logical interoffice communications
   c. Enforces routing
   d. Protects sensitive data
   e. Provides cost savings, connects secure logical interoffice communications, and protects sensitive data
   f. All of the above

5. A network administrator can protect access to sensitive data from internal users by _____.

   a. Setting up a private network for the users in question
   b. Implementing a private network segment using VLANs
   c. "Air gapping" the connection
   d. Using complex routing techniques
   e. Having a detailed network policy
   f. None of the above

6. Which of the following is a segmented address?

   a. Public DMZ
   b. Site-to-site VPN
   c. Remote-access DMZ
   d. Publicly NAT'd private range
   e. Public DMZ and remote-access DMZ
   f. All of the above

7. A host-based IDS/IPS is installed for the following reasons, except it _____.

   a. Limits vulnerabilities
   b. Tracks user activity
   c. Generates alerts
   d. Takes the place of patching a host
   e. Analyzes data for hostile intent

8. Which of the following are types of host-based IDS/IPS?

   a. Malicious and analytical
   b. Predictive and functional
   c. Signature and anomaly
   d. Restrict and respond
   e. Open and trap

9. Taking extensive and complex data from multiple devices and sources and generating alerts and actions is called _____.

   a. Detailed analysis
   b. Trending
   c. Monitoring
   d. Correlation
   e. Consolidation

10. Which of the following is not an effective security process?

    a. Securing the network
    b. Monitoring and responding to threats
    c. Testing and verifying
    d. Documenting the environment
    e. Managing and improving network and security
    f. All of the above

The answers to the "Do I Know This Already?" quiz are found in the appendix. The suggested choices for your next step are as follows:

- **8 or less overall score**—Read the entire chapter. This includes the "Foundation Topics" and "Foundation Summary" sections and the "Q&A" section.

- **9 or 10 overall score**—If you want more review on these topics, skip to the "Foundation Summary" section and then go to the "Q&A" section. Otherwise, move on to Chapter 4, "Basic Router Management."

Foundation Topics

Defense in depth is a term that describes the multiple layers of defense required to secure today's computer networks. This chapter describes the concept of defense in depth and explains why it is a critical consideration in any network design.

Overview of Defense in Depth

Internetworking refers to connecting different networks so that they can communicate, share resources, and so on. Many organizations consider their perimeter to be the connection to the Internet; however, with the liberal use of intranet, extranet, and remote user connections, the true perimeter has faded and is difficult to determine. This issue is further complicated by the security posture of the organizations on the far end of your intranet, extranet, and remote user connections. It is no longer possible to secure your network just by placing security devices (such as firewalls) at the Internet gateway.

Think of a network as a fortress that is under siege. You need to implement multiple layers of defense and try to use different types of defenses at each layer. Doing so will enable the network to handle a diverse range of attacks. A common example of this is an attack that successfully penetrates the firewalls and gets to the targeted server but is terminated by host-based intrusion detection/prevention systems installed on the server. Network attacks have become more complex and can now target multiple areas of the network simultaneously. Table 3-2 lists and describes some of the many targets on a network.

Table 3-2 *Potential Targets of Attacks*

Target	Description
Routers	The type of attack used against a router depends on the attacker's intent. An access attack is used if the intent is to gain access to the router or network. A denial-of-service (DoS) or distributed denial-of-service (DDoS) attack is used to bring down the router or to introduce routing changes to redirect traffic and deny access to the network.
Firewalls	Attacks against firewalls are virtually the same as routers; however, the techniques might differ depending on the size and type of firewall being attacked.
Switches	Any attack on a network component will affect how traffic flows across that segment. Because network traffic concentrates at the switches, it is important to ensure that switches are secure. This issue has become even more important with the increased utilization of Layer 3 switching in place of routing.

continues

Table 3-2 *Potential Targets of Attacks (Continued)*

Target	Description
Networks	Traffic flow on the network can be drastically affected by successful attacks against routers, firewalls, and switches.
Hosts	A host can be compromised to gain specific data that might reside on that system, or that system might be used to launch attacks against other network resources. Often, hosts are attacked because the attacker has discovered a vulnerability on a host and wants to exploit it.
Applications	An attacker will normally exploit a vulnerability within an application to compromise a host. As technologies advance, the number and type of attacks increase.
Data	Data can be intercepted and manipulated, but the data itself does not have any vulnerabilities. Normally, attacks are launched to access specific data. When access is gained, that data might be copied, altered, or destroyed.
Management components	Because management components are used to manage the different network components, it is important to ensure that they are secured to prevent an attacker from gaining control of the entire network.

Components Used for a Defense-in-Depth Strategy

The number and combination of different components used to secure today's networks changes continuously as new threats and threat-mitigation techniques arise. The following list identifies some of the many components used for a defense-in-depth strategy:

- **Security policy**—An effective network security policy is the centerpiece of any organization's security implementation. As described in Chapter 1, "Network Security Essentials," and Chapter 2, "Defining and Detailing Attack Threats," the security policy defines the who, what, when, where, why, and how of every aspect of an organization's operations. Although many aspects are not defined in technical detail, the overall functionality is defined in the policy. All the following elements are merely the implementation of the security policy.

- **Use of authentication, authorization, and accounting**—The implementation of authentication, authorization, and accounting (AAA) helps ensure that only authorized users access resources necessary to perform their job functions. Additionally, accounting and audit logs can be used to determine whether users are performing tasks that expose the network to unnecessary security risks.

- **VPN connectivity**—Virtual private network (VPN) technology is normally considered to be a cost-saving measure because it allows organizations to interconnect offices over the Internet. The use of VPN technology is not limited to the Internet and is normally determined by the organization, its business function, the type and value of its data, and the perceived threat. Many organizations that maintain sensitive data use VPN technology to secure dedicated circuits between their offices even though both endpoints are known.

The use of VPN technology is a major cost-savings factor for many organizations because it enables them to get rid of expensive dedicated connections and securely interconnect their different offices over public networks.

Another significant advantage is the ability to secure connections for remote users. With the increase in the availability of broadband Internet connectivity, many users are able to work from home using an encrypted connection to their corporate network.

- **Network segmentation**—Store owners would not normally leave their most valuable merchandise in an unlocked cabinet in the front of the store. In the same manner, a network administrator should not store sensitive company data on systems that allow unrestricted access from the Internet. The best way to protect assets is to segregate them by their value and restrict access to specific users, groups, and so on, as well as require AAA control gates to gain access. Network segmentation can be completed with routers, firewalls, switches, VPN devices, AAA systems, and operating system settings combined in a "solution" designed to protect the assets. A rule of thumb is the more valuable the data, the more complex the solution.

Assets that require specific access from a specific audience can be grouped together and placed on the same network segments. For example, all servers that host websites for public access should reside on a public demilitarized zone (DMZ) segment, should be assigned public address space (non-RFC 1918), and should allow access from the Internet via the standard Hypertext Transfer Protocol (HTTP) ports. Assets that require limited access should be placed further within the network and can use RFC 1918 addressing to prevent access from the Internet. Access to these assets can be restricted to specific sources and include the use of nonstandard ports by using Network Address Translation (NAT) and Port Address Translation (PAT). As the values of assets increase, the control measures required to access those assets increase, too. Figure 3-1 shows a simplified version of network segmentation.

As Figure 3-1 illustrates, all network resources are segregated by type and value. Assets with a greater value to the organization are located further within the network and are, therefore, protected at multiple layers within the network. The use of RFC 1918 addressing on internal networks prevents attacks that originate from the Internet unless those segments are NAT'd at the network perimeter. Additionally, network intrusion prevention systems (IPS) should be implemented liberally at all critical points of the network, host-based intrusion detection systems (IDS) and IPSs should be implemented on critical hosts, and virus protection should be implemented on *all* hosts.

Figure 3-1 *Network Segmentation*

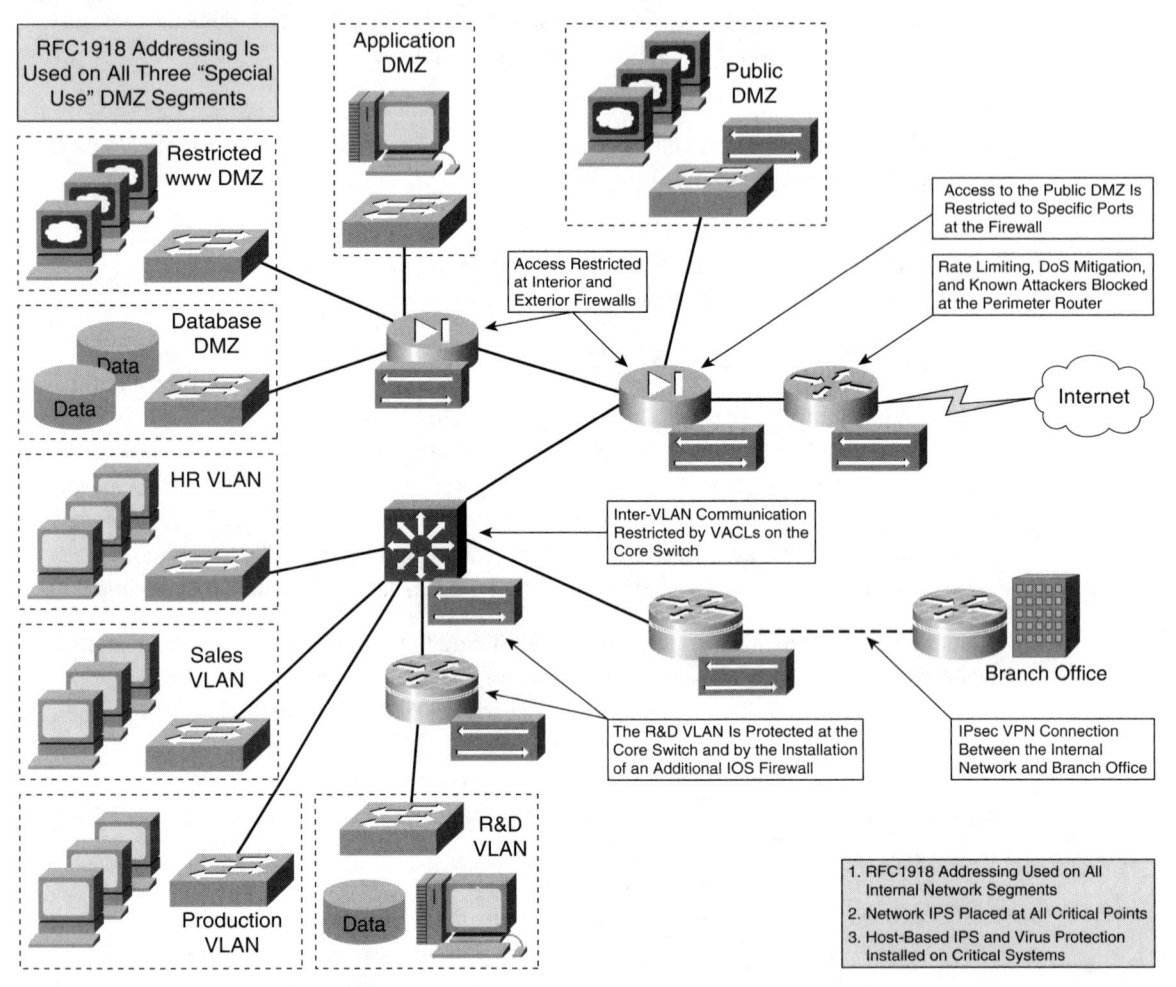

NOTE This access "from the Internet" does not include known hostile hosts and networks. Known hostile entities are addresses that have performed reconnaissance or other attacks and are identified by firewall and IDS log correlation and trending. These IP addresses are normally blocked at the perimeter. See the section "Intruder Motivations" in Chapter 2 for more information about intruders and what drives them. Some organizations will identify other organizations that have an agenda contrary to their own as "hostile" and either monitor them closely or completely block their inbound network access.

- **Dynamic perimeter security**—It would not be wise to think that a statically configured firewall or router could protect your network against attacks in an environment that is as dynamic as today's Internet. A statically configured device can only protect against known attacks. Because technology continues to change at such a rapid pace, the challenge is to protect against the unknown. The best way to do this is through the effective use of firewalls, routers, and IPSs. Cisco has developed technologies that facilitate the communication between firewalls, routers, IPSs, and AAA solutions, allowing them to interoperate and react in a more dynamic environment. This topic is discussed in greater detail throughout this book.

- **Host-based defense**—The prelude to any host-based defense is for the host to be as secure as possible. The developers of operating systems and applications produce service packs and software patches as soon as vulnerability has been identified. To limit the number of vulnerabilities that can be exploited by an attacker, it is important to ensure that all systems are always up-to-date. If an attacker were able to penetrate multiple lines of defense to get to the target host, the attack would still be ineffective if the attacker were unable to access its target. A host-based IPS is installed between the operating system and the kernel and can detect and prevent unauthorized activity on the host system. Additionally, these systems normally generate an alarm to identify that the system is under attack. There are two different types of host-based IPS:

    — **Signature based**—Signature-based IPSs watch the system and match instruction sets with the signatures of known attack profiles.

    — **Anomaly based**—Anomaly-based IPSs require time to establish a baseline of approved activities. Any instruction that is not part of the approved baseline is considered to be an attack and is blocked. Of course, you can configure the anomaly-based IPS to perform specific functions when it encounters instructions that are not within the baseline instead of just blocking the instruction. The major advantage of anomaly-based IPS is that it enables you to protect against the known and unknown threat.

- **Effective monitoring**—Why do burglar alarms normally include a siren? So people will know that someone is trying to break in. Firewalls, routers, switches, IPSs, and almost every other piece of network equipment produce an incredible amount of log data. It is important that critical systems are monitored to accurately determine the state of the network.

- **Consolidation, correlation, and trending**—This is the next step in effective monitoring. Correlation and trending enable you to determine what is "normal" network access. By identifying what is normal, you can determine what is abnormal and which type of attack you need to react to. Additionally, correlation products enable you to correlate data from multiple devices to get a better picture of the situation. This enables you to see the data from a possible attack from many different sources (firewall, router, IPS, and so on).

- **Effective security process**—Remember that the process is the effective implementation of the policy. The process is ongoing and is the driving force behind the constant improvement of your network's security posture. The security wheel discussed in Chapter 1 shows how potential threats are identified and mitigated as part of the ongoing evolution of the network. Two of the four steps of the security wheel have been modified slightly here to address more of a "real-world" scenario:

    — **Secure**—Secure the network against all known vulnerabilities by implementing equipment, processes, and system configurations.

    — **Monitor and respond**—Monitor the network to ensure that the preceding changes have the desired result and respond to any adverse effects or newly discovered issues.

    — **Test**—Test the network to verify that the components, processes, and configuration changes have secured the network.

    — **Improve and manage**—Continue to manage the network and implement improvements as necessary.

The use of defense in depth enables you to move from the old analogy of a network being a chain and only as strong as its weakest link. Today's networks should function more as vines that are constantly growing, improving, and evolving. The vines are dynamic and can adapt to changing environments, which is exactly how a secure network should function. All the components of defense in depth are discussed in great detail in the Cisco "SAFE: A Security Blueprint for Enterprise Networks" article, which you can find at http://www.cisco.com/en/US/netsol/ns340/ns394/ns171/ns128/networking_solutions_white_paper09186a008009c8b6.shtml.

Approximately 70 percent of network attacks originate from within the network. This fact, and the relatively new threat of dynamic attacks, is the driving force behind the concept of multiple layers of defense. It is crucial to ensure that your core networks are adequately protected from all attacks without regard to the source of the attack.

Physical Security

Much of this chapter is dedicated to logically securing your systems and networks. If an attacker can gain physical access to your systems, the amount of damage the attack can cause is limited only by the amount of time available. Physical access to facilities and networks should always be restricted only to necessary personnel performing a business role that requires such access.

Foundation Summary

The "Foundation Summary" section of each chapter lists the most important facts from the chapter. Although this section does not list every fact from the chapter that will be on your SNRS exam, a well-prepared candidate should at a minimum know all the details in each "Foundation Summary" before going to take the exam.

In today's dynamic environment, it is not enough just to secure the network perimeter. With the increasing use of intranets, extranets, remote users, and wireless technology, it is becoming difficult to determine where the network perimeter is actually located. Attackers now have more potential points of access and a greater selection of tools at their disposal when trying to breach a network. The best possible solution is to implement a defense-in-depth concept with multiple layers of defense that each compensates for a possible weakness of another layer. The key to defense in depth is that all components work together; failure of a single component does not necessarily mean that an attack was a success, because the attack can be mitigated at many different points.

Anything on the network, including the network itself, can be considered the target of a potential attack. Several components combine to form defense in depth, as shown in Table 3-3.

Table 3-3 *Defense-in-Depth Components*

Term	Definition
Security policy	The centerpiece of an organization's implementation. Defines everything about who/what is allowed on the network, how systems are to be protected, and who is responsible for which functions. The security policy is a policy document and does not include specific technical information.
Effective use of AAA	Enables you to ensure that only authorized users access network resources, as well as provides accounting information for trend analysis.
VPN WAN connectivity	Ensures secure connectivity between locations and network segments.
Network segmentation	Enables you to separate resources by asset value and type. Provides for multiple layers of security for resources of greater value. Includes the use of RFC 1918 addressing with NAT/PAT.
Dynamic perimeter security	Dynamic configurations that require the use of firewalls and routers with network-based IPSs. This combination allows the perimeter to detect and respond to attacks.

continues

Table 3-3 *Defense-in-Depth Components (Continued)*

Term	Definition
Host-based defense	The last line of defense. In the event that an attacker is able to breach the network and get to his target, the attacker's actions will be terminated by host-based IDSs/IPSs. An anomaly-based IDS/IPS can prevent unknown attacks because it reacts to anything that deviates from the allowed baseline.
Effective monitoring	It is important to ensure that you monitor the log output from your various network devices to spot a network attack.
Correlation and trending	Correlation and trending enable you to effectively manage the large amounts of log data produced by the many network devices. Additionally, you can correlate the data from different devices (for example, a firewall and IPS) to get a better picture of the attack.
Effective security process	This is an ongoing process that drives constant improvement, greater accuracy, and a greater understanding of the environment.

Q&A

As mentioned in the section "How to Use This Book" in the Introduction, you have two choices for review questions: the Q&A questions here or the exam simulation questions on the CD-ROM. The questions that follow present a bigger challenge than the exam itself because they use an open-ended question format. By using this more difficult format, you can exercise your memory better and prove your conceptual and factual knowledge of this chapter. You can find the answers to these questions in the appendix.

1. Setting up a layered security model for an organization helps to_____.
2. A DoS or DDoS attack against a router or firewall is used to _____.
3. An _____ is the centerpiece of any organizations security implementation.
4. In addition to providing controlled access to resources, AAA implementations provide detailed _____ and _____ to help expose and mitigate risks.
5. The use of RFC 1918 addressing on internal networks prevents attacks that originate from the Internet unless _____.
6. What is the fundamental premise behind an anomaly-based IDS/IPS?
7. Why is monitoring important?
8. Performing a network assessment would be considered part of which step of the effective security process for an organization?
9. From where do an estimated 70 percent of the network attacks originate?
10. Where should network IDS/IPS be implemented?

Part II: Managing Cisco Routers

Chapter 4 Basic Router Management

Chapter 5 Secure Router Administration

This chapter covers the following subjects:

- Router Configuration Modes
- Accessing the Cisco Router CLI
- Cisco IOS Firewall Features

CHAPTER 4

Basic Router Management

The Cisco IOS router and Cisco IOS Firewall are actually the same hardware. The difference is a low-cost, advanced Firewall feature set that was integrated into Cisco IOS Software. All the basic functionality of Cisco IOS Software remains on the Cisco IOS Firewall with additional features added, called the *Firewall feature set*. The Cisco IOS router is commonly referred to as the Cisco IOS Firewall if any of the Firewall feature set components are used. This chapter discusses access to and management of the Cisco IOS Firewall.

"Do I Know This Already?" Quiz

The purpose of the "Do I Know This Already?" quiz is to help you decide whether you really need to read the entire chapter. If you already intend to read the entire chapter, you do not necessarily need to answer these questions now.

The 10-question quiz, derived from the major sections in the "Foundation Topics" portion of the chapter, helps you determine how to spend your limited study time.

Table 4-1 outlines the major topics discussed in this chapter and the "Do I Know This Already?" quiz questions that correspond to those topics.

Table 4-1 *"Do I Know This Already?" Foundation Topics Section-to-Question Mapping*

Foundation Topics Section	Questions Covered in This Section
Router Configuration Modes	1–3
Accessing the Cisco Router CLI	4–7
Cisco IOS Firewall Features	8–10

CAUTION The goal of self-assessment is to gauge your mastery of the topics in this chapter. If you do not know the answer to a question or are only partially sure of the answer, you should mark this question wrong for purposes of the self-assessment. Giving yourself credit for an answer you correctly guess skews your self-assessment results and might provide you with a false sense of security.

1. What are the command modes on a Cisco router?
   a. ROM read-only, privileged EXEC, and configuration EXEC
   b. ROM read-only, privileged EXEC, and super EXEC
   c. ROM monitor, user EXEC, and privileged EXEC
   d. READ monitor, user EXEC, and super EXEC
   e. ROM monitor, user EXEC, and configuration EXEC

2. All of the following are various configuration modes and submodes except _____.
   a. Global configuration mode
   b. Terminal configuration mode
   c. Interface configuration mode
   d. Line configuration mode
   e. All of the above

3. If users want to control access to the console port, they make changes in which configuration mode?
   a. Global configuration mode
   b. Terminal configuration mode
   c. Console configuration mode
   d. Interface configuration mode
   e. Line configuration mode

4. What features does the question mark (?) provide on the CLI?
   a. Command syntax and general information on system
   b. Version information and configuration information
   c. List of available commands and command syntax
   d. List of available commands and configuration information
   e. Command details and customer support information
   f. All of the above

5. What is not a method for accessing the Cisco CLI?

   a. Console
   b. Auxiliary
   c. Telnet
   d. SSH
   e. Modem
   f. All of the above

6. When the Autoinstall feature is run on a new router, it _____.

   a. Completes the installation of the router configuration
   b. Allows basic communication, routing, and access
   c. Hardens the router after it is configured
   d. Generates new passwords
   e. Auto-detects new connections and routes
   f. None of the above

7. What command enables you to change the symbols that represent the configuration modes on the CLI?

   a. Symbols unchangeable
   a. The **command prompt** command
   b. The **configuration prompt** command
   c. The **prompt** command
   d. The **change symbol** command
   e. None of the above

8. What is not a feature of the Cisco IOS Firewall feature set?

   a. CBAC lists
   b. System auditing
   c. ActiveX blocking
   d. Firewall IPS
   e. Authentication proxy
   f. Alerting

9. Which type of access list provides "lock-and-key" traffic filtering?

   a. Standard access list
   b. Extended access list
   c. Reactive access list
   d. Reflective access list
   e. Dynamic access list
   f. System access list

10. Which type of AAA authentication servers does the Firewall feature set support?

    a. LDAP, RADIUS, and TACACS+
    b. TACACS+ and RADIUS
    c. Active Directory and LDAP
    d. TACACS+, RADIUS, and Kerberos
    e. Active Directory, LDAP, and Kerberos
    f. All of the above

The answers to the "Do I Know This Already?" quiz are found in the appendix. The suggested choices for your next step are as follows:

- **8 or less overall score**—Read the entire chapter. This includes the "Foundation Topics" and "Foundation Summary" sections and the "Q&A" section.

- **9 or 10 overall score**—If you want more review on these topics, skip to the "Foundation Summary" section and then go to the "Q&A" section. Otherwise, move on to Chapter 5, "Secure Router Administration."

Foundation Topics

This chapter covers the basics of managing a Cisco router. It discusses the command-line interface (CLI) and router configuration modes, and introduces the features of the Cisco IOS Firewall.

Router Configuration Modes

Before jumping into the CLI of the Cisco router, it is important to understand the different command modes available. Consider the command mode to be a level where you can perform specific functions. If you are not at the correct level, you cannot perform the correct function (to configure the router). This simplified explanation will make more sense as each mode is discussed. The following are command modes on a Cisco router:

- **ROM monitor mode**—The ROM monitor mode is the mode the router boots to if it cannot find a valid system image. You need to use this mode if you need to change only the system boot parameters to include resetting the system password. If the router has a working image installed, you need to press the **Break** key during the first 60 seconds of the router boot sequence.

- **User EXEC mode**—The user EXEC mode is the mode that you connect to by default. If the router is configured with a password, you are prompted for the password and given three attempts to provide the correct password. You will know that you are in the user EXEC mode because the router displays the host name followed by a right-angle bracket (>) symbol:

    RouterA>

    In the user EXEC mode, you can perform limited functions to check the status of the router but cannot change the router configuration. To exit the user EXEC mode, use the **logout** command.

- **Privileged EXEC mode**—To get from the user EXEC mode to privileged EXEC mode, use the **enable** command. If an enable password (or better yet, enable secret password) has been configured (and it should have), you are again prompted for a password and given three attempts. In this mode, the router displays its host name followed by the hash (#) symbol:

    RouterA>**enable**
    RouterA#

    In the privileged EXEC mode, you can perform all the functions that were available in the user EXEC mode but still cannot make any configuration changes. You do, however, have access to **show** and **debug** commands that are not available in the user EXEC mode. The privileged EXEC mode is the path to the global configuration mode. To return to the user EXEC mode, use the **disable** command:

    RouterA#**disable**
    RouterA>

- **Configuration modes and submodes**—There are many different configuration modes. Each of these makes changes to the device configuration. To ensure that those configuration changes are not lost if the router reboots, you must copy the running configuration to the startup configuration. The type and number of configuration submodes depends on the type of router, the Cisco IOS version, and the components installed on the router:

- **Global configuration mode**—The command for accessing the global configuration mode is **configure terminal**. In the global configuration mode, the router continues to display its host name followed by (config) and the # symbol:

  ```
  RouterA#configure terminal
  RouterA(config)#
  ```

  The global configuration mode is where you can make "global" changes to the configuration of the router. A common example of a global configuration is the creation of an access list. From the global configuration mode, you can move to a position that enables you to configure specific components of the router, such as the router interfaces; virtual private network (VPN) components (isakmp, crypto, and so on); CLI connections (line); authentication, authorization, and accounting (AAA) server groups; and many more. To exit to the privileged global configuration mode, use the key combination **Ctrl-Z** or type the command **end**:

  ```
  RouterA(config)#end
  RouterA#
  ```

- **Interface configuration mode**—From the global configuration mode, the command for accessing the interface configuration submode is **interface** *interface-type interface-number*. The router displays its host name followed by (config-if) indicating that it is in the interface configuration mode:

  ```
  RouterA(config)#interface Ethernet 0
  RouterA(config-if)#
  ```

  At this point, you are ready to configure the interface that you have selected. If you are configuring subinterfaces, you need to use the same command to enter the subinterface configuration mode:

  ```
  RouterA(config)#interface Ethernet 0.1
  RouterA(config-subif)#
  ```

  You can also enter the subinterface configuration mode from the interface configuration mode due to a backward command link:

  ```
  RouterA(config-if)#interface Ethernet 0.1
  RouterA(config-subif)#
  ```

- **Line configuration mode**—Another configuration submode that must be configured is the line configuration. This mode is used to configure the CLI access to the router. The command for accessing the line configuration mode from the global configuration mode is **line** *type number*:

  ```
  RouterA(config)#line con 0
  RouterA(config-line)#
  ```

  To exit the configuration mode and return to the privileged EXEC mode, use the key combination **Ctrl-Z** or type the command **end**:

  ```
  RouterA(config-subif)#end
  RouterA#
  ```

Router Configuration Modes

To return to the global configuration mode, type the **exit** command. This command works for both interface configuration and subinterface configuration modes:

```
RouterA(config-subif)#exit
RouterA(config)#
```

> **NOTE** The command prompts used in these examples are the default prompts. It is possible to change the system prompt (>, #) by using the **prompt** command while in the global configuration mode.

You can use abbreviated commands in Cisco IOS Software. Abbreviated commands allow the system to recognize the command based on the first few letters. If several commands start with the same first letters, the device responds with a list of possible choices. You can also use the **Tab** key to complete a command for you.

For a list of the available commands in each configuration, you can use the question mark (**?**). This is perhaps the most useful command written into the Cisco IOS Software. If you are not completely familiar with the correct syntax of a command, you can input a portion of the command followed by **?**. The router then provides you with the correct syntax. Table 4-2 shows how to navigate between modes when configuring the router.

Table 4-2 *Configuration Modes on the Cisco Router*

Command	Description
Router>**enable** Password: ********	Enter the router in the user EXEC mode. Use the **enable** command and the correct password to enter the privileged EXEC mode.
Router#**configure terminal**	Enter the global configuration mode.
Router(config)#**ip routing**	Enable IP routing on the router.
Router(config)#**hostname RouterA**	Configure the router host name to RouterA.
RouterA# (config) interface Ethernet 0/0	Enter the interface configuration mode.
RouterA(config-if)#**ip address 10.10.10.254 255.255.255.0**	Configure eth0/0 for 10.10.10.254/24.
RouterA(config-if)#**no shutdown**	The **no shutdown** command ensures that the interface is enabled.
RouterA(config)#**exit**	Configuration complete, exit to global configuration mode.
RouterA(config)#**end** or RouterA(config)#**Ctrl-Z**	Configuration complete, exit to privileged EXEC mode.

Chapter 4: Basic Router Management

> **NOTE** Some commands change depending on the router series and Cisco IOS Software version.

Figure 4-1 depicts the commands used to in the CLI to access the router, configure the host name, assign an IP address to Ethernet interface 0/0, and then exit the configuration.

Figure 4-1 *Using the CLI*

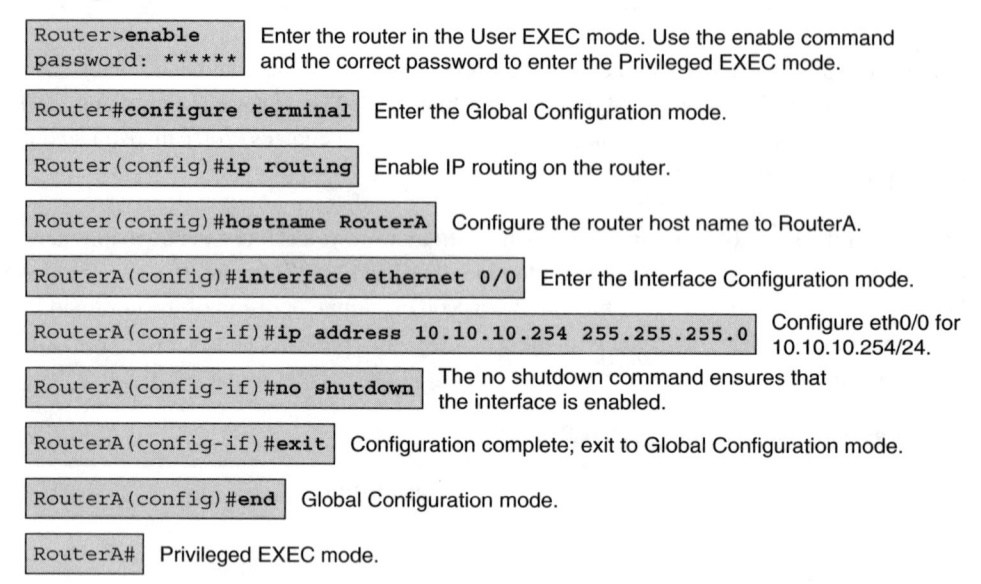

Command	Description
`Router>enable` `password: ******`	Enter the router in the User EXEC mode. Use the enable command and the correct password to enter the Privileged EXEC mode.
`Router#configure terminal`	Enter the Global Configuration mode.
`Router(config)#ip routing`	Enable IP routing on the router.
`Router(config)#hostname RouterA`	Configure the router host name to RouterA.
`RouterA(config)#interface ethernet 0/0`	Enter the Interface Configuration mode.
`RouterA(config-if)#ip address 10.10.10.254 255.255.255.0`	Configure eth0/0 for 10.10.10.254/24.
`RouterA(config-if)#no shutdown`	The no shutdown command ensures that the interface is enabled.
`RouterA(config-if)#exit`	Configuration complete; exit to Global Configuration mode.
`RouterA(config)#end`	Global Configuration mode.
`RouterA#`	Privileged EXEC mode.

Accessing the Cisco Router CLI

You can access the Cisco router CLI via any of three methods:

- **Console**—The console connection requires a direct connection to the console port of the router using a rollover cable normally from the serial interface of a computer. This method is considered the most secure for administration of the router because it requires a physical connection to the router. This method can prove impractical for enterprise networks.

- **Auxiliary**—The auxiliary connection is normally a remote dialup connection completed by connecting a modem to the aux port of the router. The administrator just dials in to the attached modem to initiate the connection to the modem. This method is commonly used for administering large networks or as a backup method to Telnet.

- **Telnet**—The Telnet connection occurs via the network interface. Telnet connections can be completed using Telnet or Secure Shell (SSH). Telnet is a clear-text protocol and should be restricted to internal (protected) network segments only. SSH is a protocol that uses encryption and can be used for remote management across public networks. This method is the most common for remote administration because it allows for administration of an entire enterprise from a central location. Additional steps are required to configure the router to accept SSH connections:

— **Enable the SSH server**—To enable the SSH server on the router, you must enter the global configuration mode and configure the domain name for the device. The domain name is important because it is used when generating the SSH key, which is used to authenticate the router when making the connection:

```
RouterA(config)#ip domain-name secur-example.com
```

Next, you should use the **crypto key generate rsa** command followed by the key length.

— **Configure the SSH parameters on the router**—The optional command **ip ssh** {[**timeout** *seconds*]|[**authentication-retries** *interger*]} enables you to configure the authentication *parameters* for the SSH connection to the router. It tells the router how long to wait for a response from the client and how many attempts to allow before terminating the connection.

If you want to restrict the router to only SSH connections, you must add the command **transport input ssh** from the line configuration mode.

The implementation of SSH is discussed in greater detail in Chapter 5.

> **NOTE** SSH requires Cisco IOS Software versions that support DES (56-bit) or 3DES (168-bit) encryption.

Figure 4-2 depicts the physical connections to a Cisco 2600 series router through which you can access the CLI.

Figure 4-2 *CLI Connections to a 2600 Router*

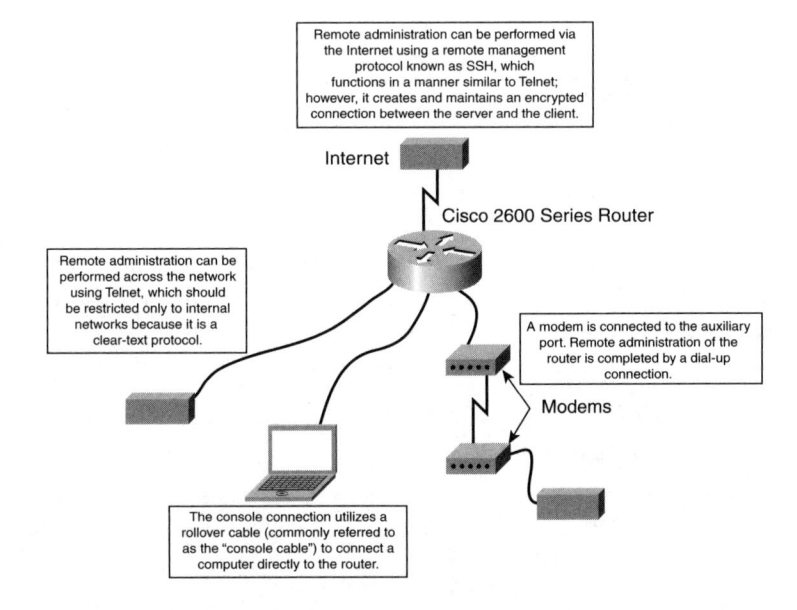

Configuring CLI Access

A new router is delivered without passwords. You must configure passwords for the access method that you intend to use and disable the methods that you do not intend to use. You can complete the initial configuration of a router in different ways. The Autoinstall feature installs a configuration on the router that is sufficient to get the router on the network and allow Telnet connections to complete the configuration. The other configuration methods usually require a physical connection to the router using a console cable. After connecting with the router, you need to enter the line configuration mode and configure the CLI access. After connecting to the router, you must configure the passwords for each type of connection. It is a good idea to configure a different password for each connection type.

Example 4-1 shows the configuration of RouterA for a console connection using the password N3wY0rk$$.

Example 4-1 *Configuring the Console Password*

```
RouterA>enable
RouterA#configure terminal
RouterA(config)#line con 0
RouterA(config-line)#login
RouterA(config-line)#password N3wY0rk$$
RouterA(config-line)#end
RouterA#
```

Example 4-2 shows the configuration of RouterA for an auxiliary connection using the password Ch1cag0!.

Example 4-2 *Configuring the Auxiliary Password*

```
RouterA>enable
RouterA#configure terminal
RouterA(config)#line aux 0
RouterA(config-line)#login
RouterA(config-line)#password Ch1cag0!
RouterA(config-line)#end
RouterA#
```

Example 4-3 shows the configuration of RouterA for a Telnet or virtual terminal (vty) connection using the password M1aM1@. Remember that you can use SSH to complete a Telnet connection if the SSH server has been enabled on the router.

Example 4-3 *Configuring the vty Password*

```
RouterA>enable
RouterA#configure terminal
RouterA(config)#line vty 0 4
RouterA(config-line)#login
```

Example 4-3 *Configuring the vty Password (Continued)*

```
RouterA(config-line)#password M1aM1@
RouterA(config-line)#transport input ssh    << This is an optional command to disallow
  telnet and only permit SSH connections
RouterA(config-line)#exit
RouterA(config)#end
RouterA#
```

Cisco IOS Firewall Features

As mentioned in the beginning of this chapter, the Cisco IOS Firewall feature is an enhancement to the Cisco IOS Software that incorporates additional security-related features. The Cisco IOS Firewall provides an additional level of security for the network without the expense of purchasing dedicated hardware. The Cisco IOS Firewall feature set was first introduced as *CiscoSecure Integrated Software* (CSIS). The Cisco IOS Firewall overview lists the following features:

- **Standard and extended access lists**—The router can be configured to perform basic traffic filtering by using standard or extended access lists.

- **Dynamic access lists**—Dynamic access lists are used to configure lock-and-key traffic filtering. This is the capability to generate dynamic temporary access through the firewall for specific predefined circumstances. This feature provides a more flexible method for filtering network access.

- **Reflexive access lists**—Reflexive access lists allow only specific traffic to maintain state. In other words, traffic that is generated on an internal network is allowed through the firewall only until the initial session state has terminated. Reflexive access lists are used only if Context-Based Access Control (CBAC) is not implemented.

- **System auditing**—The firewall maintains a record of all transactions in an orderly format that can be used for detailed reports.

- **TCP intercept**—TCP intercept is used to prevent a type of denial-of-service (DoS) attack known as the TCP SYN flood. This feature is used only if CBAC is not implemented.

- **Java blocking**—The firewall scans Java code and can block code that is unsigned or determined to be malicious.

- **Context-Based Access Control**—CBAC examines traffic passing through the firewall at all layers (up to the application layer). CBAC is used to generate dynamic access lists.

- **Cisco IOS Firewall IPS**—The Cisco IOS Firewall Intrusion Prevention System (IPS) is a signature-based device that you can configure to take the following actions when an intrusion is detected:

    — Send an alarm to the director

    — Drop the packet

    — Send a TCP reset

- **DoS mitigation**—The system is designed to detect and react to DoS attacks.

- **Authentication proxy**—Authentication proxy is used to proxy authentication requests to a AAA server. This allows authentication to occur on a per-user basis.

- **Port-to-application mapping (PAM)**—PAM enables administrators to specify which ports can be used for which services. This allows for the configuration of services on nonstandard ports when traversing the firewall.

- **Security server support**—The Cisco IOS Firewall supports the following AAA servers:

  — TACACS+

  — Remote Authentication Dial-In User Service (RADIUS)

  — Kerberos

- **Network Address Translation (NAT)**—The Cisco IOS Firewall supports NAT and Port Address Translation (PAT). NAT enables administrators to translate RFC 1918 addressing to public addressing on a one-for-one basis, and PAT enables administrators to hide an entire internal network behind a single public address. The Cisco IOS Firewall supports the use of both NAT and PAT on the same device.

- **IPsec network security**—The Cisco IOS Firewall supports all the standard IPsec protocols. This allows for the configuration of VPNs.

- **Neighbor router authentication**—The Cisco IOS Firewall can authenticate its peer routers to ensure that all routing updates are legitimate.

- **Event logging**—The Cisco IOS Firewall logs all error messages and system events to the console terminal by default. These messages can be redirected to a syslog server for easy storage and recovery. The system logs are commonly used for troubleshooting connection issues and for network forensics.

- **User authentication and authorization**—Integration with AAA servers allows for authentication and authorization on a per-user basis. This functionality enables administrators to designate specific permissions for users and groups.

- **Real-time alerts**—You can configure the firewall to perform alert functions in the event of a known attack or other event that is determined to be a severe security risk.

The individual Cisco IOS Firewall components are addressed in greater detail in individual chapters throughout this book.

Foundation Summary

The "Foundation Summary" section of each chapter lists the most important facts from the chapter. Although this section does not list every fact from the chapter that will be on your SNRS exam, a well-prepared candidate should at a minimum know all the details in each "Foundation Summary" before going to take the exam.

Router Configuration Modes

Table 4-3 lists the configuration modes of the Cisco router along with a brief description. It is important to understand that except for the ROM monitor mode you must navigate from one mode to another to view and edit different components of the router configuration.

> **NOTE** The terms *specific configuration mode* and *very specific configuration mode* in Table 4-3 are not actual Cisco terms but are used to describe how the configuration of each component becomes more granular on the router.

Table 4-3 *Router Configuration Modes*

Configuration Mode	Description
ROM monitor	The ROM monitor mode is used to change the system boot configuration only.
User EXEC	The user EXEC mode is the mode entered by default. It is possible to view the general condition of the router in this mode.
Privileged EXEC	The privileged EXEC mode is entered by using the **enable** command. It is possible to view much of the router configuration but changes cannot be made.
Global configuration	General configuration changes are made in the global configuration mode. The **configure terminal** command is used to enter this mode.

continues

Table 4-3 *Router Configuration Modes (Continued)*

Configuration Mode	Description
Specific configuration	Multiple specific items must be configured, each in its own configuration mode. Examples of specific configuration items include but are not limited to the following: • Router interfaces • Line configuration • Crypto map • ISAKMP policy • Modem pool For a list of specific configuration modes available, just enter the ? when in the global configuration mode.
Very specific configuration	Some configuration items require even more specific configuration. A common example is the subinterface configuration mode, which is used to bind additional virtual interfaces to a single physical interface.

Accessing the Cisco Router CLI

Table 4-4 describes the three connection methods used for management of the router and the command used to enter the line configuration mode to configure each access method.

Table 4-4 *CLI Connection Methods*

Connection Method	Connection Type	Command
Console port	Direct connection from a computer to the router using a console cable.	**line con 0**
Auxiliary port	Dialup connection. The receiving modem is connected to the router auxiliary port.	**line aux 0**
Telnet	Connection across the network, accessing the router via the network interface.	**line vty 0 4**

Cisco IOS Firewall Features

Table 4-5 lists the features that differentiate the Cisco router from the Cisco IOS Firewall. Note that all these features are software-based and are available in the current versions of Cisco IOS Software that include the specific features.

Table 4-5 *OS Firewall Features*

Feature	Explanation
Standard and extended access lists	Standard and extended access lists are used for static filtering of traffic passing through the firewall.
Dynamic access lists	Dynamic access lists are used to temporarily open ports to allow specific traffic through the firewall. These ports are closed as soon as the session is completed.
Reflexive access lists	Reflexive access lists only allow access as long as the connection state remains active. Reflexive access lists cannot be used in conjunction with CBAC.
System auditing	The Cisco IOS Firewall maintains an audit log of all changes made to the router.
TCP intercept	TCP intercept is used to prevent a SYN flood attack. It cannot be used in conjunction with CBAC.
Java blocking	The Cisco IOS Firewall can detect and block malicious Java code.
Context-Based Access Control	CBAC inspects traffic up to the application layer and can affect the traffic based on the configured policy.
Cisco IOS Firewall IPS	Cisco IOS Firewall IPS compares traffic to predefined attack signatures to detect and react to malicious traffic. The firewall IPS can react in any of the following manners: • Send an alert • Drop the packet • Reset the connection
DoS mitigation	The Cisco IOS Firewall can detect and react to potential DoS attacks.
Authentication proxy	Authentication proxy is used to proxy authentication requests to a AAA server. This allows for per-user or per-group policies.
Port-to-application mapping (PAM)	Port-to-application mapping enables administrators to configure applications to pass through the firewall using nonstandard ports.
Security server support	The Cisco IOS Firewall supports the following AAA servers: • TACACS+ • RADIUS • Kerberos

continues

Table 4-5 *OS Firewall Features (Continued)*

Feature	Explanation
Network Address Translation (NAT)	The Cisco IOS Firewall can translate source and destination addresses. This allows for the use of RFC 1918 addresses on internal and DMZ segments, greatly reducing the attacker's ability to route attacks across public networks.
IPSec network security	The Cisco IOS Firewall supports IPsec standards and can be used to configure VPNs.
Neighbor router authentication	Neighbor router authentication is used to ensure that the Cisco IOS Firewall receives updated routing information from only authenticated sources.
Event logging	The Cisco IOS Firewall can be configured to log all traffic that passes through it. The firewall logs can prove helpful for troubleshooting and network forensics.
User authentication and authorization	Authentication and authorization allow for the configuration of per-user and per-group policies.
Real-time alerts	The Cisco IOS Firewall can generate alerts in real time, which greatly increases the ability to react to an attempted attack.

Q&A

As mentioned in the section "How to Use This Book" in the Introduction, you have two choices for review questions: the Q&A questions here or the exam simulation questions on the CD-ROM. The questions that follow present a bigger challenge than the exam itself because they use an open-ended question format. By using this more difficult format, you can exercise your memory better and prove your conceptual and factual knowledge of this chapter. You can find the answers to these questions in the appendix.

1. To access the ROM monitor mode, you have to press the _____ key during the first 60 seconds of the router boot sequence.
2. From which command mode can a user access the global configuration mode and which command is utilized?
3. Can abbreviated commands be utilized in Cisco IOS for all commands?
4. What mode does the prompt **router(config)#** command depict?
5. What is considered the most secure method for administering a router and why?
6. What additional steps must you take to configure the router to accept SSH connections?
7. If you want to configure the console port with a password of @bc123, which commands would you type at the RouterA(config-line)# prompt?
8. Which type of access list allows specific traffic to maintain state?
9. Which type of code does the firewall feature set block?
10. What is PAM and why is it a useful feature of the Firewall feature set?

This chapter covers the following subjects:

- Privilege Levels
- Securing Console Access
- Configuring the enable Password
- **service password-encryption** Command
- Configuring Multiple Privilege Levels
- Warning Banners
- Interactive Access
- Securing vty Access
- SSH Protocol
- Port Security for Ethernet Switches
- AutoSecure

CHAPTER 5

Secure Router Administration

The Cisco IOS Firewall helps secure the trusted network from unauthorized users. The security of the network also involves the security of the Cisco IOS Firewall itself. In addition to physical security of the Cisco IOS Firewall, it is important to secure administrative accesses to interfaces on the Cisco IOS Firewall. This chapter discusses the different ways to secure administrative access to the Cisco IOS Firewall.

"Do I Know This Already?" Quiz

The purpose of the "Do I Know This Already?" quiz is to help you decide whether you really need to read the entire chapter. If you already intend to read the entire chapter, you do not necessarily need to answer these questions now.

The 14-question quiz, derived from the major sections in "Foundation Topics" section of the chapter, helps you determine how to spend your limited study time.

Table 5-1 outlines the major topics discussed in this chapter and the "Do I Know This Already?" quiz questions that correspond to those topics.

Table 5-1 *"Do I Know This Already?" Foundation Topics Section-to-Question Mapping*

Foundation Topics Section	Questions Covered in This Section
Privilege Levels	1, 2
Securing Console Access	3
Configuring the enable Password	4
service password-encryption Command	5
Configuring Multiple Privilege Levels	6
Warning Banners	7
Interactive Access	8, 9

continues

Chapter 5: Secure Router Administration

Table 5-1 *"Do I Know This Already?" Foundation Topics Section-to-Question Mapping (Continued)*

Foundation Topics Section	Questions Covered in This Section
Securing vty Access	10
SSH Protocol	11
Port Security for Ethernet Switches	12, 13
AutoSecure	14

> **CAUTION** The goal of self-assessment is to gauge your mastery of the topics in this chapter. If you do not know the answer to a question or are only partially sure of the answer, you should mark this question wrong for purposes of the self-assessment. Giving yourself credit for an answer you correctly guess skews your self-assessment results and might provide you with a false sense of security.

1. What are the default levels of access commands for the Cisco IOS Software command-line interface?

   a. Level 0 and level 15

   b. Level 1 and level 15

   c. Level 0, level 5, and level 10

   d. Level 1, level 5, and level 16

   e. Level 0 and level 16

   f. None of the above

2. Up to how many levels of password protected privileges can be configured using the CLI?

   a. 10

   b. 16

   c. 15

   d. 3

   e. 20

   f. None of the above

3. Which of the following are current mechanisms for authenticating to the router console port?

   a. Smart card

   b. Password

   c. Cookies

   d. CSACS

4. Which of the following is not a valid composition of the enable password?

   a. Must contain 1 to 25 uppercase and lowercase alphanumeric characters

   b. Cannot contain the word *enable*

   c. Must not have a number as the first character

   d. Can have leading spaces, which are ignored

   e. Can have trailing and intermediate spaces, which are recognized

   f. Can contain the question mark (?)

5. Which of the following does the **service-password encryption** command encrypt?

   a. SNMP

   b. BGP neighbor password

   c. Username password

   d. TACACS

6. To configure multiple privilege levels, use the following syntax:

   a. **privilege mode [all] {level** *level* **| reset}** *command-string*

   b. **privilege set mode {range | all} {level** *level* **| reset}** *command-string*

   c. **privilege mode** {level} [reset] *command-string*

   d. **privilege access mode {level** *level* **| reset}** *command-string*

   e. **set privilege mode** {level} **{reset}** *command-string*

   f. **set privilege mode [all] {level** *level* **| reset}** *command-string*

7. Warning banners can be set to be displayed on all the following instances except

   a. When an EXEC process is created

   b. When an incoming connection to an asynchronous line is created

   c. When a login prompt is generated

   d. Before a user logs out

   e. After a user logs in

8. Which of the following protocols are supported for interactive access to a router?

   a. FTP

   b. Telnet

   c. ICMP

   d. SSH

9. What types of TTY lines are used in an interactive access session?

   a. Local and remote
   b. Standard and virtual
   c. Single and double
   d. Sticky and dialup

10. To protect the router from DoS attacks against login attempts, which feature can you activate on the Cisco router?

    a. Login attempt log messaging
    b. Delays between login attempts
    c. Login shutdown
    d. Auto device shutdown
    e. Notch attempts
    f. None of the above

11. What command enables the secure copy feature?

    a. **ip scopy server enable**
    b. **ssh scpserver enable**
    c. **ip scp server enable**
    d. **ip scp enable**
    e. **ssh server enable**
    f. None of the above

12. What is not a feature of port security for switches?

    a. Block input to Ethernet, Fast Ethernet, and Gigabit ports.
    b. Functions by analyzing MAC address and comparing it to authorized MACs.
    c. The global resource on a system is 2048 MAC addresses plus a default MAC address.
    d. The port can dynamically configure itself by the attached device's MAC address.
    e. MACs are stored in NVRAM.

13. Which of the following statements is true when configuring port security?

   a. Port security cannot be configured on a trunk port.

   b. Port security cannot be enabled on a SPAN port.

   c. Dynamic, static, or permanent CAM entries cannot be configured on a secured port.

   d. When port security is enabled, any static or dynamic CAM entries associated with the port are cleared.

   e. All of the above.

14. What does the AutoSecure feature do within Cisco IOS?

   a. Reads configuration and makes recommendations to secure platform

   b. Enables IP security services and disables common vulnerabilities

   c. Secures communications with other systems

   d. Enables detailed audit logging and tracking

   e. Turns on heuristic analysis to auto-block new threats

The answers to the "Do I Know This Already?" quiz are found in the appendix. The suggested choices for your next step are as follows:

- **12 or less overall score**—Read the entire chapter. This includes the "Foundation Topics" and "Foundation Summary" sections and the "Q&A" section.

- **13 or 14 overall score**—If you want more review on these topics, skip to the "Foundation Summary" section and then go to the "Q&A" section. Otherwise, move on to Chapter 6, "Authentication."

Foundation Topics

An important concern when securing the network is to prevent unauthorized users from gaining access to router and switch administrative access interfaces. If an intruder were to gain console or terminal access into a networking device, such as a router, switch, or network access server, that person could significantly damage your network—perhaps by reconfiguring the device, or even by just viewing the device's configuration information.

Typically, you want administrators to have access to your networking device; you do not want other users on your LAN or those dialing in to the network to have administrative access to the router.

You can take steps to securely configure your administrative access to your network devices. Password protection enables you to restrict access to a network or a network device. Privilege levels enable you to define which commands users can issue after they have logged in to a network device.

Privilege Levels

By default, the Cisco IOS Software command-line interface (CLI) has two levels of access to commands:

- User EXEC mode (level 1)
- Privileged EXEC mode (level 15)

However, you can configure additional levels of access to commands, called *privilege levels*, to meet the needs of your users while protecting the system from unauthorized access. You can configure up to 16 privilege levels, from level 0, which is the most restricted level, to level 15, which is the least-restricted level.

Access to each privilege level is enabled through separate passwords, which you specify when configuring the privilege level. If you want a certain set of users to be able to configure only certain interfaces, but not allow them access to other configuration options, for instance, you could create a separate privilege level for only specific interface configuration commands and distribute the password for that level to those users.

Securing Console Access

Unauthorized users can access the console administrative interface by attaching a terminal (for instance, a laptop) directly to a router. Physical security has to be put in place for the router to prevent unauthorized users from gaining access to routers and the console interface. You also have to configure the router to require a password when users try to access it via the console port. The router or switch can authenticate users locally or via a remote security database, such as Cisco Secure Access Control Server (CSACS).

The console password can have from 1 to 25 uppercase and lowercase alphanumeric characters. Example 5-1 shows the configuration of a console password for a router.

Example 5-1 *Simple Console Interface Configuration for a Router*

```
Router(config)#line console 0
Router(config-line)#login
Router(config-line)#password Mer91n2
```

After a password has been configured for the console, the user might enter the password for the console interface to enter the EXEC mode. The EXEC mode is identified with the greater than sign (>) after the router name, as shown in Example 5-2.

Example 5-2 *Accessing the EXEC Mode*

```
User access Verification
Password:Mer91n2
Router>
```

However, you cannot make configuration changes to the router unless you are in the privileged mode. You can access the privileged mode by typing the **enable** command and the enable password, if one is already configured.

Another method of accessing the console is to use centralized authentication via authorization, authentication, and accounting (AAA) services discussed later in this chapter and in Chapters 9, "Cisco Secure Access Control Server," and 10, "Administration of Cisco Secure Access Control Server for Windows."

Configuring the enable Password

To set a local password to control access to various privilege levels, use the **enable password** command in global configuration mode:

```
enable password [level level] {password | [encryption-type] encrypted-password}
```

Table 5-2 shows the different options of the **enable** command.

Table 5-2 *enable Command Options**

level *level*	Level for which the password applies. You can specify up to 15 privilege levels, using numbers 1 through 15. This is an optional parameter that provides not only authentication but also authorization.
password	Password user types to enter enable mode.
encryption-type	Cisco proprietary algorithm used to encrypt the password. Currently the only encryption type available is 7.
encrypted-password	Encrypted password user types to enter enable mode. The configuration file shows an encrypted string for this parameter.

* This table has been reproduced by Cisco Press with the permission of Cisco Systems Inc. Copyright © 2003 Cisco Systems, Inc. All Rights Reserved.

The use of the privilege level in the **enable** command helps administrators and managers better manage user access to the routers. After you specify the level and the password, give the password to the users who need to access this level. Use the **privilege level** configuration command to specify commands accessible at various levels.

You will not ordinarily enter an encryption type. Typically, you enter an encryption type only if you copy and paste into this command a password that has already been encrypted by a Cisco router.

> **CAUTION** If you specify an encryption type and then enter a clear-text password, you will not be able to reenter enable mode. You cannot recover a lost password that has been encrypted by any method; however, you can enter the router in ROMMON mode. Use **config-reg 0x2142** (depending on the router) and change the enable password but do not recover the old password. For more information on password recovery, refer to Cisco.com.

You can enable or disable password encryption with the **service password-encryption** command.

An enable password is defined as follows:

- It must contain from 1 to 25 uppercase and lowercase alphanumeric characters.
- It must not have a number as the first character.
- It can have leading spaces, but they are ignored; however, intermediate and trailing spaces are recognized.
- It can contain the question mark (?) character if you precede the question mark with the Ctrl-V key combination when you create the password; for example, to create the password gen?X, do the following:

    Step 1 Enter **gen**.

    Step 2 Press **Ctrl-V**.

    Step 3 Enter **?X**.

When the system prompts you to enter the enable password, you need not precede the question mark with the Ctrl-V; you can just enter **gen?X** at the password prompt. The following example enables the password Mer0n for privilege level 4:

```
enable password level 4 Mer0n
```

enable secret Command

The **enable secret** command provides better security by storing the enable secret password using a nonreversible cryptographic function. The added layer of security encryption proves useful in environments where the password crosses the network or is stored on a Trivial File Transfer Protocol (TFTP) server:

```
enable secret [level level] {password | [encryption-type] encrypted-password}
```

You do not ordinarily enter an encryption type. Typically, you enter an encryption type only if you paste into this command an encrypted password that you copied from a router configuration file.

> **NOTE** It is important to mention that encrypting the password only helps if the configuration file of the router is configured by an unauthorized party. However, even with password encryption enabled, it is still passed in the clear of a network wire between the user workstation and the router or switch. You should consider other means of encryption such as Secure Shell (SSH) if you have concerns about passwords being sniffed or captured over the network path.

If you use the same password for the **enable password** and **enable secret** commands, you receive an error message warning that this practice is not recommended, but the password is accepted. By using the same password, however, you undermine the additional security the **enable secret** command provides.

> **NOTE** After you set a password using the **enable secret** command, a password set using the **enable password** command works only if the enable secret is disabled or an older version of Cisco IOS Software is being used, such as when running an older rxboot image. In addition, you cannot recover a lost password that has been encrypted.

An enable password is defined as follows:

- It must contain from 1 to 25 uppercase and lowercase alphanumeric characters.
- It must not have a number as the first character.
- It can have leading spaces, but they are ignored. However, intermediate and trailing spaces are recognized.
- It can contain the question mark (?) character if you precede the question mark with the key combination Ctrl-V when you create the password.

Example 5-3 specifies the enable secret password of ladyhawk.

Example 5-3 *Enable Secret Password Configuration*

```
Router(config)#enable secret ladyhawk
```

After you specify an enable secret password, users must enter this password to gain access. Any passwords set through **enable password** will no longer work.

In addition to the enable secret password, the **username secret** command provides an additional layer of security over the username password. It also provides better security by encrypting the password using nonreversible Message Digest 5 (MD5) encryption and storing the encrypted text. The added layer of MD5 encryption proves useful in environments in which the password crosses the network or is stored on a TFTP server. This command was introduced in Cisco IOS Software Release 12.0(18)S. The syntax to encrypt a user password with MD5 is as follows:

```
username name secret {[0] password | 5 encrypted-secret}
```

Example 5-4 illustrates the use of the **username secret** command.

Example 5-4 *username secret Command*

```
Router(config)# username Aida secret 0 ysf600
```

service password-encryption Command

The **service password-encryption** command stores passwords in an encrypted manner in router configuration:

```
Router(config)#service password-encryption
```

The actual encryption process occurs when the current configuration is written or when a password is configured. Password encryption is applied to all passwords, including username passwords, authentication key passwords, privileged command passwords, console and virtual terminal line access passwords, and BGP neighbor passwords. This command is primarily useful for keeping unauthorized individuals from viewing your password in your configuration file; however, it does not provide the highest level of network security. When password encryption is enabled, the encrypted form of the passwords displays.

> **NOTE** The **service password-encryption** command does not encrypt SNMP or TACACS strings.

Configuring Multiple Privilege Levels

To configure a new privilege level for users and associate commands with a privilege level, use the **privilege** command syntax as follows:

```
privilege mode [all] {level level | reset} command-string
```

Table 5-3 shows the different options that the **privilege** command provides.

Table 5-3 privilege *Command Options*

Option	Description
all	(Optional) Changes the privilege level for all the suboptions to the same level.
level *level*	Specifies the privilege level you are configuring for the specified command or commands. The *level* argument must be a number from 0 to 15.
reset *command-string*	Resets the privilege level of the specified command or commands to the default and removes the privilege level configuration from the running-config file.

The password for a privilege level defined using the **privilege** command is configured using the **enable secret** command.

Level 0 can be used to specify a more-limited subset of commands for specific users or lines. For example, you can allow user user1 to use only the **show users** and **exit** commands.

> **NOTE** Five commands are associated with privilege level 0: **disable**, **enable**, **exit**, **help**, and **logout**. If you configure AAA authorization for a privilege level greater than 0, these five commands are not included.

When you set the privilege level for a command with multiple words, note that the commands starting with the first word will also have the specified access level. If you set the **show ip route** command to level 15, for example, the **show** commands and **show ip** commands are automatically set to privilege level 15—unless you set them individually to different levels. This is necessary because you cannot execute, for instance, the **show ip** command unless you have access to **show** commands.

To change the privilege level of a group of commands, use the **all** keyword. When you set a group of commands to a privilege level using the **all** keyword, all commands that match the beginning string are enabled for that level, and all commands available in submodes of that command are enabled for that level. If you set the **show ip** keywords to level 5, for example, **show** and **ip** are changed to level 5 and all the options that follow the **show ip** string, such as **show ip accounting**, **show ip aliases**, **show ip bgp**, and so on, are available at privilege level 5.

Example 5-5 shows how to set axsforL14 as the password users must enter to use level 14 commands.

Example 5-5 *Sample Configuration Showing Different Passwords Set for Different Privilege Levels*

```
Router(config)#enable secret level 14 axsforL14
```

Example 5-6 shows how to set the **show** and **ip** keywords to level 6. The suboptions coming under **ip** are also allowed to users with privilege level 6 access.

Example 5-6 *Assignment of the* **show ip** *Command with the Privilege Level 6*

```
Router(config)# privilege exec all level 6 show ip
```

Warning Banners

In some jurisdictions, civil and criminal prosecution of crackers who break into your systems is made much easier if you provide a banner informing unauthorized users that their use is, in fact, unauthorized. In other jurisdictions, you might be forbidden to monitor the activities of even unauthorized users unless you have taken steps to notify them of your intent to do so. One way to provide this notification is to put it into a banner message configured with the Cisco IOS **banner login** command.

Legal notification requirements are complex and vary in each jurisdiction and situation. Even within jurisdictions, legal opinions vary, and you should discuss this issue with your own legal counsel. In cooperation with counsel, consider which of the following information you should include in your banner:

- A notice that the system is to be logged in to or used only by specifically authorized personnel, and perhaps information about who might authorize use.

- A notice that any unauthorized use of the system is unlawful and might be subject to civil and criminal penalties.

- A notice that any use of the system might be logged or monitored without further notice, and that the resulting logs might be used as evidence in court.

- Specific notices required by specific local laws.

From a security, rather than a legal, point of view, your login banner usually should not contain any specific information about your router, its name, its model, what software it is running, or who owns it; crackers might abuse such information.

The banner messages can be displayed when a user enters privileged EXEC mode, upon line activation, on an incoming connection to a virtual terminal, or as a message of the day. To create a banner message, use the following command:

   banner {exec | incoming | login | motd} *d message d*

Figure 5-1 shows the banner displayed when a user connects to a router via its console connection.

Figure 5-1 *Displayed Banner When a Connection Is Made*

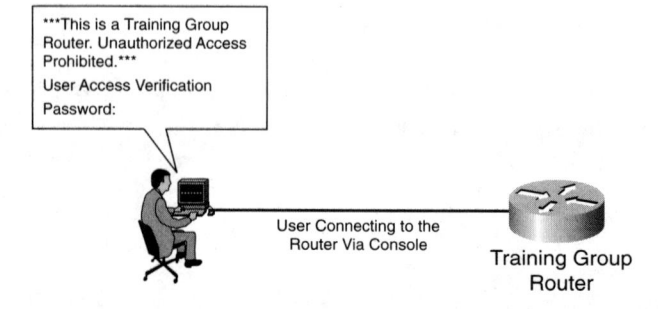

Table 5-4 shows the different command options of the **banner** command.

Table 5-4 banner *Command Options*

Command Syntax	Description
exec	Specifies a message to display when an EXEC process is created.
incoming	Specifies a banner to display when an incoming connection to a line (asynchronous) from a host on the network is made.
login	Identifies a message to display during a login before the username and password login prompts.
motd	Specifies the display of the message-of-the-day (MOTD) banner. This banner displays at login and is useful for sending messages that affect all network users.
d	You must use a delimiting character of your choice, such as a pound sign (#). You cannot use the delimiting character in the banner message.
message	The actual message text.

Interactive Access

Besides those already discussed, there are additional ways to get interactive connections to routers. Cisco IOS Software, depending on the configuration and software version, might support connections via Telnet; rlogin; SSH; non-IP–based network protocols, such as Local Area Transport (LAT), Maintenance Operation Protocol (MOP), X.29, V.120, and possibly other protocols as well as via local asynchronous connections and modem dial ins. More protocols for interactive access are always being added. Interactive Telnet access is available not only on the standard Telnet Transmission Control Protocol (TCP) port (port 23) but also on a variety of higher-numbered ports.

All interactive access mechanisms use the Cisco IOS tty abstraction. (In other words, they all involve sessions on "lines" of one sort or another.) Local asynchronous terminals and dialup modems use standard lines, known as *ttys*. Remote network connections, regardless of the protocol, use virtual ttys, or *vtys*. The best way to protect a system is to make certain that appropriate controls are applied on all lines, including both TTY lines and vty lines.

Because it is difficult to be certain that all possible modes of access have been blocked, make sure that logins on all lines are controlled using some sort of authentication mechanism, even on machines that are supposed to be inaccessible from untrusted networks. Doing so is especially important for vty lines and for lines connected to modems or other remote-access devices.

Securing vty Access

Any vty should be configured to accept connections only with the protocols actually needed. You can do this with the **transport input** command. A vty expected to receive only Telnet sessions could be configured with **transport input telnet**, for example, whereas a vty permitting both Telnet and

SSH sessions would have **transport input telnet ssh**. Not configuring a transport input for vty access is also an option if you want to disable the service.

One way to reduce this exposure is to configure an access list on all vty lines. An access list restricts the router to accept connections only from a single, specific administrative workstation. Example 5-7 shows a sample configuration of an access list configured on a vty line.

Example 5-7 *Access List Configured on a vty Line*

```
Router(config)#access-list 10 permit 192.168.100.0 0.0.0.255
Router(config)#line vty 1 5
Router(config-line)#access-class 10 in
```

Another useful tactic is to configure vty timeouts using the **exec-timeout** command. This command prevents an idle session from consuming a vty indefinitely. Although its effectiveness against deliberate attacks is relatively limited, it does provide some protection against sessions accidentally left idle.

Passwords sent over Telnet sessions are in clear text, which makes Telnet an insecure method. SSH is a more secure method of interactive access to the router.

Cisco IOS Software Release 12.3T has new functionalities available to better secure the virtual login connection. The following is a list of these features:

- **Delays between login attempts**—This feature protects the router from username and password dictionary attacks by limiting the number of successive login attempts within a certain period of time. To specify a delay between login attempts, use the **login delay** *seconds* command. The *seconds* variable is an integer between 1 and 10. The default number of seconds between login attempts is 1 second.

- **Login shutdown**—This feature protects the router from denial-of-service (DoS) attacks by preventing continuous login attempts in a specified period of time. Use the **login block-for** *seconds* **attempts** *tries* **within** *seconds* global configuration command to specify the amount of time the router must wait before allowing another login attempt. This wait time is called the *quiet period*. The *seconds* variable is an integer between 1 and 65535. The *tries* variable specifies the maximum number of failed login attempts that trigger the quiet period. During the quiet period, all attempts from all types of login protocols (that is, Telnet, SSH, and Hypertext Transfer Protocol [HTTP]) are denied. You can exclude specific hosts or subnets from the quiet period via the **login quiet-mode access-class** {*acl-name* | *acl-number*} global configuration mode. The *acl-name* and *acl-number* parameter specifies a standard, extended, or named access list.

- **Login attempt logging messages**—Successful and failed login attempts might be logged via the **login on-failure log** [every *login*] and **login on-success log** [every *login*] commands. The *login* parameter is an integer between 1 and 65535 specifying the number of attempts prior to the generation of logging messages. Future releases of Cisco IOS Software will provide support for Simple Network Management Protocol (SNMP) traps.

Example 5-8 shows a sample router configuration to block all login requests for 120 seconds if the 30 failed login attempts are exceeded within 180 seconds. A delay of 3 seconds is configured between each login attempt. Failed login attempts are logged for every third login attempt.

Example 5-8 login *Command Sample Configuration*

```
Router1#configure terminal
Router1(config)#login block-for 120 attempts 30 within 180
Router1(config)#login delay 3
Router1(config)#login on-failure log every 3
```

To verify the configuration, use the **show login** command. Example 5-9 provides a sample output for this command.

Example 5-9 *Sample* **show login** *Output*

```
Router1#show login
A login delay of 3 seconds is applied.
No Quiet-Mode access list has been configured.
Every 3 failed login is logged.
Router enabled to watch for login Attacks.
If more than 30 login failures occur in 180 seconds or less,
logins will be disabled for 120 seconds.
Router presently in Normal-Mode.
Current Watch Window remaining time 163 seconds.
Present login failure count 0.
```

SSH Protocol

SSH was originally intended to replace Telnet and the UNIX **r-** commands. Both of these session types have vulnerabilities, such as spoofing, man-in-the-middle attacks, and session hijacking, which SSH addresses and mitigates for the most part. For maintaining confidentiality and integrity in accessing a router, it is recommended to deploy SSH rather than Telnet.

SSH protects against the following:

- Attacks from machines pretending to be another server, router, or a domain name server

- Internet Protocol (IP) spoofing, where a remote host sends out packets that pretend to come from another trusted host

- IP source routing, where a host can pretend that an IP packet comes from another trusted host

- Domain Name System (DNS) spoofing, where an attacker forges name server records

- Interception of clear-text passwords or data on the network

- Manipulation of data by people in control of intermediate hosts

There are two versions of SSH, SSHv1 and SSHv2. The difference between them is that they are completely different protocols. SSHv2 is an entire rewrite of SSH1. Each version encrypts different parts of the packet, and SSH1 uses server and host keys, whereas just host keys are used in SSHv2. SSHv2 is also more secure and has better performance and portability. SSHv2 mitigates the man-in-the-middle attack vulnerability of SSHv1.

Cisco IOS Software Release 12.3(4)T introduces limited SSH server support for SSHv2. Support for Execution Shell and Secure Copy Protocol (SCP) is provided in this new version. SSHv2 clients are currently not supported in Cisco IOS Software.

NOTE SSHv1 has been supported on Cisco IOS Software since Cisco IOS Software Release 12.1T.

Setting Up SSH on a Cisco IOS Router or Switch

Before you start with the SSH configuration, download the required image on your router. The SSH server requires you to have an IPsec (DES or 3DES) encryption software image from Cisco IOS Software Release 12.1(1)T downloaded on your router. The SSH client requires you to have an IPsec (DES or 3DES) encryption software image from Cisco IOS Software Release 12.1(3)T downloaded on your router.

Here is an excerpt from 12.3 command reference documentation: "SSH Version 2 supports only the following crypto algorithms: aes128-cbc, aes192-cbc, aes256-cbc, and 3des-cbc. SSH Version 2 is supported only in 3DES images."

Configuring a Router for SSHv2 Using a Host Name and Domain Name

To enable SSH support via host name and domain name on a Cisco IOS router or switch, follow these steps:

Step 1 Configure a host name on the device using the **hostname** *hostname* global configuration command.

Step 2 Configure a domain name using the **ip domain-name** *name* global configuration command.

Step 3 Generate encryption keys for local and remote authentication using the **crypto key generate rsa** command.

Step 4 If desired, configure SSH control variables via the **ip ssh** {*timeout seconds* | *authentication-retries integer*} in global configuration mode.

Step 5 Specify the version of SSH to run on the device using the **ip ssh version** {**1**|**2**} command. This is an optional command. If the version is not specified, SSH will run in compatibility mode, where both versions are supported.

Example 5-10 shows a sample configuration of SSH using host and domain names.

Example 5-10 *Configuring SSH Using Host and Domain Names*

```
Router1#configure terminal
Router1(config)#hostname Router-ssh
Router-ssh(config)#ip domain-name mycompany.com
Router-ssh(config)#crypto key generate rsa
The name for the keys will be: Router-ssh.mycompany.com
Choose the size of the key modulus in the range of 360 to 2048 for your
General Purpose Keys. Choosing a key modulus greater than 512 may take
  a few minutes.

How many bits in the modulus [512]: 768
% Generating 768 bit RSA keys ...[OK]
Router-ssh(config)#
Router-ssh(config)#ip ssh time-out 60
Router-ssh(config)#ip ssh version 2
```

Configuring a Router for SSHv2 Using RSA Key Pairs

You can also enable SSH using an RSA key pair name. (RSA is named for its inventors: Rivest, Shamir, and Adelman.) Cisco IOS routers might have many RSA key pairs. The key pair used for SSH must be specified during the SSH configuration.

To enable SSH support via RSA key pairs on a Cisco IOS router or switch, follow these steps:

Step 1 Specify the RSA key pair to be used for SSH via the **ip ssh rsa keypair-name** *keypair-name* global configuration command.

Step 2 Generate encryption keys for local and remote authentication using the **crypto key generate rsa usage-keys label** *key-label* modulus *modulus-size* global configuration command. Modulus size must be 768 bits for SSH.

Step 3 If desired, configure SSH control variables via the **ip ssh {timeout** *seconds* | **authentication-retries** *integer*} global configuration command.

Step 4 Specify the version of SSH to run on the device using the **ip ssh version {1 | 2}** command. This is an optional command. If the version is not specified, SSH runs in compatibility mode, where both versions are supported.

Example 5-11 shows a sample configuration of SSH using RSA key pairs.

Example 5-11 *SSH Configuration Using RSA Key Pairs*

```
Router-ssh(config)# crypto key generate rsa usage-keys label rsakey modulus 768
The name for the keys will be: rsakey
% The key modulus size is 768 bits
% Generating 768 bit RSA keys ...[OK]
```

continues

Example 5-11 *SSH Configuration Using RSA Key Pairs (Continued)*

```
Router-ssh(config)#
*Mar  1 02:38:06.283: %SSH-5-ENABLED: SSH 2.0 has been enabled
Router-ssh(config)#ip ssh rsa keypair-name rsakey
Router-ssh(config)#ip ssh time-out 60
Router-ssh(config)#ip ssh version 2
```

SSH might be used for terminal-line access rather than reverse Telnet. Replacing reverse Telnet with SSH adds encryption and decryption processing that might affect the performance of the tty lines.

Example 5-12 shows the configuration commands necessary to configure SSH reverse Telnet. SSH must already be configured on the router prior to these steps. The line must be configured in its own rotary group. The **transport input ssh** command is used to specify SSH protocol for terminal access. The **ip ssh port 2001 rotary 1** command is used to enable secure network access to the tty line.

Example 5-12 *SSH Reverse Telnet Configuration*

```
line 1
   no exec
   login authentication default
rotary 1
transport input ssh
ip ssh port 2001 rotary 1
```

Secure Copy

One of the benefits of using SSH is the SCP feature. SCP uses a security and authentication mechanisms for copying router configuration or software image files. SCP works similarly to the Berkeley **r-** tools remote copy (**rcp**) command. Authorized users are permitted to copy Cisco IOS files to and from the router. Beginning in Cisco IOS Software Release 12.2T, SCP features are supported.

SCP uses SSH RSA key pairs as its transport and authentication. Therefore, SSH must be enabled on the router prior to using SCP. Activating local- or network-based AAA is also required to validate the user.

Use the **ip scp server enable** command to configure SCP.

Example 5-13 shows the commands necessary to configure SCP server-side functionality using local authentication.

Example 5-13 *SCP Server-Side Configuration Using Local Authentication*

```
Router-ssh#configure terminal
Router-ssh(config)#aaa new-model
Router-ssh(config)#aaa authentication login default local
Router-ssh(config)#aaa authorization exec default local
Router-ssh(config)#username usescp password usescp
Router-ssh(config)#ip ssh time-out 180
Router-ssh(config)#ip scp server enable
```

Port Security for Ethernet Switches

The port security feature enables you to block input to an Ethernet, Fast Ethernet, or Gigabit Ethernet port when the MAC address of the station attempting to access the port differs from any of the MAC addresses specified for that port. This functionality is also referred to as *MAC address lockdown*.

The global resource for the system varies based on the switch platform and amount of memory available. For example, the 2900 series XL switch has a global resource limitation of 1024 MAC addresses. In addition to this global resource space, there is space for one default MAC address per port to be secured. The total number of MAC addresses that can be specified per port is limited to the global resource of 1024 plus 1 default MAC address. The total number of MAC addresses on any port cannot exceed 1025.

The maximum number of MAC addresses for each port is determined by your network configuration. The following combinations are examples of valid allocations:

- 1025 (1 + 1024) addresses on 1 port and 1 address each on the rest of the ports
- 513 (1 + 512) addresses each on 2 ports in a system and 1 address each on the rest of the ports
- 901 (1 + 900) addresses on 1 port, 101 (1 + 100) addresses on another port, 25 (1 + 24) addresses on the third port, and 1 address each on the rest of the ports

After you allocate the maximum number of MAC addresses on a port, you can either specify the secure MAC address for the port manually or you can have the port dynamically configure the MAC address of the connected devices. Out of an allocated number of maximum MAC addresses on a port, you can manually configure all, allow all to be autoconfigured, or configure some manually and allow the rest to be autoconfigured. After addresses have been manually configured or autoconfigured, they are stored in nonvolatile RAM (NVRAM) and maintained after a reset.

After you allocate a maximum number of MAC addresses on a port, you can specify how long addresses on the specified port will remain secure. After the age time expires, the MAC addresses on the port become insecure. By default, all addresses on a port are secured permanently.

In the event of a security violation, you can configure the port to go into shutdown, protect, or restrictive mode. The *shutdown* mode option enables you to specify whether the port is permanently disabled or disabled for only a specified time. The default is for the port to shut down permanently. The *protect* mode option sets a maximum number of allowed MAC addresses per port. If the maximum number of MAC addresses is reached, the switch drops the packets with unknown source addresses until the number drops below the maximum value. The *restrictive* mode option enables you to configure the port to remain enabled during a security violation and drop only packets that are coming in from insecure hosts.

When a secure port receives a packet, the source MAC address of the packet is compared to the list of secure source addresses that were manually configured or autoconfigured (learned) on the port. If a MAC address of a device attached to the port differs from the list of secure addresses, the port shuts down permanently (default mode), shuts down for the time you have specified, or drops incoming packets from the insecure host. The port's behavior depends on how you configure it to respond to a security violation.

When a security violation occurs, the light-emitting diode (LED) link for that port turns orange, and a link-down trap is sent to the SNMP manager. An SNMP trap is not sent if you configure the port for restrictive violation mode. A trap is sent only if you configure the port to shut down during a security violation.

Configuring Port Security

Consider the following when configuring port security:

- You cannot configure port security on a trunk port.

- You cannot enable port security on a Switched Port Analyzer (SPAN) destination port and vice versa.

- You cannot configure dynamic, static, or permanent CAM entries on a secure port.

- When you enable port security on a port, any static or dynamic content-addressable memory (CAM) entries associated with the port are cleared; any currently configured permanent CAM entries are treated as secure.

> **NOTE** The following port security configuration is for Catalyst OS (CatOS)—it differs for switches running native IOS (for example, Cat 3550 and 2950XL).

Example 5-14 shows how to enable port security using the learned MAC address on a port and verify the configuration.

Example 5-14 *Enabling Port Security on Port 2/19*

```
ATL-SWITCH> (enable) set port security 2/19 enable
Port 2/19 port security enabled with the learned MAC address.
Trunking disabled for Port 2/19 due to security mode
ATL-SWITCH> (enable) show port 2/19
Port  Name               Status     Vlan       Level  Duplex Speed Type
----- ------------------ ---------- ---------- ------ ------ ----- ------------
 2/19                    connected  2          normal full    100  100BaseTX
Port  Security Secure--Addr    Last-Src-Addr     Shutdown Trap    IfIndex
----- -------- --------------- ----------------- -------- ------- -------
```

Example 5-14 *Enabling Port Security on Port 2/19 (Continued)*

```
2/19   enabled   00-90-2b-03-CC-08 00-90-2b-03-CC-08 No         disabled 1081
Port       Broadcast-Limit Broadcast-Drop
--------   --------------- --------------
2/19              -              0
Port   Align-Err   FCS-Err     Xmit-Err    Rcv-Err    UnderSize
-----  ----------  ----------  ----------  ----------  ---------
2/19       0           0           0           0           0
Port   Single-Col Multi-Coll Late-Coll   Excess-Col Carri-Sen Runts      Giants
-----  ---------- ---------- ----------  ---------- --------- ---------- ---------
2/19       0          0          0           0          0         0          0
Last-Time-Cleared
-------------------------
Fri June 8 2003, 19:31:10
```

Example 5-15 shows how to enable port security on a port and manually specify the secure MAC address.

Example 5-15 *Manually Assigning a MAC Address to a Port*

```
ATL-SWITCH> (enable) set port security 3/8 enable 00-A0-3C-03-2D-03
Port 3/8 port security enabled with 00-A0-3C-03-2D-03 as the secure mac address
Trunking disabled for Port 3/8 due to Security Mode
ATL-SWITCH> (enable)
ATL-SWITCH> (enable) show port security 3/8
Port  Security Violation Shutdown-Time Age-Time Max-Addr Trap     IfIndex
----- -------- --------- ------------- -------- -------- -------- -------
3/8   enabled  shutdown        300         60       5    disabled   921
Port  Num-Addr Secure--Addr      Age-Left Last-Src-Addr    Shutdown/Time-Left
----- -------- ----------------- -------- ---------------- ------------------
 3/8      1    00-A0-3C-03-2D-03    60    00-A0-3C-03-2D-03 no            -
```

AutoSecure

One of the new features in Cisco IOS Software Release 12.3T is AutoSecure. AutoSecure provides a mechanism to disable vulnerable services and enhance router security posture with a single command.

The AutoSecure feature enables IP security services, such as SSH, and disables common vulnerable IP services, such as HTTP.

> **NOTE** Using AutoSecure does not guarantee immunity from exploitation, but an enhancement to the default out-of-the-box router configuration.

You can use Cisco AutoSecure in interactive or noninteractive modes. The interactive mode requires user input for feature activation and de-activation. The noninteractive mode uses the Cisco recommended settings to activate and de-activate features.

Use the EXEC **auto secure** command to configure AutoSecure. The syntax for this command is as follows:

```
auto secure [management | forwarding | firewall | login | SSH | NTP} [no-interact]
```

Table 5-5 describes the **auto secure** command parameters.

Table 5-5 *auto secure Command Parameters*

Parameter	Description
management	Specifies to secure only the management plane.
forwarding	Specifies to secure only the forwarding plane.
firewall	Specifies the activation of Cisco IOS router Context-Based Access Control (CBAC) firewall features only.
login	Specifies the activation of only the login security features.
SSH	Specifies the activation of only the SSH security features.
NTP	Specifies the activation of only the Network Time Protocol (NTP) security features.
no-interact	Specifies an automatic implementation of recommended security policies. Using this option means the user will not be prompted to make configuration decisions.

In case of Cisco AutoSecure interactive mode, a user dialog menu is used to set security configuration parameters.

> **NOTE** Currently, no rollback feature is available for Cisco AutoSecure. To remove the AutoSecure configuration, save the running configuration, erase the configuration, and re-apply the saved configuration to the router.

AutoSecure

Example 5-16 shows a partial AutoSecure configuration for enabling the SSH security features.

Example 5-16 *AutoSecure SSH Configuration*

```
Router1#auto secure ssh
--- AutoSecure Configuration ---
*** AutoSecure configuration enhances the security of
the router, but it will not make it absolutely resistant ***

AutoSecure will modify the configuration of your device.
All configuration changes will be shown. For a detailed
explanation of how the configuration changes enhance security
and any possible side effects, please refer to Cisco.com for
Autosecure documentation.
At any prompt you may enter '?' for help.
Use Ctrl-C to abort this session at any prompt.
Gathering information about the router for AutoSecure

Is this router connected to internet? [no]: no

Configure SSH server? [yes]: yes

This is the configuration generated:

crypto key generate rsa general-keys modulus 1024
ip ssh time-out 60
ip ssh authentication-retries 2
line vty 0 15
transport input ssh telnet
!
end
```

Foundation Summary

The "Foundation Summary" section of each chapter lists the most important facts from the chapter. Although this section does not list every fact from the chapter that will be on your SNRS exam, a well-prepared candidate should at a minimum know all the details in each "Foundation Summary" section before going to take the exam.

Securing the administrative accesses to the Cisco router is one of the many tasks that needs to be completed to secure the network. Administrative access points, such as the console and vty access, should require, at minimum, a password. Creating privilege levels on the Cisco routers for the type of commands authenticated users can execute provides an additional level of security. New features, such as login security, SSHv2, and Cisco AutoSecure provide enhanced security capabilities to Cisco routers.

Q&A

As mentioned in the section "How to Use This Book" in the Introduction, you have two choices for review questions: the Q&A questions here or the exam simulation questions on the CD-ROM. The questions that follow present a bigger challenge than the exam itself because they use an open-ended question format. By using this more difficult format, you can exercise your memory better and prove your conceptual and factual knowledge of this chapter. You can find the answers to these questions in the appendix.

1. Which steps can be taken to secure administrative access to a network device?

2. What is level 15 access called on the Cisco command-line interface?

3. Access to the Cisco device console can be secured using which two methods?

4. Can the same password be used for the enable password and enable secret password?

5. If you set the privilege level of the **show ip route** command, what other effects need to be considered?

6. What are the two modes of interactive access to lines on a Cisco router, and what is a good rule of thumb for these connections?

7. Which command enables you to implement secured vty access using only the **telnet** and **ssh** commands?

8. Describe the two methods for configuring SSH on a Cisco router or switch.

9. If an incorrect MAC address is detected on a switch that has port security enabled, the port can be configured to go to _____.

10. What does the command **set port security 2/4 enable 00-B0-34-03-2C-01** do on a Cisco switch?

Part III: AAA

Chapter 6   Authentication

Chapter 7   Authentication, Authorization, and Accounting

Chapter 8   Configuring RADIUS and TACACS+ on Cisco IOS Software

Chapter 9   Cisco Secure Access Control Server

Chapter 10 Administration of Cisco Secure Access Control Server for Windows

This chapter covers the following subjects:

- Authentication

- PAP, CHAP, and EAP Authentication

CHAPTER 6

Authentication

The identification and verification of users requesting access to a device or network is one of the core objectives of security. Although several methods of authentication are available, it is essential that one or a combination of authentication methods be used to secure the device or network. This chapter covers the different types of authentication methods that you can use for Cisco devices and networks.

"Do I Know This Already?" Quiz

The purpose of the "Do I Know This Already?" quiz is to help you decide whether you really need to read the entire chapter. If you already intend to read the entire chapter, you do not necessarily need to answer these questions now.

The 10-question quiz, derived from the major sections in "Foundation Topics" section of the chapter, helps you determine how to spend your limited study time.

Table 6-1 outlines the major topics discussed in this chapter and the "Do I Know This Already?" quiz questions that correspond to those topics.

Table 6-1 *"Do I Know This Already?" Foundation Topics Section-to-Question Mapping*

Foundation Topics Section	Questions Covered in This Section
Authentication	1–5, 7–9
PAP, CHAP, and EAP Authentication	6, 10

CAUTION The goal of self-assessment is to gauge your mastery of the topics in this chapter. If you do not know the answer to a question or are only partially sure of the answer, you should mark this question wrong for purposes of the self-assessment. Giving yourself credit for an answer you correctly guess skews your self-assessment results and might provide you with a false sense of security.

Chapter 6: Authentication

1. Which of the following statements are true? (Select two.)

    a. Authentication provides a method for verifying the identity of users.

    b. NAS cannot provide authentication.

    c. Usernames and passwords can be stored on NAS.

    d. Cisco does not support RADIUS.

2. Which of the following provide the least secure methods of authentication? (Select two.)

    a. Username/password static

    b. Username/password aging

    c. Session key one-time password

    d. Token cards

3. Which of the following security authentication protocols is not supported by Cisco network devices?

    a. TACACS+

    b. RADIUS

    c. LDAP

    d. Kerberos

    e. TLS

4. Which of the following command syntax is correct for creating a username and password locally on the NAS?

    a. Router(config)#**username** *Elvis* **password** *k0nj0*

    b. Router#**username Elvis password k0nj0**

    c. Router(config)#**set username Elvis set password k0nj0**

    d. Router#**set username Elvis password k0nj0**

5. Which port is reserved for TACACS+?

    a. UDP 1645

    b. TCP 1645

    c. TCP 49

    d. UDP 49

6. TACACS+ cannot forward password types for which of the following:

  a. ARA
  b. SLIP
  c. HAP
  d. PAP
  e. Standard Telnet

7. Which of the following best describes PAP?

  a. It involves a two-way handshake where the username and password are sent across the link in clear text.
  b. It sends username and passwords in encrypted format.
  c. It involves a one-way handshake.
  d. It is not supported by Cisco network devices.

8. Which of the following port does the router use for RADIUS authentication?

  a. UDP 49
  b. TCP 1645
  c. TCP 49
  d. UDP 1645

9. How many levels of protection can be configured in Kerberos?

  a. 3
  b. 5
  c. 1
  d. 10
  e. None of the above

10. Which of the following best describes the CHAP authentication protocol?

  a. Involves a three-way handshake
  b. Involves a one-way handshake
  c. Is not supported by Cisco network devices
  d. Sends password in clear text

The answers to the "Do I Know This Already?" quiz are found in the appendix. The suggested choices for your next step are as follows:

- **8 or less overall score**—Read the entire chapter. This includes the "Foundation Topics" and "Foundation Summary" sections and the "Q&A" section.

- **9 or 10 overall score**—If you want more review on these topics, skip to the "Foundation Summary" section and then go to the "Q&A" section. Otherwise, move on to Chapter 7, "Authentication, Authorization, and Accounting."

Foundation Topics

Authentication provides the method for verifying the identity of users and administrators who are requesting access to network resources, through username and password dialog boxes, challenge and response, token cards, and other methods. The following sections describe the most common authentication tools and mechanisms in the industry.

Authentication

Various types of authentication methods are available today. They range from the simple username and password databases to stronger implementation of token cards and one-time passwords (OTPs). Table 6-2 lists the authentication methods, from the strongest and most complex methods to the weakest and simple methods.

Table 6-2 *Authentication Methods*

Method	Description
Token cards and soft tokens	Token cards are small electronic devices. A personal identification number (PIN) is given to users. The user authenticates with a combination of the token card and the PIN.
OTPs	OTP systems are based on a secret pass-phrase that generates passwords. These are only good for one-time use, and thus guard against eavesdropping attacks, playback attacks, and password attacks.
Username and passwords (with expiration date)	The user must change the password because it expires (usually every 30 to 60 days).
Static username and password database	The password is the same unless changed by the system administrator. Vulnerable to password-cracking programs and other password attacks.
No username and password	This is usually an open invitation to attackers who discover the access method to gain access to the network system.

Configuring Line Password Authentication

You can provide access control on a terminal line by entering the password and establishing password checking. To do so, use the following commands in line configuration mode:

```
Router(config)#line console 0
Router(config-line)#password password
```

The password checker is case-sensitive and can include spaces; for example, the password Secret differs from the password secret, and you can use two words for an acceptable password. You can disable line password verification by disabling password checking. To do so, use the following command in line configuration mode:

```
Router(config-line)#no login authentication
```

> **NOTE** A password for a vty line must be configured for Telnet access to work.

Configuring Username Authentication

You can create a username-based authentication system in which a user is prompted for a username and password when attempting to access the network access server (NAS) or router. The username and password database is stored locally on the Cisco NAS device.

To establish username authentication, use the following commands in global configuration mode:

```
Router(config)#username name {nopassword | password password | password encryption-type encrypted-password}
```

The following example shows the creation of a user named Elvis with a password D0wnUnd3r:

```
Router(config)#username Elvis password D0wnUnd3r
```

> **NOTE** Passwords display in clear text in your configuration unless you enable the **service password-encryption** command.

Local username and passwords work very well for administrative access authentication on small networks. For remote-access dial-in users and large enterprises, however, using an external database to do authentication might be a good choice.

Remote Security Servers

A remote security database provides uniform remote-access security policies throughout the enterprise. It centrally manages all remote user profiles. Cisco network devices support the following three primary security server protocols:

- TACACS+
- RADIUS
- Kerberos

TACACS Overview

TACACS provides a way to centrally validate all users individually before they can gain access to a router or access server. TACACS was derived from the U.S. Department of Defense and is described in RFC 1492. TACACS is an open protocol and can be ported to most username or password databases. Figure 6-1 shows a TACACS+ server supporting a dialup client.

Figure 6-1 *TACACS+ Server Supporting a Dialup Client*

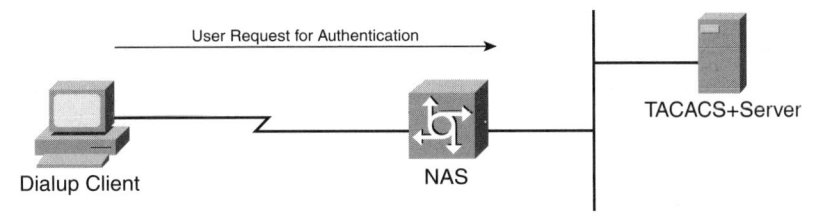

Cisco IOS Software implements TACACS to allow centralized control over who can access routers and access servers. Authentication also can be provided for Cisco IOS administration tasks on the routers' and access servers' user interfaces. With TACACS enabled, the router and access server prompts the user for a username and a password. Then, the router or access server queries a TACACS server to determine whether the user provided the correct corresponding password. TACACS was originally designed to run on UNIX workstations but can now run on Microsoft Windows, too. The three current versions of TACACS security server application are as follows:

- **TACACS**—An older access protocol, incompatible with the newer TACACS+ protocol. It provides password checking and authentication and notification of user actions for security and accounting purposes.

- **XTACACS**—An extension to the older TACACS protocol, supplying additional functionality to TACACS. Extended TACACS (XTACACs) provides information about protocol translator and router use. This information is used in UNIX auditing trails and accounting files.

- **TACACS+**—An improved protocol providing detailed accounting information and flexible administrative control over authentication and authorization processes. TACACS+ is a Cisco proprietary implementation and is facilitated and enabled only through authentication, authorization, and accounting (AAA) commands.

The TACACS and XTACACS protocols in Cisco IOS Software are officially considered end-of-maintenance and are no longer maintained by Cisco for bug fixes or enhancement.

TACACS+ provides for separate and modular authentication, authorization, and accounting facilities. TACACS+ allows for a single access control server (the TACACS+ daemon) to provide AAA services independently. Each service can be tied into its own database to take advantage of other services available on that server or on the network, depending on the capabilities of the daemon.

The TACACS+ protocol provides authentication between the NAS and the TACACS+ daemon, and it ensures confidentiality because all protocol exchanges between a NAS and a TACACS+ daemon are encrypted, typically using Message Digest 5 (MD5) algorithms. TACACS+ can forward the password types for AppleTalk Remote Access (ARA), Serial Line Internet Protocol (SLIP), Password Authentication Protocol (PAP), Challenge Handshake Authentication Protocol (CHAP), and standard Telnet. Therefore, clients can use the same username password for different protocols. TCP port 49 is reserved for TACACS+.

RADIUS Overview

RADIUS is a distributed client/server protocol that secures networks against unauthorized access. RADIUS includes two pieces: an authentication server and client protocols. A NAS operates as a client of RADIUS. The client is responsible for passing user information to designated RADIUS servers, and then acting on the response that is returned. RADIUS servers are responsible for receiving user connection requests, authenticating the user, and then returning all configuration information necessary for the client to deliver service to the user. RADIUS servers can act as proxy clients to other kinds of authentication servers. RADIUS uses User Datagram Protocol (UDP) as the communication protocol between the client and the server on port UDP 1645. Figure 6-2 shows a RADIUS server supporting a dialup client.

Figure 6-2 *Dialup Client Supported by a RADIUS Server*

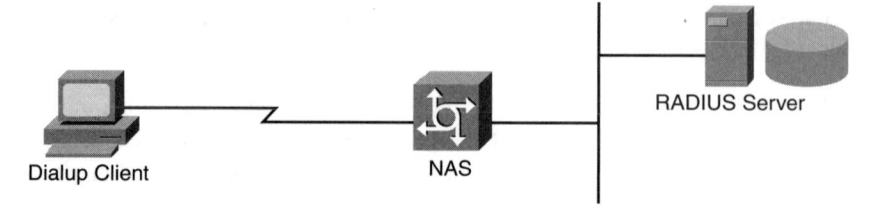

The RADIUS packet format includes the following fields:

- **Code**—Identifies the type of RADIUS command/response packet. There are five types of packets: Access-Request, Access-Accept, Access-Reject, Accounting-Request, and Accounting Response. This field is 1 octet.

- **Identifier**—Correlates a request packet to response packet and detects duplicate requests. This field is 1 octet.

- **Length**—Specifies the length of the entire RADIUS packet. This field is 2 octets.

- **Authenticator**—Authenticates the reply from the RADIUS server and is used in the password-hiding algorithm. The authenticator field is 16 octets long. The two types of authenticators are the Request Authenticator and the Response Authenticator:

- **Request Authentication**—Available in Access-Request and Accounting-Request packets
- **Response Authenticator**—Available in Access-Accept, Access-Reject, Access-Challenge, and Accounting-Response packets

- **Attributes**—Carries the specific AAA details for the request and response.

RADIUS attributes are divided into two categories: RADIUS Internet Engineering Task Force (IETF) attributes and the vendor-specific attributes (VSA). The IETF attributes are a predefined set of standard attributes that are used in the AAA communication between client and server. RADIUS VSAs are a subset of the IETF attribute: vendor-specific (attribute 26). Vendors may use this attribute to create up to 255 additional custom attributes and encapsulate them behind attribute 26. You can find a complete list of IETF attributes at Cisco.com. You can find a list of IETF attributes in RFC 2865, RFC 2866, RFC 2867, RFC 2868, and RFC 2869.

When a user attempts to log in and authenticate to an access server using RADIUS, the following steps occur:

1. The user is prompted for and enters a username and password.
2. The username and encrypted password are sent over the network to the RADIUS server.
3. The user receives one of the following responses from the RADIUS server:
   - **ACCEPT**—The user is authenticated.
   - **REJECT**—The user is not authenticated and is prompted to reenter the username and password, or access is denied.
   - **CHALLENGE**—A challenge is issued by the RADIUS server. The challenge collects additional data from the user.
   - **CHANGE PASSWORD**—A request is issued by the RADIUS server, asking the user to select a new password.

The ACCEPT or REJECT response is bundled with additional data that is used for EXEC or network authorization. You must first complete RADIUS authentication before using RADIUS authorization.

RADIUS encrypts only the password in the Access-Request packet from the client to the server. The remainder of the packet is unencrypted. Other information, such as username, authorized services, and accounting, could be captured by a third party.

The RADIUS server supports a variety of methods to authenticate a user. When it is provided with the username and original password given by the user, it can support PPP, PAP, CHAP, UNIX login, Extensible Authentication Protocol (EAP), and other authentication mechanisms.

RADIUS combines authentication and authorization. The Access-Accept packets sent by the RADIUS server to the client contain authorization information, which makes it difficult to decouple authentication and authorization. RADIUS does perform accounting separately.

RADIUS and TACACS+ have several key differences. Table 6-3 briefly compares TACACS+ and RADIUS.

Table 6-3 *Features of TACACS+ and RADIUS Protocols*

Functionality	TACACS+ Protocol	RADIUS Protocol
AAA support	AAA services are separate.	Authentication and authorization are combined, but accounting services are separate.
Transport protocol	TCP port 49.	The original RFC ports (still supported) are as follows: UDP port 1645 authentication/authorization UDP port 1646 accounting New (additional) ports are as follows: UDP port 1812 authentication/authorization UDP port 1813 accounting
Challenge/response	Bidirectional.	Unidirectional.
Protocol support	Multiproctocol support.	No NetBIOS Extended User Interface (NetBEUI), ARA, Netware Asynchronous Services Interface (NASI), and X.25 PAD.
Data integrity	The entire TACACS+ packet is encrypted in MD5.	Only the user password is encrypted.
Accounting	Limited.	Extensive.
Multilevel authorization	Per user or per group.	No privilege level control.
Interoperability	Cisco specific.	Industry-standard RFC.

Kerberos Overview

The Kerberos protocol was designed by the Massachusetts Institute of Technology (MIT) to provide strong authentication for client/server applications by using secret-key cryptography. Kerberos keeps a database of its clients and their private keys. The private key is a large number known only to Kerberos and the client to which it belongs. If the client is a user, it is an encrypted password. Network services requiring authentication register with Kerberos, as do clients wanting to use those services. The private keys are negotiated at registration.

Because Kerberos knows these private keys, it can create messages that convince one client that another is really who it claims to be. Kerberos also generates temporary private keys, called *session keys*, which are given to clients and no one else. A session key can be used to encrypt messages between two parties.

Kerberos provides three distinct levels of protection. The application programmer determines which is appropriate, according to the requirements of the application. For example, some applications require only that authenticity be established at the initiation of a network connection and can assume that further messages from a given network address originate from the authenticated party.

Other applications require authentication of each message but do not care whether the content of the message is disclosed. For these, Kerberos provides safe messages. Yet a higher level of security is provided by private messages, where each message is not only authenticated but also encrypted. Private messages are used, for example, by the Kerberos server itself for sending passwords over the network.

You can find more information on Kerberos at http://web.mit.edu/kerberos/www/.

PAP, CHAP, and EAP Authentication

Traditionally, remote users dial in to an access server to initiate a PPP session. PPP is the standard encapsulation protocol for the transport of different network protocols across Integrated Services Digital Network (ISDN), serial, or Public Switched Telephone Network (PSTN) connections.

PPP currently supports three authentication protocols: PAP and CHAP at the network layer, and EAP authentication at the link layer. These protocols are specified in RFCs 1334 and 2284. They are supported on synchronous and asynchronous interfaces. Authentication via PAP or CHAP is equivalent to typing in a username and password when prompted by the server. CHAP is considered to be more secure because the remote user's password is never sent across the connection.

EAP is an authentication framework that runs directly over an IEEE 802 or PPP data link layer, allowing the use of many different authentication types.

PAP

Password Authentication Protocol (PAP) involves a two-way handshake with the username and password sent across the link in clear text. When PAP is enabled, the remote client attempting to connect to the access server is required to send an authentication request. If the username and password specified in the authentication request are accepted, the access server sends an authentication acknowledgment. Figure 6-3 shows the two-way handshake process of PAP.

Figure 6-3 *Two-Way Handshake Process of PAP*

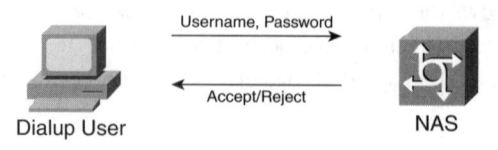

An example of a PAP authentication on a NAS follows:

```
Router(config-if)#ppp authentication pap
```

PAP provides no protection from playback and password attacks. A protocol analyzer could easily capture the password. Although many support PAP, CHAP is the preferred method of authentication because it is more secure.

CHAP

Challenge Handshake Authentication Protocol (CHAP) is a more secure authentication method than PAP because the password is never sent over the wire. CHAP periodically verifies the identity of the peer using a three-way handshake. This is done upon initial link establishment and may be repeated any time after the link has been established. Figure 6-4 shows the three-way handshake of CHAP.

Figure 6-4 *Three-Way Handshake of CHAP*

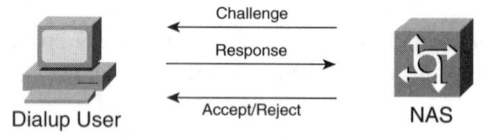

After the link-establishment phase is complete, the access server sends a "challenge" message to the remote peer. The remote peer responds with a value calculated using a "one-way hash" function (typically MD5). The access server checks the response against its own calculation of the expected hash value. If the values match, the authentication is acknowledged.

CHAP provides protection against playback attacks through the use of an incrementally changing identifier and a variable challenge value. The use of repeated challenges is intended to limit the time of exposure to any single attack. The access server is in control of the frequency and timing of the challenges.

MS-CHAP

Microsoft Challenge Handshake Authentication Protocol (MS-CHAP) is the Microsoft version of CHAP and is an extension of RFC 1994. Like the standard version of CHAP, MS-CHAP is used for PPP authentication; in this case, authentication occurs between a PC using Microsoft Windows NT or Microsoft Windows 95 and a Cisco router or access server acting as a NAS.

MS-CHAP differs from the standard CHAP as follows:

- MS-CHAP is enabled by negotiating CHAP algorithm 0x80 in Link Control Protocol (LCP) option 3 (Authentication Protocol).

- The MS-CHAP response packet is in a format designed to be compatible with Windows. This format does not require the authenticator to store a clear or reversibly encrypted password.

- MS-CHAP provides an authenticator-controlled authentication retry mechanism.

- MS-CHAP provides an authenticator-controlled change-password mechanism.

- MS-CHAP defines a set of reason-for-failure codes returned in the failure packet's Message field.

MS-CHAP Version 2

Microsoft Windows 2000 and Windows XP clients use MS-CHAP version 2 as the default authentication method for remote-access connections. MS-CHAP version 2 resolves some of the problems encountered with MS-CHAP version 1.

New features available in MS-CHAP version 2 include the following:

- **Mutual authentication between peers**—Remote-access client receives verification that the remote access server has access to the user's password.

- **Change-password feature**—Allows the client to change the account password if the password has expired.

- **Cryptographic keys**—Separate cryptographic keys are used for sent and received data.

MS-CHAP version 2 is enabled by negotiating CHAP algorithm 0x81 in LCP option 3.

Cisco supports MS-CHAP version 2 in Cisco IOS Software Release 12.2T or later code. This feature allows a remote PPP Windows client to communicate with the NAS without having to first specify an authentication method or store passwords locally.

When a user attempts to log in and authenticate using MS-CHAP version 2, the following steps occur:

Step 1 The remote access server sends a challenge to the client. The challenge contains an arbitrary string and a session.

Step 2 The client sends a response containing the username, an arbitrary challenge string, encrypted received challenge string, session identifier, user password, and peer challenge string.

Step 3 The authenticator checks the response from the client and sends back a response containing an authenticated response based on the response from the client and an indication of the success or failure of the attempted connection.

Step 4 The client verifies the authentication. The client terminates the connection if the authentication response is not correct.

Several limitations apply to the MS-CHAP version 2 capabilities in Cisco IOS Software. The change-password feature is not supported for local authentication. This feature must be used with RADIUS authentication. Also, MS-CHAP version 2 authentication is not compatible with MS-CHAP version 1 authentication.

EAP

Extensible Authentication Protocol (EAP) is an authentication protocol that runs over data link layers without requiring IP. EAP may be used on different type of links, such as dedicated point-to-point (PPP), wireless, and wired links. Cisco uses EAP for several types of environments:

- **Wireless LAN**—Using EAP to authenticate wireless clients to a centralized authentication server, such as RADIUS

- **IEEE 802.1x**—Using EAP for port-based network access control of clients to a centralized authentication server

- **Remote access**—Using EAP to authenticate remote-access users using PPP

EAP has internal support for packet retransmission and duplication but is reliant on lower-layer ordering guarantees.

EAP provides support for various authentication mechanisms, including MD5 challenge, identity, OTPs, and generic token cards. EAP also supports the use of a back-end authentication server (for example, RADIUS, Microsoft Active Directory, and so on) that the authenticator (for example, Cisco router or switch) uses to pass through the authentication exchange from the client. This capability allows enhanced compatibility between the authenticator and the third-party authentication servers, thus reducing the need for changes to the client or the NAS. Figure 6-5 shows the EAP authentication via back-end server.

Figure 6-5 *EAP Back-End Authentication*

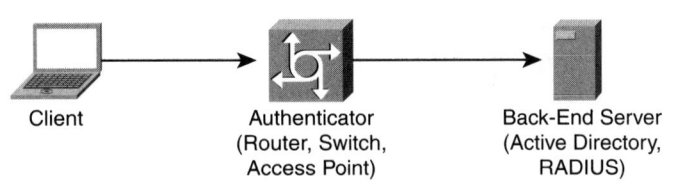

EAP has two modes of operation, proxy and local. In proxy mode, EAP allows the authentication process to be negotiated by the back-end server. In local mode, the authenticator handles the authentication via MD5 locally, instead of passing it to a back-end server.

NOTE There may be a significant increase in authentication time when using EAP in proxy mode. This increase is due to the length of time to send each packet to and from the authentication server.

Foundation Summary

The "Foundation Summary" section of each chapter lists the most important facts from the chapter. Although this section does not list every fact from the chapter that will be on your SNRS exam, a well-prepared candidate should at a minimum know all the details in each "Foundation Summary" before going to take the exam.

- Authentication methods vary from strong to weak:
    - One-time passwords using token cards
    - The session key OTP systems
    - Expiring or aging username and passwords
    - Static username and password
    - No username and password
- TACACS+ separates AAA services. RADIUS combines authentication and authorization but separates accounting services.
- CHAP periodically verifies the identity of the peer using a three-way handshake.
- PAP involves a two-way handshake with the username and password sent across the link in clear text. PAP provides no protection from playback and password attacks.
- MS-CHAP is primarily used in Microsoft client remote PPP access authentication.
- EAP is a data link layer authentication protocol used in wireless and Layer 2 environments.

Q&A

As mentioned in the section "How to Use This Book" in the Introduction, you have two choices for review questions: the Q&A questions here or the exam simulation questions on the CD-ROM. The questions that follow present a bigger challenge than the exam itself because they use an open-ended question format. By using this more difficult format, you can exercise your memory better and prove your conceptual and factual knowledge of this chapter. You can find the answers to these questions in the appendix.

1. Which port is reserved for TACACS+ use?
2. Which versions of the TACACS protocol in Cisco IOS Software have officially reached end-of-maintenance?
3. In the RADIUS security architecture, what is the network access server?
4. Which method of strong authentication for client/server applications utilizes secret-key cryptography.
5. Who developed and designed the Kerberos authentication protocol?
6. Which two popular authentication methods does PPP support?
7. Why is PAP considered insecure compared to other authentication protocols, such CHAP and MS-CHAP?
8. Which type of encryption algorithm does CHAP uses during the three-way handshake?
9. Give one difference between CHAP and MS-CHAP?
10. Where is EAP used in Cisco devices?

This chapter covers the following subjects:

- AAA Overview
- Configuring AAA Services
- Troubleshooting AAA

CHAPTER 7

Authentication, Authorization, and Accounting

An access control system needs to be in place to manage and control access to network services and resources. Authentication, authorization, and accounting (AAA) network security services provide the primary framework through which you set up access control on your router or network access server (NAS).

"Do I Know This Already?" Quiz

The purpose of the "Do I Know This Already?" quiz is to help you decide whether you really need to read the entire chapter. If you already intend to read the entire chapter, you do not necessarily need to answer these questions now.

The 10-question quiz, derived from the major sections in "Foundation Topics" section of the chapter, helps you determine how to spend your limited study time.

Table 7-1 outlines the major topics discussed in this chapter and the "Do I Know This Already?" quiz questions that correspond to those topics.

Table 7-1 *"Do I Know This Already?" Foundation Topics Section-to-Question Mapping*

Foundation Topics Section	Questions Covered in This Section
AAA Overview	1–3, 6, 8
Configuring AAA Services	4, 5, 9, 10
Troubleshooting AAA	7

> **CAUTION** The goal of self-assessment is to gauge your mastery of the topics in this chapter. If you do not know the answer to a question or are only partially sure of the answer, you should mark this question wrong for purposes of the self-assessment. Giving yourself credit for an answer you correctly guess skews your self-assessment results and might provide you with a false sense of security.

1. Which of the following best describes AAA authentication?

   a. Authentication is the last defense against hackers.//
   b. Authentication can only work with firewalls.//
   c. Authentication is the way a user is identified prior to being allowed into the network.//
   d. Authentication is a way to manage what a user can do on a network.//
   e. Authentication is a way to track what a user does when logged on to a network.

2. Which of the following best describes AAA authorization?

   a. Authorization cannot work without accounting.//
   b. Authorization provides the means of tracking and recording user activity on the network.//
   c. Authorization is the way a user is identified.//
   d. Authorization determines which resources the user is permitted to access and which operations the user is permitted to perform.

3. Which of the following best describes AAA accounting?

   a. Accounting is the way that users are identified before they log on to a network.//
   b. Accounting enables you to track the services users are accessing, as well as the amount of network resources they are consuming.//
   c. Accounting cannot be used for billing.//
   d. Accounting is a way to curtail where users can go on a NAS.//
   e. AAA accounting is used only to track users logging on to a network.

4. What command enables AAA on a NAS or a router?

   a. **aaa in**//
   b. **aaa on**//
   c. **aaa new-model**//
   d. **enable aaa**//
   e. **start aaa services**

5. Which of the following is the correct syntax to specify RADIUS as the default method for a user authentication during login?

   a. **authentication radius login**
   b. **login radius aaa authentication**
   c. **aaa login authentication group radius**
   d. **aaa authentication login default group radius**
   e. **radius authentication login**

6. Which of the following authorization methods does AAA support? (Select two.)

   a. TACACS+
   b. RADIUS
   c. SQL
   d. NDS
   e. Cisco

7. Which command enables you to troubleshoot and debug AAA authentication problems?

   a. **debug authentication**
   b. **debug aaa authentication**
   c. **authentication debug aaa**
   d. **show authentication**
   e. **show aaa authentication**

8. How do you track user activity on your NAS?

   a. You cannot track user activities on your NAS.
   b. Use AAA authorization only.
   c. Use AAA authentication only.
   d. Use AAA activity log only.
   e. Configure AAA accounting.

9. Which of the following commands requires authentication for dialup users via async or ISDN connections?

   a. **ppp authentication default radius**

   b. **aaa authentication ppp default local**

   c. **authentication line isdn**

   d. **aaa authentication login remote**

   e. **aaa ppp authentication radius**

10. After an authentication method has been defined, what is the next step to make AAA authentication work on the access server?

    a. Set up AAA accounting.

    b. Do nothing.

    c. Apply the authentication method to the desired interface.

    d. Reload the NAS or router.

The answers to the "Do I Know This Already?" quiz are found in the appendix. The suggested choices for your next step are as follows:

- **8 or less overall score**—Read the entire chapter. This includes the "Foundation Topics" and "Foundation Summary" sections and the "Q&A" section.

- **9 or 10 overall score**—If you want more review on these topics, skip to the "Foundation Summary" section and then go to the "Q&A" section. Otherwise, move on to Chapter 8, "Configuring RADIUS and TACACS+ on Cisco IOS Software."

Foundation Topics

The following sections provide general information on AAA components, implementation, and troubleshooting. The configuration command syntax provided is based on Cisco IOS Software Release 12.3(11)T.

AAA Overview

Access control is the cornerstone to ensuring the integrity, confidentiality, and availability of a network and its resources. Enforcing identification and verification of users, permitting, and then reporting or auditing their activity provides a solid framework for security. You can think of it as accessing some secure buildings today. When you first walk in front door, you are asked to provide identification. Your name is logged in, and then you are permitted to go beyond the lobby into the building. After you have access through the front door, it does not necessarily mean that you are permitted to access all the floors or offices within the building. You only have access to the rooms and floors to which you are given permission. At the end of the day when you leave, your departure from the building is logged.

This is a high-level overview of what you would like to accomplish with users accessing your network and resources. You would like to first identify and verify who they are, then give them permission to necessary resources on the network, and also have the capability to audit their activity while they are on your network. You can accomplish these functions by configuring AAA on the Cisco IOS Software.

AAA provides a modular way to perform the authentication, authorization, and accounting through the use of method lists, as discussed in the following sections.

Authentication

Authentication is the verification that the user's claimed identity is valid. Mechanisms used to verify authentication include usernames and passwords, challenge and response, and token cards. Chapter 6, "Authentication," discusses these mechanisms in more detail.

Most Cisco products support AAA authentication that uses local authentication, such as on the router, or that uses a remote security server database, such as a Cisco access control server or RADIUS server. The local authentication method is an effective solution for a small user community, whereas the separate remote security server is scalable and appropriate for a larger community of users.

AAA authentication service is implemented by first defining the authentication method, also known as *method list*, and then applying the method list to the interface desired. Having more than one method of authentication ensures a continuity of the authentication service should one of the authentication methods fail. In addition to defining the type of authentication to be performed, a method list also defines the sequence in which the authentication will be performed. If no method lists are defined for an interface, the default method list applies. With the exception of local, line password, and enable password, all authentication methods must be defined through AAA.

> **NOTE** When a user submits an incorrect username and password combination, a FAIL response occurs. On the other hand, an ERROR occurs when the security server fails to respond to an authentication request. Secondary and tertiary authentication method lists, such as local authentication, are used only when an error is detected and not when a FAIL response occurs.

Authorization

Authorization determines which resources the user is permitted to access and which operations the user is permitted to perform after being successfully authenticated. Just like AAA authentication, authorization information for each user is either stored locally on the routers or remotely on TACACS+ or RADIUS security servers. AAA authorization works by comparing attributes that describe the authorization of the user to information stored in the database. Like AAA authentication method lists, authorization method lists have to be first defined and then applied to the desired interface. All authorization methods must be defined through AAA.

Accounting

User activity reporting, such as start and stop times, executed commands, number of packets, user identities, and number of bytes, are logged by the AAA accounting service. Information collected, including the amount of resources consumed by users, can be used for billing and auditing.

Configuring AAA Services

AAA configuration includes four mandatory steps and two optional steps. It involves enabling AAA, providing security server information, defining the method list, and then applying the method list to the interface of interest. The following steps describe the configuration process:

Step 1 Activate AAA services by using the **aaa new-model** command.

Step 2 Select the type of security protocols, such as RADIUS, TACACS+, or Kerberos.

Step 3 Define the method list's authentication by using the **aaa authentication** command.

Step 4 Apply the method list to a particular interface or line, if required.

Step 5 (Optional) Configure authorization using the **aaa authorization** command.

Step 6 (Optional) Configure accounting using the **aaa accounting** command.

Configuring AAA Authentication

Administrative and remote LAN access to routers and NASs can be secured using AAA. To configure AAA authentication, perform the following steps:

Step 1 Activate AAA by using the **aaa new-model** command.

Step 2 Create a list name or use default. A list name is alphanumeric and can have one to four authentication methods.

Step 3 Specify the authentication method lists for the **aaa authentication** command. You may specify up to four.

Step 4 Apply the method list to an interface (for example, sync, async, and virtual-configured PPP, SLIP, and NASI) or to lines (for example, vty, TTY, console, and aux).

Several **aaa authentication** commands are available in Cisco IOS Software Release 12.2, including the following:

- **aaa authentication arap**
- **aaa authentication login**
- **aaa authentication ppp**
- **aaa authentication enable default**
- **aaa authentication banner**
- **aaa authentication username-prompt**
- **aaa authentication password-prompt**
- **aaa authentication attempts login**
- **aaa authentication fail-message**
- **aaa authentication dot1x**
- **aaa authentication eou**
- **aaa authentication sgbp**

The following sections discuss these four **aaa authentication** commands:

- **aaa authentication login**
- **aaa authentication ppp**
- **aaa authentication enable default**
- **aaa authentication attempts login**

See Chapter 18, "Configuring 802.1x Port-Based Authentication," for a discussion on the **aaa authentication dot1x** command.

Configuring Login Authentication Using AAA

Multiple login authentication methods are available in the AAA security services. The **aaa authentication login** command is used to enable AAA authentication. With this command, you create one or more lists of authentication methods that are tried at login. These lists are then applied to interfaces you are interested in. Table 7-2 describes the steps for applying the **aaa authentication login** command.

Step 1   Enable AAA:

```
Router(config)#aaa new-model
```

Step 2   Create a local authentication list:

```
Router(config)#aaa authentication login {default | list-name} method1
    [method2...]
```

Step 3   Apply the authentication list to a line or set of lines:

```
Router(config-line)#login authentication {default | list-name}
```

The *list-name* argument can be any name that you give to describe the list. The *method* argument is the name of the method the authentication algorithm tries. The additional methods of authentication are used only if the preceding method returns an ERROR. The **none** argument lets the authentication succeed if all the authentication methods return an ERROR.

Table 7-2 lists the supported login authentication.

Table 7-2 *Supported Login Authentication**

Keyword	Description
default	Uses the listed authentication methods that follow this argument as the default list of methods when a user logs in
list-name	Character string used to name the list of authentication methods activated when a user logs in

Table 7-2 *Supported Login Authentication* (Continued)*

Keyword	Description
method	Specifies at least the following keywords: — **enable** Uses the enable password for authentication — **krb5-telnet** Uses the Kerberos 5 Telnet authentication protocol when using Telnet to connect to the router — **line** Uses the line password for authentication — **local** Uses the local username database for authentication — **local-case** Uses case-sensitive local username authentication — **none** Uses no authentication — **group radius** Uses the list of all Remote Authentication Dial-In User Service (RADIUS) servers for authentication — **group tacacs+** Uses the list of all TACACS+ servers for authentication — **group** *group-name* Uses a subset of RADIUS or TACACS+ servers for authentication as defined by the **aaa group server radius** or **aaa group server tacacs+** command

*This table has been reproduced by Cisco Press with the permission of Cisco Systems Inc. Copyright © 2003 Cisco Systems, Inc. All Rights Reserved.

Example 7-1 shows the use of the local username database as the method of user authentication at the console interface.

Example 7-1 *Sample Configuration for Console Interface Access Using AAA Authentication Login*

```
Router(config)#aaa new-model
Router(config)#username Elvis password abc123
Router(config)#aaa authentication login conaccess local
Router(config)#line console 0
Router(config-line)#login authentication conaccess
```

Enabling Password Protection at the Privileged Level

Use the **aaa authentication enable default** command to create a series of authentication methods that are used to determine whether a user can access the privileged EXEC command level. The following shows the syntax for **aaa authentication enable**:

```
Router(config)#aaa authentication enable default method1 [method2...]
```

Table 7-3 shows **aaa authentication enable default** methods.

Table 7-3 *aaa authentication enable default Methods*

Keyword	Description
enable	Uses the enable password authentication
line	Uses the line password for authentication
none	Uses no authentication
group radius	Uses the list of all RADIUS hosts for authentication
group tacacs+	Uses the list of all TACACS+ hosts for authentication
group *group-name*	Uses a subset of RADIUS or TACACS+ servers for authentication as defined by the **aaa group server radius** or **aaa group server tacacs+** command

Example 7-2 shows a configuration for privileged EXEC access authentication using AAA.

Example 7-2 *Configuring Privileged EXEC Access Authentication Using AAA*

```
Router(config)#aaa new-model
Router(config)#aaa authentication enable default enable
```

Configuring PPP Authentication Using AAA

Users who dial in to your network need to be authenticated. Dialup configuration (PPP ARA) is typically configured on serial interfaces on your router. AAA provides a range of authentication methods for use on the serial interfaces configured for PPP. The **aaa authentication ppp** command enables AAA authentication. The syntax for **aaa authentication ppp** is as follows:

  aaa authentication ppp {default | list-name} method1 [method2...]

Table 7-4 shows **aaa authentication ppp** methods.

Table 7-4 *aaa authentication ppp Methods**

Keyword	Description
if-needed	Does not authenticate if user has already been authenticated on TTY line
krb5	Uses Kerberos 5 for authentication (can only be used for Password Authentication Protocol [PAP] authentication)
local	Uses the local username database for authentication
local-case	Uses case-sensitive local username authentication
none	Uses no authentication

Table 7-4 *aaa authentication ppp Methods* (Continued)*

Keyword	Description
group radius	Uses the list of all RADIUS servers for authentication
group tacacs+	Uses the list of all TACACS+ servers for authentication
group *group-name*	Uses a subset of RADIUS or TACACS+ servers for authentication as defined by the **aaa group server radius** or **aaa group server tacacs+** command

*This table has been reproduced by Cisco Press with the permission of Cisco Systems Inc. Copyright © 2003 Cisco Systems, Inc. All Rights Reserved.

Use the **ppp authentication** command in interface configuration mode to enable PPP authentication. The syntax for the command is as follows:

```
ppp authentication {protocol1 [protocol2...]} [if-needed] [list-name | default] [callin]
[one-time] [optional]
```

Table 7-5 shows a description of some of the syntax of this command that is relevant to this chapter.

Table 7-5 *ppp authentication Partial Command Parameters**

Keyword	Description
protocol1 [*protocol2*...]	Authentication protocol used for Point-to-Point Protocol (PPP). Available authentication protocols are Challenge Handshake Authentication Protocol (CHAP), Extensible Authentication Protocol (EAP), MS-CHAP version 1, and PAP. To use MS-CHAP version 2 authentication, use the **ppp authentication ms-chap-v2** interface configuration command.
if-needed	This optional command is used with Terminal Access Controller Access Control System (TACACS) and extended TACACS. It does not perform CHAP or PAP authentication if the user has already provided authentication. It is available only on asynchronous interfaces.
list-name	This optional command is used with AAA and specifies the name of a list of methods of authentication to use. If no list name is specified, the system uses the default. The list is created with the **aaa authentication ppp** command.
default	This optional command specifies the name of the method list created with the **aaa authentication ppp** command.

*This table has been reproduced by Cisco Press with the permission of Cisco Systems Inc. Copyright © 2003 Cisco Systems, Inc. All Rights Reserved.

The following steps outline the configuration procedure for AAA authentication methods for serial lines using PPP:

Step 1   Enable AAA globally.

Step 2   Create a local authentication list.

Step 3 Enter the configuration mode for the interface to which you want to apply the authentication list.

Step 4 Apply the authentication list to a line or set of lines.

The configuration shown in Example 7-3 has AAA authentication configured for PPP connections to use the local username database as the default method for user authentication. The PPP authentication protocol used is EAP.

Example 7-3 *Configuration of* **aaa authentication ppp**

```
Router(config)#aaa new-model
Router(config)#username Elvis password abc123
Router(config)#aaa authentication ppp default local
Router(config)#interface async 1
Router(config-if)#ppp authentication eap default
```

Limiting the Number of Login Attempts

To limit the number of AAA login attempts, use the **aaa authentication attempts login** command in global configuration mode. The syntax for this command is as follows:

> aaa authentication attempts login *number-of-attempts*

The *number-of-attempts* parameter is an integer ranging from 1 to 25. The default value is 3 login attempts.

Example 7-4 shows a sample configuration, limiting the number of login attempts to 10.

Example 7-4 *Configuring* **aaa authentication attempts login** *Command*

```
Router1#configure terminal
Router1(config)#aaa authentication attempts login 10
```

Configuring AAA Authorization

You can restrict the type of operation users can perform or the network resources they can access by using the AAA authorization service. After AAA authorization is enabled and customized, user profiles are stored on the local database or in a remote security server. From information in these profiles, users' sessions are configured after they have been authenticated.

AAA supports six different methods of authorization:

- **TACACS+**—User profile information is stored on a remote security server that has TACACS+ services running. The network access server communicates with the TACACS+ service to configure the user's session.

- **If-authenticated**—Successful authentication is required first before the user is allowed to access the requested function.

- **None**—Authorization is not performed over this line or interface.

- **Local**—User information is stored locally on the router or access server.

- **RADIUS**—User profile information is stored on a remote security server. The router or access server requests authorization information from the RADIUS security server.

- **Kerberos instance map**—The router or access server uses the instance defined by the **kerberos instance map** command to authorize.

AAA authorization controls the user's activity by permitting or denying access to which type of network access a user can start (PPP, Serial Line Internet Protocol [SLIP], AppleTalk Remote Access Protocol [ARAP]), what type of commands the user can execute, and more. Cisco IOS Software supports the following 11 types of AAA authorization:

- **auth-proxy**—Applies specific security policies on a per-user basis. It requires the Cisco IOS Firewall feature set.

- **commands**—Applies to the EXEC mode commands a user issues. Command authorization attempts authorization for all EXEC mode commands, including global configuration commands, associated with a specific privilege level.

- **exec**—Applies to a user EXEC terminal session.

- **network**—Applies to network connections. This can include a PPP, SLIP, or ARAP connection.

- **reverse-access**—Applies to reverse Telnet sessions.

- **configuration**—Applies to downloading configurations from the AAA server.

- **ipmobile**—Applies to authorization for IP mobile services.

- **cache filterserver**—Applies to the authorization for caches and the downloading of access control list (ACL) configurations from a RADIUS filter server.

- **config-commands**—Applies to configuration mode commands a user issues. Identical commands in configuration and EXEC mode can lead to some confusion in the authorization process. This command clears the intended authorization to the specific mode.

- **console**—Applies to the authorization of console line. This command must be used in conjunction with the **authorization** command on the line console port.

- **template**—Applies to the use of customer templates for virtual private network (VPN) routing and forwarding (VRF). The templates may be local or remote.

The syntax for the **aaa authorization** command is as follows:

```
aaa authorization {auth-proxy | network | cache filterserver | exec | commands level |
    config-commands | console | reverse-access | configuration | ipmobile | template}
    default | list-name} [method1 [method2]]
```

Table 7-6 shows **aaa authorization** command parameters.

Table 7-6 *aaa authorization Command Parameters**

Keyword	Description
network	Enables authorization for all network-related service requests, including SLIP, PPP, PPP Network Control Programs (NCP), ARAP and 802.1X virtual LAN (VLAN) assignments.
auth-proxy	Enables authorization that applies specific security policies on a per-user basis. For detailed information on the authentication proxy feature, see Chapter 16, "Authentication Proxy and the Cisco IOS Firewall."
cache	AAA cache configuration.
exec	Enables authorization to determine whether a user is allowed to run an EXEC shell.
commands *level*	Enables authorization for specific, individual EXEC commands associated with a specific privilege level. This enables you to authorize all commands associated with a specified command level from 0 to 15.
config-commands	Enables authorization for configuration mode commands.
console	Enables authorization for console line commands.
reverse-access	Enables authorization for reverse Telnet functions.
configuration	Downloads the configuration from the AAA server.
default	Uses the listed authentication methods that follow this argument as the default list of methods for authorization.
template	Enables template authorization.
ipmobile	Enables authorization for mobile IP services.
list-name	Character string used to name the list of authentication methods.
method	Specifies at least one of the keywords that follow.
group *group-name*	Uses a subset of RADIUS or TACACS+ servers for authentication as defined by the **aaa group server radius** or **aaa group server tacacs+** command.

*This table has been reproduced by Cisco Press with the permission of Cisco Systems Inc. Copyright © 2003 Cisco Systems, Inc. All Rights Reserved.

The following steps outline the configuration procedure for AAA authorization methods:

1. Create an authorization method list for a particular authorization type and enable authorization:

   ```
   Router(config)# aaa authorization {auth-proxy | network | cache filterserver | exec | commands
       level | config-commands | console | reverse-access | configuration | ipmobile | template}
       default | list-name} [method1 [method2...]]
   ```

2. Enter the line configuration mode for the lines to which you want to apply the authorization method list:

   ```
   Router(config)# line [aux | console | tty | vty] line-number [ending-line-number]
   ```

3. Apply the authorization list to a line or set of lines:

   ```
   Router(config-line)# authorization {arap | commands level | exec | reverse-access}
       {default | list-name}
   ```

Example 7-5 shows a sample configuration of a NAS for AAA services to be provided by the RADIUS server. If the RADIUS server fails to respond, the local database is queried for authentication and authorization information.

Example 7-5 *Configuring a NAS for AAA Services Provided by the RADIUS Server*

```
Router(config)#aaa new-model
Router(config)#aaa authentication dialins group radius local
Router(config)#aaa authorization network la-users group radius local
Router(config)#username mark password whatisthema7r1x
Router(config)#radius-server host 10.2.1.17
Router(config)#radius-server key ToPs3cret
Router(config)#interface group-async 1
Router(config-line)#group-range 1 16
Router(config-line)#encapsulation ppp
Router(config-line)#ppp authentication chap dialins
Router(config-line)#ppp authorization la-users
Router(config)#line 1 16
Router(config-line)#autoselect ppp
Router(config-line)#modem dialin
```

The lines in this sample RADIUS AAA configuration are defined as follows:

- The **aaa new-model** command enables AAA network security services.

- The **aaa authentication ppp dialins group radius local** command defines the authentication method list dialins, which specifies that RADIUS authentication and then (if the RADIUS server does not respond) local authentication is used on serial lines using PPP.

- The **aaa authorization network la-users group radius local** command defines the network authorization method list named la-users, which specifies that RADIUS authorization will be used on serial lines using PPP. If the RADIUS server fails to respond, local network authorization is performed.

- The **username** command defines the username and password to be used for the PPP PAP caller identification.

- The **radius-server host** command defines the name of the RADIUS server host.

- The **radius-server key** command defines the shared secret text string between the NAS and the RADIUS server host.

- The **interface group-async** command selects and defines an asynchronous interface group.

- The **ppp authentication chap dialins** command selects CHAP as the method of PPP authentication and applies the admins method list to the specified interfaces.

- The **ppp authorization la-users** command applies the la-users network authorization method list to the specified interfaces.

Configuring AAA Accounting

Enabling the AAA accounting feature helps you log user activity, including network resource utilization, which you can use for billing and auditing. Like authentication and authorization, the AAA accounting feature has method lists. The two methods used by the AAA accounting feature are RADIUS and TACACS+.

You can configure the following nine types of accounting on the Cisco IOS Software:

- **network**—Provides information for all PPP, SLIP, or ARAP sessions, including packet and byte counts.

- **exec**—Provides information about user EXEC terminal sessions of the NAS.

- **commands**—Provides information about the EXEC mode commands that a user issues. Command accounting generates accounting records for all EXEC mode commands, including global configuration commands, associated with a specific privilege level.

- **connection**—Provides information about all outbound connections made from the NAS, such as Telnet.

- **system**—Provides information about system-level events.

- **resource**—Provides start and stop records for calls that have passed user authentication and provides stop records for calls that fail to authenticate.

- **auth-proxy**—Provides accounting information about all authenticated hosts that use the authentication proxy service.

- **tunnel**—Provides accounting records for Virtual Private Dialup Network (VPDN) tunnel status changes. This includes Tunnel-Start, Tunnel-Stop, and Tunnel-Reject.

- **tunnel-link**—Provides accounting records for VPDN tunnel-link status changes. This includes Tunnel-Link-Start, Tunnel-Link-Stop, and Tunnel-Link-Reject.

The syntax for the **aaa accounting** command is as follows:

```
aaa accounting {auth-proxy | system | network | exec | connection | commands level} {default
   | list-name} [vrf vrf-name]{start-stop | stop-only | none} [broadcast] group groupname
```

Table 7-7 explains the keywords and arguments for the **aaa accounting** command.

Table 7-7 aaa accounting *Command Syntax Explanation*

Keywords	Description
auth-proxy	Provides information about all authenticated proxy user events.
system	Performs accounting for all system-level events not associated with users, such as reloads.
network	Runs accounting for all network-related service requests, including SLIP, PPP NCPs, and ARAP.
exec	Runs accounting for EXEC shell session. This keyword might return user profile information, such as what is generated by the **autocommand** command.
connection	Provides information about all outbound connections made from the NAS, such as Telnet, Local Area Transport (LAT), TN3270, Packet Assembler/Disassembler (PAD), and rlogin.
commands *level*	Runs accounting for all commands at the specified privilege level. Valid privilege level entries are integers from 0 through 15.
default	Uses the listed accounting methods that follow this argument as the default list of methods for accounting services.
list-name	Character string used to name the list of at least one of the accounting methods.
vrf *vrf-name*	This optional command specifies the VRF configuration. This command may be used only when working with system accounting.
start-stop	Sends a "start" accounting notice at the beginning of a process and a "stop" accounting notice at the end of a process.
wait-start	Similar to the **start-stop** parameter, the **wait-start** parameter sends "start" and "stop" accounting notices. The difference is that the requested user process does not begin until the accounting server receives the "start" accounting notice.

continues

Table 7-7 aaa accounting *Command Syntax Explanation (Continued)*

Keywords	Description
stop-only	Sends a "stop" accounting notice at the end of the requested user process.
none	Disables accounting services on this line or interface.
broadcast	This optional command enables sending accounting records to multiple AAA servers. Simultaneously sends accounting records to the first server in each group. If the first server is unavailable, failover occurs using the backup servers defined within that group.
group *groupname*	radius or tacacs+.

Use the **stop-only** keyword to receive minimal amount of accounting records. To collect the maximum accounting records, use the **start-stop** keyword. Accounting is stored only on the RADIUS server.

The following example shows **aaa accounting** configuration for users accessing the NAS via PPP:

```
Router(config)#aaa new-model
Router(config)#aaa authentication login neteng group radius local
Router(config)#aaa authentication ppp default group radius local
Router(config)#aaa authorization exec neteng group radius
Router(config)#aaa authorization network neteng group radius
Router(config)#aaa accounting exec neteng start-stop group radius
Router(config)#aaa accounting network neteng start-stop group radius
```

Troubleshooting AAA

After configuring AAA services, you must test and monitor your configuration. The **show** and **debug** commands are useful commands to verify, troubleshoot, and test your AAA configuration. The following **show** and **debug** commands enable you to troubleshoot and test your AAA configuration:

- show aaa user
- debug aaa authentication
- debug aaa authorization
- debug aaa accounting

The **show aaa servers** displays information about the number of packets the AAA servers sent or received for all transaction types. Currently, only RADIUS servers are supported by this command.

Example 7-6 provides sample output of the **show aaa servers** command displaying the RADIUS server information.

Example 7-6 show aaa servers *Command Output*

```
Router#show aaa servers

RADIUS: id 1, priority 1, host 10.200.200.1, auth-port 1812, acct-port 1813
    State: current UP, duration 18140s, previous duration 0s
    Dead: total time 0s, count 0
    Authen: request 4, timeouts 0
            Response: unexpected 0, server error 0, incorrect 0, time 8759038ms
            Transaction: success 4, failure 0
    Author: request 0, timeouts 0
            Response: unexpected 0, server error 0, incorrect 0, time 0ms
            Transaction: success 0, failure 0
    Account: request 8, timeouts 7
            Response: unexpected 0, server error 0, incorrect 0, time 0ms
            Transaction: success 3, failure 4
    Elapsed time since counters last cleared: 21m
Router#
```

The **show aaa user** displays attribute information about AAA authentication users. The attributes may include the IP address of the user, the protocol being used, AAA call events related to the session, and the speed of the connection.

Example 7-7 provides partial output of the **show aaa user** command displaying the RADIUS server information.

Example 7-7 show aaa user *Command Output*

```
Router1#show aaa user 14
Unique id 14 is currently in use.
Accounting:
  log=0x98021
  Events recorded :
    CALL START
    ATTR ADD
    INTERIM START
    INTERIM STOP
    OUTB_TELNET_START
  update method(s) :
. . . .
. . . .
. . . .
  Component = OUTB_TELNET
Authen: service=LOGIN type=ASCII method=RADIUS Fallover-from= 0
. . . .
. . . .
. . . .
```

continues

Example 7-7 show aaa user *Command Output (Continued)*

```
   Attribute List:
     82577618 0 00000009 interface(157) 5 tty66
     8257762C 0 00000001 port-type(161) 4 Virtual Terminal
     82577640 0 00000009 clid(27) 14 10.200.200.254
PerU: No data available
```

Example 7-8 provides sample output of the **debug aaa authentication** command. A single EXEC login that uses the default method list and the first method, TACACS+, is displayed. The TACACS+ server sends a GETUSER request to prompt for the username, then a GETPASS request to prompt for the password, and finally a PASS response to indicate a successful login. The number 35149617 is the session ID, which is unique for each authentication. Use this ID number to distinguish different authentications if several are occurring concurrently.

Example 7-8 debug aaa authentication *Command Output*

```
Router# debug aaa authentication
2:13:51: AAA/AUTHEN: create_user user='' ruser='' port='tty19'
  rem_addr='192.168.100.14' authen_type=1 service=1 priv=1
2:13:51: AAA/AUTHEN/START (0): port='tty19' list='' action=LOGIN service=LOGIN
2:13:51: AAA/AUTHEN/START (0): using "default" list
2:13:51: AAA/AUTHEN/START (35149617): Method=TACACS+
2:13:51: TAC+ (35149617): received authen response status = GETUSER
2:13:51: AAA/AUTHEN (35149617): status = GETUSER
2:13:51: AAA/AUTHEN/CONT (35149617): continue_login
2:13:51: AAA/AUTHEN (35149617): status = GETUSER
2:13:51: AAA/AUTHEN (35149617): Method=TACACS+
2:13:51: TAC+: send AUTHEN/CONT packet
2:13:51: TAC+ (35149617): received authen response status = GETPASS
2:13:51: AAA/AUTHEN (35149617): status = GETPASS
2:13:51: AAA/AUTHEN/CONT (35149617): continue_login
2:13:51: AAA/AUTHEN (35149617): status = GETPASS
2:13:51: AAA/AUTHEN (35149617): Method=TACACS+
2:13:51: TAC+: send AUTHEN/CONT packet
2:13:51: TAC+ (35149617): received authen response status = PASS
2:13:51: AAA/AUTHEN (35149617): status = PASS
```

Example 7-9 shows sample output from the **debug aaa authorization** command. In this display, an EXEC authorization for user Howard is performed. On the first line, the username is authorized. On the second and third lines, the attribute value (AV) pairs are authorized. The **debug** output displays a line for each AV pair that is authenticated. Next, the display indicates the authorization method used. The final line in the display indicates the status of the authorization process, which in this case has failed.

Example 7-9 debug aaa authorization *Command Output*

```
Router# debug aaa authorization
12:41:21: AAA/AUTHOR (0): user='Howard'
12:41:21: AAA/AUTHOR (0): send AV service=shell
12:41:21: AAA/AUTHOR (0): send AV cmd*
12:41:21: AAA/AUTHOR (642335165): Method=TACACS+
12:41:21: AAA/AUTHOR/TAC+ (642335165): user=Chris
12:41:21: AAA/AUTHOR/TAC+ (642335165): send AV service=shell
12:41:21: AAA/AUTHOR/TAC+ (642335165): send AV cmd*
12:41:21: AAA/AUTHOR (642335165): Post authorization status = FAIL
```

The **aaa authorization** command causes a request packet containing user profile information to be sent to the TACACS+ services daemon as part of the authorization process. The service responds in one of the following three ways:

- Accepts the request as is
- Makes changes to the request
- Refuses the request, thereby refusing authorization

Example 7-10 demonstrates sample output from the **debug aaa accounting** command.

Example 7-10 debug aaa accounting *Command Output*

```
Router# debug aaa accounting
16:49:21: AAA/ACCT: EXEC acct start, line 10
16:49:32: AAA/ACCT: Connect start, line 10, glare
16:49:47: AAA/ACCT: Connection acct stop:
task_id=70 service=exec port=10 protocol=telnet address=172.31.3.78
 cmd=glare bytes_in=308 bytes_out=76 paks_in=45 paks_out=54 elapsed_time=14
```

The information displayed by the **debug aaa accounting** command is independent of the accounting protocol used to transfer the accounting information to a server. You also can use the **show accounting** command to step through all active sessions and to print all the accounting records for actively accounted functions.

The **show accounting** command enables you to display the active accountable events on the system. It provides systems administrators a quick look at what is happening and may also prove useful for collecting information in the event of a data loss of some kind on the accounting server.

Foundation Summary

The "Foundation Summary" section of each chapter lists the most important facts from the chapter. Although this section does not list every fact from the chapter that will be on your SNRS exam, a well-prepared candidate should at a minimum know all the details in each "Foundation Summary" before going to take the exam.

To configure security on a Cisco router or access server using AAA, follow these steps:

Step 1	Activate AAA services by using the **aaa new-model** command.
Step 2	Select the type of security protocols, such as RADIUS, TACACS+, or Kerberos.
Step 3	Define the method list's authentication by using an **aaa authentication** command.
Step 4	Apply the method lists to a particular interface or line, if required.
Step 5	(Optional) Configure authorization using the **aaa authorization** command.
Step 6	(Optional) Configure accounting using the **aaa accounting** command.

- **show aaa server**—Displays RADIUS servers used for AAA authentication
- **show aaa user**—Displays AAA authenticated users statistics
- **debug aaa authentication**—Displays debugging messages on authentication functions
- **debug aaa authorization**—Displays debugging messages on authorization functions
- **debug aaa accounting**—Displays debugging messages on accounting functions

Q&A

As mentioned in the section "How to Use This Book" in the Introduction, you have two choices for review questions: the Q&A questions here or the exam simulation questions on the CD-ROM. The questions that follow present a bigger challenge than the exam itself because they use an open-ended question format. By using this more difficult format, you can exercise your memory better and prove your conceptual and factual knowledge of this chapter. You can find the answers to these questions in the appendix.

1. Which command enables AAA on a router/NAS?
2. Which of the AAA services can be used for billing and auditing?
3. What is the difference between console and network AAA authorization supported on the Cisco IOS Software?
4. Which AAA command would you use to configure authentication for login to an access server?
5. Where is authorization information stored for each user?
6. Which command enables you to troubleshoot a AAA accounting problem?
7. What types of auditing methods can be specified in a AAA configuration?
8. What is the difference between a FAIL response and an ERROR response in a AAA configuration?
9. How do you display all the detailed accounting records for actively accounted functions?
10. What command disables AAA functionality on your access server?

This chapter covers the following subject:

- Configuring TACACS+ on Cisco IOS Software

- Configuring RADIUS on Cisco IOS Software

CHAPTER 8

Configuring RADIUS and TACACS+ on Cisco IOS Software

Remote Authentication Dial-In User Service (RADIUS) and Terminal Access Controller Access Control System (TACACS+) provide a way to centrally validate users attempting to gain access to a router or access server. This chapter discusses the basic configuration of a network access server (NAS) and router to work with RADIUS and TACACS+ servers.

"Do I Know This Already?" Quiz

The purpose of the "Do I Know This Already?" quiz is to help you decide whether you really need to read the entire chapter. If you already intend to read the entire chapter, you do not necessarily need to answer these questions now.

The 10-question quiz, derived from the major sections in the "Foundation Topics" portion of the chapter, helps you determine how to spend your limited study time.

Table 8-1 outlines the major topics discussed in this chapter and the "Do I Know This Already?" quiz questions that correspond to those topics.

Table 8-1 *"Do I Know This Already?" Foundation Topics Section-to-Question Mapping*

Foundation Topics Section	Questions Covered in This Section
Configuring TACACS+ on Cisco IOS Software	1, 2, 4
Configuring RADIUS on Cisco IOS Software	3, 5–10

CAUTION The goal of self-assessment is to gauge your mastery of the topics in this chapter. If you do not know the answer to a question or are only partially sure of the answer, you should mark this question wrong for purposes of the self-assessment. Giving yourself credit for an answer you correctly guess skews your self-assessment results and might provide you with a false sense of security.

Chapter 8: Configuring RADIUS and TACACS+ on Cisco IOS Software

1. Which of the following is the command to specify the TACACS+ server on the access server?

    a. **tacacs-server host**

    b. **tacacs host**

    c. **server tacacs+**

    d. **server host**

2. Which is the default port reserved for TACACS?

    a. UDP 49

    b. TCP 49

    c. UDP 1046

    d. TCP 1046

3. Which of the following commands enables you to verify or troubleshoot a RADIUS configuration on a NAS?

    a. **show radius**

    b. **debug radius**

    c. **debug radius-server**

    d. **verify radius**

4. What is the significance of the **tacacs-server key** command?

    a. It specifies an encryption key that will be used to encrypt all exchanges between the access server and the TACACS+ server.

    b. It is used to specify a special text when the user logs in to the access server.

    c. It is an optional configuration and not required in the TACACS+ configuration.

    d. It uniquely identifies the TACACS+ server.

5. Which of the following commands identifies a RADIUS server in a RADIUS configuration?

    a. **radius-server host**

    b. **radius-host**

    c. **server radius+**

    d. **server host**

6. Which of the following are the basic steps required to configure RADIUS on Cisco IOS Software? (Select three.)

    a. Enable AAA.
    b. Create an access list.
    c. Identify RADIUS server.
    d. Define the method list using AAA authentication.

7. Which of the following commands displays an abbreviated **debug** output for a RADIUS server?

    a. debug radius
    b. show radius abbreviate
    c. debug radius brief
    d. display abrev server

8. Which of the following is the default port used by RADIUS?

    a. TCP 1685
    b. UDP 1645
    c. TCP 1645
    d. UDP 1685

9. Which two commands test and verify a RADIUS configuration?

    a. debug radius
    b. debug server radius
    c. display radius server
    d. debug aaa authentication

10. Which of the following commands deletes the RADIUS server with IP address 10.2.100.64 from a router configuration?

    a. del radius-server host 10.2.100.64
    b. remove radius-server host 10.2.100.64
    c. no radius-server host 10.2.100.64
    d. disable radius-server host 10.2.100.64

The answers to the "Do I Know This Already?" quiz are found in the appendix. The suggested choices for your next step are as follows:

- **8 or less overall score**—Read the entire chapter. This includes the "Foundation Topics" and "Foundation Summary" sections and the "Q&A" section.

- **9 or 10 overall score**—If you want more review on these topics, skip to the "Foundation Summary" section and then go to the "Q&A" section. Otherwise, move on to Chapter 9, "Cisco Secure Access Control Server."

Foundation Topics

The following sections provide basic descriptions and examples of configuring TACACS+ and RADIUS on devices running Cisco IOS Software Release 12.3(11)T.

Configuring TACACS+ on Cisco IOS Software

To configure the Cisco access server to support TACACS+, you must complete the following steps:

Step 1 Enable AAA. Use the **aaa new-model** command to enable AAA.

Step 2 Identify the TACACS+ server. Use the **tacacs-server host** command to specify the IP address or name of one or more TACACS+ servers.

Step 3 Configure AAA services. Use the **aaa authentication** command to define method lists that use TACACS+ for authentication.

Step 4 Apply the method lists to the interfaces. Use **line** and **interface** commands to apply the defined method lists to various interfaces.

If needed, you can configure authorization using the **aaa authorization** command to authorize user-specific functions. Unlike authentication, which can be configured per line or per interface, authorization is configured globally for the entire NAS. Similarly, accounting can be configured using the **aaa accounting** command to enable accounting for TACACS+ connections.

TACACS+ authorization and accounting functions are implemented by transmitting and receiving TACACS+ attribute-value (AV) pairs for each user session. All TACACS+ values are strings. For a list of supported TACACS+ AV pairs, refer to the Cisco.com online documentation.

The **tacacs-server host** command enables you to specify the names of the IP host or hosts maintaining a TACACS+ server. Because the TACACS+ software searches for the hosts in the order specified, this feature can prove useful for setting up a list of preferred servers. The following syntax is used to specify a TACACS server:

```
tacacs-server host host-name [single-connection] [port integer] [timeout integer]
   [key string]
```

Table 8-2 shows the command parameters and the description of the **tacacs-server host** command.

Table 8-2 *Parameters for the* **tacacs-server host** *Command*

Parameter	Description
hostname	Name or IP address of the host.
single-connection	(Optional) Used to specify a single connection. Rather than have the router open and close a TCP connection to the daemon each time it must communicate, the **single-connection** option maintains a single open connection between the router and the daemon. This proves more efficient because it allows the daemon to handle a higher number of TACACS operations.
port	(Optional) Used to identify the server port number. If this option is used, it overrides the default, which is port 49.
timeout	(Optional) Specifies a timeout value. This overrides the global timeout value set with the **tacacs-server timeout** command for this server only.
key	(Optional) Specifies an authentication and encryption key. This must match the key used by the TACACS+ daemon. Specifying this key overrides the key set by the global command **tacacs-server key** for this server only.
string	(Optional) Character string specifying authentication and encryption key.
nat	(Optional) Sends the client port Network Address Translation (NAT) address to the TACACS+ server.

It is also possible to configure the encryption key used for TACACS+ separately using the **tacacs-server key** command. Specifying the encryption key with the **tacacs-server host** command overrides the default key set by the **tacacs-server key** command in global configuration mode for this server only. Note that this key is not encrypted when viewing the configuration text, therefore making it more important to protect any copies of the configuration file in printed or saved format.

The following example specifies a TACACS+ server with an IP address of 192.168.1.10:

```
NAS(config)#tacacs-server host 192.168.1.10
```

The following example specifies that, for AAA confirmation, the access server consults the TACACS+ server with IP address 192.168.1.10 on port number 49. The timeout value for requests on this connection is 3 seconds; the encryption key is **seferea**:

```
NAS(config)#tacacs-server host 192.168.1.10 port 49 timeout 3 key seferea
```

> **NOTE** The key string must be configured as the last item in the preceding command syntax because any leading spaces in between and after the key are used as part of the key string. This could cause mismatched key strings between the RADIUS server and authenticator device.

TACACS+ Authentication Example

Example 8-1 shows a sample configuration for TACACS+ to be used for PPP authentication. Figure 8-1 shows a remote client being authenticated via the TACACS+ server.

Example 8-1 *Sample Configuration for PPP Authentication Using TACACS+*

```
NAS(config)#aaa new-model
NAS(config)#aaa authentication ppp meist group tacacs+ local
NAS(config)#tacacs-server host 192.168.1.15 key Elvis
NAS(config)#interface serial 0
NAS(config-if)#ppp authentication chap meist
```

Figure 8-1 *TACACS+ Server Being Used as a Security Server for AAA Authentication*

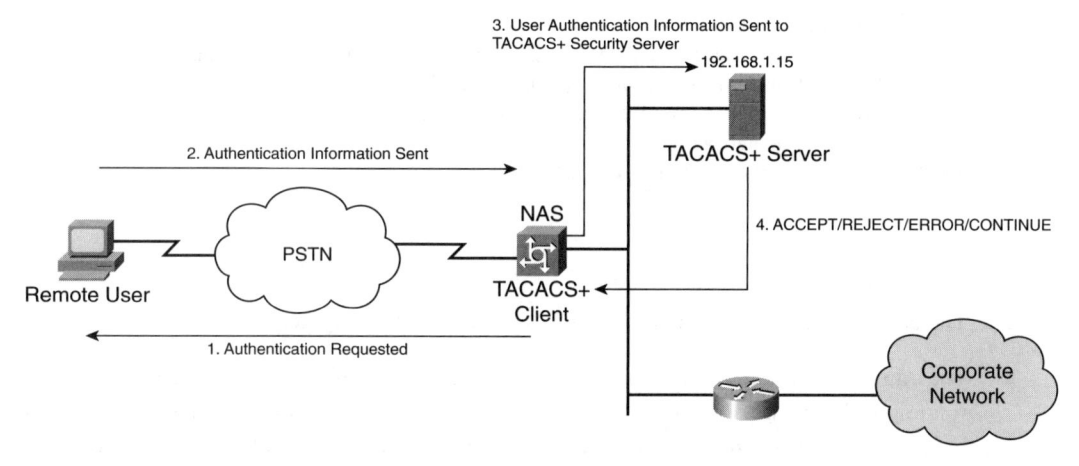

In this example

- The **aaa new-model** command enables the AAA services globally.

- The **aaa authentication** command defines a method list, meist, to be used on serial interfaces running PPP. The keyword **tacacs+** means that authentication will be done through TACACS+. If TACACS+ returns an error of some sort during authentication, the keyword **local** indicates that authentication will be attempted using the local database on NAS.

- The **tacacs-server host** command identifies the TACACS+ server as having an IP address of 192.168.1.15. The **tacacs-server key** command defines the shared encryption key to be Elvis.

- The **interface** command selects the line or port, and the **ppp authentication** command indicates CHAP protocol and applies the meist method list to this interface.

> **NOTE** The local database is chosen if the TACACS+ server does not respond; it is not used if authentication does not pass because of an incorrect username/password combination.

TACACS+ Authorization Example

Example 8-2 shows a sample TACACS+ configuration to be used for PPP authentication using the default method list and network authorization.

Example 8-2 *Sample Configuration for PPP AAA Authentication and Network Authorization*

```
NAS(config)#aaa new-model
NAS(config)#aaa authentication ppp default if-needed group tacacs+ local
NAS(config)#aaa authorization network group tacacs+
NAS(config)#tacacs-server host 192.168.1.15 key Elvis
NAS(config)#interface serial 0
NAS(config-if)#ppp authentication default
```

In this example

- The **aaa new-model** command enables the AAA security services.

- The **aaa authentication** command defines a method list, default, to be used on serial interfaces running PPP. The keyword **default** means that PPP authentication is applied by default to all interfaces. The **if-needed** keyword means that if the user has already authenticated by going through the ASCII login procedure, PPP authentication is not necessary and can be skipped. If authentication is needed, the keyword **tacacs+** means that authentication will be done through TACACS+. The keyword **local** indicates that local authentication will be used using the local database if the TACACS+ returns an error (that is, if the TACACS+ server is not accessible).

- The **aaa authorization** command configures network authorization via TACACS+. Unlike authentication lists, this authorization list always applies to all incoming network connections made to NAS.

- The **tacacs-server host** command identifies the TACACS+ server as having an IP address of 192.168.1.15. The **tacacs-server key** command defines the shared encryption key to be Elvis.

- The **interface** command selects the line or port, and the **ppp authentication** command applies the default method list to this interface.

TACACS+ Accounting Example

Example 8-3 shows a sample TACACS+ configuration to be used for PPP authentication using the default method list and accounting.

Example 8-3 *Sample Configuration for AAA Authentication and Accounting with TACACS+ Security Server*

```
NAS(config)#aaa new-model
NAS(config)#aaa authentication ppp default if-needed group tacacs+ local
NAS(config)#aaa accounting network default stop-only group tacacs+
NAS(config)#tacacs-server host 192.168.1.15
NAS(config)#tacacs-server key Elvis
NAS(config)#interface serial 0
NAS(config-if)#ppp authentication default
```

In this example

- The **aaa new-model** command enables the AAA security services.

- The **aaa accounting** command configures network accounting via TACACS+. Accounting records describing the session that just terminated will be sent to the TACACS+ server whenever a network connection terminates.

- The **tacacs-server host** command identifies the TACACS+ daemon as having an IP address of 192.168.1.15. The **tacacs-server key** command defines the shared encryption key to be Elvis.

- The **interface** command selects the line or port, and the **ppp authentication** command applies the default method list to this line.

AAA TACACS+ Testing and Troubleshooting

To verify and test a TACACS+ configuration, use the following commands:

- **debug aaa authentication**
- **debug tacacs**
- **debug tacacs events**

The following examples show sample output from the commands listed in the preceding sections.

debug aaa authentication

Example 8-4 shows a sample output from the **debug aaa authentication** command for a TACACS login attempt that was successful. The information indicates that TACACS+ is the authentication method used.

Example 8-4 *Sample Output from a* **debug aaa authentication** *Command Showing a Successful Authentication*

```
NAS#debug aaa authentication
17:42:03: AAA/AUTHEN (182481354): Method=TACACS+
17:42:03: TAC+: send AUTHEN/CONT packet
17:42:03: TAC+ (182481354): received authen response status = PASS
17:42:03: AAA/AUTHEN (182481354): status = PASS
```

An AAA session has three possible results: PASS, FAIL, or ERROR.

debug tacacs

Example 8-5 shows a sample output from the **debug tacacs** command for a TACACS login attempt that was successful, as indicated by the status PASS.

Example 8-5 *Sample Output from a* **debug tacacs** *Command Showing a Successful Authentication*

```
NAS#debug tacacs
18:03:33: TAC+: Opening TCP/IP connection to 192.168.1.15 using source 10.100.14.184
18:03:33: TAC+: Sending TCP/IP packet number 858503507-1 to 192.168.1.15 (AUTHEN/START)
18:03:33: TAC+: Receiving TCP/IP packet number 858503507-2 from 192.168.1.15
18:03:33: TAC+ (858503507): received authen response status = GETUSER
18:03:34: TAC+: send AUTHEN/CONT packet
18:03:34: TAC+: Sending TCP/IP packet number 858503507-3 to 192.168.1.15 (AUTHEN/CONT)
18:03:34: TAC+: Receiving TCP/IP packet number 858503507-4 from 192.168.1.15
18:03:34: TAC+ (858503507): received authen response status = GETPASS
18:03:38: TAC+: send AUTHEN/CONT packet
18:03:38: TAC+: Sending TCP/IP packet number 858503507-5 to 192.168.1.15 (AUTHEN/CONT)
18:03:38: TAC+: Receiving TCP/IP packet number 858503507-6 from 192.168.1.15
18:03:38: TAC+ (858503507): received authen response status = PASS
18:03:38: TAC+: Closing TCP/IP connection to 192.168.1.15
```

debug tacacs events

Example 8-6 shows a sample output from the **debug tacacs events** command. This example shows the opening and closing of a TCP connection to a TACACS+ server (192.168.100.24), the bytes read and written over the connection, and the TCP status of the connection.

Example 8-6 *Sample Output from a* **debug tacacs events** *Command Showing a Successful Communication to the TACACS+ Server*

```
NAS#debug tacacs events
13:22:21: TAC+: Opening TCP/IP to 192.168.100.24/1049 timeout=15      <--
13:22:21: TAC+: Opened TCP/IP handle 0x48A87C to 192.168.100.24/1049
13:22:21: TAC+: periodic timer started
13:22:21: TAC+: 192.168.100.24 req=3BD868 id=-1242409656 ver=193
   handle=0x48A87C (ESTAB)
```

Example 8-6 *Sample Output from a* **debug tacacs events** *Command Showing a Successful Communication to the TACACS+ Server (Continued)*

```
expire=14 AUTHEN/START/SENDAUTH/CHAP queued
13:22:22: TAC+: 192.168.100.24 ESTAB 3BD868 wrote 46 of 46 bytes
13:22:27: TAC+: 192.168.100.24 CLOSEWAIT read=12 wanted=12 alloc=12 got=12
13:22:27: TAC+: 192.168.100.24 CLOSEWAIT read=61 wanted=61 alloc=61 got=49
13:22:27: TAC+: 192.168.100.24 received 61 byte reply for 3BD868
13:22:27: TAC+: req=3BD868 id=-1242409656 ver=193 handle=0x48A87C (CLOSEWAIT)
  expire=9
AUTHEN/START/SENDAUTH/CHAP processed
13:22:27: TAC+: periodic timer stopped (queue empty)
13:22:27: TAC+: Closing TCP/IP 0x48A87C connection to 192.168.100.24/1049
13:22:27: TAC+: Opening TCP/IP to 192.168.100.24/1049 timeout=15
13:22:27: TAC+: Opened TCP/IP handle 0x489F08 to 192.168.100.24/1049
13:22:27: TAC+: periodic timer started
13:22:27: TAC+: 192.168.100.24 req=3BD868 id=299214410 ver=192 handle=0x489F08
  (ESTAB)
expire=14 AUTHEN/START/SENDPASS/CHAP queued
00:03:23: TAC+: 192.168.100.24 ESTAB 3BD868 wrote 41 of 41 bytes
00:03:23: TAC+: 192.168.100.24 CLOSEWAIT read=12 wanted=12 alloc=12 got=12
00:03:23: TAC+: 192.168.100.24 CLOSEWAIT read=21 wanted=21 alloc=21 got=9
00:03:23: TAC+: 192.168.100.24 received 21 byte reply for 3BD868
00:03:23: TAC+: req=3BD868 id=299214410 ver=192 handle=0x489F08 (CLOSEWAIT)
  expire=13
AUTHEN/START/SENDPASS/CHAP processed
00:03:23: TAC+: periodic timer stopped (queue empty)
```

Among the messages shown in Example 8-6 are the following:

- Line 1 indicates that a TCP open request to host 192.168.100.24 on port 1049 will time out in 15 seconds if it gets no response.

- Line 2 indicates a successful open operation and provides the address of the internal TCP handle for this connection.

- Line 7 indicates that all 46 bytes were written to address 192.168.100.24 for request 3BD868.

- Line 8 indicates that 12 bytes were read in reply to the request.

- Line 26 indicates that the TACACS+ server helper process switched itself off when it had no more work to do.

The **debug tacacs events** command generates a substantial amount of output and has to be used with caution. For more detailed information, refer to the "Debug Command Reference" documentation at Cisco.com.

Configuring RADIUS on Cisco IOS Software

To configure RADIUS on your Cisco router or access server, you must complete the following steps:

Step 1 Enable AAA. Use the **aaa new-model** global configuration command to enable AAA.

Step 2 Identify the RADIUS server. Use the **radius-server host** command to specify the IP address. Use the **radius-server key** command to specify an encryption key that will be used to encrypt all exchanges between NAS and the RADIUS server.

Step 3 Configure AAA services. Use the **aaa authentication** global configuration command to define method lists that use RADIUS for authentication.

Step 4 Apply the method lists to the interfaces. Use **line** and **interface** commands to apply the defined method lists to various interfaces.

If needed, you can configure authorization using the **aaa authorization** command to NAS. Similarly, you can configure accounting using the **aaa accounting** command to enable accounting for RADIUS connections.

To configure RADIUS to use the AAA security commands, you must specify the host running the RADIUS server and a secret text string that it shares with the access server. To specify a RADIUS server host and shared secret text string, use the following commands in global configuration mode:

```
radius-server host {hostname | ip-address} [auth-port port-number]
  [acct-port port-number] [timeout seconds] [retransmit retries] [key string]
  [alias {hostname | ip-address}]
```

Table 8-3 shows the **radius-server host** command parameters and their description.

Table 8-3 *Parameters for the* **radius-server host** *Command*

Parameter	Description
*hostname	ip-address*	Name or IP address of the RADIUS server host.
auth-port *port-number*	(Optional) Specifies the UDP destination port for authentication requests. The default port number is 1645.
acct-port *port-number*	(Optional) Specifies the UDP destination port for accounting requests. The default port number is 1646.

Table 8-3 *Parameters for the* **radius-server host** *Command (Continued)*

Parameter	Description
timeout *seconds*	(Optional) The time interval (in seconds) that the router waits for the RADIUS server to reply before retransmitting. This setting overrides the global value of the **radius-server timeout** command. If no timeout value is specified, the global value is used. Enter a value in the range from 1 to 1000.
retransmit *retries*	(Optional) The number of times a RADIUS request is re-sent to a server if that server is not responding or responding slowly. This setting overrides the global setting of the **radius-server retransmit** command.
key *string*	(Optional) Specifies the authentication and encryption key used between the router and the RADIUS daemon running on this RADIUS server. This key overrides the global setting of the **radius-server key** command. If no key string is specified, the global value is used. The key is a text string that must match the encryption key used on the RADIUS server.
alias	(Optional) Allows up to eight aliases per line for any given RADIUS server.

Multiple RADIUS hosts may be specified on the router or NAS. This increases the availability of the RADIUS during periods of high load and server failure. The following are the steps that occur in a RADIUS server failure scenario when multiple RADIUS servers are specified:

1. A new request for RADIUS transaction is performed by the router or NAS.

2. A RADIUS packet is sent to the first server in the group that is not marked dead (as per the configured **deadtime**). The packet is retransmitted for the configured number of retransmissions.

3. If the packet retransmits time out (as per the configured **timeout**), the router will transmit the packet to the next nondead server in the list and retransmit for the configured number of time.

4. Step 3 is repeated until the specified maximum number of transmissions has been made. If the end of the list is reached before the maximum number of transmissions has been reached, the router goes back to the beginning of the list and continues from there.

This scenario does not allow for multiple RADIUS servers to share transaction load because each RADIUS server is used until marked dead. To balance the load between various servers and specify the initial server for the router or access server, use the **radius-server retry method reorder** command in global configuration mode.

If this command is configured, the decision about which RADIUS server to use will be as follows:

- The router or NAS maintains the status of the first server to which a transmission is sent. This server is identified as the "flagged" server. At boot time, the flagged server is the first server listed in the server group. If the flagged server is marked as dead, the next nondead server listed after the flagged server is designated for this role. If the flagged server is the last server in the list and it is marked dead, the transaction fails, and the first server on the list becomes the flagged server.

- The transmission is sent to the flagged server for the configured number of retransmissions.

- NAS then sequentially sends the transmission through the list of nondead servers in the server group until a response is received or the maximum retries (a configurable parameter) is reached.

A server is marked dead only if both of the following conditions are met:

1. The server has not responded to the configured number of retransmission. The number of retransmission is configurable via the **radius-server transaction max-tries** command.

2. The server has not responded to any requests for the configured period of time.

RADIUS Authentication Example

To use a RADIUS server for AAA authentication at login or PPP, AAA must be enabled. Then, specify the RADIUS server IP or host name and key.

Example 8-7 shows a sample configuration to authenticate using RADIUS.

Example 8-7 *Sample Configuration Using RADIUS*

```
NAS(config)#aaa new-model
NAS(config)#radius-server host 192.168.100.15
NAS(config)#radius-server key ladyhawk
NAS(config)#username Elvis password k0nj0
NAS(config)#aaa authentication login test group radius local
NAS(config)#aaa authentication ppp test if-needed group radius
NAS(config)#interface group-async 1
NAS(config-line)#ppp authentication pap test
```

The configuration lines in this sample RADIUS authentication configuration mean the following:

- The **radius-server host** command defines the IP address of the RADIUS server host.

- The **radius-server key** command defines the shared secret text string between NAS and the RADIUS server host.

- The **aaa authentication login test group radius local** command configures the router to use RADIUS for authentication at the login prompt. If RADIUS returns an error, the user is authenticated using the local database. In this example, test is the name of the method list, which specifies RADIUS and then local authentication.

- The **aaa authentication ppp test if-needed group radius** command configures the Cisco IOS Software to use RADIUS authentication for lines using PPP if the user has not already been authenticated. If the EXEC facility has authenticated the user, RADIUS authentication is not performed. In this example, **test** is the name of the method list defining RADIUS as the if-needed authentication method.

- The **ppp authentication pap test** command applies the **test** method list to the lines specified.

RADIUS Authorization Example

To use a RADIUS server for AAA authorization, AAA must be enabled. Then, specify the RADIUS server IP or host name and key.

Example 8-8 shows a sample configuration to authorize using RADIUS.

Example 8-8 *Sample Configuration Using RADIUS*

```
NAS(config)#aaa new-model
NAS(config)#radius-server host 192.168.100.15
NAS(config)#radius-server key ladyhawk
NAS(config)#username Elvis password k0nj0
NAS(config)#aaa authorization exec list group radius
NAS(config)#aaa authorization network list group radius
```

- The **aaa authorization exec radius** command sets the RADIUS information that is used for EXEC authorization, autocommands, and access lists.

- The **aaa authorization network radius** command sets RADIUS for network authorization, address assignment, and access lists.

Example 8-9 shows the tasks performed to direct traffic to another server in the server group.

Example 8-9 *Configuring* **radius-server retry method reorder** *Command*

```
Router1(config)#aaa new-model
Router1(config)#radius-server retry method reorder
Router1(config)#radius-server retransmit 1
Router1(config)#radius-server transaction max-tries 5
Router1(config)#radius-server host 192.168.100.15 key ladyhawk1
Router1(config)#radius-server host 192.168.100.16 key ladyhawk2
Router1(config)#radius-server host 192.168.100.17 key ladyhawk3
```

The configuration lines in this sample **radius-server retry method reorder** configuration mean the following:

- The reorder is configured as host 192.168.100.15 as the flagged server and the 192.168.100.16 as a second server.

- If both RADIUS servers are not responding to RADIUS packets but are not marked dead yet, the transmission for the first transaction from NAS is as follows:

    192.168.100.15

    192.168.100.16

    192.168.100.16

    192.168.100.17

- If RADIUS server 192.168.100.17 responds, all transactions initiated after that point will be sent to this host.

> **NOTE** An additional 4 bytes of memory are required per server group. However, because most server configurations have only a small number of server groups configured, the additional 4 bytes should minimally impact performance.

To improve the RADIUS response times if a server fails, use the **radius-server deadtime command** in global configuration mode. The following is the syntax for this command:

```
radius-server deadtime minutes
```

The *minutes* parameter is an integer between 1 and 1440.

RADIUS Accounting Example

Example 8-10 is the RADIUS accounting configuration using RADIUS with the AAA command set. Figure 8-2 shows this configuration.

Configuring RADIUS on Cisco IOS Software 165

Figure 8-2 *General Configuration Using RADIUS*

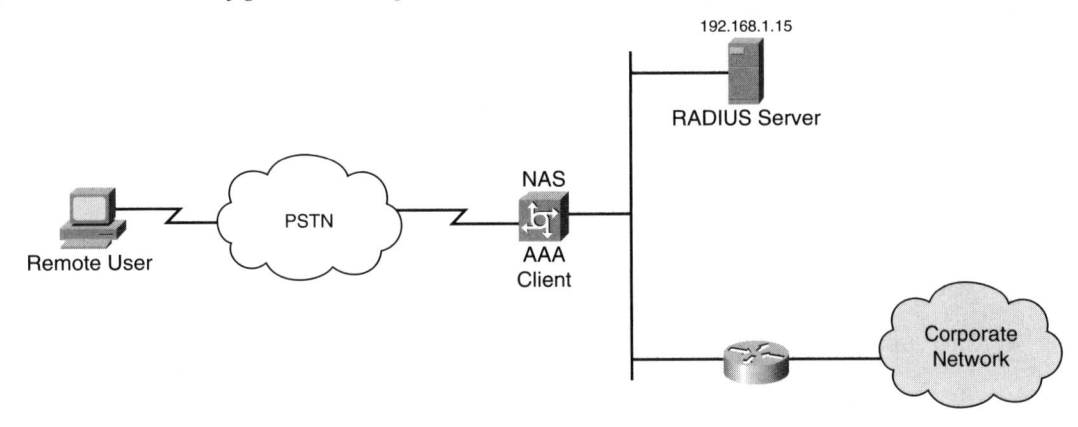

Example 8-10 *Sample RADIUS Configuration*

```
NAS(config)#aaa new-model
NAS(config)#radius-server host 192.168.100.15
NAS(config)#radius-server key ladyhawk
NAS(config)#username Elvis password k0nj0
NAS(config)#aaa authentication ppp test1 radius local
NAS(config)#aaa authorization network default group radius local
NAS(config)#aaa accounting network default start-stop group radius
NAS(config)#aaa authentication login admins local
NAS(config)#aaa authorization exec default local
NAS(config)#line 1 8
NAS(config-line)#login authentication admins
NAS(config)#interface group-async 1
NAS(config-line)#ppp authentication pap test1
```

The last four lines in this example RADIUS AAA configuration mean the following:

- The **aaa authorization network radius local** command is used to assign an address and other network parameters to the RADIUS user.

- The **aaa accounting network start-stop radius** command tracks PPP usage.

- The **aaa authentication login admins local** command defines another method list, admins, for login authentication.

- The **login authentication admins** command applies the admins method list for login authentication from lines 1 to 8.

RADIUS Configuration Testing and Troubleshooting

The commands to test and troubleshoot a RADIUS configuration are similar to ones used by TACACS and AAA:

- **debug radius**
- **debug radius brief**
- **debug radius hex**
- **debug aaa authentication**
- **debug aaa authorization**
- **debug aaa accounting**
- **show accounting**

The **debug radius** command provides information related to RADIUS. Prior to Cisco IOS Software Release 12.2(11)T, the output of this command was only available in an expanded, hexadecimal string format. This format made it difficult to interpret and analyze the output. The new options for **debug radius**, brief and hex, offer a user-friendly ASCII format and a complete display of attribute values.

Example 8-11 shows a sample output from the **debug radius** command.

Example 8-11 *Sample Output from the* **debug radius** *Command*

```
NAS#debug radius
RADIUS protocol debugging is on
RADIUS packet hex dump debugging is off
Router#
14:51:04: RADIUS: ustruct sharecount=3
14:51:04: Radius: radius_port_info() success=0 radius_nas_port=1
14:51:04: RADIUS: Initial Transmit ISDN 0:D:23 id 0 192.168.100.64:1824,
 Accounting-Request, len 358
14:51:04: RADIUS: NAS-IP-Address    [4]  6  10.100.10.0
14:51:04: RADIUS: Vendor, Cisco     [26] 19 VT=02 TL=13 ISDN 0:D:23
14:51:04: RADIUS: NAS-Port-Type     [61] 6  Async
14:51:04: RADIUS: User-Name         [1]  12 "7358827539"
14:51:04: RADIUS: Called-Station-Id [30] 7  "74762"
14:51:04: RADIUS: Calling-Station-Id[31] 12 "7358827539"
14:51:04: RADIUS: Acct-Status-Type  [40] 6  Start
14:51:04: RADIUS: Service-Type      [6]  6  Login
14:51:04: RADIUS: Vendor, Cisco     [26] 27 VT=33 TL=21 h323-gw-id=5300_43.
14:51:04: RADIUS: Vendor, Cisco     [26] 55 VT=01 TL=49
h323-incoming-conf-id=8F3A3163 B4980003 0 29BD0
14:51:04: RADIUS: Vendor, Cisco     [26] 31 VT=26 TL=25 h323-call-origin=answer
```

Example 8-11 *Sample Output from the* **debug radius** *Command (Continued)*

```
14:51:04: RADIUS: Vendor, Cisco [26] 32 VT=27 TL=26 h323-call-type=Telephony
14:51:04: RADIUS: Vendor, Cisco [26] 57 VT=25 TL=51
  h323-setup-time=*13:14:02.222 EST sun mar 22 2003
14:51:04: RADIUS: Vendor, Cisco [26] 46 VT=24 TL=40 h323-conf-id=8F3A3163
  B4980003 0 29BD0
14:51:04: RADIUS: Acct-Session-Id [44] 10 "00000004"
14:51:04: RADIUS: Delay-Time [41] 6 0
```

Example 8-12 is sample output from the **debug aaa accounting** command.

Example 8-12 *Sample Output from the* **debug aaa accounting** *Command*

```
NAS#debug aaa accounting
08:22:12: AAA/ACCT: EXEC acct start, line 10
08:22:22: AAA/ACCT: Connect start, line 10, glare
08:22:37: AAA/ACCT: Connection acct stop:
task_id=11 service=exec port=10 protocol=telnet address=192.168.100.17 cmd=glare
  bytes_in=283 bytes_out=84 paks_in=32 paks_out=41 elapsed_time=12
```

The **show accounting** command enables you to display the active accountable events on the network. It provides system administrators with a quick look at what is going on, and it can help collect information in the event of a data loss on the accounting server, as shown in Example 8-13.

Example 8-13 *Sample Output from the* **show accounting** *Command*

```
NAS#show accounting
Active Accounted actions on Interface Serial4:16, User Elvis Priv 1
Task ID 27, Network Accounting record, 00:00:13 Elapsed
task_id=27 timezone=EDT service=ppp mlp-links-max=4 mlp-links-current=4
protocol=ip addr=192.168.100.12 mlp-sess-id=6
Active Accounted actions on Interface Serial4:17, User Elvis Priv 1
Task ID 29, Network Accounting record, 00:00:44 Elapsed
task_id=29 timezone=EDT service=ppp mlp-links-max=4 mlp-links-current=4
protocol=ip addr=192.168.100.12mlp-sess-id=1
Active Accounted actions on Interface Serial4:18, User Elvis Priv 1
Task ID 17, Network Accounting record, 00:01:19 Elapsed
task_id=17 timezone=EDT service=ppp mlp-links-max=4 mlp-links-current=4
protocol=ip addr=192.168.100.12mlp-sess-id=1
Active Accounted actions on Interface Serial4:20, User Elvis Priv 1
Task ID 14, Network Accounting record, 00:01:03 Elapsed
task_id=14 timezone=EDT service=ppp mlp-links-max=4 mlp-links-current=4
mlp-sess-id=1 protocol=ip addr=192.168.100.12
Active Accounted actions on , User (not logged in) Priv 0
```

continues

Example 8-13 *Sample Output from the* **show accounting** *Command (Continued)*

```
Task ID 1, Resource-management Accounting record, 04:32:21 Elapsed
task_id=1 timezone=EDT rm-protocol-version=1.0
service=resource-management
protocol=nas-status event=nas-start reason=reload
Overall Accounting Traffic
Starts Stops Updates Active Drops
Exec 0 0 0 0 0
Network 8 4 0 4 0
Connect 0 0 0 0 0
Command 0 0 0 0 0
R-mgmt 1 0 0 1 0
System 0 0 0 0 0
User creates:21, frees:9, Acctinfo mallocs:15, frees:6
Users freed with accounting unaccounted for:0
Queue length:0
```

Foundation Summary

The "Foundation Summary" section of each chapter lists the most important facts from the chapter. Although this section does not list every fact from the chapter that will be on your SNRS exam, a well-prepared candidate should at a minimum know all the details in each "Foundation Summary" before going to take the exam.

The following are the steps required to configure TACACS+ or RADIUS server on a router:

Step 1   Enable AAA. Use the **aaa new-model** global configuration command to enable AAA.

Step 2   Identify the server. Use the **radius-server host** or **tacacs-server host** command to specify the IP address. Use the **radius-server key** or **tacacs- server key** command to specify an encryption key that will be used to encrypt all exchanges between the router and authentication servers.

Step 3   Configure AAA services. Use the **aaa authentication** global configuration command to define method lists that use RADIUS for authentication.

Step 4   Apply the method lists to the interfaces. Use **line** and **interface** commands to apply the defined method lists to various interfaces.

Multiple RADIUS hosts may be specified on the router or NAS. This increases the availability of the RADIUS during periods of high load and server failure.

To balance the load between various servers and specify the initial server for the router or access server, use the **radius-server retry method reorder** command.

If needed, you can configure authorization using the **aaa authorization** command for NAS. Similarly, you can configure accounting using the **aaa accounting** command to enable accounting for RADIUS connections.

The following troubleshooting commands enable you to test and verify RADIUS and TACACS+ server configurations:

- **debug radius**
- **debug radius brief**
- **debug radius hex**
- **debug aaa authentication**
- **debug tacacs**
- **debug tacacs events**

Q&A

As mentioned in the section "How to Use This Book" in the Introduction, you have two choices for review questions: The questions that follow present a bigger challenge than the exam itself because they use an open-ended question format. By using this more difficult format, you can exercise your memory better and prove your conceptual and factual knowledge of this chapter. You can find the answers to these questions in the appendix.

1. What command specifies a TACACS server?
2. What is the purpose of the **radius-server retry method reorder** command?
3. What purpose does the **tacacs-server key** command serve?
4. What is the purpose of the keyword **local** in the following configuration line?

    aaa authentication ppp test1 tacacs local

5. Is it possible to change the default port used by RADIUS authentication?
6. What command deletes a RADIUS server configuration?
7. What command enables network-level authorization to use a TACACS+ server?
8. Which testing and verifying command used for TACACS+ produces a substantial amount of output?
9. What default port is reserved for TACACS?
10. Is it possible to have both RADIUS and TACACS configured on a single router/NAS?

This chapter covers the following subjects:

- Cisco Secure ACS for Microsoft Windows

- Cisco Secure ACS for Microsoft Windows Architecture

CHAPTER 9

Cisco Secure Access Control Server

Cisco Secure Access Control Server (Cisco Secure ACS) provides authentication, authorization, and accounting (AAA) services for dialup access, dial-out access, wireless, VLAN access, firewalls, virtual private network (VPN) concentrators, administrative controls, and more. The list of external databases supported has also continued to grow, and the use of multiple databases, as well as multiple Cisco Secure ACSs, has become more common.

This chapter describes the features and architectural components of the Cisco Secure ACS.

"Do I Know This Already?" Quiz

The purpose of the "Do I Know This Already?" quiz is to help you decide whether you really need to read the entire chapter. If you already intend to read the entire chapter, you do not necessarily need to answer these questions now.

The 12-question quiz, derived from the major sections in "Foundation Topics" section of the chapter, helps you determine how to spend your limited study time.

Table 9-1 outlines the major topics discussed in this chapter and the "Do I Know This Already?" quiz questions that correspond to those topics.

Table 9-1 *"Do I Know This Already?" Foundation Topics Section-to-Question Mapping*

Foundation Topics Section	Questions Covered in This Section
Cisco Secure ACS for Windows	1, 2, 4–6
Administration	11
Replicating, Synchronizing, and Backing Up Databases	9, 10
Cisco Secure ACS for Windows Architecture	7, 8
Authenticating Users	3
Enabling User Changeable Passwords	12

> **CAUTION** The goal of self-assessment is to gauge your mastery of the topics in this chapter. If you do not know the answer to a question or are only partially sure of the answer, you should mark this question wrong for purposes of the self-assessment. Giving yourself credit for an answer you correctly guess skews your self-assessment results and might provide you with a false sense of security.

1. Which of the following devices does Cisco Secure ACS support? (Select two.)

    a. Cisco PIX Firewall

    b. Cisco NAS

    c. Cisco 412

    d. Cisco 550

2. Which of the following is true about Cisco Secure ACS?

    a. Centralizes access control and accounting

    b. Centralizes configuration management for routers and switches

    c. Is a distributed security application only for firewalls

    d. Only supports Cisco products

3. Which of the following user repository systems does Cisco support? (Select three.)

    a. Microsoft Windows NT/2000 user database

    b. Generic LDAP

    c. Novell NetWare Directory Services

    d. CipherTec database

4. Which of the following password protocols does Cisco Secure ACS not support?

    a. EAP-CHAP

    b. EAP-TLS

    c. LEAP

    d. ERTP

5. Which of the following is a feature of the Cisco Secure ACS authorization feature?

    a. Denying logins based on time of day and day of week

    b. Denying access based on operating system of the client

    c. Permitting access based on packet size

    d. Permitting access based on the type of encryption used

6. Which of the following are the types of accounting logs that can be generated by Cisco Secure ACS? (Select three.)

   a. Administrative accounting

   b. PAP accounting

   c. TACACS+ accounting

   d. RADIUS accounting

7. Which of the following is not part of the main services/modules that are installed for Cisco Secure ACS for Microsoft Windows?

   a. CSMon

   b. CSAdmin

   c. CSAuth

   d. CSACS

8. What does the CSMon services do?

   a. Provides logging services for both accounting and system activity

   b. Provides the HTML interface for administration

   c. Provides recording and notification of Cisco Secure ACS performance

   d. Monitors firewall activities

9. Cisco Secure ACS servers running different versions of ACS software may be configured to replicate to each other.

   a. True

   b. False

10. What are the main components of RDBMS synchronization? (Select two.)

    a. TACACS+

    b. CSDBSync

    c. CSAdmin

    d. CSMon

    e. AccountActions table

11. What administrative capabilities does the Cisco Secure ACS provide? (Select three.)

    a. Log administrator activities

    b. Configure a router's AAA settings

    c. Restore Cisco Secure ACS configuration, user accounts, and group profiles from a backup file

    d. CSMon service, providing monitoring, notification, logging, and limited automated failure response

12. Which of the following is a prerequisite for UCP?

    a. Microsoft Windows XP user desktop

    b. Microsoft IIS 5.0 or 6.0 web server running

    c. Linux

    d. Token

The answers to the "Do I Know This Already?" quiz are found in the appendix. The suggested choices for your next step are as follows:

- **10 or less overall score**—Read the entire chapter. This includes the "Foundation Topics" and "Foundation Summary" sections and the "Q&A" section.

- **11 or 12 overall score**—If you want more review on these topics, skip to the "Foundation Summary" section and then go to the "Q&A" section. Otherwise, move on to Chapter 10, "Administration of Cisco Secure Access Control Server for Windows."

Foundation Topics

This chapter provides an overview of Cisco Secure (ACS) for Windows version 3.3 and components related to it. It does not cover the Cisco Secure ACS Solution Engine or Cisco Secure ACS for UNIX. Cisco Secure ACS for UNIX has been discontinued.

Cisco Secure ACS for Windows

Cisco Secure ACS is a highly scalable and operates as a centralized RADIUS server or TACACS+ server system. It controls the AAA of users who access corporate resources through a network.

Cisco Secure ACS for Windows provides AAA services to network devices that function as AAA clients, such as a network access servers (NAS), PIX Firewalls, and routers. The AAA client in Figure 9-1 represents any such device that provides AAA client functionality and uses one of the AAA protocols supported by Cisco Secure ACS.

Figure 9-1 *AAA Client Being Supported by a Cisco Secure ACS*

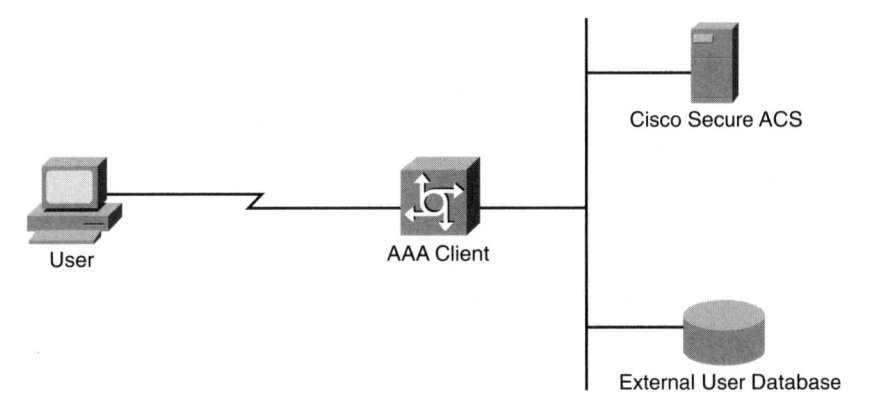

Cisco Secure ACS supports a broad set of networking access products, including all Cisco IOS routers, VPN access products, VoIP solutions, cable broadband access, content networks, wireless solutions, storage networks, and 802.1X-enabled Cisco Catalyst switches. It also supports third-party devices that can be configured with RADIUS or TACACS+. Cisco Secure ACS treats all such devices as AAA clients.

Cisco Secure ACS centralizes access control and accounting. With Cisco Secure ACS, network administrators can quickly administer accounts and globally change levels of service offerings for entire groups of users. Although the external user database shown in Figure 9-1 is optional, support for many popular user repository implementations enables companies to put to use the working knowledge gained from and the investment already made in building their corporate user repositories, such as Microsoft Windows Active Directory.

To maintain reliability and security in your network, the AAA features of the Cisco Secure ACS application help you monitor and control the following:

- **Authentication**—Who is logging in to the system
- **Authorization**—Whether a particular user should be using the requested service
- **Accounting**—What each user has been doing while on the network

The NAS directs all dial-in user access requests for authentication and authorization to Cisco Secure ACS using the RADIUS or TACACS+ protocol. If the user's request is authenticated, Cisco Secure ACS sends the user's authorizing attributes, and the accounting function is then started. Figure 9-2 shows an overview of how Cisco Secure ACS for Microsoft Windows works.

Figure 9-2 *Cisco Secure ACS Overview*

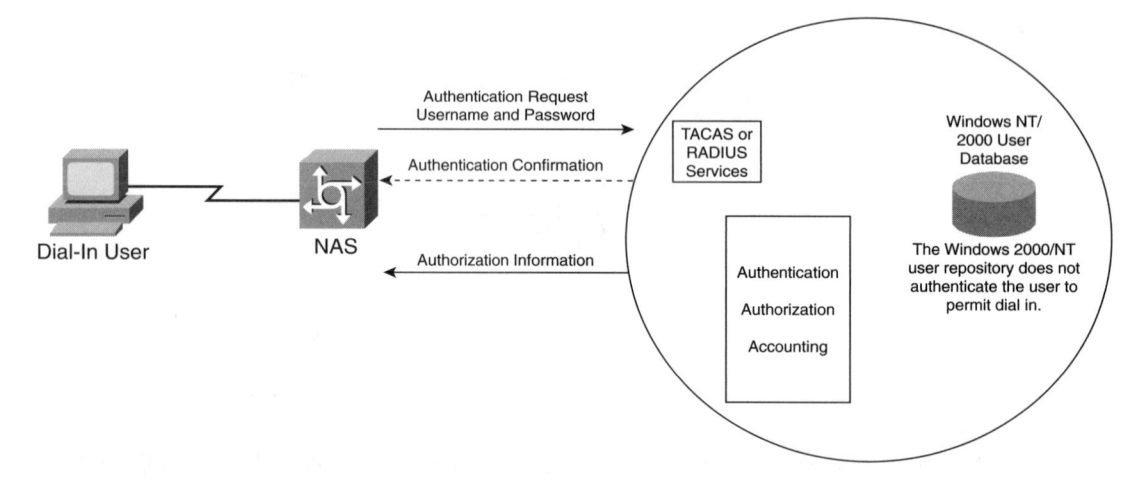

Authentication

Cisco Secure ACS supports a variety of user databases for authentication. It supports the Cisco Secure user database and the following external user databases:

- Microsoft Windows NT/2000 user database
- Generic Lightweight Directory Access Protocol (LDAP)

- LEAP Proxy RADIUS servers
- Novell NetWare Directory Services (NDS)
- Open Database Connectivity (ODBC)-compliant relational databases
- CRYPTOCard token server
- SafeWord token server
- PassGo token server
- Rivest, Shamir, and Adelman (RSA) SecureID token server
- Symantec (AXENT) Defender token servers
- LEAP proxy agent
- Secure Computing Safeword
- ActivCard token server
- Vasco token server

You can configure Cisco Secure ACS to forward authentication of users to one or more external user databases, which means that different levels of security can be concurrently used with Cisco Secure ACS for different requirements. The basic user-to-network security level is Password Authentication Protocol (PAP). Although it represents the unencrypted security, PAP does offer convenience and simplicity for the client. PAP allows authentication against the Windows NT/2000 database. With this configuration, users need to log in only once.

Challenge Handshake Authentication Protocol (CHAP) allows a higher level of security than PAP for encrypting passwords when communicating from an end-user client to the AAA client. You can use CHAP with the Cisco Secure user database. AppleTalk Remote Access Protocol (ARA Protocol) support is included to support Apple clients.

Cisco Secure ACS supports many common password protocols, including EAP-CHAP, EAP Transport Layer Security (EAP-TLS), Light EAP (LEAP), ARA Protocol, ASCII/PAP, CHAP, and MS-CHAP.

With Cisco Secure ACS, you can choose whether and how you want to use password aging. Control for password aging may reside either in the Cisco Secure user database or in a Windows NT/2000 user database. Each password-aging mechanism differs as to requirements and setting configurations.

The password-aging feature controlled by the Cisco Secure user database enables you to force users to change their passwords under any of the following conditions:

- After a specified number of days
- After a specified number of logins
- The first time a new user logs in

The Windows NT/2000-based password-aging feature enables you to control the following password-aging parameters:

- Maximum password age in days
- Minimum password age in days

The methods and functionality of Windows password aging differ according to whether you are using Windows NT or Windows 2000 and whether you use Active Directory (AD) or Security Accounts Manager (SAM).

Authorization

Cisco Secure ACS can send user profile policies to a AAA client to determine the network services the user can access. You can configure authorization to give different users and groups different levels of service. For example, standard dialup users might not have the same access privileges as premium customers and users. You can also differentiate by levels of security, access times, and services.

The Cisco Secure ACS access-restrictions feature enables you to permit or deny logins based on time of day and day of week. For example, you could create a group for temporary accounts that can be disabled on specified dates. This would make it possible for a service provider to offer a 14-day free trial. The same authorization could be used to create a temporary account for a consultant with login permission limited to Monday through Friday, 9 a.m. to 5 p.m.

You can restrict users to a service or combination of services such as PPP, ARA, or Serial Line Internet Protocol (SLIP), or EXEC. After a service is selected, you can restrict Layer 2 and Layer 3 protocols, such as IP and IPX, and you can apply individual access lists. Access lists on a per-user or per-group basis can restrict users from reaching parts of the network where critical information is stored or prevent them from using certain services, such as FTP or SNMP.

Cisco Secure ACS can provide information to the network device for a specific user to configure a secure tunnel through a public network such as the Internet. The information can be for the access server, such as the home gateway for that user, or for the home gateway router to validate the user at the customer premises. In either case, Cisco Secure ACS can be used for each end of the Virtual Private Dialup Network (VPDN).

Additional authorization-related features of Cisco Secure ACS features include the following:

- Group administration of users, with support for up to 500 groups
- The capability to map a user from an external user database to a specific Cisco Secure ACS group
- Restricting access by time-of-day and day-of-week access
- Support for VoIP, including configurable logging of accounting data
- Disabling an account after a number of failed attempts, specified by the administrator
- Disabling an account on a specific date
- Restricting network access based on remote address caller line identification (CLID) and dialed number identification service (DNIS)
- Per-user and per-group RADIUS or TACACS+ attributes
- Define usage quotas by duration or total number based on daily, monthly, or weekly periods

In addition to support for standard IETF attributes, Cisco Secure ACS includes support for RADIUS vendor-specific attributes (VSA). Some of the predefined VSAs in Cisco Secure ACS include Cisco Ascend, Juniper, Microsoft, and Nortel VSAs.

Cisco Secure ACS also supports up to 10 user-definable VSAs.

Accounting

AAA clients use the accounting functions provided by the RADIUS and TACACS+ protocols to communicate relevant data for each user session to the AAA server for recording. Cisco Secure ACS writes accounting records to a comma-separated value (CSV) log file or ODBC database, depending on your configuration. You can easily import these logs into popular database and spreadsheet applications for billing, security audits, and report generation. You can generate the following types of accounting:

- **Administrative accounting**—Lists commands entered on a network device with TACACS+ command authorization enabled
- **RADIUS accounting**—Lists when sessions stop and start; records AAA client messages with username; provides CLID information; and records the duration of each session
- **TACACS+ accounting**—Lists when sessions start and stop; records AAA client messages with username; provides CLID information; and records the duration of each session

The Cisco Secure ACS provides the following accounting features:

- Configurable supplementary user ID fields for capturing additional information in logs
- Centralized logging, allowing several Cisco Secure ACS servers to forward their accounting data to a remote Cisco Secure ACS server
- Customizable logs, enabling you to capture as much information as needed

Administration

The web administration interface is platform independent so that it can configure, maintain, and protect its AAA functionality. Almost all the configuration of the Cisco ACS server is done via the web interface post installation. The HTML interface enables you to easily modify Cisco Secure ACS configuration from any connection on your LAN or WAN.

The administration interface primarily uses HTML, along with some Java functions, to enhance ease of use. This design keeps the interface responsive and straightforward. The inclusion of Java requires that the browser used for administrative sessions supports Java.

Through the web interface, you can do the following:

- View and edit user and group information
- Restart services
- Add remote administrators
- View reports from anywhere on the network
- Back up the system
- Change AAA client information

The Cisco Secure ACS provides the following administrative capabilities:

- Define different privileges per administrator
- Log administrator activities
- View a list of logged-in users
- Restore Cisco Secure ACS configuration, user accounts, and group profiles from a backup file

- CSMon service, providing monitoring, notification, logging, and limited automated failure response

- Replication of Cisco Secure user database components to other Cisco Secure ACS servers

- Automatic configuration of users, groups, network devices, and custom RADIUS VSAs

- Scheduled and on-demand Cisco Secure ACS system backups

Replicating, Synchronizing, and Backing Up Databases

Several database utilities automate the maintenance, update, and backup of the Cisco Secure ACS database and network configuration. This section discusses the following three utilities:

- Database replication

- Relational database management system (RDBMS)

- Database backup

Database Replication

The database replication feature is designed to duplicate parts of the primary Cisco Secure ACS setup to one or more secondary servers, providing fault-tolerant AAA services. In the event the primary ACS server fails or is out of service, clients may be configured to use these secondary servers as backup for AAA services.

Database replication provides the following features:

- Replicate parts of the primary Cisco Secure ACS configuration

- Create schedules and timing for the replication process

- Export selected configuration from the primary Cisco Secure ACS

- Transport selected configuration data from primary to secondary Cisco Secure ACS servers securely

Some items in the Cisco Secure ACS database may not be replicated:

- Unknown user group mapping configuration

- IP pool definitions

- Cisco Secure ACS certificate and private key files
- External user database configurations
- Logging configurations
- User-defined RADIUS dictionaries
- RDBMS synchronization settings
- System configurations settings in the ACS Service Management page
- Third-party software such as RSA ACE client software

Figure 9-3 depicts a replication scenario. Server 1 acts as the primary Cisco Secure ACS, replicating to servers 2 and 3, which act as secondary Cisco Secure ACS devices.

Figure 9-3 *Cisco Secure ACS Database Replication Scenario*

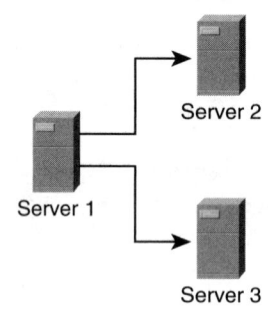

The primary Cisco Secure ACS sends replicated database components to secondary Cisco Secure servers. The secondary Cisco Secure ACS receives the replicated database components from a primary Cisco Secure ACS.

The replication is one way from the primary server to the secondary servers. Bidirectional replication, to and from servers to each other, is not supported. For example, if server 2 were configured to replicate to server 1 in addition to receiving replication from server 1, replication to server 2 would fail.

In a cascading multiserver ACS environment, it is possible for a server to be both primary and secondary ACS to different servers. Figure 9-4 shows a cascading ACS replication scenario.

Figure 9-4 *Cascading ACS Replication Scenario*

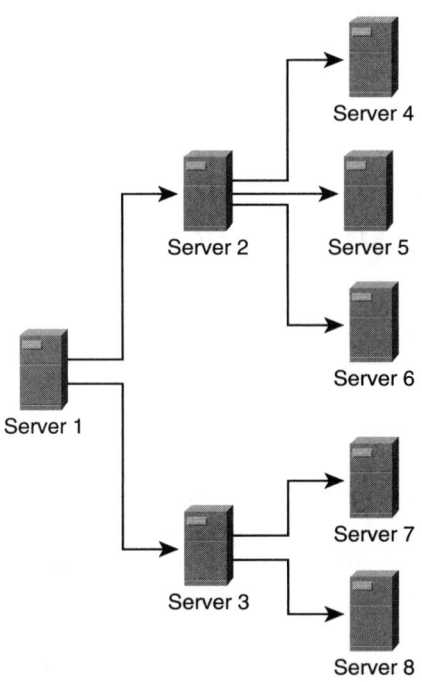

In this scenario, the server performing primary ACS functions, server 1, may not act as a secondary ACS to any of the other servers.

In a cascading replication environment, the primary Cisco Secure ACS must be configured with all the servers that will receive replication database components directly or indirectly. In Figure 9-4, server 1 must have an entry in its AAA Servers table for each of the other seven servers. If this is not done, after replication, servers 2 and 3 do not have servers 4 through 8 in their AAA Servers tables and replication will fail.

> **NOTE** Cisco Secure ACS involved in replication must run the same release of the Cisco Secure ACS software and patch level. For example, if the primary Cisco Secure ACS is running Cisco Secure ACS version 3.3, all secondary Cisco Secure ACS must be running Cisco Secure ACS version 3.3.

The replication frequency is manually configurable. Shorter replication frequency allows for more current secondary Cisco Secure ACS databases; however, there is a cost to having frequent replication. In addition to additional network traffic, the processing load on the server itself is increased. Replication will not occur if there are no changes in the primary ACS.

Cisco Secure ACS logs all replication events in the Windows Event Log and the Database Replication Report.

It is important to archive logs on a regular basis. In environments for which the *crash on audit fail* option is turned on, when the logs fill up, the server stops responding to all network requests.

RDBMS Synchronization

The RDBMS synchronization feature allows the automated synchronization with a Sybase, SQL, or Oracle RDBMS data source. It specifies the ODBC data source to use for Cisco Secure ACS and the other RDBMS application and controls the timing of the import/synchronization process. The RDBMS synchronization has two main components:

- **CSDBSync**—Performs automated user and group account management services. It is a dedicated Windows NT or Windows 2000 service for Cisco Secure ACS.

- **AccountActions table**—Specifies a set of rows that define actions CSDBSync is to perform in the Cisco Secure user database. Each row holds user, group, or AAA client information corresponding to the data stored for each user in the Cisco Secure ACS database. Other information in the row includes the action code (for example, adding or deleting a user). CSDBSync reads from the table and takes actions on each record based on the action code specified. CSDBSync uses these rows to update Cisco Secure user database. The AccountActions table database must support a multithreaded ODBC driver to prevent any problems with multiple systems attempting to access the table simultaneously.

Database Backup

System backup archives data into a format that can be later used to restore the configuration in case of corrupted ACS database. Several generations of database backup files may be stored.

Note that database replication is not a replacement for database backup. Both types of services are used in an operational environment to ensure data integrity and availability.

The following are two methods for backing up Cisco Secure ACS:

- **ACS system backup**—Backs up the Cisco Secure ACS system information to a file on the local hard drive. The backup may be performed manually or automatically at regular intervals or at selected days of the week and times.

- **CSUtil.exe**—A command-line utility that provides database backup and restore functionality. It also allows for adding, changing, and deleting users from a colon-delimited text file.

Cisco Secure ACS for Windows Architecture

Cisco Secure ACS is modular and flexible to fit the needs of both simple and large networks. Cisco Secure ACS for Windows operates as a set of Windows 2000 services and controls the AAA, for users accessing networks.

When you install Cisco Secure ACS on your server, the installation adds several Windows services. These services provide the core of the Cisco Secure ACS functionality and are as follows:

- **CSAdmin**—Provides the HTML interface for administration of Cisco Secure ACS
- **CSAuth**—Provides authentication and authorization services
- **CSDBSync**—Provides synchronization of the Cisco Secure user database with an external RDBMS application
- **CSLog**—Provides logging services, both for accounting and system activity
- **CSMon**—Provides monitoring, recording, and notification of Cisco Secure ACS performance, and includes automatic response to some scenarios
- **CSTacacs and CSRadius**—Provides communication between RADIUS or TACACS+ AAA clients and the CSAuth service

Figure 9-5 shows the cores services in the Cisco ACS for Windows.

Figure 9-5 *Core Services of Cisco ACS for Windows*

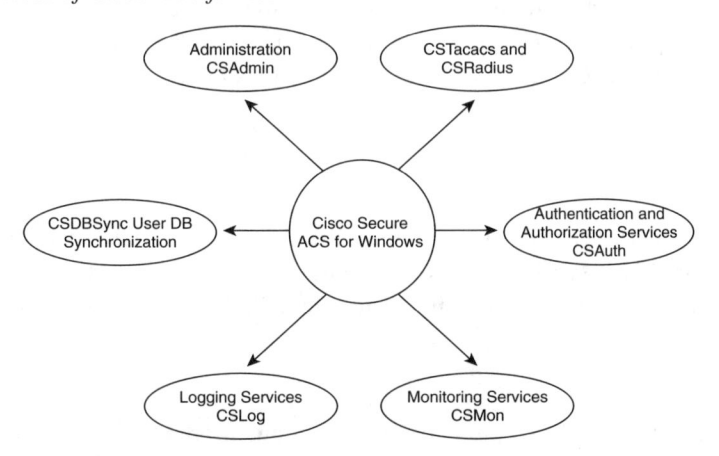

CSAdmin

CSAdmin provides the web server for the Cisco Secure ACS HTML interface. After installing Cisco Secure ACS, you must configure it from its HTML interface; therefore, CSAdmin must be running when you configure Cisco Secure ACS.

Cisco Secure ACS has a built-in web server for ACS administration. The web server uses port 2002 rather than the standard port 80 usually associated with HTTP traffic. CSAdmin is multithreaded, which enables several Cisco Secure ACS administrators to access it at the same time. Therefore, CSAdmin is well-suited for distributed, multiprocessor environments.

CSAuth

CSAuth is the authentication and authorization service. It permits or denies access to users by processing authentication and authorization requests. CSAuth determines whether access should be granted and defines the privileges for a particular user. It is the Cisco Secure ACS database manager.

To authenticate users, Cisco Secure ACS can use the internal user database or one of many external databases. When a request for authentication arrives, Cisco Secure ACS checks the database that is configured for that user. If the user is unknown, Cisco Secure ACS checks the database configured for unknown users.

When a user has authenticated, Cisco Secure ACS obtains a set of authorizations from the user profile and the group to which the user is assigned. This information is stored with the username in the Cisco Secure user database. Some of the authorizations included are the services to which the user is entitled, such as IP over PPP, IP pools from which to draw an IP address, access lists, and password-aging information. The authorizations, with the approval of authentication, are then passed to the CSTacacs or CSRadius modules to be forwarded to the requesting device.

CSDBSync

CSDBSync is the service used to synchronize the Cisco Secure ACS database with third-party RDBMSs. CSDBSync synchronizes AAA client, AAA server, network device groups, and proxy table information with data from a table in an external relational database.

CSLog

CSLog is the service used to capture and place logging information. CSLog gathers data from the RADIUS or TACACS+ packet and CSAuth and then manipulates the data to be placed into the CSV files. CSV files can be imported into spreadsheets that support this format.

CSMon

CSMon is a service that helps minimize downtime in a remote-access network environment. It provides monitoring, recording, and notification of Cisco Secure ACS performance and includes automatic responses to some scenarios. CSMon works for both RADIUS and TACACS+ and automatically detects which protocols are in use. You can use the Cisco Secure ACS HTML interface to configure the CSMon service. The Cisco Secure ACS Active Service Management feature provides the options for configuring CSMon behavior.

CSTacacs and CSRadius

The CSTacacs and CSRadius services communicate between the CSAuth module and the access device that is requesting authentication and authorization services. CSTacacs is used to communicate with TACACS+ devices and CSRadius to communicate with RADIUS devices. Both services can run at the same time. When only one security protocol is used, only the applicable service needs to be running; however, the other service does not interfere with normal operation and does not need to be disabled. However, for improved security, it is recommended that all unnecessary ports are disabled.

Authenticating Users

Users are authenticated against the Cisco Secure ACS database. This database may be the internal or external to Cisco Secure Server, allowing the flexibility to authenticate users based on information collected in different locations. The following are the types of user databases Cisco Secure ACS supports:

- Cisco Secure user database
- Windows AD and SAM database
- Generic LDAP
- Novell NDS
- LEAP Proxy RADIUS servers
- Token servers (RSA SecurID)
- RADIUS-compliant token servers
- ODBC-compliant relational databases

The Cisco Secure ACS method of interaction with external user databases varies with the database type. In the case of a Windows user database and generic LDAP, the authentication application program interface (API) is provided via the Windows operating system. With other external databases, such as NDS and LDAP, TCP connections are used, and additional software installation is required. When the communication between the external data set and ACS is established, the user may be authenticated in one of two ways:

- **Specific user assignment**—Cisco Secure ACS may be configured to authenticate specific users with an external user database. To do this, the user must exist in the Cisco Secure user database, and the password authentication list in the user setup must be set to the external user database that contains the user's credentials. The user may be placed in the desired Cisco Secure ACS group to receive the applicable access profile.

- **Unknown user policy**—Cisco Secure ACS may attempt to authenticate users not found in the Cisco Secure user database via an external user database. In this method, users do not need to exist in the Cisco Secure user database to get authenticated.

A common configuration is to use a Windows user database for standard network users and a token server for network administrators.

NOTE Whereas authentication may be determined via external sources, authorization privileges are only granted via the Cisco Secure ACS group to which the user is assigned.

The databases supported by Cisco Secure ACS support multiple password protocols. Tables 9-2 and 9-3 list the database types and the Extensible Authentication Protocol (EAP) and non-EAP password protocols each database supports.

Table 9-2 *Non-EAP Authentication Protocol and User Database Compatibility*

Database	ASCII/PAP	CHAP	ARAP	MS-CHAP v.1	MS-CHAP v.2
Cisco Secure ACS	Yes	Yes	Yes	Yes	Yes
Windows SAM	Yes	No	No	Yes	Yes
Windows AD	Yes	No	No	Yes	Yes
LDAP	Yes	No	No	No	No
Novell NDS	Yes	No	No	No	No
ODBC	Yes	Yes	Yes	Yes	Yes
LEAP proxy RADIUS server	Yes	No	No	Yes	Yes
All token servers	Yes	No	No	No	No

Table 9-3 *EAP Authentication Protocol and User Database Compatibility*

Database	LEAP	EAP-MD5	EAP-TLS	PEAP (EAP-MS CHAPv2)	PEAP (EAP-GTC)	EAP-FAST Phase Zero	EAP-FAST Phase Two
Cisco Secure ACS	Yes	Yes	Yes	Yes	Yes	Yes	Yes
Windows SAM	Yes	No	No	Yes	Yes	Yes	Yes
Windows AD	Yes	No	Yes	Yes	Yes	Yes	Yes
LDAP	No	No	Yes	No	Yes	No	Yes
Novell NDS	No	No	No	No	Yes	No	Yes
ODBC	Yes	Yes	Yes	Yes	Yes	Yes	Yes
LEAP proxy RADIUS server	Yes	No	No	Yes	Yes	Yes	Yes
All token servers	No	No	No	No	Yes	No	No

The following sections describe the user authentication process using Cisco Secure internal and external database mechanisms using the internal Cisco Secure user database, Windows NT/2000, generic LDAP, and third-party token server. For information regarding other ACS user database capabilities, such as ODBC and NDS, visit Cisco.com.

Local Database

The Cisco Secure local database is the internal database to Cisco Secure ACS. The authentication types supported are as follows:

- ASCII
- PAP
- CHAP
- MS-CHAP
- ARAP
- LEAP
- EAP-MD5
- EAP-TLS
- PEAP (EAP-GTC)
- PEAP (EAP-MSCHAPv2)
- EAP-FAST (phase zero and phase two)

192 Chapter 9: Cisco Secure Access Control Server

The Cisco Secure user database extracts information from several files, including the varsdb.MDB and the Windows registry. The varsdb.MDB file, a Microsoft Jet database file, uses an index and tree structure for searches to occur logarithmically. This structure provides faster lookups than the linear searching of data. Figure 9-6 shows the authentication process using the local ACS user database.

Figure 9-6 *Authentication Process Using Local ACS Database*

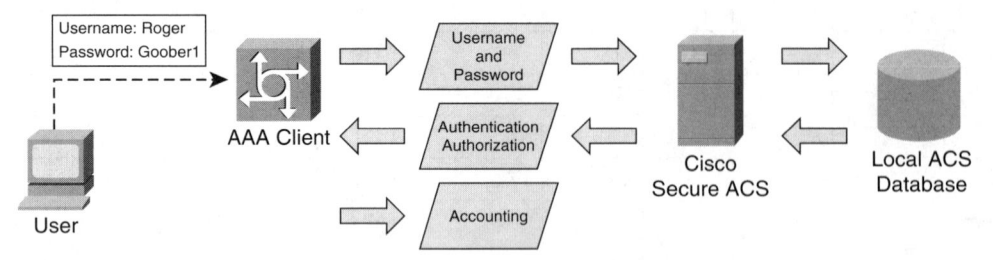

Cisco Secure ACS stores user passwords in an encrypted format, using RC2 encryption with a 40-bit key.

The disadvantage of using the local database is the duplication of user databases in an organization that already has an existing user database.

Windows NT/2000 AD

Cisco Secure ACS provides integration to Windows NT/2000 data stores, taking advantage of the work an organization has already invested in building these databases. By using the existing user databases, ACS eliminates the need for separate databases.

Cisco Secure ACS forwards user credentials to the Windows operating system of the computer running ACS. The Windows database checks its own user database and forwards the credentials to other domains with which it has trust relationships established. When it completes its check of the user, it either passes or fails the authentication request. Upon receiving a response from the Windows database, ACS instructs the requesting AAA client to grant or deny the user access, depending on the response from the Windows database. Figure 9-7 shows the authentication process using an external AD user database.

Figure 9-7 *ACS User Authentication via AD*

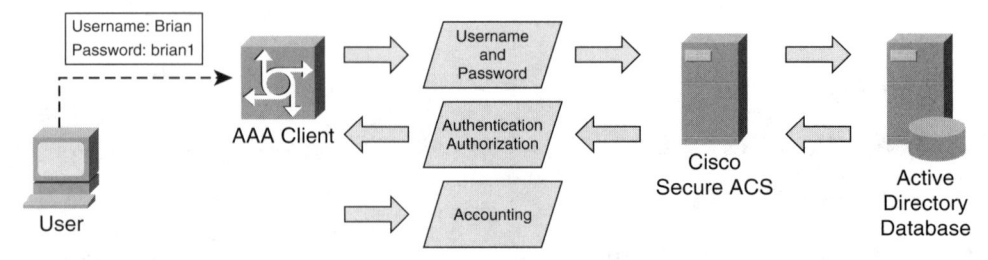

Cisco Secure ACS can also control access via checking the setting for "Grant dial-in permission to user" in the Windows domain database or across a trusted domain. If this feature is disabled for the user, access to a dial-in user is denied.

One of the limitations of using Windows 2000 database in a Cisco Catalyst switch environment is that Cisco Secure ACS is unable to support EAP-MD5 (CHAP) authentication at this time. Windows NT/2000 AD does not currently support MD5-Digest authentication for LDAP version 3 compliance. However, EAP-TLS support is currently available, allowing for digital certificate and smart card deployments as another authentication mechanism.

Generic LDAP User Database

The ACS authentication process and capabilities using generic LDAP is similar to that used for authenticating users via external Windows AD database.

Cisco Secure ACS supports ASCII, PAP, EAP-TLS, PEAP (EAP-GTC), and EAP-FAST (phase two only) authentication via generic LDAP.

ACS forwards the username and password to an LDAP database using a TCP connection on a manually specified port. The LDAP database validates or fails the authentication request based on the credentials received. When Cisco Secure ACS receives a response from LDAP, it instructs the AAA client to grant or deny the user access.

Cisco Secure ACS can authenticate multiple LDAP instances. These LDAP instances may have different IP addresses or port settings but could be pointing to the same LDAP database, which is useful when the LDAP database contains more than one subtree for users and groups. Using separate LDAP instances, one user subtree and one group subtree instance may be configured for authentication requests.

Token Server

ASCII, PAP, and PEAP (EAP-GTC) are the token server authentication protocols supported by Cisco Secure ACS.

In a scenario that ACS has been configured to authenticate against a token server, the username, PIN, and one-time password (OTP) provided by the token is forwarded to the token server for validation. If the token server verifies the authentication request, the user is successfully authenticated, and the appropriate authorizations are granted. In this type of configuration, the ACS acts as a client to the token server. For all token servers except RSA SecurID, ACS accomplishes this using the RADIUS interface. For RSA SecurID, ACS uses an RSA proprietary API. For information about configuration and support for each type of token server, refer to Cisco.com.

Figure 9-8 shows the user authentication to a network device using a token server.

Figure 9-8 *ACS User Authentication via Token Server*

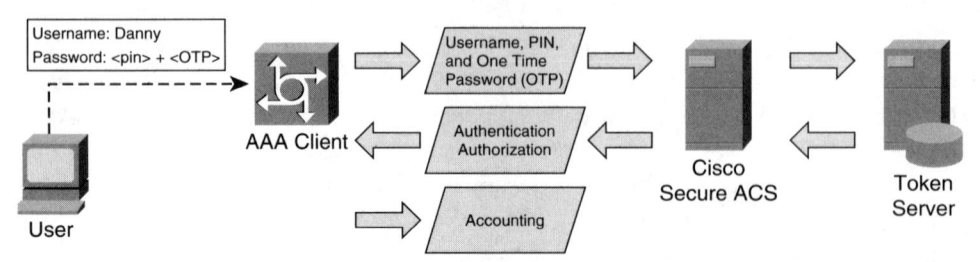

Enabling User Changeable Passwords

User Changeable Passwords (UCP) is an application that enables users to change their Cisco Secure ACS passwords with a web-based utility. When users need to change passwords, they can access the UCP web page using a web browser. The UCP web page requires users to log in. The password required is the user's existing PAP password. After UCP authenticates the users to Cisco Secure ACS, it allows them to change their PAP/CHAP passwords.

To install UCP, a Microsoft IIS 5.0 or 6.0 web server must be running. Communications between UCP and Cisco Secure ACS is protected with 128-bit encryption. It is also possible to secure communications between the user and UCP by using Secure Sockets Layer (SSL). SSL is the recommended method for UCP.

For users authenticated against a Windows user database, ACS user passwords may be changed upon password expiration. This feature may be enabled in the Cisco Secure ACS server in the MS-CHAP and Windows EAP settings tables in the external user database section. The compatible password protocols that may be used are MS-CHAP, PEAP (EAP-GTC), PEAP (EAP-MSCAPv2), and EAP-FAST.

For installation and configuration information regarding UCP, visit Cisco.com.

Foundation Summary

The "Foundation Summary" section of each chapter lists the most important facts from the chapter. Although this section does not list every fact from the chapter that will be on your SNRS exam, a well-prepared candidate should at a minimum know all the details in each "Foundation Summary" section before going to take the exam.

Important points to remember about Cisco Secure ACS for Windows 2000/NT include the following:

- Helps centralize access control and accounting
- Can authenticate against many popular token servers
- Uses the RADIUS and TACACS+ protocols to provide AAA services that ensure a secure environment
- Provides AAA services to network devices that function as AAA clients, such as NASs, switches, PIX Firewalls, VPN concentrators, and routers
- Enables network administrators to quickly administer accounts and globally change levels of service for service offerings for entire group of users
- Supports many popular user repository implementations, including Windows AD

Several database utilities automate the maintenance, update, and backup of the Cisco Secure ACS database and network configuration, including the following:

- Database replication
- RDBMS
- Database backup

The major services that make up Cisco Secure ACS for Windows are as follows:

- **CSAdmin**—Provides the HTML interface for administration of Cisco Secure ACS
- **CSAuth**—Provides authentication services
- **CSDBSync**—Provides synchronization of the Cisco Secure user database with an external RDBMS application
- **CSLog**—Provides logging services, both for accounting and system activity

- **CSMon**—Provides monitoring, recording, and notification of Cisco Secure ACS performance, and includes automatic response to some scenarios
- **CSTacacs and CSRadius**—Provides communication between RADIUS or TACACS+ AAA clients and the CSAuth service

Users are authenticated against the Cisco Secure ACS database. This database may be the internal or external to Cisco Secure server, allowing the flexibility to authenticate users based on information collected in different locations. The following are the types of user databases Cisco Secure ACS supports:

- Cisco Secure user database
- Windows AD and SAM database
- Generic LDAP
- Novell NDS
- LEAP proxy RADIUS servers
- Token servers (RSA SecurID)
- RADIUS-compliant token servers
- ODBC-compliant relational databases

UCP is an application that enables users to change their Cisco Secure ACS passwords with a web-based utility.

Q&A

As mentioned in the section "How to Use This Book" in the Introduction, you have two choices for review questions: the Q&A questions here or the exam simulation questions on the CD-ROM. The questions that follow present a bigger challenge than the exam itself because they use an open-ended question format. By using this more difficult format, you can exercise your memory better and prove your conceptual and factual knowledge of this chapter. You can find the answers to these questions in the appendix.

1. Where does the Cisco Secure ACS write its accounting records?
2. Give one example of the user repository that Cisco Secure ACS supports?
3. Give one advantage of using Cisco Secure ACS?
4. Give two examples of the password protocols that Cisco Secure ACS supports?
5. Roger is a network administrator at an engineering firm. He would like to restrict access to consultants during the weekend. Can Cisco Secure ACS help him?
6. What are the core services of Cisco Secure ACS 3.3?
7. What is the function of CSAdmin?
8. What are the benefits of database replication?
9. Which core service of the Cisco Secure ACS for Windows provides synchronization with external RDBMS applications?
10. Name two types of accounting logs generated by Cisco Secure ACS?

This chapter covers the following subjects:

- Basic Deployment Factors for Cisco Secure ACS

- Installing Cisco Secure ACS for Microsoft Windows

- Troubleshooting Cisco Secure ACS for Microsoft Windows

CHAPTER 10

Administration of Cisco Secure Access Control Server for Windows

Authentication, authorization, and accounting (AAA) was conceived originally to provide a centralized point of control for user access via dialup services. As user databases grew, more capability was required of the AAA server. Regional, and then global, requirements became common.

This chapter provides insight into the deployment process and presents a collection of factors that you should consider before deploying Cisco Secure Access Control Server (Cisco Secure ACS).

"Do I Know This Already?" Quiz

The purpose of the "Do I Know This Already?" quiz is to help you decide whether you really need to read the entire chapter. If you already intend to read the entire chapter, you do not necessarily need to answer these questions now.

The 10-question quiz, derived from the major sections in "Foundation Topics" section of the chapter, helps you determine how to spend your limited study time.

Table 10-1 outlines the major topics discussed in this chapter and the "Do I Know This Already?" quiz questions that correspond to those topics."

Table 10-1 *"Do I Know This Already?" Foundation Topics Section-to-Question Mapping*

Foundation Topics Section	Questions Covered in This Section
Basic Deployment Factors for Cisco Secure ACS	1, 2, 5, 7
Installing Cisco Secure ACS for Microsoft Windows	3, 4, 6
Troubleshooting Cisco Secure ACS for Windows	8–10

> **CAUTION** The goal of self-assessment is to gauge your mastery of the topics in this chapter. If you do not know the answer to a question or are only partially sure of the answer, you should mark this question wrong for purposes of the self-assessment. Giving yourself credit for an answer you correctly guess skews your self-assessment results and might provide you with a false sense of security.

1. Which of the following points do you have to consider before deploying Cisco Secure ACS?

    a. Dialup topology

    b. Number of users

    c. Remote access policy

    d. Number of Linux servers

2. Which of the following is the minimum CPU requirement for a Cisco Secure ACS?

    a. At least a Pentium II 330 MHz

    b. At least a Pentium III 550 MHz

    c. Will work on any Pentium platform

    d. Both items, including at least a Pentium II 330 MHz and at least a Pentium III 550 MHz

3. Which of the following are task buttons that are present on the web administrative interface of Cisco Secure ACS?

    a. User Setup

    b. Group Setup

    c. Network Configuration

    d. System Configuration

4. Which of the following are checklist items that come up during the installation of Cisco Secure ACS? (Select two.)

    a. Microsoft Windows server can successfully ping AAA clients.

    b. End users can successfully connect to AAA clients.

    c. Users have at least Netscape version 6.02.

    d. Users have a T1 connection.

5. What is the minimum browser version that is supported by Cisco ACS version 3.3?

   a. Netscape 6.02 and Microsoft Internet Explorer 6.0

   b. Mosaic 3.0 and Microsoft Internet Explorer 5.5

   c. Netscape 7.1 and Microsoft Internet Explorer 6.0 with SP1

   d. Mosaic 3.0 and Netscape 7.02

6. What are the default ports RADIUS uses for authentication?

   a. TCP 21, 22

   b. UDP 1646, 1813

   c. UDP 1645, 1812

   d. TCP 1645, 1812

7. What is the maximum number of AAA devices Cisco Secure ACS can support?

   a. 1500 network devices running AAA client

   b. 2000 network devices running AAA client

   c. 10,000 network devices running AAA client

   d. 80,000 network devices running AAA client

8. How would network latency affect the deployment of Cisco Secure ACS?

   a. It would not affect anything.

   b. It would cause the ACS installation to abort.

   c. It would cause authentication problems if the **tacacs-server timeout** is set too low for the environment.

   d. It would cause RADIUS authentication failure.

9. Which of the following is *not* a troubleshooting step for ACS authentication failure?

   a. Verify whether Cisco Secure ACS is configured to authenticate to the Windows 2000 user database.

   b. Verify whether the correct username and password is being used.

   c. Confirm the existence of the username.

   d. Verify the proper network services checked in the group settings.

10. Which of the following **debug** command is best suited to troubleshooting unsuccessful TACACS login commands to the router?

   a. **debug ip access**

   b. **debug tacacs**

   c. **debug ip tacacs**

   d. **debug aaa authentication**

The answers to the "Do I Know This Already?" quiz are found in the appendix. The suggested choices for your next step are as follows:

- **8 or less overall score**—Read the entire chapter. This includes the "Foundation Topics" and "Foundation Summary" sections and the "Q&A" section.

- **9 or 10 overall score**—If you want more review on these topics, skip to the "Foundation Summary" section and then go to the "Q&A" section. Otherwise, move on to Chapter 11, "Securing Networks with Cisco Routers."

Foundation Topics

The following sections provide general information on the installation prerequisites, process, and troubleshooting for Cisco Secure ACS 3.3.

Basic Deployment Factors for Cisco Secure ACS

Generally, the ease in deploying Cisco Secure ACS is directly related to the complexity of the implementation planned and the degree to which you have defined your network's policies and requirements. Deployment factors include the following:

- Number of users
- Network topology
- Access policy
- Network latency and reliability
- Remote-access policy
- Administrative-access policy

The following factors are just a few things to consider when deploying Cisco Secure ACS. In addition, minimum hardware and operating system requirements apply when you are installing Cisco Secure ACS. The following sections detail these specifications.

Hardware Requirements

The computer running Cisco Secure ACS must meet the following minimum hardware requirements:

- Pentium III processor, 550 MHz or faster.
- 256 MB of RAM.
- At least 250 MB of free disk space if you are running an external database (if not, disk space of 2 GB or more is required).
- Minimum graphics resolution of 256 colors at 800 × 600 lines.

Operating System Requirements

The Cisco Secure ACS should be installed on a computer running an English language version of Microsoft Windows 2000 Server with Service Pack (SP) 4.

Microsoft Windows 2000 Advanced Server is supported with SP 4 installed, provided that the Advanced Server–specific features, including Microsoft clustering service, are not installed.

Microsoft Windows Server Enterprise and Standard editions are also supported.

For up-to-date information regarding tested operating systems and service pack, visit Cisco.com.

Browser Compatibility

Your Cisco Secure ACS must have a compatible browser installed. Cisco Secure ACS 3.3 has been tested with English language versions of the following browsers on Microsoft Windows operating systems:

- Microsoft Internet Explorer 6.0, SP 1 and Sun Java Plug-in 1.4.2_04 or Microsoft Java Virtual Machine (JVM)

- Netscape Communicator 7.1 and Sun Java Plug-in 1.4.2_04

To use a web browser to access the Cisco Secure ACS Hypertext Markup Language (HTML) interface, you must enable both Java and JavaScript in the browser. Also, the web browser must not be configured to use a proxy server. If the browser used for an administrative session is configured to use a proxy server, Cisco Secure ACS sees the administrative session originating from the IP address of the proxy server rather than from the actual address of the computer.

Performance Considerations

Several factors determine the performance capabilities of Cisco Secure ACS. The size of the Windows server on which ACS is installed, the type of AAA clients, the selection of user databases and network topology are a few of these factors.

The following are some of the considerations when sizing the Cisco Secure ACS platform:

- **Maximum number of users supported by the Cisco Secure user database**—The Cisco Secure database does not have a theoretical limit to the number of users supported. The software has been successfully tested with databases in excess of 100,000 users. The limit for a single Cisco Secure ACS server authenticating against internal and external databases is approximately 300,000 to 500,000 users. If the authentication load is distributed across a number of replicated Cisco Secure ACS servers, the number increases significantly.

- **Transactions per second per number of users**—A single processor 300-MHz Pentium II server containing a 10,000-user Cisco Secure ACS database provides 80 RADIUS full login cycles (authentication, accounting start, and accounting stop) per second and approximately 40 TACACS+ logins per second. As the database grows, this performance declines approximately proportionately.

- **Maximum number of AAA clients supported**—Depending on the platform, Cisco Secure ACS can support AAA services for approximately 10,000 network devices running AAA client.

The performance of the Cisco Secure ACS server varies depending on the hardware on which it is installed and running. For example, Cisco Secure ACS can perform significantly more authentications per second if it is running on a 1.2-GHz Pentium IV server with Windows 2000 Advanced Server hosted on a local 1-GB Ethernet backbone than it could if it is running on a 550-MHz Pentium III server with Microsoft Windows 2000-hosted on a 100-MB LAN located remotely across the WAN.

The specific environment and AAA requirements also impact the performance of the Cisco Secure ACS. For instance, the location of the Cisco Secure ACS system is a factor when using an external Microsoft Windows Active Directory. Location of the Active Directory domain controller with respect to the Cisco Secure ACS might cause authentication delays. Cisco Secure ACS initially requests authentication service from the Active Directory domain controller that serves the local domain in which the Cisco Secure ACS resides. If the user does not exist in the domain controller, it requests authentication from its trusted neighbors, a situation that could result in longer delays.

AAA Clients

The AAA clients are the devices on the network through which or to which service access is being attempted. An example of a AAA client is a dialup user who obtains network access from a network access server. Another example is an administrator who is trying to log in to a network device for administrative purposes. When a user attempts to access a AAA-enabled device, the device contacts the AAA server (Cisco Secure ACS) for authentication and authorization services. The communication between the AAA client and AAA server is either TACACS+ or RADIUS. Therefore, for full TACACS+ and RADIUS support on Cisco IOS devices, make sure that the Cisco AAA clients are running Cisco IOS Software Release 11.2 or later. Cisco Secure ACS can also provide AAA services for third-party devices. The third-party device must support RADIUS or TACACS authentication mechanisms and be configured to use one of these authentication types.

Installing Cisco Secure ACS for Microsoft Windows

After confirming your network's system requirements for Cisco Secure ACS for Windows, run the setup program to install the software. Figure 10-1 shows a checklist window that comes up during the first part of the installation process.

Figure 10-1 *Checklist Window That Appears During the Installation Process for Cisco ACS 3.3*

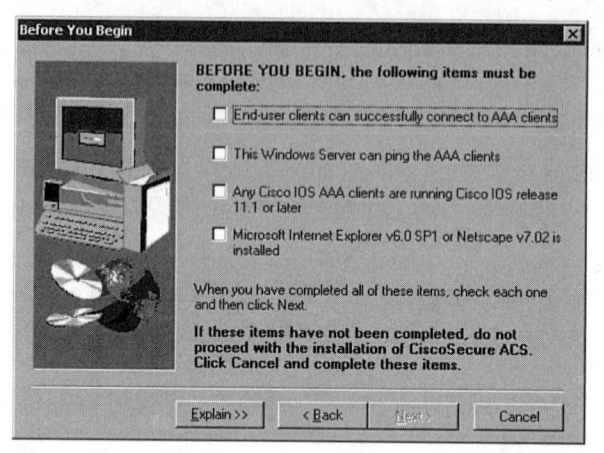

As shown in Figure 10-1, the installation process wants you to test and confirm a few things before moving forward with the installation process:

Step 1   Ensure that end users can successfully connect to the AAA client, such as a network access server (NAS).

Step 2   Verify connections between Microsoft Windows NT/2000 server and other network devices. Make sure that the AAA devices on your network can reach (ping) your Windows NT/2000 server. This will make the installation smoother and help reduce any troubleshooting issues later on when it is time to configure the Cisco Secure ACS and the devices that use it.

Step 3   Check the Cisco IOS Software version on the AAA client and make sure that it is later than Cisco IOS Software Release 11.1.

Step 4   Make sure that your browser version complies with the required version for Microsoft Windows (Internet Explorer 6.0 and Netscape 7.0).

When the installation is complete, the HTML interface appears, in which you make administration and configuration changes to the Cisco Secure ACS.

At the left of the browser, the HTML interface provides a navigation bar with task buttons. Figure 10-2 shows the HTML interface with the navigation bar containing the task buttons.

Figure 10-2 *HTML Interface for Cisco Secure ACS*

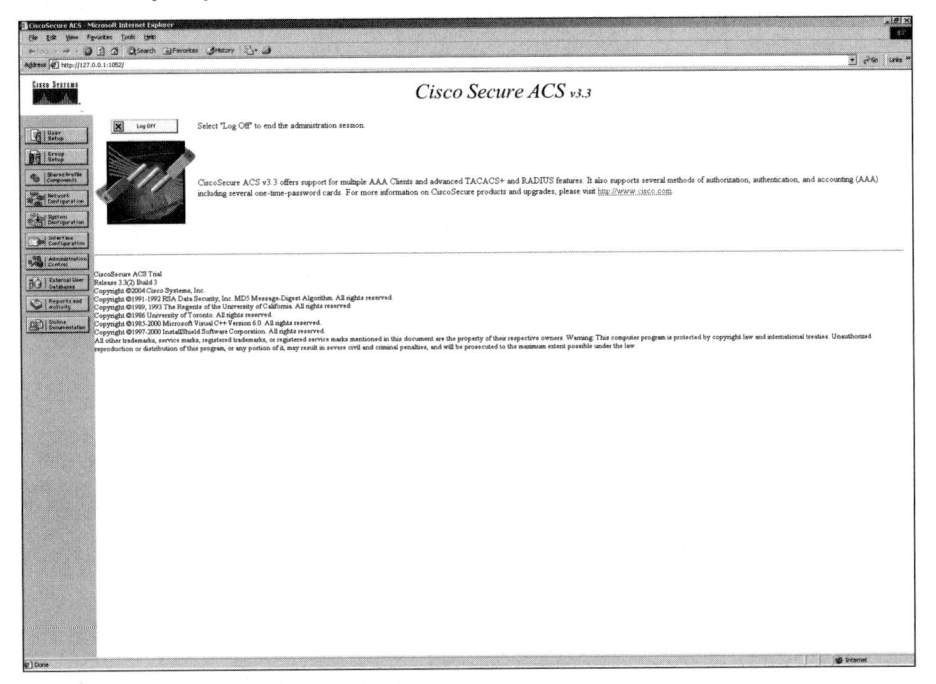

The task buttons are as follows:

- **User Setup**—Add and edit user profiles.

- **Group Setup**—Configure network services and protocols for groups of users.

- **Shared Profile Components**—Add and edit network access restriction and command authorization sets to be applied to users and groups.

- **Network Configuration**—Add and edit network access devices and configure distributed systems.

- **System Configuration**—Configure database information and accounting.

- **Interface Configuration**—Display or hide product features and options to be configured.

- **Administration Control**—Define and configure access policies.

- **External User Databases**—Configure external databases for authentication.

- **Reports and Activity**—Display accounting and logging information.

- **Online Documentation**—View the user guide.

Table 10-2 shows the ports that Cisco Secure ACS uses.

Table 10-2 *Cisco Secure ACS Ports Requirement*

Service Name	UDP	TCP
RADIUS authentication and authorization	1645, 1812	N/A
RADIUS accounting	1646, 1813	N/A
TACACS+ AAA	N/A	49
Replication and relational database management system (RDBMS) synchronization	N/A	2000
ACS remote logging	N/A	2001
Hypertext Transfer Protocol (HTTP) administrative access	N/A	2002
HTTP administrative port for new sessions	N/A	Configurable; default 1024 through 65535
User-changeable password web application	N/A	2000

Cisco Secure ACS Deployment Sequence

Although there is no single process for all Cisco Secure ACS deployments, you should consider following the sequence keyed to the high-level functions represented in the navigation toolbar. Also remember that many of these deployment activities are iterative in nature; you might find that you repeatedly return to such tasks as interface configuration as your deployment proceeds.

- **Configure administrators**—You should configure at least one administrator at the beginning of the deployment. This will provide remote administrative access.

- **Configure the Cisco Secure ACS HTML interface**—Select the features and controls that you intend to use through the HTML interface. This makes using Cisco Secure ACS easier than it would be if you had to contend with multiple parts of the HTML interface that you do not plan to use.

- **Configure the system**—There are more than a dozen functions within the System Configuration section to be considered, from setting the format for the display of dates and password validation to configuring settings for database replication and RDBMS synchronization.

- **Configure the network**—You control distributed and proxied AAA functions in the Network Configuration section of the HTML interface. From here, you establish the identity, location, and grouping of AAA clients and servers and determine which authentication protocols each is to use.

- **Configure the external user database**—During this phase of deployment, you must decide whether and how you intend to implement an external database to establish and maintain user authentication accounts. Typically, this decision is made according to your existing network administration mechanisms.

- **Configure the shared profile components**—With most aspects of network configuration already established and before configuring user groups, you should configure your shared profile components. When you set up and name the network access restrictions and command authorization sets you intend to use, you lay out an efficient basis for specifying user group and single-user access privileges.

- **Configure groups**—Having previously configured any external user databases you intend to use and before configuring your user groups, you should decide how to implement two other Cisco Secure ACS features related to external user databases: unknown user processing and database group mapping.

- **Configure users**—With groups established, you can establish user accounts. It is useful to remember that a particular user can belong to only one user group, and that settings made at the user level override settings made at the group level.

- **Configure reports**—Using the Reports and Activities section of the Cisco Secure ACS HTML interface, you can specify the nature and scope of logging that Cisco Secure ACS performs.

Troubleshooting Cisco Secure ACS for Microsoft Windows

A good place to start troubleshooting Cisco Secure ACS related AAA problems is the Failed Attempts Report under Reports and Activity. The report displays several types of failures. If no entry is found in the Failed Attempts Report, it could be that there is a misconfiguration between the Cisco Secure ACS and the router/client. In this case, do the following:

- Verify that the router can ping the server and that the server can ping the router.

- Verify that the TACACS+ host IP address is correctly configured on the router.

- Verify that the TACACS+ host key entered on both the router and the Cisco Secure ACS is the same.

Authentication Problems

If there are entries of authentication failure in the Failed Attempts Report and you are authenticating against a Windows 2000 user database, check the following items:

- Verify whether Cisco Secure ACS is configured to authenticate to the Windows 2000 user database.

- Verify whether the correct username and password is being used.

- Confirm the existence of the username.

- Check whether the user account has User Must Change Password at Next Login selected. If this option is selected, deselect it.

- Confirm that Cisco Secure ACS is configured to reference the Grant Dial-In Permission to User.

- Verify whether the retry interval is too brief. (The default is 5 seconds). Increase the retry interval (**tacacs-server timeout** *20*) on the AAA client to 20 or greater.

Troubleshooting Authorization Problems

If the authentication is working but the authorization is failing, check the following items:

- Are the proper network services checked in the group settings?

- If IP is checked, how is the dial-in user obtaining an address?

- Is there an IP pool configured on the NAS?

- Has the radio button for the command been permitted?

- Has the radio button for the argument been permitted?

Administration Issues

Table 10-3 details how to approach some of the problems that might arise in a Cisco Secure ACS installation.

Table 10-3 *Cisco Secure ACS Installation Troubleshooting*

Issue	Troubleshooting Tasks
Remote administrator cannot bring up the Cisco Secure ACS HTML interface in a browser or receives a warning that access is not permitted.	Ping Cisco Secure ACS to confirm connectivity. Verify that the remote administrator is using a valid administrator name and password that has already been added in Administration Control. Verify that Java functionality is enabled in the browser. Determine whether the remote administrator is trying to administer Cisco Secure ACS through a firewall, through a device performing Network Address Translation, or from a browser configured to use an HTTP proxy server. Ensure that the HTTP port of the ACS HTML interface is specified in the browser address. For example, if port 8080 is the ACS administration port for ACS server at acs.mycompany.com, the HTML link would be http://acs.mycompany.com:8080.
Unauthorized users can log in.	Ensure that Reject Listed IP Addresses is selected, but no start or stop IP addresses are listed. Go to Administrator Control: Access Policy and specify the start IP address and stop IP address.
No remote administrators can log in.	Ensure that Allow Only Listed IP Addresses to Connect is selected, but no start or stop IP addresses are listed. Go to Administrator Control: Access Policy and specify the start IP address and stop IP address.

Other useful troubleshooting commands for NAS devices (IOS and CatOS) include the following:

- **debug tacacs+**—Displays TACACS related information such as unsuccessful login attempts to a router configured for TACACS authentication

- **debug radius**—Displays RADIUS client/server-related information in ASCII format

- **debug aaa authentication**—Displays information related to AAA TACACS+ authentication

- **debug aaa authorization**—Displays information related to AAA TACACS+ authorization

Foundation Summary

The "Foundation Summary" section of each chapter lists the most important facts from the chapter. Although this section does not list every fact from the chapter that will be on your SNRS exam, a well-prepared candidate should at a minimum know all the details in each "Foundation Summary" section before going to take the exam.

Cisco Secure ACS deployment factors that you need to consider include the following:

- Number of users
- Network topology
- Access policy
- Network latency and reliability
- Remote-access policy
- Administrative-access policy

The following are the minimum hardware requirements for installing Cisco Secure ACS for Windows:

- Pentium III processor, 550 MHz or faster.
- 256 MB of RAM.
- At least 250 MB of free disk space. If you are running your database on the same machine, more disk space is required.
- Minimum graphics resolution of 256 colors at 800 × 600 lines.

A good place to start troubleshooting Cisco Secure ACS related AAA problems is the Failed Attempts Report under Reports and Activity. The report displays several types of failures. Other useful troubleshooting commands include the following:

- **debug tacacs+**—Displays TACACS related information such as unsuccessful login attempts to a router configured for TACACS authentication
- **debug radius**—Displays RADIUS client/server-related information in ASCII format
- **debug aaa authentication**—Displays information related to AAA TACACS+ authentication
- **debug aaa authorization**—Displays information related to AAA TACACS+ authorization

Q&A

As mentioned in the section "How to Use This Book" in the Introduction, you have two choices for review questions: the Q&A questions here or the exam simulation questions on the CD-ROM. The questions that follow present a bigger challenge than the exam itself because they use an open-ended question format. By using this more difficult format, you can exercise your memory better and prove your conceptual and factual knowledge of this chapter. You can find the answers to these questions in the appendix.

1. Which factors should you consider when deploying Cisco Secure ACS?
2. What are the minimum hardware requirements to install Cisco Secure ACS?
3. How does Cisco Secure ACS provide control for remote-access policies?
4. Where would be a good place to start to troubleshoot Cisco Secure ACS related AAA problems?
5. Does a browser using a proxy server have any effect in the administration of a Cisco Secure ACS remotely?
6. What should you check when there are authentication failures in the failed attempts report against a Windows 2000 user database?
7. Which protocol and port number does UCP use for the web application?

Part IV: IOS Firewall Feature Set

Chapter 11   Securing Networks with Cisco Routers

Chapter 12   The Cisco IOS Firewall and Advanced Security Feature Set

Chapter 13   Cisco IOS Intrusion Prevention System

Chapter 14   Mitigating Layer 2 Attacks

Chapter 15   Context-Based Access Control

Chapter 16   Authentication Proxy and the Cisco IOS Firewall

Chapter 17   Identity-Based Networking Services

Chapter 18   Configuring 802.1x Port-Based Authentication

This chapter covers the following subjects:

- Designing ACLs
- Simple Network Management Protocol
- Disabling Directed Broadcasts

CHAPTER 11

Securing Networks with Cisco Routers

This chapter addresses how Cisco IOS Software uses access control lists (ACL) to secure a network, and discusses how to configure routers so that they do not present potential vulnerabilities through services or features that might be exploited by intruders. Disabling unnecessary services helps reduce the vulnerability of your network, and this chapter identifies some of the common services and features that you should consider disabling if not in use.

"Do I Know This Already?" Quiz

The purpose of the "Do I Know This Already?" quiz is to help you decide whether you really need to read the entire chapter. If you already intend to read the entire chapter, you do not necessarily need to answer these questions now.

The 10-question quiz, derived from the major sections in "Foundation Topics" section of the chapter, helps you determine how to spend your limited study time.

Table 11-1 outlines the major topics discussed in this chapter and the "Do I Know This Already?" quiz questions that correspond to those topics.

Table 11-1 *"Do I Know This Already?" Foundation Topics Section-to-Question Mapping*

Foundation Topics Section	Questions Covered in This Section
Defining ACLs	1–6
Simple Network Management Protocol	7, 8
Disabling Directed Broadcasts	9, 10

CAUTION The goal of self-assessment is to gauge your mastery of the topics in this chapter. If you do not know the answer to a question or are only partially sure of the answer, you should mark this question wrong for purposes of the self-assessment. Giving yourself credit for an answer you correctly guess skews your self-assessment results and might provide you with a false sense of security.

1. In what order is an access list reviewed for a packet entering the system?
   a. All at once
   b. Top to bottom
   c. Bottom to top
   d. Configured by the user

2. What are the ACL criteria by which packet analysis may be implemented? (Select four.)
   a. Source address of the traffic
   b. Destination address of the traffic
   c. Data
   d. MAC
   e. Port

3. Which kind of an ACL allows information to be filtered based on session information?
   a. Reflexive
   b. Context-based Access Control
   c. Standard
   d. Session
   e. Transport
   f. Port

4. What is the wildcard mask for 255.248.0.0?
   a. 255.248.0.0
   b. 0.7.0.0
   c. 0.7.255.255
   d. 255.255.255.255
   e. 255.255.7.0

5. What does the following ACL command do and which type of protocol is it?

```
#access-list 101 deny icmp any 192.168.150.0 0.0.0.255
#access-list 101 permit ip any 192.168.150.0 0.0.0.255
```

   a. Allows all other networks other than 192.168.150.0/25 to use ICMP, and it is of type Standard

   b. Denies ICMP originating from any source to destination 192.168.150.0/24, and it is of type Extended

   c. Denies ICMP originating from 192.168.150.0/24, and it is of type Standard

   d. Allows all IP traffic originating from 192.168.150.0/24, and it is of type Extended

   e. None of the above

6. What are the two main tasks associated with using an ACL?

   a. Create ACL.

   b. Identify ACL.

   c. Apply ACL.

   d. Enable ACL mode on the device.

   e. All of the above.

7. What version of SNMP is not supported in the latest version of Cisco IOS Software?

   a. SNMPv1

   b. SNMPv2

   c. SNMPv3

   d. SNMPv0

   e. All of the above

8. What is not a security feature of SNMPv3? (Select two.)

   a. Message integrity

   b. MD5 hash sum checks

   c. Authentication

   d. Three-phase authentication

   e. Encryption

9. A Smurf attack generates traffic on the network that targets what?

   a. Networks

   b. Broadcasts

   c. Hosts

   d. Routers and switches

   e. All of the above

10. What is the best reason to disable finger on a Cisco device?

   a. Causes information leakage

   b. Shows who is logged in to the device

   c. Divulges network information

   d. Grants access to users

   e. All of the above

The answers to the "Do I Know This Already?" quiz are found in the appendix. The suggested choices for your next step are as follows:

- **8 or less overall score**—Read the entire chapter. This includes the "Foundation Topics" and "Foundation Summary" sections and the "Q&A" section.

- **9 or 10 overall score**—If you want more review on these topics, skip to the "Foundation Summary" section and then go to the "Q&A" section. Otherwise, move on to Chapter 12, "The Cisco IOS Firewall and Advanced Security Feature Set."

Foundation Topics

ACLs tell the router which traffic to allow and which traffic to deny. They can be configured to be general or specific. Unused but enabled services on routers represent a potential vulnerability for your network. Every network is unique, and therefore, every network requires a different type of configuration on its routers. This chapter covers some of the Cisco IOS Software services that should be turned off in most network settings to prevent security breaches or network downtime. It also discusses some commonly configured management services and how to securely operate them.

Defining ACLs

ACLs are rules that deny or permit packets coming in or out of an interface. An ACL typically consists of multiple ACL entries (ACE), organized internally by the router. When a packet is subjected to access control, the router searches this linked list in order from top to bottom to find a matching element. The matching element is then examined to determine whether the packet is allowed or denied. Figure 11-1 shows the behavior of a router that has an ACL configured on its interfaces.

Figure 11-1 *High-Level Overview of How an ACL Is Processed by a Router*

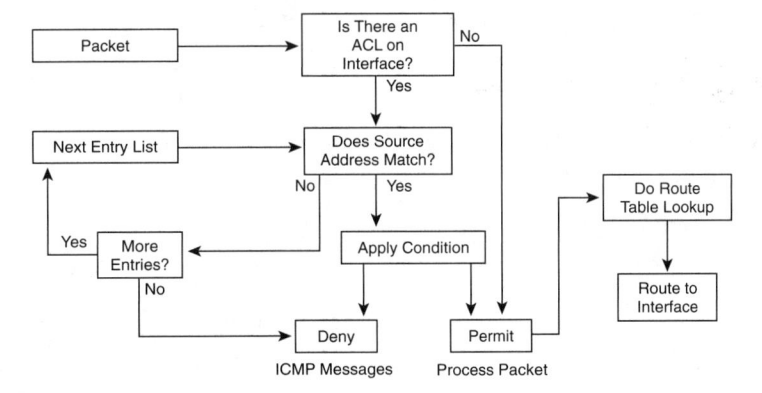

ACL functions include the following:

- To control the transmission of packets coming into or out from an interface
- To control virtual terminal line access
- To restrict contents of routing updates
- To define interesting traffic

Numbered ACEs are entered one line at a time, and the list is scanned for a match in that same order. If you must make a change, you have to reenter the entire list.

ACL criteria can be the source address of the traffic, the destination address of the traffic, the upper-layer protocol, or other information.

There is an implied deny for traffic that is not permitted. A single-entry ACL with only one deny entry has the effect of denying all traffic. You must have at least one **permit** statement in an ACL; otherwise, all traffic will be blocked.

Determining When to Configure Access Lists

To provide the security benefits of ACLs, you should, at a minimum, configure ACLs on border routers, which are routers situated at the edges of your networks. This setup provides a basic buffer from the outside network or from a less-controlled area of your own network into a more sensitive area of your network.

You can configure ACLs so that inbound traffic or outbound traffic, or both, are filtered on an interface. ACLs must be defined on a per-protocol basis. In other words, you should define ACLs for every protocol enabled on an interface if you want to control traffic flow for that protocol.

Types of IP ACLs

Cisco IOS Software supports the following types of ACLs for IP:

- **Standard IP ACLs**—Use source addresses for matching operations.

- **Extended IP ACLs**—Use source and destination addresses for matching operations and optional protocol type information for finer granularity of control.

- **Reflexive ACLs**—Allow IP packets to be filtered based on session information. Reflexive ACLs contain temporary entries and are nested within an extended, named IP ACLs.

- **Time-based ACLs**—Time-based ACLs, as the name intuitively indicates, are triggered by a time function.

- **Context-Based Access Control (CBAC)**—Introduced in Cisco IOS Software Release 12.0.5.T and requires the Cisco IOS Firewall feature set. CBAC inspects traffic that travels through the firewall to discover and manage state information for TCP and UDP sessions. This state information is used to create temporary openings in the firewall's ACLs. This topic is discussed in greater detail in Chapter 15, "Context-Based Access Control."

Standard IP ACLs

Standard IP ACLs are the oldest type of ACLs, dating back as early as Cisco IOS Software Release 8.3. Standard IP ACLs control traffic by comparing the source address of the IP packets to the addresses configured in the ACL.

The following is the command syntax format of a standard IP ACL:

```
access-list access-list-number {permit | deny} {host | source source-wildcard | any}
    log
```

In all software releases, the *access-list-number* can be anything from 1 to 99. Table 11-2 shows the protocol and the corresponding number range for the ACL identification. In Cisco IOS Software Release 12.0.1, standard IP ACLs began using additional numbers (1300 to 1999). These additional numbers are referred to as *expanded IP ACLs*. In addition to using numbers to identify ACLs, Cisco IOS Software Release 11.2 and later added the ability to use the list *name* in standard IP ACLs.

The **log** option enables you to monitor how many packets are permitted or denied by a particular ACL, including the source address of each packet. The logging message includes the ACL number, whether the packet was permitted or denied, the source IP address of the packet, and the number of packets from that source permitted or denied in the prior 5-minute interval.

Table 11-2 *Protocols and Their Corresponding Number Identification for an ACL*

Protocol	Range
Standard IP	1–99 and 1300–1999
Extended IP	100–199 and 2000–2699
Ethernet type code	200–299
Ethernet address	700–799
Transparent bridging (protocol type)	200–299
Transparent bridging (vendor code)	700–799
Extended transparent bridging	1100–1199
DECnet and extended DECnet	300–399
Xerox Network Systems (XNS)	400–499
Extended XNS	500–599
AppleTalk	600–699
Source-route bridging (protocol type)	200–299
Source-route bridging (vendor code)	700–799

continues

Table 11-2 *Protocols and Their Corresponding Number Identification for an ACL (Continued)*

Protocol	Range
Internetwork Packet Exchange (IPX)	800–899
Extended IPX	900–999
IPX Service Advertising Protocol (SAP)	1000–1099
Standard Virtual Integrated Network Service (VINES)	1–100
Extended VINES	101–200
Simple VINES	201–300

Wildcard masks in conjunction with IP addresses are used to identify the source address in an ACL. Wildcard masks are also known as *reverse netmasks*. If your netmask normally is 255.255.255.0, for example, in binary that is as follows:

```
11111111 11111111 11111111 00000000
```

Swapping the bits, that yields the following:

```
00000000 00000000 00000000 11111111
```

or 0.0.0.255 (your wildcard mask).

Another way to calculate your wildcard mask is to take your network mask and subtract each octet from 255. If your network mask is 255.255.248.0, for example, you calculate your wildcard by subtracting 255 from each octet, yielding a 0.0.7.255 wildcard mask.

After defining an ACL, you must apply it to the interface (inbound or outbound):

```
interface interface
ip access-group number {in | out}
```

Example 11-1 shows the use of a standard IP ACL to block all traffic except that from source 192.168.100.x.

Example 11-1 *Sample ACL Configuration Permitting Network 192.168.100.0 and Implicitly Denying All Other IP Traffic*

```
Firewall(config)#access-list 1 permit 192.168.100.0 0.0.0.255
Firewall(config)#interface Ethernet0/0
Firewall(config-if)#ip address 192.168.100.1 255.255.255.0
```

The terms *in*, *out*, *source*, and *destination* are used as referenced by the router. Traffic on the router could be compared to traffic on the highway. If you were a law enforcement officer in the United States and wanted to stop a truck coming from Mexico and traveling to Canada, the truck's source would be Mexico, and the truck's destination would be Canada. The roadblock could be applied at the U.S./Canadian border (out) or the U.S./Mexican border (in). See Figures 11-2 and 11-3.

Figure 11-2 *Border Patrol (ACL) Stopping a Truck (Packet) from Entering the United States from Mexico*

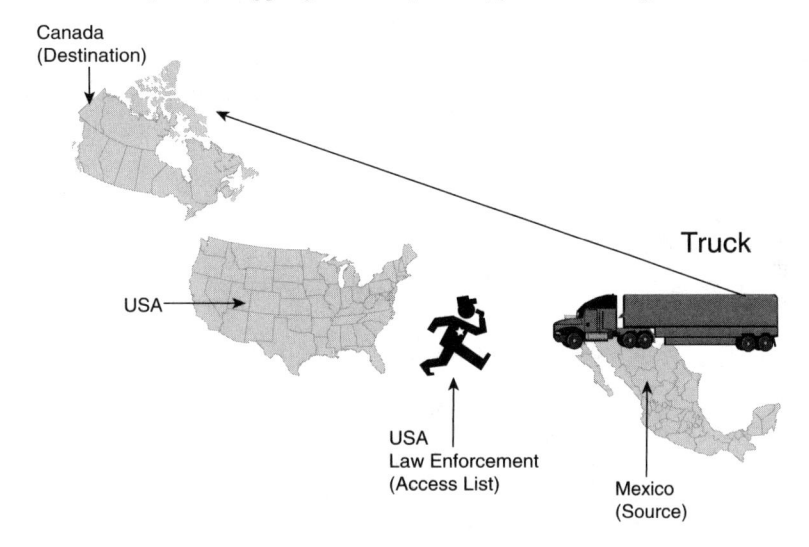

Figure 11-3 *Border Patrol (ACL) Stopping a Truck from Mexico (Packet) Leaving the United States*

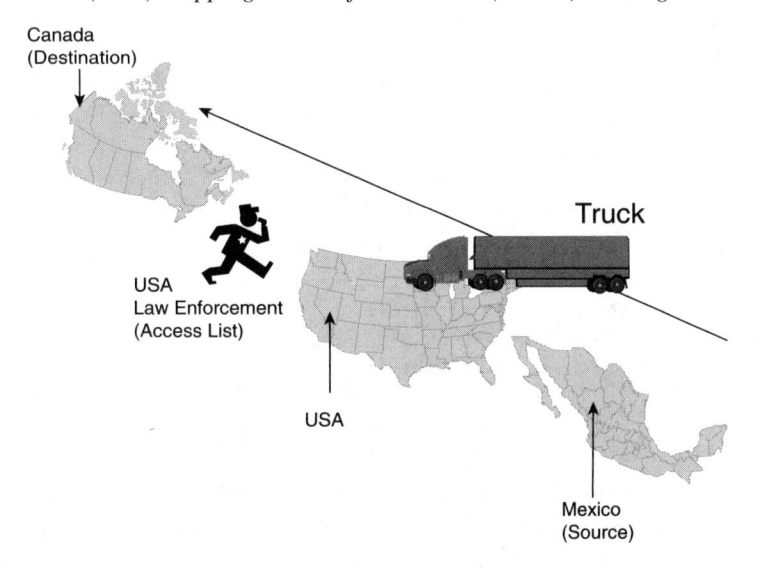

With regard to a router, these terms mean the following:

- **In**—Traffic that is arriving on the interface and which will go through the router; the source is where it has been, and the destination is where it is going (on the other side of the router).

- **Out**—Traffic that has already been through the router and is leaving the interface; the source is where it has been (on the other side of the router), and the destination is where it is going.

The in ACL has a source on a segment of the interface to which it is applied and a destination off of any other interface. The out ACL has a source on a segment of any interface other than the interface to which it is applied and a destination off of the interface to which it is applied.

Extended IP ACLs

Extended IP ACLs were introduced in Cisco IOS Software Release 8.3. Extended IP ACLs control traffic by not only comparing the source and destination IP addresses but also comparing the source and destination port numbers of the IP packets to those configured in the ACL.

The following is the command syntax format of extended IP ACLs:

```
ip access-list access-list-number [dynamic dynamic-name [timeout minutes]]
   {deny | permit} protocol source source-wildcard destination destination-wildcard
   [precedence precedence] [tos tos] [log | log-input] [time-range time-range-name]
```

In all software releases, the *access-list-number* can be 101 to 199. In Cisco IOS Software Release 12.0.1, extended IP ACLs began using additional numbers (2000 to 2699). These additional numbers are referred to as *expanded IP ACLs*. Cisco IOS Software Release 11.2 added the ability to use the list *name* in extended IP ACLs.

Example 11-2 shows an extended IP ACL used to permit traffic on the 192.168.100.x network (inside) and to receive ping responses from the outside while preventing unsolicited pings from people outside (permitting all other traffic).

Example 11-2 *Sample Configuration for an Extended IP ACL*

```
Firewall(config)#access-list 101 deny icmp any 192.168.100.0 0.0.0.255 echo
Firewall(config)#access-list 101 permit ip any 192.168.100.0 0.0.0.255
Firewall(config)#interface fastethernet0/1
Firewall(config-if)#ip address 172.16.8.1 255.255.255.0
Firewall(config-if)#ip access-group 101 in
```

Reflexive ACLs

Cisco IOS Software Release 11.3 introduced reflexive ACLs. Reflexive ACLs enable IP packets to be filtered based on upper-layer session information. They are generally used to allow outbound traffic and to limit inbound traffic in response to sessions originating inside the router.

Reflexive ACLs can be defined only with extended, named IP ACLs. They cannot be defined with numbered or standard, named IP ACLs or with other protocol ACLs. Reflexive ACLs can be used in conjunction with other standard and static extended IP ACLs.

Example 11-3 demonstrates the process of permitting ICMP outbound and inbound traffic. TCP traffic that had been initiated from inside is permitted in, whereas all other traffic is denied.

Example 11-3 *Sample Configuration for a Reflexive ACL*

```
Firewall(config)#ip access-list extended incoming
Firewall(config-ext-nacl)#permit icmp
 192.168.100.0 0.0.0.255 10.1.1.0 0.0.0.255
Firewall(config-ext-nacl)#evaluate traffic
Firewall(config-ext-nacl)#exit
Firewall(config)#ip access-list extended outgoing
Firewall(config-ext-nacl)#permit icmp
 10.10.10.0 0 0.0.0.255 192.168.1.0 0.0.0.255
Firewall(config-ext-nacl)#permit tcp
 10.10.10.0 0.0.0.255 192.168.1.0 0.0.0.255 reflect traffic
Firewall(config)#ip reflexive-list timeout 90
Firewall(config)#interface Ethernet0/0
Firewall(config-if)#ip address 192.168.100.1 255.255.255.0
Firewall(config-if)#ip access-group incoming in
Firewall(config-if)#ip accesss-group outgoing out
Firewall(config)#exit
```

Time-Based ACLs

Cisco IOS Software Release 12.0.1.T introduced time-based ACLs. Although similar to extended IP ACLs in function, they allow for access control based on time. To implement time-based ACLs, a time range is created that defines specific times of the day and week. The time range is identified by a name and then referenced by a function. Therefore, the time restrictions are imposed on the function itself. The time range relies on the router's system clock. The router clock can be used, but the feature works best with Network Time Protocol (NTP) synchronization.

Time-based ACL commands require the following syntax:

```
time-range time-range-name
periodic days-of-the-week hh:mm to [days-of-the-week] hh:mm
absolute [start time date] [end time date]
ip access-list name | number extended_definition time-range name_of_time-range
```

Example 11-4 shows a Telnet connection permitted from the inside to the outside network on Monday, Tuesday, and Thursday during the hours of 7 a.m. until 6 p.m.

Example 11-4 *Sample Configuration for Time-Range ACL*

```
Firewall(config)#interface Ethernet0/0
Firewall(config-if)#ip address 192.168.100.253 255.255.255.0
Firewall(config-if)#ip access-group 111 in
Firewall(config)#access-list 111 permit tcp
 10.1.1.0 0.0.0.255 172.16.1.0 0.0.0.255
eq telnet time-range TelnetAccess
Firewall(config)#time-range TelnetAccess
Firewall(config-time-range)#periodic Monday Tuesday Thursday 7:00 to 18:00
```

Time ranges offer many possible benefits, including the following:

- The network administrator has more control over permitting or denying a user access to resources. These resources include an application (identified by an IP address/mask pair and a port number), policy routing, or an on-demand link (identified as interesting traffic to the dialer).

- When provider access rates vary by time of day, it is possible to automatically reroute traffic cost-effectively.

- Service providers can dynamically change a committed access rate (CAR) configuration to support the quality-of-service (QoS) service level agreements (SLA) that are negotiated for certain times of day.

- Network administrators can control logging messages. ACL entries can log traffic at certain times of the day but not constantly. Therefore, administrators can just deny access without analyzing the many logs generated during peak hours.

- Policy-based routing and queuing functions are enhanced.

Certificate-Based ACLs

Cisco IOS Software Release 12.2(15)T introduced certificate-based ACLs. To implement certificate-based ACLs, you need a certificate authority (CA) on the network that supports Public Key Infrastructure (PKI) protocol and Simple Certificate Enrollment Protocol (SCEP), or manual "cut and paste" certificate enrollment. You can find more information about certificate-based ACLs at http://www.cisco.com/univercd/cc/td/doc/product/software/ios122/122newft/122t/122t15/ftcrtacl.htm. A discussion of Certificate-based ACLs is beyond the scope of this book.

Configuring ACLs on a Router

When creating an ACL, you define criteria that is applied to each packet processed by the router; the router decides whether to forward or block each packet based on whether the packet matches the criteria.

Defining ACLs

In ACLs, you generally define criteria such as packet source addresses, packet destination addresses, or upper-layer protocol of the packet. However, each protocol has its own specific set of criteria that you can define.

For a single ACL, you can define multiple criteria in multiple, separate ACL statements. Each of these statements should reference the same identifying name or number to tie the statements to the same ACL. You can have as many criteria statements as you want, limited only by the available memory. Of course, the more statements you have, the more difficult it will be to comprehend and manage your ACLs.

The two main tasks involved in using ACLs are as follows:

Step 1   Create an ACL by specifying an ACL number or name and access conditions.

Step 2   Apply the ACL to interfaces or terminal lines.

Figure 11-4 depicts the application of an ACL to the serial interface of a router providing access to the server 10.40.100.5.

Figure 11-4   *ACL webserver2 Being Applied to the Serial Interface of a Router*

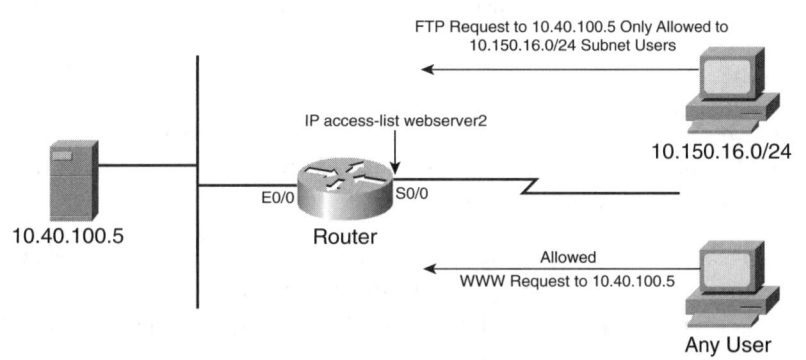

In Example 11-5, an extended IP ACL called webserver2 is created. ACL webserver2 has ACL entries that permit TCP port 80 (WWW) from any source to 10.40.100.5 and FTP access to 10.40.100.5 10.150.16.0.0/24 only.

Example 11-5 shows an example of ACL configuration.

Example 11-5   *ACL Configuration Called webserver2 Permitting Web Access*

```
Firewall# configure terminal
Firewall(config)#ip access-list extended webserver2
Firewall(config-ext-nacl)#permit tcp any host 10.40.100.5 eq www
Firewall(config-ext-nacl)#permit tcp 10.150.16.0 0.0.0.255 host 10.40.100.5 eq ftp
```

You can define ACLs without applying them. However, the ACLs will have no effect until they are applied to the router's interface. Example 11-6 applies the ACL server2 to the serial0 interface creating an ACL.

Example 11-6 *Applying the ACL server2 to the serial0 Interface*

```
Firewall(config)#interface serial0
Firewall(config-if)# ip access-group webserver2 in
```

Simple Network Management Protocol

Simple Network Management Protocol (SNMP) is widely used for router monitoring and configuration changes. If not configured properly, SNMP could provide a wealth of information about the device to intruders running SNMP discovery tools.

Cisco IOS Software Release 12.1 supports the following versions of SNMP:

- **SNMPv1**—Version 1 of SNMP is a full Internet standard, defined in RFC 1157. Security is based on community strings.

- **SNMPv2c**—The community string-based administrative framework for SNMPv2. SNMPv2c (the *c* stands for *community*) is an experimental IP defined in RFC 1901, RFC 1905, and RFC 1906.

- **SNMPv3**—Version 3 of SNMP is an interoperable standards-based protocol defined in RFCs 2273 through 2275. SNMPv3 provides secure access to devices by a combination of authenticating and encrypting packets over the network.

Unfortunately, SNMPv1, which is the most commonly used, uses a very weak authentication scheme based on a community string, which amounts to a fixed password transmitted over the network without encryption. SNMPv1 is ill suited for use across the public Internet for the following reasons:

- It uses clear-text strings.

- Most SNMP implementations send clear-text strings repeatedly as part of periodic polling.

- SNMP strings are readable in the router/switch configuration.

You should carefully consider the implications before using it that way. If it is absolutely necessary that you use SNMP, it is recommended you use SNMPv3, which supports a Message Digest 5 (MD5)-based digest authentication scheme and allows for restricted access to various management data.

Another way to mitigate the potential threats posed when using SNMPv1 is to avoid using the same community strings for all network devices. Use a different string or strings for each device, or at least for each area of the network. Use strong community names. Do not make a read-only string the same as a read-write string. If possible, periodic SNMPv1 polling should be done with a read-only community string; read-write strings should be used only for actual write operations.

In most networks, legitimate SNMP messages come only from certain management stations. If this is true in your network, you should probably use the access list *number* option on the **snmp-server community** command to restrict SNMPv1 access to only the IP addresses of the management stations. Do not use the **snmp-server community** command for any purpose in a pure SNMPv2 environment; this command implicitly enables SNMPv1. SNMP access is restricted using standard ACLs.

SNMPv3 uses a secure form of communication. The security features provided in SNMPv3 are as follows:

- **Message integrity**—Ensuring that a packet has not been tampered with in transit by an unauthorized source

- **Authentication**—Determining the message is from a valid source

- **Encryption**—Scrambling the contents of a packet to prevent it from being seen by an unauthorized source

SNMP management stations often have large databases of authentication information, such as community strings. This information may provide access to many routers and other network devices. This concentration of information makes the SNMP management station a natural target for attack, and so it should be secured accordingly.

Controlling Interactive Access Through a Browser

Administrative access support via a browser is supported by Cisco IOS Software Release 11.0(6) and later. This feature is disabled by default but can be enabled via the **ip http server** command.

However, the use of HTTP to manage a router presents some inherent vulnerabilities. The Cisco IOS HTTP server provides authentication, but not encryption, for client connections. The data that the client and server transmit to each other is not encrypted. This leaves communication between clients and servers vulnerable to interception and attack. To reduce the risk posed by this vulnerability, you can use the **ip http access-class** command to restrict the hosts that connect to the router via HTTP or use the Cisco secure HTTP server.

The Secure HTTP (HTTPS) feature enables you to connect to the Cisco IOS HTTPS server securely. You enable it with the **ip http secure-server** command.

It uses Secure Sockets Layer (SSL) and Transport Layer Security (TLS) to provide device authentication and data encryption. HTTPS authenticates the client and the server with each other before establishing a connection. HTTPS supports the following standards:

- MD5
- Secure Hash Algorithm 1 (SHA-1)
- Public key-based encryption:
    - Rivest, Shamir, and Adelman (RSA)—Encryption/decryption/generation
    - Directory System Agent (DSA)—Encryption/decryption/generation
    - Diffie-Hellman—Key exchange/key generation
    - X.509 digital certificates

Before you start configuring the HTTPS server, generate an RSA usage key pair with a length of 1024 bits or greater for the device using the **crypto key generate rsa usage 1024** command. If you do not generate an RSA usage key pair manually, an RSA usage key pair with a length of 768 bits is generated automatically when you connect to the HTTPS server for the first time. Unless you save these automatically generated keys manually to NVRAM, they will be lost when the device is rebooted. The following is an example of enabling HTTPS on a router:

```
FW1(config)#ip http secure-server
```

Disabling Directed Broadcasts

On IP networks, a packet can be directed to an individual machine or broadcast to an entire network. When a packet is sent to an IP broadcast address from a machine on the local network, that packet is delivered to all machines on that network. When a packet is sent to that IP broadcast address from a machine outside of the local network, it is broadcast to all machines on the target network.

IP broadcast addresses are usually network addresses with the host portion of the address having all 1 bits. For example, the IP broadcast address for the network 192.168.100.0 is 192.168.100.255. Network addresses with all 0s in the host portion, such as 192.168.100.0, can also produce a broadcast response. In a Smurf attack, attackers are using Internet Control Message Protocol (ICMP) echo request packets directed to IP broadcast addresses from remote locations to generate denial-of-service (DoS) attacks. There are three parties in these attacks: the attacker, the intermediary, and the victim.

> **NOTE** These attacks have been referred to as Smurf attacks because the name of one of the exploit programs attackers used to execute this attack is called Smurf.

The intermediary receives an ICMP echo request packet directed to the IP broadcast address of its network. If the intermediary does not filter ICMP traffic directed to IP broadcast addresses, many of the machines on the network receive this ICMP echo request packet and send an ICMP echo reply packet back. When (potentially) all the machines on a network respond to this ICMP echo request, the result can be severe network congestion or outages.

When the attackers create these packets, they do not use the IP address of their own machine as the source address. Instead, they create forged packets that contain the spoofed source address of the attacker's intended victim. The result is that when all the machines at the intermediary's site respond to the ICMP echo requests, they send replies to the victim's machine. The victim is subjected to network congestion that could potentially make the network overrun and unusable.

Unless applications or other explicit requirements need the router interfaces to have IP directed broadcasts, they should be turned off to suppress the effects of this attack. You can use the **no ip directed-broadcast** command to do so on the Cisco IOS Software.

The **no ip directed-broadcast** interface command is the default in Cisco IOS Software Release 12.0 and later. In earlier versions, the command should be applied to every LAN interface that is not known to forward legitimate directed broadcasts.

Routing Protocol Authentication

One of the ways that routers update their routing tables is by route updates they receive from other routers via routing protocols. Routing protocols are vulnerable to spoofing of route updates. A mechanism for receiving reliable routing information from a trusted source router should be put in place to avoid getting bad updates by "rogue" or misconfigured routers. It is quite possible to have a rogue router provide bad routes to your router, which could cause the failure of your network. One way to combat this problem is to use authentication and encryption for the communication between routers that share routing updates.

When authentication is configured, neighbor authentication occurs whenever routing updates are exchanged between neighbor routers. This authentication ensures that a routing protocol receives reliable routing information from a trusted source.

The authentication process works by requiring a unique key to first verify the source (neighbor router) before a routing update is accepted by a routing protocol. This way "rogue" routers will not be able to participate in the route update process. The process of authentication occurs as follows (in summary):

1. A router sends a routing update with a key to the neighbor router.

2. The receiving (neighbor) router compares the received key against the key stored in its own memory.

3. If the two keys match, the receiving router accepts the routing update packet. If the two keys do not match, the routing update packet is rejected.

MD5 authentication works much like plain-text authentication, except that the key is never sent over the wire. Instead, the router uses the MD5 algorithm to produce a "message digest" of the key (also called a *hash*). The message digest is then sent rather than the key itself. This ensures that nobody can eavesdrop on the line and learn keys during transmission. Routing protocols such as Open Shortest Path First (OSPF), Routing Information Protocol (RIP) version 2, Interior Gateway Routing Protocol (IGRP), and Border Gateway Protocol (BGP) use it.

Defining Small Server Services

TCP and UDP small servers are services that run in the router and are useful for diagnostics, including Echo, Chargen, Daytime, and Discard:

- **Echo (UDP, TCP)**—This simple port just echoes whatever is sent to it.

- **Chargen (UDP, TCP)**—This simple port generates a stream of characters (TCP) or a packet containing characters (UDP).

- **Daytime (TCP)**—This simple port responds with the current time of day. The protocol specification does not clearly define the format of the data returned, so every machine responds in a slightly different format. This can be used to fingerprint machines.

- **Discard (UDP, TCP)**—This simple port throws traffic away.

These services, especially their UDP versions, are used for diagnostic purposes but can be used to launch DoS and other attacks that would otherwise be prevented by packet filtering. It is recommended that these services not be enabled unless doing so is absolutely necessary. These services could be exploited indirectly to gain information about the target system or directly as is the case with the Fraggle attack, which uses UDP echo.

> **NOTE** *Fingerprinting* is the technique of interpreting the responses of a system to identify the operating system and systems in use. In particular, unexpected combinations of data are sometimes sent at the system to trigger these responses.

Exploitation of these services may result in symptoms such as the process table being full of error messages (%SYS-3 NOPROC) or high CPU utilization. The small services are disabled by default in Cisco IOS Software 12.0 and later. In earlier software, you can disable them by using the **no service tcp-small-servers** and **no service udp-small-servers** commands.

Disabling Finger Services

Cisco routers provide an implementation of the finger service, which is used to find out which users are logged in to a network device. This service is equivalent to issuing a remote **show users** command. Although the information gained may seem harmless, it could be valuable to an attacker. You can disable the finger service with the **no service finger** command.

Disabling Network Time Protocol

Network Time Protocol (NTP) is not especially dangerous, but any unneeded service may represent a path for penetration. If NTP is actually used, it is important to explicitly configure a trusted time source and to use proper authentication, because corrupting the time base is a good way to subvert certain security protocols. If NTP is not being used on a particular router interface, you can disable it with the **no ntp enable** interface command. One security risk of NTP is that it can be altered as a method of subverting security protocols and can compromise logging information. This is the primary reason for disabling NTP if it is not in use.

Disabling Cisco Discovery Protocol

Cisco Discovery Protocol (CDP) is a Cisco proprietary Layer 2 protocol that is media and protocol independent and runs on all Cisco-manufactured equipment, including routers, access servers, and switches. CDP is primarily used to obtain protocol addresses of neighboring devices and discover the platform of those devices. CDP can also be used to show information about the interfaces your router uses.

To compromise the neighboring device and consequently the network, an attacker can potentially use the information provided by CDP. If your network does not use CDP, you should turn it off. You can disable CDP with the global configuration **no cdp running** command. You can disable CDP on a particular interface with the **no cdp enable** command.

Foundation Summary

The "Foundation Summary" section of each chapter lists the most important facts from the chapter. Although this section does not list every fact from the chapter that will be on your SNRS exam, a well-prepared candidate should at a minimum know all the details in each "Foundation Summary" section before going to take the exam.

ACLs filter network traffic by controlling whether routed packets are forwarded or blocked at the router's interfaces. Your router examines each packet to determine whether to forward or drop the packet, based on the criteria you specified within the ACLs.

The two main tasks involved in using ACLs are as follows:

- Create an ACL by specifying an ACL number or name and access conditions.
- Apply the ACL to interfaces or terminal lines.

Cisco IOS Software supports the following types of ACLs for IP:

- Standard IP ACLs
- Extended IP ACLs
- Reflexive ACLs
- Time-based ACLs
- CBAC

At the end of every ACL is an implied "deny all traffic" criteria statement. Therefore, if a packet does not match any of your criteria statements, the packet is blocked. Table 11-3 lists the commands used to mitigate attacks on IOS routers.

Table 11-3 *Commands for Preventing Attacks Against the Cisco IOS Router*

Command	Description
no service tcp-small-servers **no service udp-small-servers**	Prevents abuse of the small services from DoS or other attacks
no service finger	Avoids releasing user information to possible attackers

continues

Table 11-3 *Commands for Preventing Attacks Against the Cisco IOS Router (Continued)*

Command	Description
no cdp running **no cdp enable**	Avoids releasing information about the router to directly connected devices
no ntp enable	Prevents attacks against the NTP service
no ip directed-broadcast	Prevents attackers from using the router as a Smurf amplifier
snmp-server party... **authentication md5** *secret* ...	Configures MD5-based SNMPv2 authentication. Enable SNMP only if it is needed in your network
ip http authentication *method*	Authenticates HTTP connection requests (if you've enabled HTTP on your router)
ip http access-class *list*	Further controls HTTP access by restricting it to certain host addresses (if you have enabled HTTP on your router)

Q&A

As mentioned in the section "How to Use This Book" in the Introduction, you have two choices for review questions: the Q&A questions here or the exam simulation questions on the CD-ROM. The questions that follow present a bigger challenge than the exam itself because they use an open-ended question format. By using this more difficult format, you can exercise your memory better and prove your conceptual and factual knowledge of this chapter. You can find the answers to these questions in the appendix.

1. Describe one of the functions of an ACL.
2. What is the difference between a standard and extended ACL?
3. Which types of ACLs do the corresponding number identifiers of 150, 750, and 1400 implement?
4. What command sets an ACL that will deny inbound traffic from 192.168.10.0/24 from interface E0/0?
5. Where is the traffic filtered when an ACL is configured with the "out" command?
6. What command enables you to set the ACL time to limit Telnet access to Sunday evenings between 8 p.m. and 10 p.m.?
7. For security reasons, why is SNMPv1 ill suited for configuring devices that are directly connected to the Internet?
8. Why would the following command be issued?

    `Firewall(config)#ip http secure-server`
9. Why is it a good practice to limit directed broadcasts on the Cisco devices?
10. Can routing protocols utilize ACLs? If so, what is the logic behind it?

This chapter covers the following subject:

- The Cisco IOS Firewall and Advanced Security Feature Set

CHAPTER 12

The Cisco IOS Firewall and Advanced Security Feature Set

The Cisco IOS Firewall is a component of the Cisco IOS Advanced Security feature set. The Cisco IOS Advanced Security feature set was introduced in Cisco IOS Software Release 12.3 and is an improvement on the Cisco IOS Firewall feature set from previous Cisco IOS Software versions.

"Do I Know This Already?" Quiz

The purpose of the "Do I Know This Already?" quiz is to help you decide whether you really need to read the entire chapter. If you already intend to read the entire chapter, you do not necessarily need to answer these questions now.

The 10-question quiz, derived from the major sections in "Foundation Topics" section of the chapter, helps you determine how to spend your limited study time.

Table 12-1 outlines the major topics discussed in this chapter and the "Do I Know This Already?" quiz questions that correspond to those topics.

Table 12-1 *"Do I Know This Already?" Foundation Topics Section-to-Question Mapping*

Foundation Topics Section	Questions Covered in This Section
The Cisco IOS Firewall and Advanced Security Feature Set	1–10

CAUTION The goal of self-assessment is to gauge your mastery of the topics in this chapter. If you do not know the answer to a question or are only partially sure of the answer, you should mark this question wrong for purposes of the self-assessment. Giving yourself credit for an answer you correctly guess skews your self-assessment results and might provide you with a false sense of security.

1. Which of the following technologies is not part of the Cisco IOS Firewall and Advanced Security feature set?

    a. VPN service
    b. Firewall services
    c. Intrusion prevention
    d. Advanced routing support
    e. VoIP
    f. None of the above

2. Which features are part of the Cisco IOS Firewall engine?

    a. Event logging
    b. Audit trail
    c. Redundancy/failover
    d. Transparency
    e. Firewall management
    f. All of the above

3. DoS protection provides all of the following except what?

    a. Protection from SYN attacks
    b. Extended logging features
    c. Packet injection protection
    d. Traffic blackholing
    e. None of the above

4. Real-time alerts sent via syslog include which of the following information?

    a. Source
    b. Destination
    c. Bytes
    d. Session
    e. Source, destination, and session
    f. All of the above

5. Which of the following are types of PAM support provided by the Cisco IOS Firewall?

   a. System-defined support

   b. Network-specific support

   c. User-defined support

   d. Host-specific support

   e. Unusual support

   f. All of the above

6. Which of the following ports is not a system-defined PAM entry?

   a. H323

   b. FTP

   c. StreamWorks

   d. SQLNet

   e. SSH

   f. NetShow

7. If you try to map multiple ports to a PAM service, what will happen?

   a. System will generate an error.

   b. Old entries for that service will be overwritten.

   c. Multiple entries for a service will be allowed.

   d. System will be rebooted.

   e. Old entries for that service will be overwritten, and multiple entries will be allowed.

8. Mapping port 8080 to Telnet for one host and mapping 8080 to HTTP for another is called what?

   a. Host-specific PAM

   b. ID-specific PAM

   c. Service-specific PAM

   d. IP-specific PAM

   e. None of the above

9. URL filtering in the Cisco IOS Firewall feature set supports which of the following methods?
   a. Global
   b. User
   c. Category
   d. Customized
   e. Keyword
   f. All of the above

10. If you want to block all websites with the word *illegal* in the URL, which kind of filter must you use?
    a. Global
    b. User
    c. Category
    d. Customized
    e. Keyword
    f. None of the above

The answers to the "Do I Know This Already?" quiz are found in the appendix. The suggested choices for your next step are as follows:

- **8 or less overall score**—Read the entire chapter. This includes the "Foundation Topics" and "Foundation Summary" sections and the "Q&A" section.

- **9 or 10 overall score**—If you want more review on these topics, skip to the "Foundation Summary" section and then go to the "Q&A" section. Otherwise, move on to Chapter 13, "Cisco IOS Intrusion Prevention System."

Foundation Topics

Firewalls are networking devices that control access to your organization's network assets. Firewalls are usually positioned at the ingress/egress points of your network. If your network has multiple entrance points, you must position a firewall at each ingress/egress point to provide effective network access control.

The most basic function of a firewall is to monitor and filter traffic. In addition to placing firewalls on the perimeter of your network, you can place firewalls within the network to control access to specific segments of your network. For example, you can position firewalls at all the entry points into a research and development network to prevent unauthorized access to proprietary information. Firewalls can be simple or elaborate, depending on the network requirements. The Cisco IOS Firewall provides firewall functionality on a Cisco router. The Cisco IOS Firewall and Advanced Security feature set were introduced with Software Release 12.3 and include significant enhancements to the IOS Firewall feature set, which was available with previous Cisco IOS Software versions. The Cisco IOS Firewall provides the following benefits to smaller networks through a single device rather than multiple devices:

- Controls access into and out of the network based on predefined parameters
- Provides secure connectivity with other networks via the public infrastructure (Internet)
- Provides a method for enforcing security policy

Cisco IOS Firewall and Advanced Security Feature Set

The Cisco IOS Firewall and Advanced Security feature set is a security-specific configuration that integrates the functionality of several devices. It is designed for smaller environments where network utilization is such that it does not require individual components for each network/security function. This ability enables you to consolidate the functions of several different components into a single platform, provides cost savings, and simplifies management.

Figure 12-1 depicts the consolidation of functions from the different components into the Cisco IOS Firewall.

Chapter 12: The Cisco IOS Firewall and Advanced Security Feature Set

Figure 12-1 *Cisco IOS Firewall Functions*

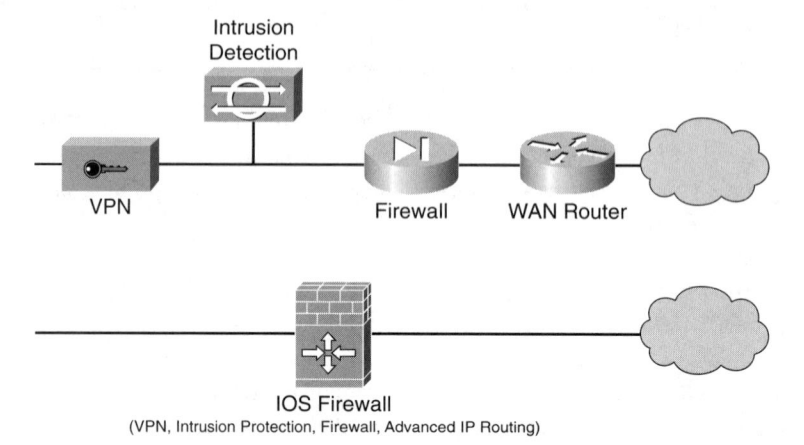

[Diagram: Top path shows VPN, Intrusion Detection, Firewall, WAN Router connecting to a cloud. Bottom path shows IOS Firewall (VPN, Intrusion Protection, Firewall, Advanced IP Routing) connecting to a cloud.]

The Cisco IOS Firewall and Advanced Security feature set integrate the following technologies:

- **VPN services**—The Cisco IOS Firewall supports dynamic multipoint virtual private networks (VPN). This allows PPP and point-to-multipoint (p2mp) VPNs to be dynamically configured supporting hub-and-spoke or full-mesh architectures with stateful IPsec failover to provide redundancy. Scalability has been increased to minimize bandwidth overhead, reduce jitter, and provide the highest quality of service (QoS) for voice and video traffic, as well as data. A low-latency queue (LLQ) for IPsec traffic has been integrated to provide QoS for encrypted voice and video. IPsec Network Address Translation (NAT) transparency has been added to all encrypted traffic to traverse NAT/PAT (Port Address Translation) devices, easing the integration of VPN connectivity across diverse network and addressing infrastructures. The Cisco IOS Firewall VPN supports Advanced Encryption Standard (AES) encryption with up to 256-bit keys.

- **Intrusion protection**—The Cisco IOS Intrusion Prevention System (IPS) supports several signatures to offer improved protection against attacks. Chapter 13 discusses the Cisco IOS IPS in detail.

- **Firewall services**—The core of the Cisco IOS Firewall is the advanced firewall engine. This engine tracks the state and context of network connections to secure traffic flow. It enhances security for TCP and UDP applications that use well-known ports, such as e-mail (SMTP) and Telnet traffic, by examining source and destination addresses. The Cisco IOS Firewall services include the following:

    — **Authentication proxy**—LAN-based, dynamic, per-user authentication and authorization via TACACS+ and RADIUS authentication servers for both inbound and outbound users. Chapter 16, "Authentication Proxy and the Cisco IOS Firewall," discusses authentication proxy in greater detail.

- **Audit trail**—Details transactions. Records time stamp, source host, destination host, ports, duration, and total number of types transmitted for detailed reporting. Can be configured on a per-application, per-feature basis.
- **Basic and advanced traffic filtering**—Standard and extended IP access control lists (ACL). Lock-and-key dynamic ACLs grant temporary access through firewalls upon user identification.
- **Context-Based Access Control (CBAC)**—Provides internal users with secure, per-application access. Chapter 15, "Context-Based Access Control" discusses CBAC in greater detail.
- **DoS detection and prevention**—Defends and protects router resources against common attacks. This is a function of the Cisco IOS IPS.
- **Dynamic port mapping**—Allows CBAC-supported applications to run on nonstandard ports.
- **Event logging**—Enables administrators to track potential security breaches or other nonstandard activities in real time by logging system error message output to a console terminal or syslog server.
- **Firewall management**—A wizard-based network configuration tool that offers step-by-step guidance through network design, addressing, and Cisco Firewall feature set implementation.
- **URL filtering support**—The Cisco IOS Firewall supports the N2H2 and Websense protocols to provide web access control and auditing.
- **Java applet blocking**—Protects against unidentified, malicious Java applets.
- **Network Address Translation (NAT)**—Hides the internal network from the outside for enhanced security.
- **Peer router authentication**—Ensures that routers receive reliable routing information from trusted sources.
- **Policy-based multi-interface support**—Provides the ability to control user access by IP address and interface as determined by the security policy.
- **Redundancy/failover**—Automatically routes traffic to a backup router if a failure occurs.
- **Firewall transparency**—Cisco IOS Software Release 12.3(7) introduced the transparent Cisco IOS Firewall. A transparent Cisco IOS Firewall can be implemented on an existing network without having to reconfigure statically defined devices because it utilizes a combination of CBAC and ACLs configured on the bridged interface. The firewall is considered to be transparent because it is intercepting packets at Layer 2 and is not restricted to traditional Layer 3 firewall limitations. You can configure both the Layer 2 and Layer 3 firewall on the same device.

- **Quality of service**—The Cisco IOS Firewall Advanced Security feature set is an add-in to Cisco IOS Software and provides support for QoS.

- **Multiprotocol support**—The Cisco IOS Firewall Advanced Security features set is an add-in to Cisco IOS Software and provides support for multiple protocols.

- **Multicast support**—The Cisco IOS Firewall Advanced Security features set is an add-in to Cisco IOS Software and provides support for multicast.

- **Advanced routing support**—The Cisco IOS Firewall Advanced Security features set is an add-in to Cisco IOS Software and provides support for advanced routing protocols.

- **VoIP support**—The Cisco IOS Firewall Advanced Security features set is an add-in to Cisco IOS Software and provides support for Session Initiation Protocol (SIP) and Skinny Client Control Protocol (SCCP). An enhancement to VoIP functionality includes support for Voice and Video over VPNs (V3PNs). Combining the multipoint VPN technology with low-latency queuing (LLQ) for IPsec provides the support. LLQ is a QoS method for providing class-based weighted fair queuing of traffic.

Authentication Proxy

When you configure the authentication proxy, you do not assign a direction at the interface because it is always applied to the inbound path. Authentication proxy intercepts the packet before it reaches the inbound ACL. Consequently, an inbound ACL can block all traffic, except for the special servers or devices that need to communicate with the Cisco IOS Firewall.

Authentication proxy dynamically opens connections on the inbound ACL of the input interface where the proxy is enabled, as well as on the outbound ACL of the output interface where the packet exits. This enables the packet to leave and lets the firewall engine intercept and take control.

DoS Protection

Cisco IOS Firewall provides protection from denial-of-service (DoS) attacks, such as SYN flood denial, port scans, and packet injection. Additional functionality has been added to the logging features introduced in Cisco IOS Software Release 12.3(7) to track the source IP of traffic flowing to a host that might be the target of a DoS attack. This features enables you to track the traffic across the entire network, determine the network entry point, and maintain logs simultaneously on each device that passes the traffic.

Logging and Audit Trail

Real-time alerts send syslog error messages to central management consoles upon the detection of suspicious activity. Enhanced audit trail features use syslog to track all transactions and to record timestamps, source host, destination host, ports used, session duration, and the total number of transmitted bytes for advanced, session-based reporting

— http://www.cisco.com/en/ US/products/sw/secursw/ps1018/products_implementation_ design_guide09186a00800fd670.html

To enable logging and send messages to a syslog server, use the following commands:

```
Firewall(config)#logging on
Firewall(config)#logging 192.168.100.6
```

To enable an audit trail of firewall messages, use the following syntax:

```
Firewall(config)#ip inspect audit-trail
```

To control the number of audit trail messages, you can enable or disable the audit trail per protocol in the firewall rules.

Port-to-Application Mapping

Port-to-application mapping (PAM) enables you to customize TCP or UDP port numbers for network services or applications. PAM uses this information to support network environments that run services using ports that differ from the registered or well-known ports associated with an application, such as those identified later in this chapter.

PAM enables CBAC-supported applications to be run on nonstandard ports. Using PAM, network administrators can customize access control for specific applications and services to meet the distinct needs of their networks.

PAM also supports host- or subnet-specific port mapping, which enables you to apply PAM to a single host or subnet using standard ACLs. Host- or subnet-specific port mapping is done using standard IP ACLs. The PAM table provides three types of mapping information, each of which is discussed in more detail in the following sections:

- System-defined port mapping
- User-defined port mapping
- Host-specific port mapping

System-Defined Port Mapping

By default, a table of system-defined mapping entries using the well-known or registered port mapping is created. The system-defined entries comprise all the services supported by CBAC, which requires the system-defined mapping information to function properly. The system-defined mapping information cannot be deleted or changed; that is, you cannot map SMTP services to port 21 (FTP) or FTP services to port 80 (HTTP). Table 12-2 lists the default system-defined services and applications in the PAM table.

Table 12-2 *Default System-Defined Services and Applications in the PAM Table*

Application Name	Protocol Description	Port Number
cuseeme	CU-SeeMe Protocol	7648
exec	Remote Process Execution	512
ftp	File Transfer Protocol (control port)	21
http	Hypertext Transfer Protocol	80
H323	H.323 Protocol (for example, Microsoft NetMeeting, Intel Video Phone)	1720
Login	Remote login	513
mgcp	Media Gateway Control Protocol	2427
msrpc	Microsoft Remote Procedure Call	135
netshow	Microsoft NetShow	1755
real-audio-video	RealAudio and RealVideo	7070
rtsp	Real Time Streaming Protocol	8559
shell	Remote command	514
sip	Session Initiation Protocol	5060
smtp	Simple Mail Transfer Protocol	25
sqlnet	SQL-NET	1521
streamworks	StreamWorks Protocol	1558
sunrpc	SUN Remote Procedure Call	111
telnet	Telnet	23
tftp	Trivial File Transfer Protocol (TFTP)	69
vdolive	VDOLive Protocol	7000

This table has been reproduced by Cisco Press with the permission of Cisco Systems Inc. Copyright © 2003 Cisco Systems, Inc. All Rights Reserved.

User-Defined Port Mapping

Network services or applications that use nonstandard ports require user-defined entries in the PAM table. For example, your network might run Telnet services on the nonstandard port 9000 rather than on the system-defined default port (port 23). In this case, you can use PAM to map port 9000 with Telnet services. If Telnet services run on other ports, use PAM to create additional port-mapping entries. After you define a port mapping, you can overwrite that entry later by just mapping that specific port with a different application.

User-defined port mapping information can also specify a range of ports for an application by establishing a separate entry in the PAM table for each port number in the range.

Host-Specific Port Mapping

In some environments, it might be necessary to override the default port-mapping information for a specific host or subnet. With host-specific port mapping, you can use the same port number for different services on different hosts. This means that you can map port 8080 with HTTP services for one host, while mapping port 8080 with Telnet services for another host.

Host-specific port mapping also enables you to apply PAM to a specific subnet when that subnet runs a service that uses a port number that differs from the port number defined in the default mapping information. For example, hosts on subnet 10.100.10.11 might run HTTP services on nonstandard port 8080, whereas other traffic through the firewall uses the default port for HTTP services, which is port 80.

Host-specific port mapping enables you to override a system-defined entry in the PAM table. If CBAC finds an entry in the PAM table that maps port 21 (the system-defined port for FTP) with SMTP for a specific host, for example, CBAC identifies port 21 as SMTP protocol traffic on that host.

To configure PAM, use the **ip port-map** command, as follows:

```
ip port-map appl_name port port_num [list acl_num]
```

Use the *list* option to associate this port mapping to the specific hosts in the ACL. (PAM uses standard IP ACLs only.) If an ACL is included, the hosts defined in that ACL have the application *appl_name* running on port *port_num*. The following example shows an HTTP mapped to port 8080 by an **ip port-map** command:

```
ip port-map http port 8080
```

URL Filtering

URL filtering enables you to control the Internet content that network users can access based on predefined policies, such as the organization's acceptable use policy. URL filtering enables you to block access to websites based on specific filters. The Cisco IOS Firewall supports the following filtering methods:

- **Global filtering**—Filters applied to all IP addresses, users, and groups.

- **User/group-based filtering**—Filters applied to specific users or groups.

- **Keyword-based filtering**—Filters applied to block access to websites that have a specific keyword as part of their URL.

- **Category-based filtering**—Filters applied to block access to websites that fall within specific categories. Websites are categorized based on their content and content filtering products, such as Websense, and enable you to block access based on category. Many organizations block access to pornography and gambling websites.

- **Customized filtering**—Filters applied to block access to specific URLs.

Foundation Summary

The "Foundation Summary" section of each chapter lists the most important facts from the chapter. Although this section does not list every fact from the chapter that will be on your SNRS exam, a well-prepared candidate should at a minimum know all the details in each "Foundation Summary" section before going to take the exam.

The Cisco IOS Firewall Advanced Security feature set enables you to consolidate the functions of several devices into a single Cisco IOS Firewall:

- Controls access into and out of the network based on predefined parameters
- Provides secure connectivity to other trusted networks via the public infrastructure (Internet)
- Provides a method for enforcing security policy

You can use the Cisco IOS Firewall feature set to configure your Cisco IOS router as a

- Network perimeter firewall or part of an perimeter firewall stack
- Firewall between groups in your internal network
- VPN firewall providing secure connections to or from branch offices
- Firewall between your company's network and your company's partners' networks

Some of the Cisco IOS Firewall features in Cisco IOS Software include the following:

- **Dynamic port mapping**—Allows CBAC-supported applications to run on nonstandard ports.
- **Intrusion protection**—Intrusion protection capability in the critical packet path provides dynamic monitoring, interception, and reporting of network attacks and misuse.
- **Authentication proxy**—LAN-based, dynamic, per-user authentication and authorization via TACACS+ and RADIUS authentication servers enables setting individual security policies.
- **Logging and audit trail**—Configurable audit trail and alerts; Cisco IOS Firewall alerts and audit trails are now configurable on a per-application basis. Java blocking is also configurable on a modular basis.
- **URL filtering**—URL filtering enables you to block browser access to websites that violate an organization's acceptable use policy. URL filtering can be applied based on several filtering methods.

Q&A

As mentioned in the section "How to Use This Book" in the Introduction, you have two choices for review questions: the Q&A questions here or the exam simulation questions on the CD-ROM. The questions that follow present a bigger challenge than the exam itself because they use an open-ended question format. By using this more difficult format, you can exercise your memory better and prove your conceptual and factual knowledge of this chapter. You can find the answers to these questions in the appendix.

1. The LLQ is used to provide QoS for what type of encrypted traffic type?
2. The Cisco IOS Firewall provides URL filtering support, Context-Based Access Control, and complete vulnerability detection and prevention. True or false?
3. The authentication proxy dynamically opens connections on the inbound ACL of the input interface where the proxy is enabled, as well as on the outbound ACL of the output interface where the packet exits. True or false?
4. What command must you use to enable the audit trail of firewall messages?
5. Can system-defined PAM entries be modified to allow Telnet through port 80?
6. Mapping port 9000 to run Telnet services via the PAM table can be done using which kind of mapping?
7. Which command must you use to configure PAM?
8. What does the *list* option in the **port mapping** command give?
9. The Cisco IOS Firewall feature set provides DoS protection. True or False?
10. The Cisco IOS Firewall was designed for which type of network?

This chapter covers the following subjects:

- Cisco IOS IPS

- Cisco IOS IPS Configuration Tasks

- Initializing the Cisco IOS IPS

- Working with Cisco IOS IPS Signatures and Rules

- Verifying the Cisco IOS IPS Configuration

- Cisco IOS IPS Deployment Strategies

CHAPTER 13

Cisco IOS Intrusion Prevention System

Intrusion detection and prevention is a key component of the Cisco Self-Defending Network solution. This technology combined with the firewall and NetFlow services provides threat defense and provides prevention and response to malicious network attacks and threats.

Cisco IOS Intrusion Prevention System (IPS) is the evolution of the Cisco IOS Intrusion Detection System (IDS) solution. Cisco IPS products go beyond IDS signature matching by incorporating features such as stateful pattern recognition, protocol analysis, traffic anomaly detection, and protocol anomaly detection. These features provide the level of detail required to accurately identify the widest range of relevant attacks.

Similar to Cisco IDS, Cisco IPS is composed of hardware- and software-deployment solutions. Cisco IPS 4200 series sensor appliances are dedicated and purpose-built devices capable of protecting multiple network segments. Integrated hardware solutions are available, too, using the Cisco Catalyst 6500 IDS modules and the network module for the Cisco access routers. Cisco IOS IPS provides a subset of IPS capabilities via Cisco IOS Software on the router.

Cisco IOS IPS enhances the features of Cisco IOS IDS from a passive device that monitors traffic, to an inline reactive and prevention device. The capability of Cisco IOS IPS to drop traffic or reset connections is the primary difference between the two solutions.

"Do I Know This Already?" Quiz

The purpose of the "Do I Know This Already?" quiz is to help you decide whether you really need to read the entire chapter. If you already intend to read the entire chapter, you do not necessarily need to answer these questions now.

The 12-question quiz, derived from the major sections in the "Foundation Topics" portion of the chapter, helps you determine how to spend your limited study time.

Table 13-1 outlines the major topics discussed in this chapter and the "Do I Know This Already?" quiz questions that correspond to those topics.

Table 13-1 *"Do I Know This Already?" Foundation Topics Section-to-Question Mapping*

Foundation Topics Section	Questions Covered in This Section
Cisco IOS IPS	1–4
Cisco IOS IPS Configuration Tasks	6
Initializing the Cisco IOS IPS	5, 7
Working with Cisco IOS IPS Signatures and Rules	8–10
Verifying the Cisco IOS IPS Configuration	11
Cisco IOS IPS Deployment Strategies	12

> **CAUTION** The goal of self-assessment is to gauge your mastery of the topics in this chapter. If you do not know the answer to a question or are only partially sure of the answer, you should mark this question wrong for purposes of the self-assessment. Giving yourself credit for an answer you correctly guess skews your self-assessment results and might provide you with a false sense of security.

1. Which types of ACLs are not supported by Cisco IOS IPS?

    a. Named

    b. Extended

    c. Standard

    d. Shorthand

2. What is an exploit attack?

    a. An activity violating corporate policy

    b. An activity to send large numbers of requests to a system or network resource with the intention to disrupt normal operations

    c. An activity to gain access to a compromised system or network resource

    d. An activity to send large numbers of requests to a system or network resource with the intention to disrupt normal operations

3. What is the SDF file format?

    a. HTTP

    b. CISL

    c. XML

    d. PEARL

4. How many signatures can a platform with 128 MB of memory support?

    a. 460

    b. 256

    c. 128

    d. 563

5. What is a standards-based format used by Cisco and other IPS vendors to send IDS notifications?

    a. SNMP

    b. SDEE

    c. TFTP

    d. PPP

6. What has to be enabled in order to use SDEE for notification?

    a. FTP

    b. TFTP

    c. HTTP

    d. TACACS

7. What command enables you to configure the maximum number of events stored in the event buffer?

    a. Router1(config)#**ip events sdee events** *events*

    b. Router1(config)#**ip sdee events** *events*

    c. Router1#**ip events sdee** *events*

    d. Router1#**ip sdee events** *events*

8. What are the two main types of signature definition files in the Cisco IOS IPS?

    a. IPS built-in signatures and Attack-drop.sdf

    b. Blocking.sdf and IPS SDF

    c. Signature micro-engines SDF and Attack-drop.sdf

    d. Atomic.sdf and Atomic.IP.sdf

9. What command enables you to load a desired SDF file from a specific location?

   a. **sdf ips location** *url*

   b. **ip sdf ips location** *url*

   c. **ip location url ips** *sdf*

   d. **ip ips sdf location** *url*

10. For what is the **ip ips signature** command used?

    a. Install new signatures

    b. Attach a policy to a signature

    c. Create an IPS rule

    d. Set IPS to forward alerts

11. What command enables you to display the number of packets audited and the number of alarms sent?

    a. **show ip ips audit**

    b. **show ip ips statistics**

    c. **show ip ips session**

    d. **show ip ips alarms**

12. Cisco IOS IPS is best suited for which types of networks? (Select two.)

    a. Enterprise

    b. Branch office

    c. Hub and spoke

    d. Telecommuter

The answers to the "Do I Know This Already?" quiz are found in the appendix. The suggested choices for your next step are as follows:

- **10 or less overall score**—Read the entire chapter. This includes the "Foundation Topics" and "Foundation Summary" sections and the "Q&A" section.

- **11 or 12 overall score**—If you want more review on these topics, skip to the "Foundation Summary" section and then go to the "Q&A" section. Otherwise, move on to Chapter 14, "Mitigating Layer 2 Attacks."

Foundation Topics

This chapter discusses the Cisco IOS IPS. Unlike the Cisco IOS IDS, which was tightly integrated with the Cisco IOS Firewall feature set, IPS is an independent component of Cisco IOS Software Release 12.3T and later.

This chapter covers six main topics that pertain to the Cisco IOS IPS, as follows:

- Cisco IOS IPS
- Cisco IOS IPS configuration
- Initializing the Cisco IOS IPS
- Working with Cisco IOS IPS signatures and rules
- Verifying the Cisco IOS IPS configuration
- Cisco IOS IPS deployment strategies

Cisco IOS IPS

This section describes the Cisco IOS features and new enhancements.

Cisco IOS IPS Features

The Cisco IOS IPS feature set provides inline packet-inspection capability as packets flow through the router. It also looks for any traffic matching a specific signature that indicates malicious traffic. If it finds traffic that matches a signature, it can quickly react and eliminate the threat before it adversely affects the network. You can configure the Cisco IOS IPS to react by notifying a syslog or management server via an alarm, by dropping the matched packets or resetting the TCP connection, or by a combination or these actions. Note that these features have always been incorporated in Cisco IOS Software via the Cisco IOS Firewall feature set. However, the new IPS-based features include several enhancements:

- **Expanded signature capability**—Cisco IOS IPS provides more than 700 signatures supported in the hardware platforms, and enables you to modify an existing signature or create new custom signatures. New signatures can be loaded without having to upgrade the Cisco IOS Software image.

- **Parallel signature scanning**—All the signatures in a single micro-engine are scanned in parallel fashion rather than serially.

- **Extended ACL support**—Support for both named and numbered extended ACLs as opposed to standard numbered ACLs.

Attacks detected by Cisco IOS IPS signatures are broken down into four types:

- **Exploit**—An activity to gain access to a compromised system or network resource
- **Denial-of-service (DoS)**—An activity to send large numbers of requests to a system or network resource with the intention to disrupt normal operations
- **Reconnaissance**—An information-gathering activity to collect data on system and network resources as targets that have a potential to be compromised later
- **Misuse**—An activity that violates corporate policy

The four signature types can apply to either of the following categories:

- **Atomic**—Atomic signatures trigger the IPS with simple patterns usually with a single packet to a single host. These signatures tend to be less memory intensive because the IPS is not required to gather large amounts of data.
- **Compound**—Compound signatures require the IPS to gather and compare greater and more complex amounts of data to trigger an event. It is usually an attack on multiple hosts, over extended time periods, and with multiple packets.

Cisco IOS IPS Functions

Cisco IOS IPS has two main components: signature definition files (SDF) and signature microengines (SME).

The SDF contains the signature definitions and configurable actions for each signature. This file is in Extensible Markup Language (XML) format. Cisco IOS IPS loads and compiles the SDF and populates its internal tables with the information necessary to detect each signature. The location of the SDF file is definable, and it can reside locally on the router's Flash file system or on a remote server capable of Trivial File Transfer Protocol (TFTP), File Transfer Protocol (FTP), Secure Copy Protocol (SCP), or Remote Copy Protocol (RCP). Access routers with Cisco IOS Software Release 12.3(8)T or higher contain a preconfigured SDF called the Attack-drop.sdf in Flash, which provides the up-to-date, highly severe worm and attack signatures. For a list of signatures provided in Attack-drop.sdf, refer to the Cisco IOS documentation at Cisco.com.

An SME loads the SDF and scans signatures for various conditions that match a defined pattern. The SME signature engine parses values from the packet and passes them to the regular-expression engine for inspection. The regular-expression engine searches for multiple patterns simultaneously. The current version of Cisco IOS IPS 12.3(11)T contains Atomic.IP, Atomic.ICMP, Atomic.IPOPTIONS, Atomic.UDP, Atomic.TCP, Service.DNS, Service.RPC, Service.SMTP, Service.HTTP, and Service.FTP SMEs. The SDF normally contains signature definitions from several of these engines.

> **NOTE** Cisco IOS Software Release 12.3(14)T will also deliver new application inspection and control capabilities, such as enhanced application security for port 80 misuse and VoIP environments.

Cisco IOS IPS Restrictions

Prior to installing and configuring Cisco IOS IPS, you need to consider a few factors, as discussed in the following sections.

Memory Considerations

One of the main concerns about the number of signatures loaded on a Cisco router is the amount of memory consumed by each signature. Based on tests conducted on Cisco access routers, a platform with 128 MB of memory can support up to 563 signatures. If the platform has 256 MB of memory, it can support 737 signatures.

Table 13-2 estimates the memory required per signature. According to these estimates, Simple Mail Transfer Protocol (SMTP) signatures require the most memory.

Table 13-2 *Estimated Memory Needed per Signature*

Signatures	Memory in MB
Atomic.IP	0.004
Atomic.IPOPTIONS	0.004
Atomic.ICMP	0.005
Atomic.UDP	0.006
Atomic.TCP	0.05
Service.HTTP	0.08
Service.DNS	0.006
Service.RPC	0.006

Unsupported Signatures

Cisco IOS IPS (effective Cisco IOS Software Release 12.3(8)T) will no longer support the following signatures:

- 1100 IP fragment attack (attack, atomic)
- 1105 broadcast source address (compound/attack)
- 1106 multicast IP source address (compound/attack)
- 8000 FTP retrieve password file (attack, atomic) SubSig ID: 2101

Unsupported CLI Features

Cisco IOS IPS actions (such as resetting the TCP connection) are no longer configurable via the command-line interface (CLI). The signatures are now preset with actions to mitigate the attack.

Cisco IOS IPS Application

Figure 13-1 shows how the Cisco IOS IPS functions when it discovers traffic that matches an attack signature considered malicious.

Figure 13-1 *Cisco IOS IPS Attack Response*

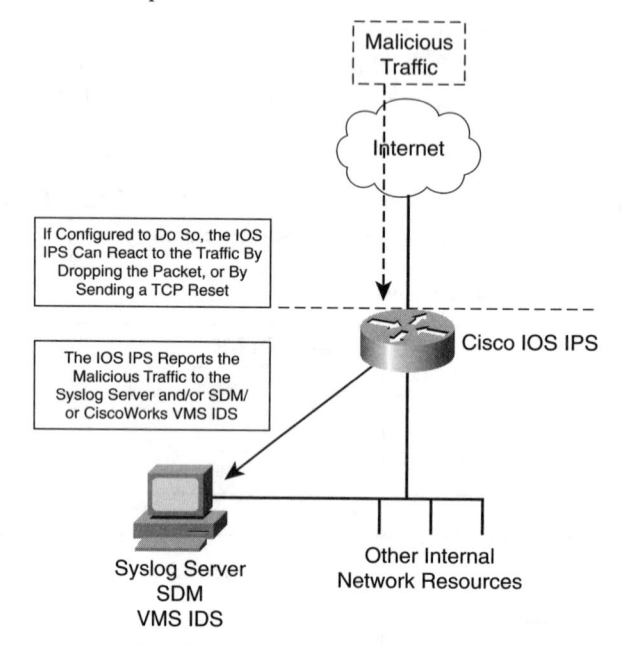

Figure 13-1 illustrates how the Cisco IOS IPS may respond to an attack. When malicious traffic is detected, the Cisco IOS IPS sends an alarm to the syslog or management server and drops the traffic or resets the session.

Cisco IOS IPS Configuration Tasks

You can configure Cisco IOS IPS in three ways:

- **Cisco router CLI**—A basic set of Cisco IOS IPS commands is available in the CLI. This set includes enabling and disabling signatures, applying rules to an interface (inbound/outbound), applying access lists to traffic, setting notification, selecting protected networks, loading an SDF file, and naming configuration policies.

- **Cisco Router and Security Device Manager (SDM)**—Cisco SDM supports Cisco IPS configuration provided in the CLI. It also has configuration options for Cisco IOS IPS actions, such as resetting the TCP connection and cloning/creating signatures, which cannot be configured via the CLI.

- **CiscoWorks VPN/Security Management Solution IDS Management Center (VMS IDS)**— VMS IDS version 2.3 will support Cisco IOS IPS. VMS IDS manages Cisco IOS IPS in the same way it manages Cisco IDS 4200 series sensors.

This book provides guidelines on configuring Cisco IOS IPS via the CLI only. For detailed information on configuration via SDM or VMS IDS, refer to Cisco.com.

To implement the Cisco IOS IPS, complete these steps:

Step 1   Initialize the Cisco IOS IPS on the router.

Step 2   Configure signatures.

Step 3   Create and apply IPS rules.

Step 4   Verify configuration.

Initializing the Cisco IOS IPS

Initializing the Cisco IOS IPS includes three subtasks. The first two tasks are configuring the notification type and configuring the router maximum queue for alarms. The last task provides configuration information for defining a protected network.

Configuring the Notification Type

The Cisco IOS IPS provides several notification methods:

- **Cisco IOS logging**—Uses a syslog format to send messages to a syslog server.

- **PostOffice Protocol (POP)**—This is a proprietary protocol used to provide Cisco IOS Firewall IDS routers and IDS sensors a mechanism to communicate with each other and the management system. POP operates by pushing alarms and queuing to the management server. Cisco IOS IPS will only support POP-related functionality up to Cisco IOS Software Release 12.3(14)T.

- **Security Device Event Exchange (SDEE)**—This is an ICSA standardized IPS communications protocol and message format. It uses pull technology initiated by the network management application to the IPS device. SDEE, also known as Remote Data Exchange Protocol (RDEP) v2, uses XML, HTTP, and Secure Sockets Layer (SSL) or Transport Layer Security (TLS) to provide a standard interface. Cisco IOS IPS uses this protocol to send notifications to its management server. The primary benefit of SDEE is that it is an open standard format allowing coexistence of multiple IPS vendors in a network environment.

Chapter 13: Cisco IOS Intrusion Prevention System

Use the **ip ips notify log | sdee** command (in global configuration mode) to configure the Cisco IOS IPS to forward alerts to the syslog server, SDM, or VMS IDS.

Table 13-3 lists the notification commands and associated options. Example 13-1 depicts the command format.

Table 13-3 *Configuring the Notification Type*

ip ips notify log	Sends messages in syslog format
ip ips notify sdee	Sends messages in SDEE

Example 13-1 *Configuring Notification Type*

```
Router1#configure terminal
Router1(config)#ip ips notify log
<<<or>>>>
Router1(config)#ip ips notify sdee
```

NOTE To use SDEE, the HTTP server must be enabled. Use the **ip http server** command to enable HTTPS on the router.

To configure the Cisco IOS IPS to forward alerts to the Cisco Unix Director, the Cisco Secure Policy Manager (CSPM), and IDS Management Center (MC), use the **ip ips notify nr-director** command (in global configuration mode). Table 13-4 lists the notification commands.

Table 13-4 *POP Setting Notification Commands*

Command	Description
ip ips notify nr-director	Sends the alarms in POP format to the central manager (Director, router MC, Event Viewer, or other IDS 3.x sensors)

Example 13-2 shows the commands used to set the notification type.

Example 13-2 *Configuring the Notification Type*

```
Router1#configure terminal
Router1(config)#ip ips notify nr-director
```

NOTE The **ip ips notify nr-director** command will be phased out in Cisco IOS Software Release 12.3(14)T. This phaseout is attributable to the replacement of the POP with the standard RDEP.

Configuring the Router Maximum Queue for Alarms

The size of an events buffer determines the amount of memory used on the router. It is important to limit the amount of memory reserved for alerts and alarms. If event notification is enabled, a maximum of 200 events can be stored in the buffer. The following command sets the SDEE notification queue size:

   `ip sdee events` *events*

The *events* keyword is an integer between 1 and 1000 that designates the maximum number of events that can remain in the queue. This command must be used in global configuration mode.

Example 13-3 shows the command used to set the maximum queue size to allow 300 events.

Example 13-3 *Configuring the Maximum Queue Size*

```
Router1#configure terminal
Router1(config)#ip sdee events 300
```

The following command sets the notification queue size for the POP:

   `ip ips po max-events` *num-of-events*

The *num-of-events* keyword is an integer between 1 and 65535 that designates the maximum number of events that can remain in the queue. The default value for **max-events** is 100.

Example 13-4 shows the command used to set the maximum queue size.

Example 13-4 *Configuring the Maximum Events*

```
Router1#configure terminal
Router1(config)#ip ips po max-events 200
```

> **NOTE** The **ip ips po max-events** command will be deprecated in Cisco IOS Software Release 12.3(14)T. This deprecation is attributable to the replacement of the POP with the standard RDEP.

Defining the Protected Network

To effectively identify an attack, the Cisco IOS IPS must be able to determine the IP address range of the protected network. This function does not affect the operation of the IDS but is used to log traffic as inbound or outbound. The following command specifies the beginning and end of the address range for the protected network:

   `ip ips protected` *ip-addr* `to` *ip-addr*

Example 13-5 shows the command used to configure the protected network.

Example 13-5 *Configuring a Protected Network*

```
Router1#configure terminal
Router1(config)#ip ips protected 10.10.10.1 to 10.10.10.254
```

> **NOTE** The **ip ips protected** command will be phased out in Cisco IOS Software Release 12.3(14)T.

Working with Cisco IOS IPS Signatures and Rules

This section describes the steps required to load IPS signatures and create and apply IPS rules.

Loading IPS-Based Signatures

The Cisco IOS IPS has two main types of signature definition files:

- **IPS SDF (default)**—Built-in 132 signatures inherited from Cisco IOS Software IDS technology to enable backward-compatibility. These signatures are hard-coded into Cisco IOS Software Release 12.3(8)T or later and set to alarm only for the default action.

- **Attack-drop.sdf**—At the time of this writing, the Attack-drop.sdf file contains 118 high-fidelity IPS signatures. This file is available in Flash on all Cisco IOS IPS-enabled routers shipped with Cisco IOS Software 12.3(8)T or later and can be further modified by adding or deleting signatures based on the network requirements. For a complete number and list of supported signatures in this file, refer to the online documentation for the specific Cisco IOS version.

The command required to load the desired SDF file from a specific location is as follows:

```
ip ips sdf location url
```

This command must be used in the global configuration mode. The available *url* options are local Flash, FTP server, RCP, and TFTP server.

If the router cannot find the file specified in the *url*, it attempts to load the built-in signature file. To prevent the router from loading the built-in signature file, execute the following command in global configuration mode:

```
no ip ips sdf built-in
```

Cisco recommends using the signatures provided in Flash for attack mitigation. If the built-in signatures do not provide adequate protection for a network, you can merge these signatures with Attack-drop.sdf. The **copy [/erase]** *url* **ips-sdf** command enables you to merge the two files. Table 13-5 briefly describes the command and its options.

Table 13-5 *Merging Signature Files*

Command	Description
copy [/erase] *url* **ips-sdf**	Specifies an IPS rule. (You must be in global configuration mode.)
copy *url* **ips-sdf**	Merges the built-in signatures with the SDF file specified with the URL.
	Available *url* locations are local Flash, FTP server, RCP, and TFTP server.

Working with Cisco IOS IPS Signatures and Rules

By default, when IOS is compiling new signatures for a particular engine, the IOS passes the packets without scanning for that particular engine. To instruct the router to drop all packets until the signature engine is built and ready to scan traffic, use the following command:

`ip ips fail closed`

Example 13-6 depicts the loading of the Attack-drop.sdf from the router Flash card using the commands described previously.

Example 13-6 *Configuring the SDF Location*

```
Router#configure terminal
Router(config)#ip ips sdf location disk1:attack-drop.sdf
Router(config)#ip ips name HQIPS
Router(config)# no ip ips sdf built-in
Router(config)#ip ips fail closed
```

NOTE If Flash is erased, the Attack-drop.sdf file might also be erased. Therefore, if you are copying a Cisco IOS image to Flash and are prompted to erase the contents of Flash before copying the new image, you risk erasing the Attack-drop.sdf file. If this occurs, the router will refer to the built-in signatures within the Cisco IOS image.

It is important to define *normal traffic* on the network and configure the signatures accordingly to ensure that the IDS reacts to traffic that is truly malicious. If the IPS reacts to normal traffic, the alert is referred to as *false positive*. A *false negative* occurs when the IPS incorrectly interprets malicious traffic as normal for the network. To further reduce the number of false positive alerts, it is important to correctly configure the signature thresholds and disable or exclude specific signatures. Disabling a signature turns the signature off completely. When a signature is excluded, it designates specific hosts on the networks that are not inspected for a signature. A signature number identifies all IPS signatures. You can find the signature numbers and explanations at Cisco.com.

Use the **ip ips signature** command in global configuration mode to attach a policy to a signature. The command syntax is as follows:

`ip ips signature signature-id[:sub-signature-id] {delete | disable | list acl-list}`

Table 13-6 describes this command syntax.

Use the **no** form of this command to reenable the signature. If the policy attached an access list to the signature, use the **no** form of this command to remove the access list:

`no ip ips signature signature-id[:sub-signature-id]`

Table 13-6 ips signature *Configuration Options*

signature-id [:subsignature-id]	Signature within the SDF that is not reported, if detected. If a subsignature is not specified, the default is 0. For example, if signature 1104 is specified without a subsignature, the router will interpret the signature as 1104:0.
delete	Deletes a specified signature.
disable	Disables a specified signature. Instructs the router to scan for a given signature but not take any action if the signature is detected.
list acl-list	(Optional) A named, standard, or extended access control list (ACL) to filter the traffic that will be scanned. If the packet is permitted by the ACL, the signature will be scanned and reported. If the packet is denied by the ACL, the signature is deemed disabled.

Example 13-7 shows an example of using the **ip ips signature** command to disable and delete signatures.

Example 13-7 *Disabling and Deleting Signatures*

```
Router1#configure terminal
Router1(config)#ip ips signature 5184 disable
Router1(config)#ip ips signature 9202 delete
```

It is also possible to exclude traffic from being scanned for the specified signature by using an access list. The access list should be created prior to applying it to the signature. Example 13-8 shows the access list and how it is applied to exclude host 10.200.53.53.

Example 13-8 *Excluding Signatures*

```
Router1#configure terminal
Router1(config)#access-list 33 deny host 10.200.53.53
Router1(config)#access-list 33 permit any
Router1(config)#ip ips signature 9211 list 33
Router1(config)#ip ips signature 1006 list 33
```

Creating and Applying IPS Rules

Enabling Cisco IOS IPS will install the latest Cisco IOS IPS signatures on a router for the first time. So, this task is used for preparation of new routers.

Use the **ip ips** command to create an IPS rule and apply it to the inbound or outbound traffic on an interface. Table 13-7 lists the **ips** command with a brief description.

Working with Cisco IOS IPS Signatures and Rules

Table 13-7 *Setting IPS IOS Rules Command*

Command	Description
ip ips name	Specifies an IPS rule. (You must be in global configuration mode.)
ip ips in|out [list acl]	Applies an IPS rule to inbound or outbound traffic on an interface. (You must be in interface configuration mode.)

Example 13-9 shows the commands necessary to enable IPS on a router.

Example 13-9 *Enabling IPS on a New Router*

```
Router1#configure terminal
Router1(config)#ip ips name HQIPS
Router1(config)#interface gigabitethernet0/1
Router1(config-if)#ip ips HQIPS in
```

> **NOTE** It might take several minutes while the default signatures load. The router prompt will not blink at this time. It is recommended that message logging be enabled to monitor the status. This step uses the built-in signatures available statically in IOS. The next section describes the commands to specify the location of a different signature definition file.

To exclude specific hosts or subnets from the IPS rule, append an access list with the **ip ips in|out** command. Example 13-10 lists the syntax for creating audit rule HQIPS and excludes scanning the internal network subnet of 10.200.53.0/24 via access list 34. This access list does not deny any traffic from the internal network but filters it from the IPS rule. The line **permit any** forces all other traffic through the audit process.

Example 13-10 *Creating a Conditional IPS Rule*

```
Router1#configure terminal
Router1(config)#access-list 34 deny 10.200.53.0 0.0.0.255
Router1(config)#access-list 34 permit any
Router1(config)#ip ips name HQIPS list 34
```

> **NOTE** Cisco IOS IPS inbound rules analyze packets prior to them being processed by the ACL to ensure that potential attacks are logged even if the router has an ACL policy that would drop the packets. On outbound rules, Cisco IOS IPS processes packets after they are acted on by the input ACL of the incoming interface. This action might result in loss of IPS event alarms even though the attack does not succeed.

Verifying the Cisco IOS IPS Configuration

It is important to ensure that your system is properly configured. You can use three commands to verify the configuration of the Cisco IOS IPS:

- **show**—You enter the **show** command in the privileged EXEC mode, and it enables you to see the current Cisco IOS IPS configuration. Table 13-8 lists the **show** commands with a brief description of each.

Table 13-8 show *Commands*

Command	Description
show ip ips all	Displays all available IPS information.
show ip ips configuration	Displays additional configuration information, including default values that might not display using the **show running-config** command.
show ip ips interfaces	Displays the interface configuration.
show ip ips name *name*	Displays information only for the specified IPS rule.
show ip ips statistics [reset]	Displays information such as the number of packets audited and the number of alarms sent. The optional **reset** keyword resets sample output to reflect the latest statistics.
show ip ips sessions [details]	Displays IPS session-related information. The optional **details** keyword shows detailed session information.
show ip ips signatures [details]	Displays signature information, such as which signatures are disabled and marked for deletion. The optional **details** keyword shows detailed signature information.
show ip sdee alerts	Displays the IDS alert buffer.
show ip sdee all	Displays all information available for IDS SDEE notifications.
show ip sdee errors	Displays IDS SDEE error messages.
show ip sdee events	Displays IDS SDEE events.
show ip sdee configuration	Displays SDEE configuration parameters.
show ip sdee status	Displays the status events that are currently in the buffer.
show ip sdee subscription	Displays IDS SDEE subscription information.

Example 13-11 displays the output from the **show ip ips configuration** Command.

Example 13-11 show ip ips configuration *Command Output*

```
Router1#show ip ips configuration
Configured SDF Locations:
 flash:attack-drop.sdf
```

Example 13-11 show ip ips configuration *Command Output (Continued)*

```
Builtin signatures are enabled but not loaded
Last successful SDF load time: 00:26:54 UTC Mar 1 2005
IDS fail closed is disabled
Fastpath ips is enabled
Quick run mode is enabled
Event notification through syslog is enabled
Event notification through Net Director is disabled
Event notification through SDEE is disabled
Total Active Signatures: 82
Total Inactive Signatures: 0
Signature 5184:0 disable
PostOffice:HostID:0 OrgID:0 Msg dropped:0
             :Curr Event Buf Size:0  Configured:100
Post Office is not enabled - No connections are active
IDS Rule Configuration
 IPS name HQIPS
    acl list 34
Interface Configuration
 Interface FastEthernet0/0
Inbound IPS rule is HQIPS
   acl list 34
   Outgoing IPS rule is not set
```

Example 13-12 displays a partial output from the **show ip ips signatures** command.

Example 13-12 show ip ips signatures *Command Partial Output*

```
Router1#show ip ips signatures
Signatures were last loaded from flash:attack-drop.sdf
SDF release version attack-drop.sdf v2
*=Marked for Deletion   Action=(A)larm,(D)rop,(R)eset   Trait=AlarmTraits
MH=MinHits              AI=AlarmInterval                CT=ChokeThreshold
TI=ThrottleInterval     AT=AlarmThrottle                FA=FlipAddr
WF=WantFrag             Ver=Signature Version


Signature Micro-Engine: SERVICE.SMTP (1 sigs)
  SigID:SubID On Action  Sev Trait   MH    AI    CT    TI AT FA WF Ver
  ----------- -- ------  --- -----   ----- ----- ----- -- -- -- -- ---
    3129:0       Y    ADR    MED   0     0     0     0     15 FA    N    S59
 . . . .
 . . . .
 . . . .
Signature Micro-Engine: SERVICE.RPC (29 sigs)
  SigID:SubID On Action  Sev Trait   MH    AI    CT    TI AT FA WF Ver
  ----------- -- ------  --- -----   ----- ----- ----- -- -- -- -- ---
    6100:0       Y    AD     HIGH  0     0     0     100   30 FA    N    1.0
```

continues

Example 13-12 show ip ips signatures *Command Partial Output (Continued)*

```
  6100:1        Y    ADR   HIGH    0      0      0     100    30 FA  N      1.0
  6101:0        Y    AD    HIGH    0      0      0     100    30 FA  N      1.0
  6195:0        Y    AD    HIGH    0      0      0     100    30 FA  N      2.2
Signature Micro-Engine: ATOMIC.L3.IP (4 sigs)
  SigID:SubID On Action  Sev Trait    MH     AI     CT    TI AT FA WF Ver
  ----------- -- ------  --- -----   -----  -----  -----  -- -- -- -- ---
  1102:0        Y    AD    HIGH    0      0      0     100    30 FA  N      2.1
  1104:0        Y    AD    HIGH    0      0      0     100    30 FA  N      2.2
  1108:0        Y    AD    HIGH    0      0      0     100    30 GS  N      S27
  2154:0        Y    AD    HIGH    0      0      0     100    30 FA  N  Y   1.0
Total Active Signatures: 82
Total Inactive Signatures: 0
```

Example 13-13 displays the output from the **show ip sdee status** command.

Example 13-13 show ip sdee status *Command Output*

```
Router1#show ip sdee stat
Event storage: 100 events maximum, using 65600 bytes of memory
                             SDEE Status Messages
        Time                     Message                      Description
    1: 000 00:49:31 UTC Mar 1 2005   ENGINE_READY: SERVICE.HTTP - 1236 ms - packets for this
engine will be scanned
    2: 000 00:49:31 UTC Mar 1 2005   ENGINE_BUILDING: ATOMIC.TCP - 6 signatures - 9 of 13
engines
    3: 000 00:49:31 UTC Mar 1 2005   ENGINE_READY: ATOMIC.TCP - 28 ms - packets for this
engine will be scanned
    4: 000 00:49:31 UTC Mar 1 2005   ENGINE_BUILDING: ATOMIC.UDP - 0 signatures - 10 of 13
engines
    5: 000 00:49:31 UTC Mar 1 2005   ENGINE_BUILD_SKIPPED: ATOMIC.UDP - there are no new
signature definitions for this engine
    6: 000 00:49:31 UTC Mar 1 2005   ENGINE_BUILDING: ATOMIC.ICMP - 0 signatures - 11 of 13
engines
    7: 000 00:49:31 UTC Mar 1 2005   ENGINE_BUILD_SKIPPED: ATOMIC.ICMP - there are no new
signature definitions for this engine
    8: 000 00:49:31 UTC Mar 1 2005   ENGINE_BUILDING: ATOMIC.IPOPTIONS - 1 signatures - 12
of 13 engines
    9: 000 00:49:31 UTC Mar 1 2005   ENGINE_READY: ATOMIC.IPOPTIONS - 4 ms - packets for this
engine will be scanned
   10: 000 00:49:31 UTC Mar 1 2005   ENGINE_BUILDING: ATOMIC.L3.IP - 4 signatures - 13 of 13
engines
   11: 000 00:49:31 UTC Mar 1 2005   ENGINE_READY: ATOMIC.L3.IP - 8 ms - packets for this
engine will be scanned
   12: 000 00:49:30 UTC Mar 1 2005   ENGINE_READY: SERVICE.RPC - 204 ms - packets for this
engine will be scanned
   13: 000 00:49:30 UTC Mar 1 2005   ENGINE_BUILDING: SERVICE.DNS - 18 signatures - 7 of 13
engines
   14: 000 00:49:30 UTC Mar 1 2005   ENGINE_READY: SERVICE.DNS - 64 ms - packets for this
engine will be scanned
   15: 000 00:49:30 UTC Mar 1 2005   ENGINE_BUILDING: SERVICE.HTTP - 23 signatures - 8 of 13
engines
```

- **clear**—The **clear** command enables you to clear statistics and configurations. Table 13-9 lists the available **clear** commands.

Table 13-9 **clear** *Commands*

Command	Description
clear ip ips configuration	Disables the Cisco IOS Firewall IPS, removes all intrusion detection configuration entries, and releases dynamic resources. You must use this command in EXEC mode.
clear ip ips statistics	Resets all IPS statistics on packets analyzed and alarms sent. You must use this command in EXEC mode.
clear ip sdee {events | subscriptions}	Clears SDEE events and subscription. Because subscriptions are properly closed by the Cisco IOS IPS client, typically this command is used only to help with error recovery.

- **debug**—The output from **debug** privileged EXEC command displays diagnostic information concerning router events and general status. Table 13-10 lists the available **debug** commands.

Table 13-10 **debug ip ips** *Command Parameters*

Command	Description
debug ip ips	Enables debug messages for Cisco IOS IPS
debug ip ips engine	(Optional) Displays debug messages only for a specific signature engine
debug ip ips detailed	(Optional) Displays detailed debug messages for the specified signature engine or for all IPS actions
debug ip sdee	Enables debug messages for SDEE notification events
debug ip sdee alerts	Displays new alerts reported to SDEE from the IPS
debug ip sdee detail	Displays detailed SDEE messages
debug ip sdee messages	Displays error and status messages reported to SDEE from the IPS
debug ip sdee requests	Displays SDEE client requests
debug ip sdee subscriptions	Displays SDEE client subscription requests

Cisco IOS IPS Deployment Strategies

As discussed earlier in this chapter, the primary advantage with the Cisco IOS IPS is that it is an integrated solution that leverages existing Cisco router infrastructure to mitigate both internal and external attacks on the network with its inline capabilities.

The Cisco IOS IPS features complement Cisco IOS Firewall and virtual private network (VPN) solutions for threat protection at all entry points into the network. The Cisco IOS IPS is the perfect solution for network segments that do not require, or might not support, the use of appliance-based IPS solutions.

Although Cisco IOS IPS is ideal for any size network, several strategies are best supported by the deployment of the Cisco IOS IPS:

- **Branch offices**—Small to medium-size branch and home offices that act as an extension of an enterprise could benefit from an integrated, cost-effective router with integrated security.

- **Telecommuter sites**—Corporate employees who connect to their company networks via VPNs using an Internet service provider (ISP). Typically, there is little to no physical security at the remote location. An integrated VPN, IPS, and firewall security solution could provide a safe and flexible solution.

Figure 13-2 depicts a typical configuration of network connectivity between a headquarters, branch offices, and telecommuters.

Figure 13-2 *Typical Network Connectivity*

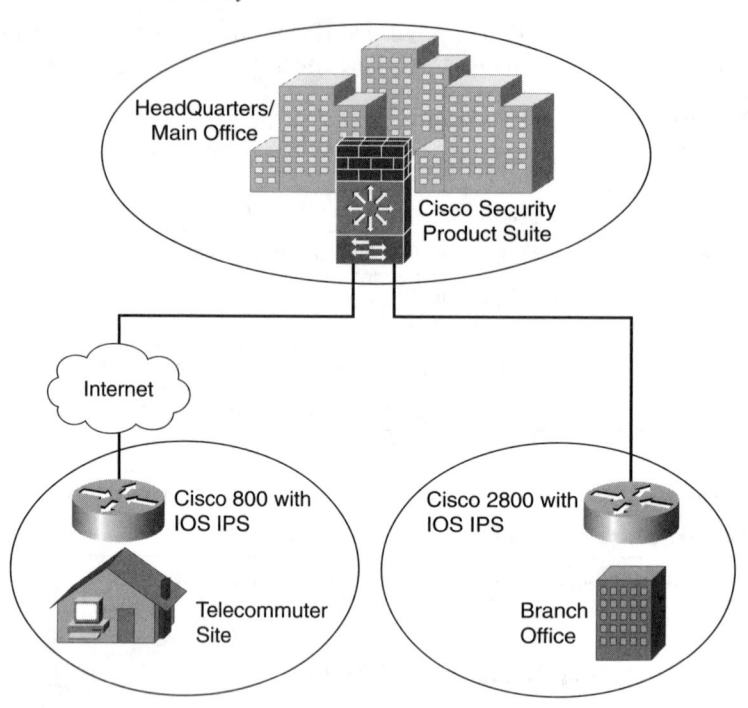

Foundation Summary

The "Foundation Summary" section of each chapter lists the most important facts from the chapter. Although this section does not list every fact from the chapter that will be on your SNRS exam, a well-prepared candidate should at a minimum know all the details in each "Foundation Summary" before going to take the exam.

The Cisco IOS IPS is the next generation of intrusion detection technology designed to complement the security infrastructure by integrating into Cisco IOS Software. This integration of technologies makes the Cisco IOS router a cost-effective and functional tool. A review of several Cisco IOS IPS concepts follows:

- The Cisco IOS IPS can communicate with Cisco SDM, CiscoWorks VMS IDS, and syslog server.

- The devices communicate using SDEE. You configure the communications by enabling HTTP services on the router and the SDEE notification parameters.

- The four categories of signatures are exploit, DoS, reconnaissance, and misuse.

- The two signature types are atomic and compound.

- Each IPS signature takes up a portion of router memory. The number of signatures supported on platforms with 128 MB of memory is 563. If the platform has 256 MB of memory, it can support 737 signatures.

- Cisco IOS IPS has two main components. SDF contains signature information. The SME parses the SDF and inspects the signature for a pattern match.

- Four actions are required to configure the IPS:

    1. Initialize the Cisco IOS IPS on the router.
        — Configure the notification type.
        — Configure the maximum queue for alarms.
    2. Configure signatures.
        — Load IPS signatures via SDF files.
        — Delete and exclude signatures to comply with "normal" network traffic.

3. Create and apply IPS rules.

— Create and enable Cisco IPS rules on an interface.

4. Verify configuration.

— Use the **show**, **clear**, and **debug** commands to verify successful configuration.

Q&A

As mentioned in the section "How to Use This Book" in the Introduction, you have two choices for review questions: the Q&A questions here or the exam simulation questions on the CD-ROM. The questions that follow present a bigger challenge than the exam itself because they use an open-ended question format. By using this more difficult format, you can exercise your memory better and prove your conceptual and factual knowledge of this chapter. You can find the answers to these questions in the appendix.

1. What three actions are performed by the Cisco IOS IPS when malicious traffic is discovered?
2. Why would you want to disable a signature?
3. What is SDEE?
4. Which type of signatures create greater load on the router performance?
5. How do you exclude a signature?
6. What is contained in SDF?
7. What notification types can you configure up to Cisco IOS Software Release 12.3(14)T?
8. What is the benefit of the new Cisco IOS IPS?
9. What default action does Cisco IOS take on incoming packets while compiling new signatures for a particular engine?
10. What tools can you use to configure the Cisco IOS IPS?

This chapter covers the following subjects:

- Layer 2 Attacks
- Mitigation of Layer 2 Attacks

CHAPTER 14

Mitigating Layer 2 Attacks

Unlike hubs, switches cannot regulate the flow of data between their ports by creating almost "instant" networks that contain only the two end devices communicating with each other. Data frames are sent by end systems, and their source and destination addresses are not changed throughout the switched domain. Switches maintain content-addressable memory (CAM) lookup tables to track the source addresses located on the switch ports. These lookup tables are populated by an address-learning process on the switch. If the destination address of a frame is not known or if the frame received by the switch is destined for a broadcast address, the switch forwards the frame out all ports. With their ability to isolate traffic and create the "instant" networks, switches can be used to divide a physical network into multiple logical or VLANs through the use of Layer 2 traffic segmentation.

VLANs enable network administrators to divide their physical networks into a set of smaller logical networks. Like their physical counterparts, each VLAN consists of a single broadcast domain isolated from other VLANs and work by tagging packets with an identification header and then restricting the ports that the tagged packets can be received on to those that are part of the VLAN. The two most prevalent VLAN tagging techniques are the IEEE 802.1q tag and the Cisco Inter-Switch Link (ISL) tag.

This chapter discusses Layer 2 attacks, mitigations, best practices, and functionality.

"Do I Know This Already?" Quiz

The purpose of the "Do I Know This Already?" quiz is to help you decide whether you really need to read the entire chapter. If you already intend to read the entire chapter, you do not necessarily need to answer these questions now.

The 11-question quiz, derived from the major sections in "Foundation Topics" section of the chapter, helps you determine how to spend your limited study time.

Table 14-1 outlines the major topics discussed in this chapter and the "Do I Know This Already?" quiz questions that correspond to those topics.

Table 14-1 *"Do I Know This Already?" Foundation Topics Section-to-Question Mapping*

Foundation Topics Section	Questions Covered in This Section
Types of Attacks	1–10
Factors Affecting Layer 2 Mitigation Techniques	11

> **CAUTION** The goal of self-assessment is to gauge your mastery of the topics in this chapter. If you do not know the answer to a question or are only partially sure of the answer, you should mark this question wrong for purposes of the self-assessment. Giving yourself credit for an answer you correctly guess skews your self-assessment results and might provide you with a false sense of security.

1. What is the default inactivity expire time period on a Cisco Catalyst switch CAM table?

    a. 1 minute

    b. 5 minutes

    c. 10 minutes

    d. 50 minutes

2. What are three methods of implementing port security?

    a. Active secure MAC addresses, fixed secure MAC address, and closed secure MAC address

    b. Static secure MAC address, dynamic secure MAC addresses, and sticky secure MAC addresses

    c. Default secure MAC address, evasive secure MAC address, and evading secure MAC address

    d. Stinky secure MAC addresses, dynamite secure MAC, and clammy secure MAC addresses

3. Which command enables port security on an interface?

    a. **switchport mode port-security**

    b. **switchport mode interface-security**

    c. **switchport interface-security**

    d. **switchport port-security**

4. What is the default action mode for security violations?

    a. Protect

    b. Restrict

    c. Shutdown

5. The DTP state on a trunk port may be set to what?

    a. Auto, on, off, undesirable, or non-negotiate

    b. Auto, on, off, desirable, or non-negotiate

    c. Auto, on, off, desirable, or negotiate

    d. Auto, on, off, undesirable, or negotiate

6. What are the two different types of VLAN hopping attacks?

    a. Switch spoofing and double tagging

    b. Switch goofing and double teaming

    c. Switch impersonation and double grouping

    d. Switch imitation and double alliance

7. Which features of Cisco IOS Software enable you to mitigate STP manipulation? (Select two.)

    a. **spanning-tree portfast bpduguard**

    b. **spanning-tree guard rootguard**

    c. **set spantree global-default loopguard enable**

    d. **set udld enable**

8. What are the three types of private VLAN ports?

    a. Neighborhood, remote, and loose

    b. Community, isolated, and promiscuous

    c. Communal, remote, and licentious

    d. Area, secluded, and wanton

9. What common tool is used to launch MAC overflow attacks?

    a. Dsniff

    b. Macof

    c. Arpspoof

    d. Tablstuf

10. Protect mode security is recommended for trunk ports.

    a. True

    b. False

11. What are the three factors when designing a Layer 2 protected network?

    a. Number of users groups

    b. Number of DNS zones

    c. Number of switches

    d. Number of buildings

    e. Number of security zones

The answers to the "Do I Know This Already?" quiz are found in the appendix. The suggested choices for your next step are as follows:

- **9 or less overall score**—Read the entire chapter. This includes the "Foundation Topics" and "Foundation Summary" sections and the "Q&A" section.

- **10 or 11 overall score**—If you want more review on these topics, skip to the "Foundation Summary" section and then go to the "Q&A" section. Otherwise, move on to Chapter 15, "Context-Based Access Control."

Foundation Topics

The following sections provide an overview of the most common Layer 2 attacks and suggested mitigations. There is also a brief description of factors that you should consider when designing Layer 2 protected networks.

Types of Attacks

For years, the focus on security has been at the network edge or the IP level (Open System Interconnection [OSI] Layer 3). As the popularity of Ethernet switching and wireless LANs grow, however, the emphasis on Layer 2 security has become more important. Yet, less public information is available regarding security risks in a Layer 2 environment and mitigating strategies of these risks. In addition, switches and wireless access points are susceptible to many of the same Layer 3 attacks as routers.

The most common types of Layer 2 attacks are as follows:

- CAM table overflow
- VLAN hopping
- Spanning Tree Protocol (STP) manipulation
- MAC address spoofing
- Private VLAN
- DHCP "starvation"

The following sections discuss the most common Layer 2 attacks and recommended methods to reduce the effects of these attacks.

CAM Table Overflow Attacks

The content-addressable memory (CAM) table in a switch stores information, such as MAC addresses and associated VLAN parameters. It is similar to a router's routing table. CAM tables have a fixed size.

A MAC address is a 48-bit hexadecimal number composed of two descriptive fields. The first 24 bits comprise the manufacturer code assigned by the IEEE. The second 24 bits comprise the specific interface number assigned by the hardware manufacturer. A MAC address of FF.FF.FF.FF.FF.FF is a broadcast address. Each MAC address is a unique series of numbers, similar to serial numbers or LAN IP addresses. A manufacturer should not have two devices with the same MAC address.

When a Layer 2 switch receives a frame, the switch looks in the CAM table for the destination MAC address. If an entry exists for that MAC address, the switch forwards the frame to the port identified in the CAM table for that MAC address. If the MAC address is not in the CAM table, the switch forwards the frame out all ports on the switch. If the switch sees a response as a result of the forwarded frame, it updates the CAM table with the port on which the communication was received.

In a typical LAN environment where there are multiple switches connected on the network, all the switches receive the unknown destination frame. Figure 14-1 shows the CAM table operation.

Figure 14-1 *CAM Table Operation*

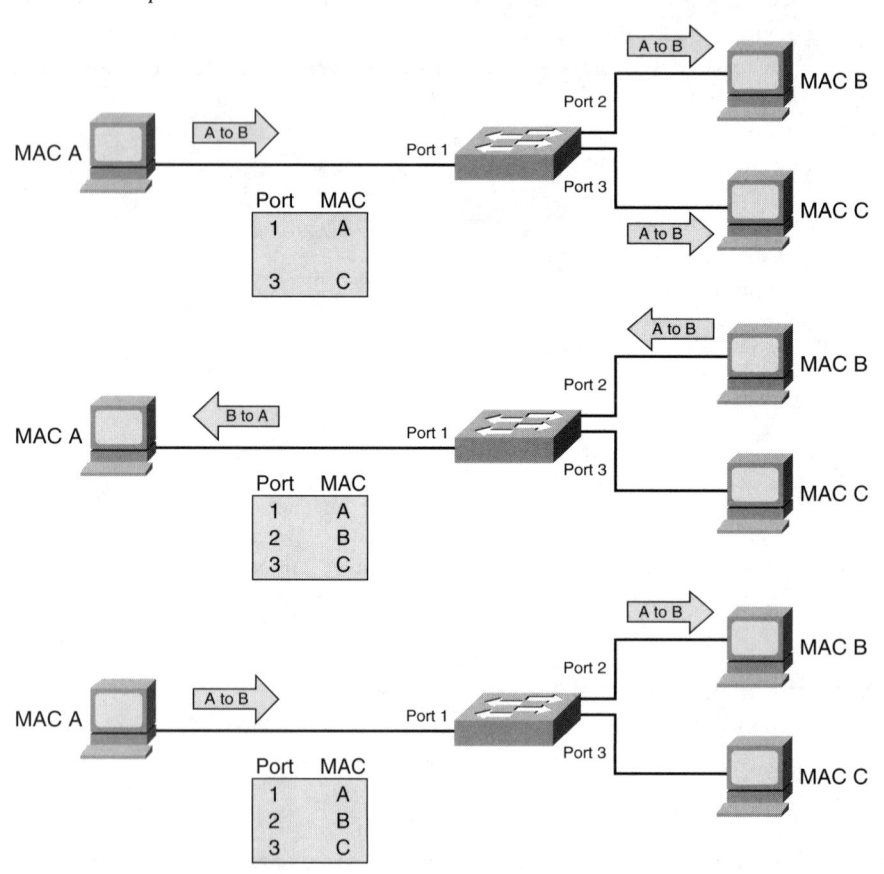

As previously mentioned, the CAM table has a limited size. Cisco Catalyst switches use the 63 bits of source (MAC, VLAN, and so on) and create a 14-bit hash value. If the value is the same, there are eight buckets in which to place CAM entries. These entries expire after a certain inactivity period. (The default on the Cisco Catalyst switch is 5 minutes.) If enough MAC addresses are flooded to a switch before existing entries expire, the CAM table fills up, and new entries are not accepted. When the CAM table is full, the switch starts flooding the packets out all ports. This scenario is called a *CAM table overflow*.

In a CAM table overflow attack, an attacker sends thousands of bogus MAC addresses from one port, which looks like valid hosts' communication, to the switch. One of the more popular tools used for launching this type of attack is called Macof, which was written using PERL code, ported to C language, and bundled into the Dsniff suite. Dsniff is a collection of tools for network auditing and penetration testing. Macof can generate 155,000 MAC entries on a switch per minute. The goal is to flood the switch with traffic by filling the CAM table with false entries. When flooded, the switch broadcasts traffic without a CAM entry out on its local VLAN, thus allowing the attacker to see other VLAN traffic that would not otherwise display.

Figure 14-2 shows a CAM table overflow attack.

Figure 14-2 *CAM Table Overflow Attack*

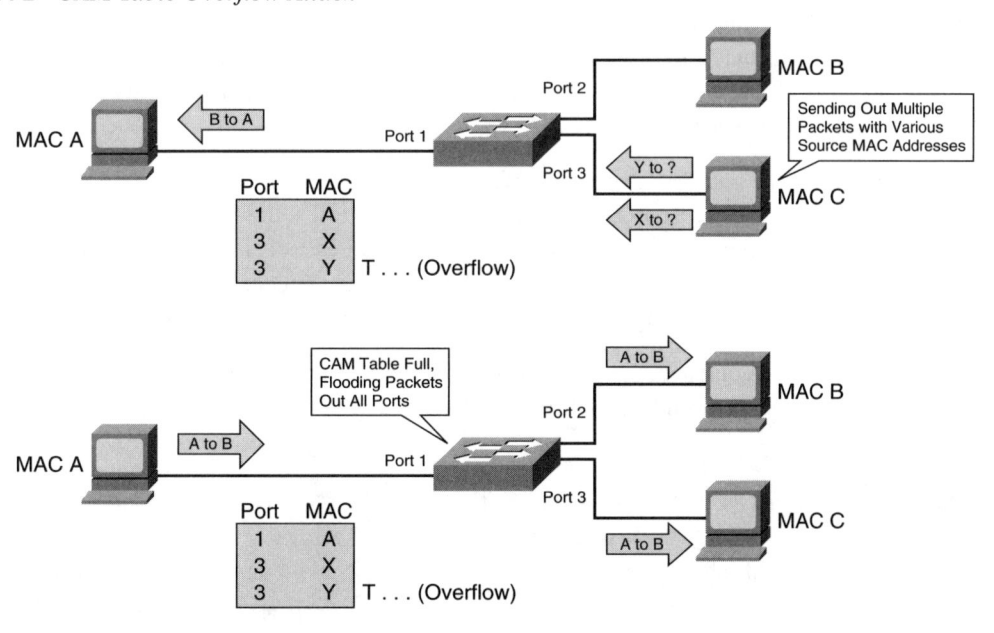

Mitigating CAM Table Overflow Attacks

You can mitigate CAM table overflow attacks in several ways. One of the primary ways is to configure port security on the switch. You can apply port security in three ways:

- **Static secure MAC addresses**—A switch port may be manually configured with the specific MAC address of the device that connects to it.

- **Dynamic secure MAC addresses**—The maximum number of MAC addresses that will be learned on a single switch port is specified. These MAC addresses are dynamically learned, stored only in the address table, and removed when the switch restarts.

- **Sticky secure MAC addresses**—The maximum number of MAC addresses on a given port may be dynamically learned or manually configured. The manual configuration is not a recommended method because of the high administrative overhead. The *sticky* addresses will be stored in the address table and added to the running configuration. If the addresses are saved in the configuration file, the interface does not need to dynamically relearn them when the switch restarts.

The type of action taken when a port security violation occurs falls into the following three categories:

- **Protect**—If the number of secure MAC addresses reaches the limit allowed on the port, packets with unknown source addresses are dropped until a number of MAC addresses are removed or the number of allowable addresses is increased. You receive no notification of the security violation in this type of instance.

- **Restrict**—If the number of secure MAC addresses reaches the limit allowed on the port, packets with unknown source addresses are dropped until some number of secure MAC addresses are removed or the maximum allowable addresses is increased. In this mode, a security notification is sent to the Simple Network Management Protocol (SNMP) server (if configured) and a syslog message is logged. The violation counter is also incremented.

- **Shutdown**—If a port security violation occurs, the interface changes to error-disabled and the LED is turned off. It sends an SNMP trap, logs to a syslog message, and increments the violation counter.

> **NOTE** On some Cisco Catalyst switch platforms, such as 3550 and 6500, you can enable port security on trunk ports. On these platforms, Protect mode security is not recommended for a trunk port. Protect mode disables learning when any VLAN reaches its maximum limit, even if the port has not reached its maximum limit.

Table 14-2 shows the commands used to configure port security.

Table 14-2 *Configuring Port Security*

Command	Description
**switchport mode {access	trunk}**	Sets the interface mode as access or trunk. An interface in the default mode of dynamic auto cannot be configured as a secure port.
switchport port-security	Enables port security on the interface.
switchport port-security [maximum value [vlan {*vlan-list***	{access	voice}}]]**	Sets the maximum number of secure MAC addresses for the interface. The active Switch Database Management (SDM) template determines the maximum number of available addresses. The default is 1.
**switchport port-security violation {protect	restrict	shutdown}**	Sets the action to be taken when a security violation is detected. The default mode for security violations is to shut down the interface.
switchport port-security [mac-address *mac-address* **[vlan {***vlan-id***	{access	voice}}]**	Sets a secure MAC address for the interface. This command may be used to enter the maximum number of secure MAC addresses. If fewer secure MAC addresses are configured than the maximum, the remaining MAC addresses are dynamically learned.
switchport port-security mac-address sticky	Enables sticky learning on the interface.
switchport port-security mac-address sticky [*mac-address*	**vlan** {*vlan-id*	{**access**	**voice**}}]	Sets a sticky secure MAC address. This command can be repeated as many times as necessary. If fewer secure MAC addresses are configured than the maximum, the remaining MAC addresses are dynamically learned, converted to sticky secure MAC addresses, and added to the running configuration.

NOTE When a secure port is in the error-disabled state, you can bring it out of this state by entering the **errdisable recovery cause psecure-violation** global configuration command, or you can manually re-enable it by entering the **shutdown** and **no shutdown** interface configuration commands.

The following example configures a switch port as an access port and sets dynamic port security with maximum number of addresses learned to 20. The violation mode is the default shutdown mode, sticky learning is enabled, and no static MAC addresses are configured. In the scenario where a twenty-first computer tries to connect, the port will be placed in error-disabled state and will send out an SNMP trap notification.

Example 14-1 *Configuring Dynamic Port Security*

```
Switch#configure terminal
Switch(config)#interface fastethernet0/0
Switch(config-if)#switchport mode access
Switch(config-if)#switchport port-security
Switch(config-if)#switchport port-security maximum 20
Switch(config-if)#switchport port-security mac-address sticky
```

VLAN Hopping Attacks

VLANs are a simple way to segment the network within an enterprise to improve performance and simplify maintenance. Each VLAN consists of a single broadcast domain. VLANs work by tagging packets with an identification header. Ports are restricted to receiving only packets that are part of the VLAN. The VLAN information may be carried between switches in a LAN using trunk ports. Trunk ports have access to all VLANs by default. They route traffic for multiple VLANs across the same physical link. Two types of trunks are used: 802.1q and ISL. The trunking mode on a switch port may be sensed using Dynamic Trunk Protocol (DTP), which automatically senses whether the adjacent device to the port may be capable of trunking. If so, it synchronizes the trunking mode on the two ends. The DTP state on a trunk port may be set to auto, on, off, desirable, or non-negotiate. The DTP default on most switches is auto.

One of the areas of concern with Layer 2 security is the variety of mechanisms by which packets that are sent from one VLAN may be intercepted or redirected to another VLAN, which is called *VLAN hopping*. VLAN hopping attacks are designed to allow attackers to bypass a Layer 3 device when communicating from one VLAN to another. The attack works by taking advantage of an incorrectly configured trunk port.

It is important to note that this type of attack does not work on a single switch because the frame will never be forwarded to the destination. But in a multiswitch environment, a trunk link could be exploited to transmit the packet. There are two different types of VLAN hopping attacks:

- **Switch spoofing**—The network attacker configures a system to spoof itself as a switch by emulating either ISL or 802.1q, and DTP signaling. This makes the attacker appear to be a switch with a trunk port and therefore a member of all VLANs.

- **Double tagging**—Another variation of the VLAN hopping attack involves tagging the transmitted frames with two 802.1q headers. Most switches today perform only one level of decapsulation. So when the first switch sees the double-tagged frame, it strips the first tag off the frame and then forwards with the inner 802.1q tag to all switch ports in the attacker's VLAN as well as to all trunk ports. The second switch forwards the packet based on the VLAN ID in the second 802.1q header. This type of attack works even if the trunk ports are set to off.

Figure 14-3 shows VLAN hopping with a double-tagging scenario.

Figure 14-3 *Double-Tagging VLAN Hopping Attack*

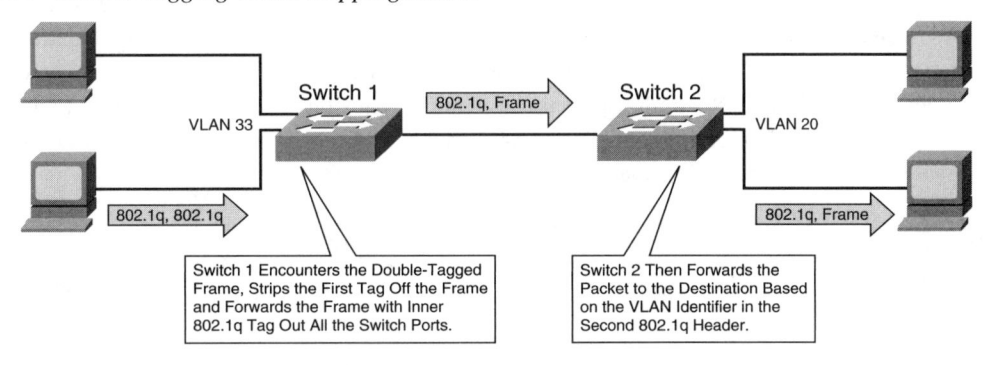

Mitigating VLAN Hopping Attacks

Mitigating VLAN hopping attacks requires the following configuration modifications:

- Always use dedicated VLAN IDs for all trunk ports.
- Disable all unused ports and place them in an unused VLAN.
- Set all user ports to nontrunking mode by disabling DTP. Use the **switchport mode access** command in the interface configuration mode.
- For backbone switch-to-switch connections, explicitly configure trunking.
- Do not use the user native VLAN as the trunk port native VLAN.
- Do not use VLAN 1 as the switch management VLAN.

STP Manipulation Attacks

STP prevents bridging loops in a redundant switched network environment. By avoiding loops, you can ensure that broadcast traffic does not become a traffic storm.

STP is a hierarchical tree-like topology with a "root" switch at the top. A switch is elected as root based on the lowest configured priority of any switch (0 through 65,535). When a switch boots up, it begins a process of identifying other switches and determining the root bridge. After a root bridge is elected, the topology is established from its perspective of the connectivity. The switches determine the path to the root bridge, and all redundant paths are blocked. STP sends configuration and topology change notifications and acknowledgments (TCN/TCA) using bridge protocol data units (BPDU).

An STP attack involves an attacker spoofing the root bridge in the topology. The attacker broadcasts out an STP configuration/topology change BPDU in an attempt to force an STP recalculation. The BPDU sent out announces that the attacker's system has a lower bridge priority. The attacker can then see a variety of frames forwarded from other switches to it. STP recalculation may also cause a denial-of-service (DoS) condition on the network by causing an interruption of 30 to 45 seconds each time the root bridge changes. Figure 14-4 shows an attacker using STP network topology changes to force its host to be elected as the root bridge.

Figure 14-4 *STP Attack*

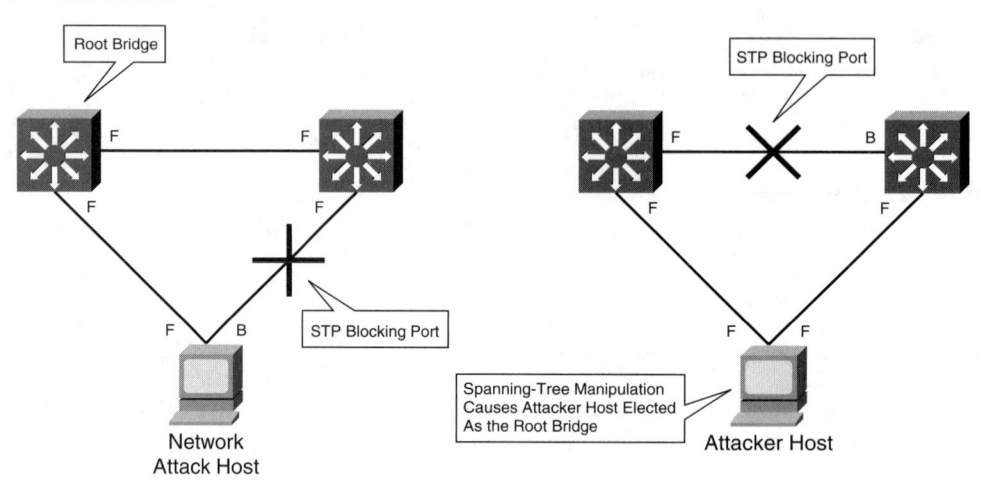

Preventing STP Manipulation Attacks

To mitigate STP manipulation, use the root guard and BPDU guard features in the Cisco IOS Software. These commands enforce the placement of the root bridge and the STP domain borders. The STP root guard feature is designed to allow the placement of the root bridge in the network. The STP BPDU guard is used to keep all active network topology predictable.

Example 14-2 shows an example of enabling BPDU guard, using **portfast**, to disable ports upon detection of a BPDU message and to disable ports that would become the root bridge because of their BPDU advertisement.

Example 14-2 *Enabling BPDU and Root Guard*

```
Switch#configure terminal
Switch(config)#spanning-tree portfast bpduguard
Switch(config)#interface fa0/10
Switch(config)#spanning-tree guard root
```

MAC Address Spoofing—Man-in-the-Middle Attacks

MAC spoofing involves the use of a known MAC address of another host that is authorized to access the network. The attacker attempts to make the target switch forward frames destined for the actual host to the attacker device instead. This is done by sending a frame with the other host's source Ethernet address with the objective to overwrite the CAM table entry. After the CAM is overwritten, all the packets destined for the actual host will be diverted to the attacker. If the original host sends out traffic, the CAM table will be rewritten again, moving the traffic back to the original host port. Figure 14-5 shows how MAC spoofing works.

Figure 14-5 *MAC Spoofing Attack*

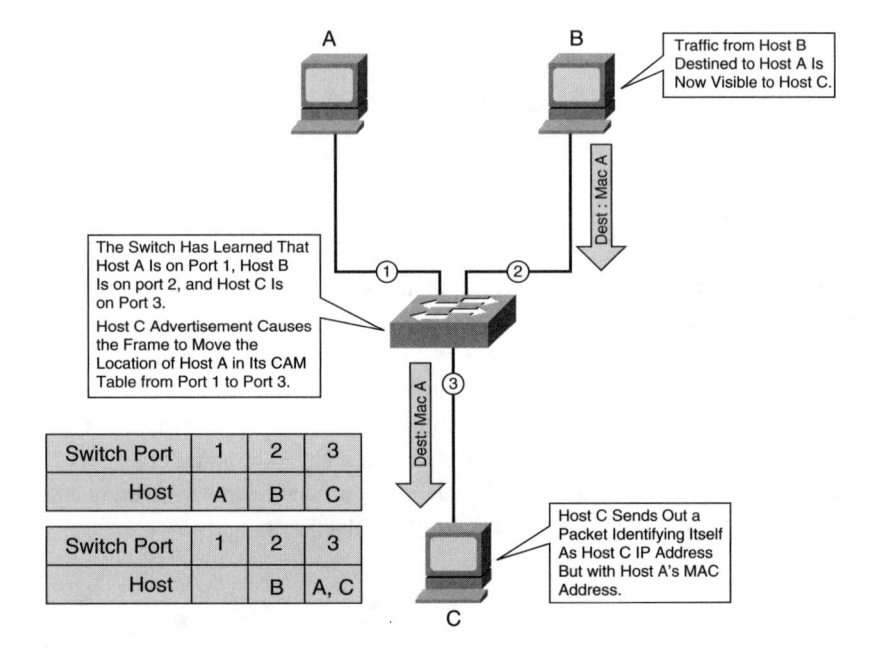

Another method of spoofing MAC addresses is to use Address Resolution Protocol (ARP), which is used to map IP addressing to MAC addresses residing on one LAN segment. When a host sends out a broadcast ARP request to find a MAC address of a particular host, an ARP response comes from the host whose address matches the request. The ARP response is cached by the requesting host. ARP protocol also has another method of identifying host IP-to-MAC associations, which is called Gratuitous ARP (GARP), which is a broadcast packet used by hosts to announce their IP address to the LAN to avoid duplicate IP addresses on the network. GARP can be exploited maliciously by an attacker to spoof the identity of an IP address on a LAN segment. This is typically used to spoof the identity between two hosts or all traffic to and from the default gateway.

One of the tools used to spoof ARP entries is called Arpspoof and is part of a collection of tools known as *Dsniff*.

Mitigating MAC Address Spoofing Attacks

Use the **port-security** command described in the "Mitigating CAM Table Overflow Attacks" section to specify MAC addresses connected to particular ports; however, this type of configuration has a high administrative overhead and is prone to mistakes. There are other mechanisms, such as hold-down timers, that you can use to mitigate ARP spoofing attacks by setting the length of time an entry will stay in the ARP cache. Hold-down timers by themselves are insufficient to mitigate attacks. It is possible to combine this with the modifications to the ARP cache expiration time for all the hosts, but this is also unmanageable. One recommended alternative is to use private VLANs to mitigate these types of network attacks. A couple of other features of Cisco IOS Software provide protection from this type of attack: dynamic ARP inspection (DAI) and DHCP snooping (as described in the following sections).

Using DHCP Snooping

DHCP snooping is a DHCP security feature that provides network security by filtering untrusted DHCP messages using a DHCP snooping binding database it builds and maintains (referred to as the *DHCP snooping binding table*). An untrusted message is a message received from outside the network or firewall. When a switch receives a packet on an untrusted interface and DHCP snooping is enabled on that interface or VLAN, the switch compares the source MAC address and the DHCP client requester hardware address. If the addresses match, the switch forwards the packet. If the addresses do not match, the switch drops the packet. DHCP snooping considers DHCP messages originating from any user port to a DHCP server as untrusted. An untrusted port should not send DHCP server type responses, such as DHCPOffer, DHCPAck, and so on.

A trusted interface is an interface configured to receive messages only from within the network.

The DHCP snooping binding table contains information such as the host MAC address (dynamic and static), IP address, lease time, binding type, and VLAN number. The database can have up to 8192 bindings.

To enable DHCP snooping on a switch, use the **ip dhcp snooping** command in global configuration mode. To enable the switch to insert and remove DHCP relay information (option 82 field) in forwarded DHCP request messages to the DHCP server, use the **ip dhcp snooping information option** global configuration command.

Example 14-3 shows the configuration for enabling DHCP snooping for VLAN 34 and how to configure a rate limit of 70 packets per second on FastEthernet port 0/0.

Example 14-3 *Configuring DHCP Snooping*

```
Switch#configure terminal
Switch(config)#ip dhcp snooping
Switch(config)#ip dhcp snooping vlan 34
Switch(config)#ip dhcp snooping information option
Switch(config)#interface fa0/0
Switch(config-if)#ip dhcp snooping limit rate 70
```

Using DAI

DAI is a security feature that intercepts and verifies IP-to-MAC address bindings and discards invalid ARP packets. DAI uses the DHCP snooping database to validate bindings. It associates a trust state with each interface on the switch. Packets arriving on trusted interfaces bypass all DAI validation checks, and those arriving on untrusted interfaces undergo the DAI validation process. In a typical network, all ports on the switch connected to host are configured as untrusted, and switch ports are considered trusted. Use the **ip arp inspection trust** interface command to configure the trust settings. When the switch is configured for DAI, it rate-limits incoming ARP packets to prevent DoS attacks. The default rate for an untrusted interface is 14 packets per second. Trusted interfaces are not rate-limited. You can change the rate limit by using the **ip arp inspection limit** interface configuration command. DAI uses the ARP access control lists (ACLs) and DHCP snooping database for the list of valid IP-to-MAC address bindings. ARP ACLs take precedence over entries in the DHCP snooping bindings database but have to be manually configured. Use the **ip arp inspection filter** global configuration command to configure ARP ACLs. The switch will drop a packet that is denied in the ARP ACLs even if the DHCP snooping database has a valid binding for it. When a switch drops a packet, it is logged in the buffer and generates a system message. Use the

ip arp inspection log-buffer to configure the number of buffers and the number of entries needed in the specified interval to generate system messages. Example 14-4 shows a sample DAI configuration for VLAN 34, setting fa0/0 to a trusted port with a rate limit of 20 packets per second and a logging buffer of 64 messages.

Example 14-4 *DAI Configuration*

```
Switch#configure terminal
Switch(config)#ip arp inspection vlan 34
Switch(config)#interface fa0/0
Switch(config-if)#ip arp inspection trust
Switch(config-if)#ip arp inspection limit rate 20 burst interval 2
Switch(config-if)#exit
Switch(config)#ip arp inspection log-buffer entries 64
```

Private VLAN Vulnerabilities

Private VLANs isolate ports within a VLAN to communicate only with other ports in the same VLAN. There are three types of private VLAN ports:

- **Community**—Communicate among themselves and with other promiscuous ports. These interfaces are isolated at Layer 2 from all other interfaces in other communities or isolated ports within their private VLAN.

- **Isolated**—Has complete Layer 2 separation from other ports within the same private VLAN except for the promiscuous port. Private VLANs block all traffic to isolated ports except traffic from promiscuous ports. Traffic received from an isolated port is forwarded only to promiscuous ports.

- **Promiscuous**—Communicates with all interfaces, including community and isolated ports within a private VLAN.

> **NOTE** To configure a private VLAN, the switch must be in Virtual Terminal Protocol (VTP) transparent mode.

A network vulnerability of private VLANs involves the use of a proxy to bypass access restrictions of the private VLAN. In a proxy attack, frames are forwarded to a host on the network connected to a promiscuous port, such as a router. The network attacker sends a packet with its source IP and MAC address and a destination IP address of the target system but a destination MAC address of the router. The switch forwards the frame to the router. The router routes the traffic, rewrites the destination MAC address as that of the target, and sends the packet out. Because the router is authorized to communicate with the private VLANs, the packet is forwarded to the target system. This type of attack allows for unidirectional traffic only because the private VLAN filter blocks the target's attempts to respond. This vulnerability is not a private VLAN vulnerability per se because all the rules of that VLAN were enforced.

Figure 14-6 shows how the private VLAN proxy vulnerability works.

Figure 14-6 *Private VLAN Proxy Attack*

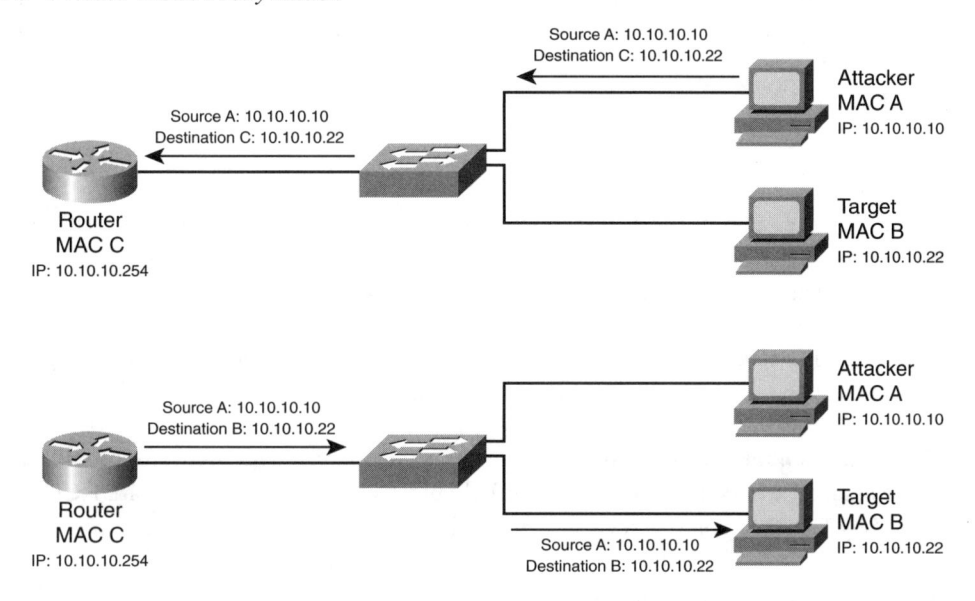

Defending Private VLANs

You can configure ACLs on the router port to mitigate private VLAN attacks. You can also use virtual ACLs on the Cisco Catalyst Layer 3 switch platforms to help mitigate the effects of private VLAN attacks. Example 14-6 shows the ACLs configured on a router port to segment the private VLAN subnet 192.168.200.0/24.

Example 14-5 *Mitigating Private VLAN Proxy Attack*

```
Router1#configure terminal
Router1(config)#access-list 150 deny ip 192.168.200.0 0.0.0.255 192.168.200.0 0.0.0.255 log
Router1(config)#access-list 150 permit ip any any
Router1(config)#interface fa0/0
Router1(config-if)#ip access-group 150 in
```

DHCP Starvation Attacks

A DHCP server dynamically assigns IP addresses to hosts on a network. The administrator creates pools of addresses available for assignment. A lease time is associated with the addresses. DHCP is a standard defined in RFC 2131.

A DHCP starvation attack works by broadcasting DHCP requests with spoofed MAC addresses. This scenario is achieved with attack tools such as gobbler, which looks at the entire DHCP scope and tries to lease all the DHCP addresses available in the DHCP scope. This is a simple resource starvation attack, similar to a SYN flood attack. The attacker can then set up a rogue DHCP server and respond to new DHCP requests from clients on the network. This might result in a "man-in-the-middle" attack.

Mitigating DHCP Starvation Attacks

The methods used to mitigate a MAC address spoofing attack may also prevent DHCP starvation by using the DHCP snooping feature. Implementation of RFC 3118, *Authentication for DHCP Message*, will also assist in mitigating this type of attack.

You can also limit the number of MAC addresses on a switch port, a mitigation strategy for CAM table flooding, to mitigate DHCP starvation attacks.

Other features on the Cisco Catalyst switch, such as IP source guard, may also provide additional defense against attacks. IP source guard initially blocks all IP traffic except for DHCP packets captured by DHCP snooping process. When a client receives a valid IP address from the DHCP server, an ACL is applied to the port. This ACL restricts the traffic from the client to those source IP addresses configured in the binding.

One way to prevent a rogue DHCP server from responding to DHCP requests is to use VLAN ACLs (VACL). You can use VACLs to limit DHCP replies to legitimate DHCP servers and deny them to all others. You should use this type of configuration if the network does not support DHCP snooping.

IEEE 802.1x EAP Attacks

IEEE 802.1x is an IEEE standard link layer (Layer 2) protocol designed to provide port-based network access control using authentication unique to a device or user. Extensible Authentication Protocol (EAP) is the transport mechanism used in 802.1x to authenticate supplicants/clients against a back-end data store, which is typically a RADIUS server. Figure 14-7 shows client authentication using 802.1x and EAP.

Figure 14-7 *Client Authentication Using 802.1x and EAP*

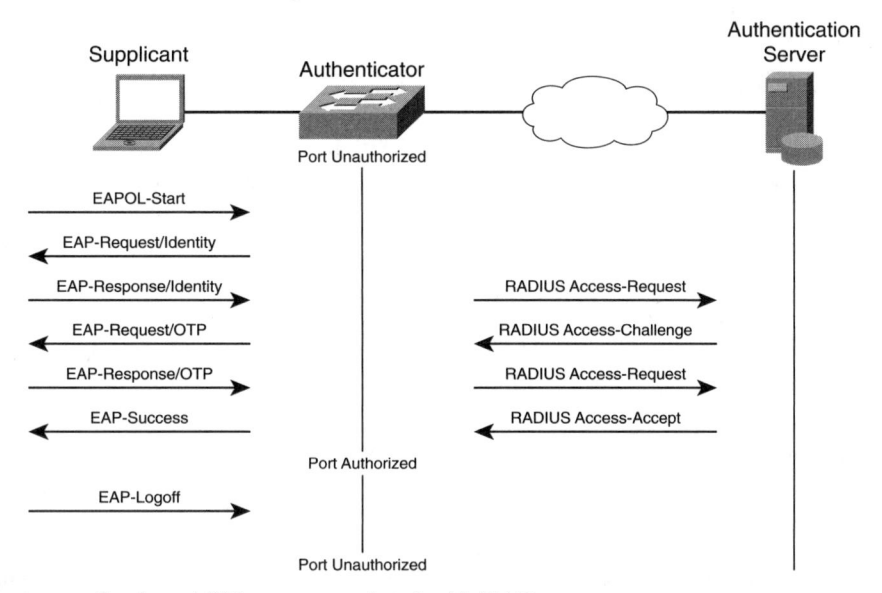

Two types of vulnerabilities are associated with EAP:

- **Man-in-the-middle attack**—At the end of the EAP Over LAN (EAPOL) authentication, the attacker sends the client an EAP-Success message that identifies the attacker as the authenticator. When this action is successful, the attacker is in the path between the client and the authenticator.

- **Session-hijacking attack**—This attack occurs after the authentication process between the client and the authentication server is complete. If the attacker sends a disassociate management frame with the authenticator's MAC address to the client, it will force the client to disconnect from the network; however, the authenticator state is still in an authenticated and associated state, which allows the attacker to access the network.

Mitigating IEEE 802.1x EAP Attacks

Cisco recommends deploying Protected EAP (PEAP) for use in a wireless LAN environment and deploying 802.1x on all access switches to limit physical access to the network. Example 14-6 shows a sample configuration enabling 802.1x authentication on a Cisco router.

Example 14-6 *Enabling 802.1x on a Cisco Router*

```
Router#configure terminal
Router(config)#aaa new-model
Router(config)#aaa authentication dot1x default group radius
Router(config)#dot1x system-auth-control
Router(config)#interface fa0/0
Router(config-if)#dot1x port-control auto
```

For more information on functionality and configuration of IEEE 802.1x and EAP, see Chapters 17, "Identity-Based Networking Services," and 18, "Configuring 802.1x Port-Based Authentication," of this book.

Factors Affecting Layer 2 Mitigation Techniques

You must consider several factors when designing a protected Layer 2 network, as divided into the following three categories:

- **The number of user groups**—Depending on the size of the network, users can be grouped by function, location, or access level. For example, users in a company's financial division can be grouped together to provide restricted access to the financial resources.

- **The number of switches**—The number of switches depends on the scale of the network. As the number of switches in the network grows, the manageability and security of the devices becomes more difficult, and the security vulnerabilities increase.

- **The number of security zones**—A security zone is a logical and physical environment under the common set of policies and procedure with visible management support. In this environment, access privileges to all information assets are defined and followed. For example, an organization's internal LAN could be considered on a security zone or it could be broken up into several depending on the types of users. If the organization has external partners it exchanges information with, it will typically have a separate security zone for this communication, sometimes referred to as an *extranet*. If the network is connected to the Internet, there is typically a demilitarized zone (DMZ), separating the internal network from the Internet. A DMZ is a security zone that might contain resources (such as web servers, mail, and so on) accessible via the public Internet without compromising the internal network security.

Based on the preceding categories, a combination of eight scenarios could be developed to address security vulnerabilities in typical networking environments. Table 14-3 briefly describes the vulnerabilities associated with these combinations.

Table 14-3 *Layer 2 Security Vulnerabilities in Different Network Combinations*

Combination #	Security Zones	Number of User Groups	Number of Switch Devices	Layer 2 Vulnerabilities
1	Single	Single	Single	MAC spoofing CAM table overflow
2	Single	Single	Multiple	MAC spoofing CAM table overflow VLAN hopping STP
3	Single	Multiple	Single	MAC spoofing CAM table overflow VLAN hopping
4	Single	Multiple	Multiple	MAC spoofing CAM table overflow VLAN hopping STP
5	Multiple	Single	Single	MAC spoofing CAM table overflow VLAN hopping
6	Multiple	Single	Multiple	MAC spoofing CAM table overflow VLAN hopping STP

continues

Table 14-3 *Layer 2 Security Vulnerabilities in Different Network Combinations (Continued)*

Combination #	Security Zones	Number of User Groups	Number of Switch Devices	Layer 2 Vulnerabilities
7	Multiple	Multiple	Single	MAC spoofing CAM table overflow VLAN hopping Private VLAN attacks
8	Multiple	Multiple	Multiple	MAC spoofing CAM table overflow VLAN hopping STP VTP attacks

Part VII of this book, "Scenarios," outlines specific examples of these combinations.

Foundation Summary

The "Foundation Summary" section of each chapter lists the most important facts from the chapter. Although this section does not list every fact from the chapter that will be on your SNRS exam, a well-prepared candidate should at a minimum know all the details in each "Foundation Summary" before going to take the exam.

The most common types of Layer 2 attacks and mitigation strategies are as follows:

CAM table overflow—In a CAM table overflow attack, an attacker sends thousands of bogus MAC addresses from one port, which looks like valid hosts' communication to the switch. You can mitigate CAM table overflow attacks in several ways. One of the primary ways is to configure port security on the switch. You can apply port security in three ways: static secure MAC addresses, dynamic secure MAC addresses, and sticky secure MAC addresses.

VLAN hopping—There are two different types of VLAN hopping attacks: switch spoofing and double tagging. Mitigating VLAN hopping attacks requires the following configuration modifications:

- Always use dedicated VLAN IDs for all trunk ports.
- Disable all unused ports and place them in an unused VLAN.
- Set all user ports to nontrunking mode by disabling DTP. Use the **switchport mode access** command in the interface configuration mode.
- For backbone switch-to-switch connections, explicitly configure trunking.
- Do not use the user native VLAN as the trunk port native VLAN.
- Do not use VLAN 1 as the switch management VLAN.

STP manipulation—An STP attack involves an attacker spoofing the root bridge in the topology. The attacker broadcasts out an STP configuration/topology change BPDU in an attempt to force an STP recalculation. The BPDU sent out announces that the attacker's system has a lower bridge priority. The attacker can then see a variety of frames forwarded from other switches to it.

To mitigate STP manipulation, use the root guard and BPDU guard features in the Cisco IOS Software.

MAC address spoofing—MAC address spoofing involves the use of a known MAC address of another host authorized to access the network. The attacker attempts to make the target switch forward frames destined for the actual host to the attacker device instead. Another way to spoof MAC addresses is by using ARP.

Use the **port-security** command described in the "Mitigating CAM Table Overflow Attacks" section to specify MAC addresses connected to particular ports. DHCP snooping could be used as a method to mitigate MAC address spoofing. Another method of mitigating MAC address spoofing is DAI.

Private VLAN—Private VLANs isolate ports within a VLAN to communicate only with other ports in the same VLAN. The three types of private VLAN ports are community, isolated, and promiscuous.

You can configure ACLs on the router port to mitigate private VLAN attacks. You can also use virtual ACLs on the Cisco Catalyst Layer 3 switch platforms to help mitigate the effects of private VLAN attacks.

DHCP "starvation"—A DHCP starvation attack works by broadcasting DHCP requests with spoofed MAC addresses.

The methods used to mitigate MAC address spoofing attack may also prevent DHCP starvation by using the DHCP snooping feature. You can limit the number of MAC addresses on a switch port, a mitigation strategy for CAM table flooding, to mitigate DHCP starvation attack. Other features on the Cisco Catalyst switch, such as IP source guard, may also provide additional defense against attacks.

IEEE 802.1x attack—IEEE 802.1x is an IEEE standard link layer (Layer 2) protocol designed to provide port-based network access control using authentication unique to a device or user.

Two types of vulnerabilities are associated with EAP: man-in-the-middle attacks and session-hijacking attacks. Cisco recommends deploying PEAP for use in a wireless LAN environment and deploying 802.1x on all access switches to limit physical access to the network.

Q&A

As mentioned in the section "How to Use This Book" in the Introduction, you have two choices for review questions. The questions that follow present a bigger challenge than the exam itself because they use an open-ended question format. By reviewing using this more difficult format, you can exercise your memory better and prove your conceptual and factual knowledge of this chapter. You can find the answers to these questions in the appendix.

For more practice with exam-like question formats, including questions using a router simulator and multiple-choice questions, use the exam engine on the CD-ROM.

1. What are the most common types of Layer 2 attacks?
2. Describe the CAM table overflow attack.
3. Explain the three categories of action that can be taken when a port security violation occurs.
4. When a secure port is in the error-disabled state, how can it be brought out of this state?
5. How can you mitigate VLAN hopping attacks?
6. What is involved in an STP attack?
7. How does MAC spoofing–man-in-the-middle attacks work?
8. How can you mitigate MAC spoofing attacks?
9. Describe how a proxy attack bypasses access restrictions of private VLANs.
10. Explain how a DHCP starvation attack is performed.

This chapter covers the following subjects:

- Context-Based Access Control Features
- Configuring CBAC
- Verifying and Debugging CBAC

CHAPTER 15

Context-Based Access Control

The Cisco IOS Firewall is a group of features that makes Cisco IOS Software more effective for perimeter security. The firewall features are based on inspection of data in IP packets, including data beyond the headers.

> **NOTE** The perimeter router running the Cisco IOS Software security feature is referred to as *Cisco IOS Firewall* in this chapter.

"Do I Know This Already?" Quiz

The purpose of the "Do I Know This Already?" quiz is to help you decide whether you really need to read the entire chapter. If you already intend to read the entire chapter, you do not necessarily need to answer these questions now.

The 12-question quiz, derived from the major sections in the "Foundation Topics" portion of the chapter, helps you determine how to spend your limited study time.

Table 15-1 outlines the major topics discussed in this chapter and the "Do I Know This Already?" quiz questions that correspond to those topics.

Table 15-1 *"Do I Know This Already?" Foundation Topics Section-to-Question Mapping*

Foundation Topics Section	Questions Covered in This Section
Context-Based Access Control Features	1–6, 12
Configuring CBAC	8–11
Verifying and Debugging CBAC	7

> **CAUTION** The goal of self-assessment is to gauge your mastery of the topics in this chapter. If you do not know the answer to a question or are only partially sure of the answer, you should mark this question wrong for purposes of the self-assessment. Giving yourself credit for an answer you correctly guess skews your self-assessment results and might provide you with a false sense of security.

1. Which of the following is *not* true about CBAC?

   a. CBAC provides secure per-application access control across network perimeters.

   b. CBAC intelligently filters TCP and UDP packets based on application layer protocol session information.

   c. The CBAC feature is only available on Cisco switches.

   d. CBAC uses state information to create temporary openings in the firewall's ACL to allow return traffic.

2. What is the advantage of using CBAC versus ACLs?

   a. CBAC examines and inspects packets at the network, transport, and application layer level, whereas ACLs do not inspect all three levels.

   b. CBAC is less complicated to configure than ACLs.

   c. CBAC works on hubs.

   d. The CBAC memory requirement is less than ACL memory requirements.

3. How does CBAC handle UDP sessions?

   a. CBAC cannot build a state table for UDP sessions because UDP is a connectionless protocol.

   b. CBAC approximates UDP sessions by examining the information in the packet and determining whether the packet is similar to other UDP packets.

   c. CBAC does not inspect UDP packets.

   d. CBAC denies suspicious UDP packets randomly.

4. Approximately how much memory per connection does CBAC require?

   a. 2 KB

   b. 6 KB

   c. 200 bytes

   d. 600 bytes

5. Which of the following is true about ACLs created by CBAC? (Select two.)

   a. ACL entries are created and deleted dynamically.

   b. After they are created, they are saved to NVRAM.

   c. CBAC does not create or delete ACLs.

   d. CBAC uses ACL bypassing feature in 12.3(4)T code.

6. Which of the following protocols are supported by CBAC? (Select three.)

   a. FTP
   b. SMTP
   c. H.323
   d. OSPF

7. Which three types of **debug** command are used to debug CBAC? (Select three.)

   a. Network-level **debug**
   b. Transport-level **debug**
   c. Application protocol **debug**
   d. Generic **debug**

8. What command enables you to define an inspection rule?

   a. **inspection rule** *name protocol*
   b. **ip inspect name** *inspection name protocol*
   c. **ip protocol inspect** *inspection name protocol*
   d. **ip protocol inspection** *name*

9. What command enables you to inspect an application-level protocol?

   a. **debug ip inspect protocol**
   b. **debug ip inspect tcp**
   c. **debug ip inspect udp**
   d. **debug up inspect app**

10. What command enables you to show existing sessions currently being tracked and inspected by CBAC?

    a. **show ip inspect session** [*detail*]
    b. **display current ip inspect**
    c. **show current ip inspect**
    d. **display ip inspect session** [*detail*]

11. Support for ESMTP is not available in CBAC.

    a. True
    b. False

12. What are the three types of CBAC mapping entries for PAM? (Select three.)

   a. System-defined port mapping

   b. User-defined port mapping

   c. Host-specific port mapping

   d. Administrator-defined port mapping

   e. Mapquest-defined port mapping

The answers to the "Do I Know This Already?" quiz are found in the appendix. The suggested choices for your next step are as follows:

- **10 or less overall score**—Read the entire chapter. This includes the "Foundation Topics" and "Foundation Summary" sections and the "Q&A" section.

- **11 or 12 overall score**—If you want more review on these topics, skip to the "Foundation Summary" section and then go to the "Q&A" section. Otherwise, move on to Chapter 16, "Authentication Proxy and the Cisco IOS Firewall."

Foundation Topics

Context-Based Access Control Features

A Context-Based Access Control (CBAC) engine provides secure, per-application access control across network perimeters. CBAC lets the router maintain a persistent state, based on information from inspected packets, and use that information to decide which traffic should be forwarded. CBAC is the centerpiece of the Cisco IOS Firewall feature set, and the other features in the set build on CBAC. The CBAC features include the following:

- Detecting and preventing denial-of-service (DoS) attacks
- Generating real-time alerts and audit trails
- Providing secure per-application access control
- Providing filtering on generic TCP and UDP packets

CBAC can be used for intranets, extranets, and the Internet because of its inherent capability to distill packets (TCP and UDP) based on application protocol session information. For example, you can configure CBAC to permit specific TCP and UDP traffic through a Cisco IOS Firewall only when the connection is initiated from within the network you want to protect. In other words, CBAC can inspect traffic for sessions that originate from the external network.

Unlike access control lists (ACL), which are limited to the examination of packets at the network level, CBAC examines not only network layer and transport layer information but also examines the application layer protocol information (such as FTP connection information) to learn about the state of the TCP or UDP session. This extended examination allows support of protocols that involve multiple channels created as a result of negotiations in the control channel. Most of the multimedia protocols—as well as some other protocols including FTP, Remote Procedure Call (RPC), and SQL*Net—involve multiple channels.

In scenarios where Network Address Translation (NAT) is applied to traffic passing through a router that has CBAC enabled, the firewall first performs the CBAC inspection and then hides the internal IP address from the outside entities. This process provides extra protection for protocols that CBAC does not support.

CBAC also allows for Java blocking, filtering HTTP traffic. Java applets may be blocked based on the server address. It is also possible to deny access to Java applets not embedded in an archived or compressed file. A CBAC inspection rule may be created to filter Java applets at the firewall to allow users to download only trusted applets.

Detecting and Protecting Against DoS Attacks

CBAC inspects traffic that travels through the firewall to discover and manage state information for TCP and UDP sessions. This state information is used to create temporary openings in the firewall's ACLs to allow return traffic and additional data connections for permissible sessions.

Inspecting packets at the application layer, and maintaining TCP and UDP session information, provides CBAC with the capability to detect and prevent certain types of network attacks, such as SYN flooding. TCP SYN messages are sent to servers from clients as a first step in a three-step process known as the TCP three-way handshake to establish a TCP session, as shown in Figure 15-1.

Figure 15-1 *Three-Way TCP Handshake Process*

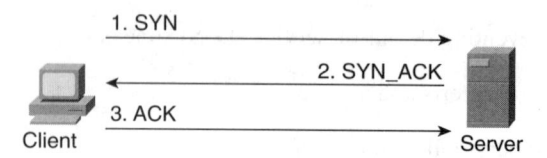

1. SYN
2. SYN_ACK
3. ACK

Client and server can now send service-specific data.

A SYN flood occurs when several hundred or thousand TCP SYN messages are sent to a server but never complete the TCP session. The resulting volume of half-open connections can overwhelm the server, causing it to deny service to valid requests. Network attacks that deny access to a network device are called DoS attacks.

CBAC helps to protect against DoS attacks in other ways. CBAC inspects packet sequence numbers in TCP connections to see whether they are within expected ranges. You can also configure CBAC to drop half-open connections. Additionally, CBAC can detect unusually high rates of new connections and issue alert messages.

Generating Alerts and Audit Trails

CBAC also generates real-time alerts and audit trails. Enhanced audit trail features use syslog to track all network transaction—recording time stamps, source host, destination host, ports used, and the total number of transmitted bytes for advanced, session-based reporting. Real-time alerts send syslog error messages to central management consoles upon detecting suspicious activity. Using CBAC inspection rules, you can configure alerts and audit trail information on a per-application protocol basis. If you want to generate audit trail information for HTTP traffic, for example, you can specify that in the CBAC rule covering HTTP inspection.

CBAC has typically been available only for IP traffic. Only TCP and UDP packets are inspected. However, beginning in Cisco IOS Software Release 12.2(15)T, limited stateful inspection of the most common types of Internet Control Message Protocol (ICMP) messages is also available. Table 15-2 lists the ICMP types supported by CBAC.

Table 15-2 *CBAC-Supported ICMP Types*

ICMP Packet Type	Name	Description
0	Echo Reply	Reply to Echo Request (Type 8).
3	Destination Unreachable	Possible reply to any request. This packet is included because it is a possible response to any ICMP packet request.
8	Echo Request	Ping or traceroute request.
11	Time Exceeded	Reply to any request if the Time-to-Live (TTL) packet is 0.
13	Timestamp Request	Request.
14	Timestamp Reply	Reply to Timestamp Request (Type 13).

Other IP traffic cannot be filtered with CBAC and should be filtered with extended IP ACLs instead.

> **NOTE** CBAC does not protect against attacks originating from within the protected network. CBAC only detects and protects against attacks that travel through the Cisco IOS Firewall.

Turn on logging and the audit trail to provide a record of network access through the firewall, including unauthorized access attempts. To configure logging and audit trail functions, use the **logging** and **ip inspect audit-trail** commands in global configuration mode. Example 15-1 shows the enabling of logging and audit trail functions on a router.

Example 15-1 *Enabling Logging and Audit Trail*

```
Router1#configure terminal
Router1(config)#service timestamps log datetime
Router1(config)#logging 192.168.20.1
Router1(config)#logging facility 5
Router1(config)#ip inspect audit-trail
```

How CBAC Works

A CBAC inspection rule is created to specify which protocols you want to be inspected. You then apply the rule to the desired interface and specify the direction (in or out). Only specified protocols are inspected by CBAC. Packets entering the Cisco IOS Firewall are inspected by CBAC only if they first pass the inbound ACL at the interface. If a packet is denied by the ACL, the packet is just dropped and not inspected by CBAC.

CBAC creates temporary openings in ACLs at Cisco IOS Firewall interfaces. These openings are created when specified traffic exits your internal network through the Cisco IOS Firewall. The openings allow returning traffic that would normally be blocked. The traffic is allowed back through the Cisco IOS Firewall only if it is part of the same session as the original traffic that triggered CBAC when exiting through the Cisco IOS Firewall. Figure 15-2 shows a Telnet user trying to access a server with CBAC enabled on the router.

Figure 15-2 *Telnet User Trying to Access a Server with CBAC Enabled on the Router*

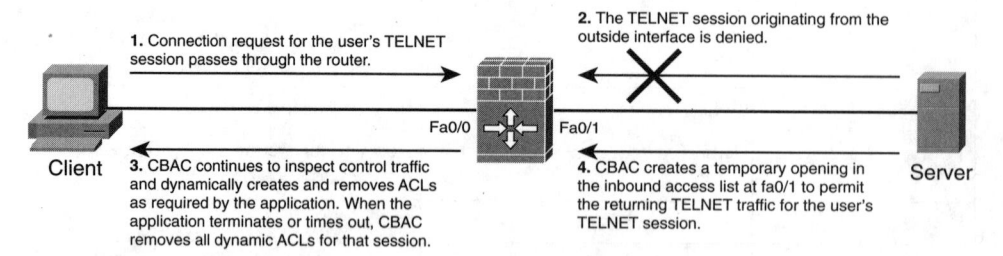

In Figure 15-2, the inbound ACL at Fa0/1 is configured to block Telnet traffic, and no outbound ACL is configured at Fa0/0. When the connection request for User1's Telnet session passes through the Cisco IOS Firewall, CBAC creates a temporary opening in the inbound ACL at Fa0/1 to permit returning Telnet traffic for User1's Telnet session.

CBAC inspects and monitors only the control channels of connections, not the data channels. For example, during FTP sessions, both the control and data channels, which are created when a data file is transferred, are monitored for state changes. However, only the control channel is inspected.

CBAC inspection recognizes application-specific commands in the control channel, and detects and prevents certain application-level attacks.

Whenever a packet is inspected, a state table is updated to include information about the state of the packet's connection. The traffic permitted back through the Cisco IOS Firewall is composed of packets that have a permissible session from the state table.

The following is an example of CBAC inspection process of passive FTP:

1. FTP client sends out a synchronize start (SYN) packet on the control channel:

    Client ----SYN----→ Server

2. The firewall validates the rule set for permitting the connection and creates a hole in the ACL:

    Client ←----SYN ACK---- Server

3. TCP handshake is completed:

    Client ----ACK----→ Server

4. When the **ls** command is entered, the FTP client sends **PASV** and **LIST** commands to the server:

   Client ----**PASV**----→ Server

   Client ←-----Address/port info---- Server

   On seeing the address/port in the reply to the **PASV** command, the firewall creates a pregenerated session and ACL holes. The holes point from the client to the server because the client will try to connect the server to create the data channel as per Passive FTP specifications.

5. The FTP client sends the SYN for this data connection:

   Client ----SYN DAT----→ Server

6. On seeing the SYN packet, the firewall creates holes that allow a synchronize acknowledge (SYN ACK) reply from the server:

   Client ←-----SYN ACK----- Server

   The holes can take from 5 to 10 seconds to create. At least three packets are exchanged between the client and the server when the user sends the **ls** command:

   — PASV

   — Reply to **PASV** with address/port information

   — SYN to the new address/port

UDP Sessions

Unlike TCP sessions, UDP sessions are connectionless. This characteristic of UDP sessions makes it harder to identify packets that belong to the same session. In these cases, CBAC uses source/destination addresses and port numbers and whether the packet was detected soon after another similar UDP packet to determine whether the packet belongs to that particular session. "Soon" means within the configurable UDP idle timeout period.

ACL Entries

Cisco IOS Software Release 12.3(4)T provides a new feature called *ACL bypassing*. Prior to this feature, a packet could go through as many as three redundant searches, one input ACL search, one output ACL search, and one inspection session search. When the ACL bypassing feature was introduced, it eliminated the necessity for input and output dynamic ACL searches. This is because every time CBAC creates a dynamic ACL for a single session, it is implied that the packets that match the ACL entry are already part of an existing session, and therefore allowed through. ACL bypassing subjects the packet to one search only, the session inspection. ACL bypassing is performed by default.

314   Chapter 15: Context-Based Access Control

The temporary ACL entries are never saved to nonvolatile RAM (NVRAM).

Handling Half-Open Sessions

A half-open session is a session open in only one direction. For TCP traffic, this might mean that the session did not complete the three-way handshake. For UDP traffic, this means that return traffic was not detected.

CBAC can measure the rate and the active number of half-open sessions several times per minute. If the number of sessions or the rate of the new connection attempts increases above the threshold, the software deletes the half-open sessions. This deletion process continues until the rate drops below the threshold.

You can set the thresholds for the number and rate of half-open sessions by using the **max-incomplete high**, **max-incomplete low**, **one-minute high**, and **one-minute low** commands.

CBAC Restrictions

CBAC has the following restrictions:

- Packets with the firewall as the source or destination address are not inspected by CBAC.

- If you reconfigure your ACLs when you configure CBAC, be aware that if your ACLs block TFTP traffic into an interface, you will not be able to netboot over that interface. (This limitation is not CBAC specific, but is part of existing ACL functionality.)

- CBAC is available for IP and limited ICMP protocol traffic. Only TCP, UDP, and most common ICMP type packets are inspected. Other IP traffic, such as routing protocol advertisements, cannot be inspected with CBAC and should be filtered with extended ACLs instead.

- H.323v2 and Real Time Streaming Protocol (RTSP) inspection supports only the following multimedia client/server applications: Cisco IP/TV, RealNetworks RealAudio G2 Player, and Apple QuickTime 4.

Supported Protocols

You can configure CBAC to inspect all TCP and UDP sessions. You can also configure CBAC to specifically inspect certain application layer protocols. You can configure all the following application layer protocols for CBAC:

- TCP sessions, regardless of the application layer protocol (sometimes called "single-channel" or "generic" TCP inspection)

- UDP sessions, regardless of the application layer protocol (sometimes called "single-channel" or "generic" UDP inspection)

- FTP

- Simple Mail Transfer Protocol (SMTP)

Context-Based Access Control Features   315

- SQL*Net
- TFTP
- UNIX R commands (such as **rlogin**, **rexec**, and **rsh**)
- RPC (Sun RPC, not DCE RPC or Microsoft RPC)
- HTTP (Java blocking)
- Java
- Microsoft NetShow
- RealAudio
- StreamWorks
- VDOLive
- CU-SeeMe (only the White Pine version)
- H.323 (such as NetMeeting, ProShare)
- RTSP

> **NOTE** Support for Extended Simple Mail Transport Protocol (ESMTP) is not available in CBAC. To determine whether a mail server is using SMTP or ESMTP, Telnet to the mail server port 25 and observe the banner for SMTP or ESMTP specification. You can also consult the mail server software vendor.

When a protocol is configured for CBAC, the protocol's traffic will be inspected, state information will be maintained, and in general, packets will be allowed back through the Cisco IOS Firewall only if they belong to a permissible session. When CBAC inspects FTP traffic, it only allows data channels with the destination port in the range of 1024 to 65,535. It will not open a data channel if the FTP client/server authentication fails.

CBAC supports multimedia application inspection of RTSP and H.323v2.

RTSP

RTSP is an application-level protocol used for control over the delivery of data with real-time properties, such as audio and video streams. As defined in Internet Engineering Task Force (IETF) RFC 2326, RTSP may run over UDP, multicast UDP, or TCP. Currently, CBAC supports only TCP-based RTSP, which is supported by a variety of multimedia vendor products, including Cisco IP/TV, RealAudio G2 Player, Apple QuickTime 4 software, and RealNetworks.

RTSP may use many different data transport modes. CBAC supports the following data transport modes:

- **RealNetworks Real Data Transport (RDT)**—RDT is a proprietary protocol developed by RealNetworks for data transport. It uses RTSP for communications control and RDT for data connection and retransmission of lost packets.

- **Interleaved (tunnel mode)**—RTSP uses the control channel to tunnel RTP or RDT traffic.

- **Synchronized Multimedia Integration Language (SMIL)**—SMIL is a simple HTML-like language that enables the creation of interactive multimedia presentations consisting of multiple elements of music, voice, images, text, video, and graphics. It is a proposed specification of the World Wide Web Consortium (W3C).

- **Real-Time Transport Protocol (RTP)**—RTP is the transport protocol for RTSP. RTP uses the UDP packet format for transmitting real-time data. It may be used in conjunction with Real-time Transport Control Protocol (RTCP) to get feedback on quality of data transmission and information about participants in a streaming session.

An RTSP client uses TCP port 554 or 8554 to open a connection to a multimedia server. The data port between the client and server is negotiated between the client and the server on UDP ports between 1024 and 65,536. RTP and RTCP ports work in pairs, with RTP using an even-numbered port and RTCP using the next consecutive port.

CBAC uses the client port and connection information to create dynamic entries in the ACL. As the connections are terminated, CBAC removes the dynamic entries from the ACL.

H.323

CBAC supports the following H.323 inspection:

- **H.323v1**—When a TCP connection is created between the client and server (H.225), a separate channel for media control (H.245) is opened through which multimedia channels for audit and video are further negotiated.

- **H.323v2**—Provides additional options, including "fast start." The fast start option minimizes delay between the time the user initiates a connection and the time that the user gets data. This version is backward-compatible with version 1 inspection. The client opens a connection to the server on port 1720. The data channel between the client and the server is dynamically negotiated using UDP ports between 1024 and 65,536.

CBAC uses the client port and connection information to create dynamic entries in the ACL. As the connections are terminated, CBAC removes the dynamic entries from the ACL.

CPU and Performance Impact

Even though CBAC uses slightly less than 600 bytes of memory per connection, it can still impact the efficiency of your router. Because of this memory usage, you should use CBAC only when you need to. In addition, a slight amount of additional processing occurs whenever packets are inspected. Cisco IOS Software Release 12.2(8)T provides improvements to the Cisco IOS Firewall performance. The following sections provide an overview of these improvements.

Throughput Performance

CBAC uses a hash table to search for packets that belong to a specific session. It is possible to have entries for a certain pattern of addresses hashed into the same bucket, hence causing collisions in the table. This results in poor performance of hash function distribution. A small hash table size might negatively impact throughput performance when a large number of sessions are active.

The **ip inspect hashtable** command enables users to dynamically change the size of the session hash table without having to reload the router. Increasing the size of the hash table allows for the reduction of the number of sessions per hash bucket, which would improve the throughput performance of the CBAC engine.

Session Connection Improvements

One of the Cisco IOS Software performance improvements is to address CBAC limitations in the number of TCP connections per second. Previously, CBAC would send all packets for a session to the process switch during the TCP connection setup and teardown, which slowed down the processing of packets significantly. The new version of Cisco IOS Software sends only the first packet of any connection to the process-switching path, while the engine in the fast path processes the rest of the packets. This arrangement speeds up the processing of the packets and leads to improved performance.

CPU Utilization Improvements

You can measure the CPU utilization of the router running CBAC while a specific throughput or connections-per-second metric is maintained. This improvement is used in conjunction with the throughput and connections-per-second metrics.

Improvement Benefits

The improvements with the new release of Cisco IOS Software provide the following benefits:

- **Layer 4 processing**—The code path for connection initiation and teardown was rewritten, and enables quicker creation of connections per second and reduces CPU utilization per connection.

- **Hash table function**—Dynamic configuration of the hash table size, from 1K to 8K, ensures better distribution of the hash function.

- **Application module tuning**—By improving the connections per seconds for a session, it enhances the performance of the associated application.

Configuring CBAC

To configure CBAC, follow these steps:

Step 1   Select an interface.

Step 2   Configure IP ACLs at the interface.

Step 3   Configure global timeouts and thresholds.

Step 4   Define an inspection rule.

Step 5   Apply the inspection rule to an interface.

Step 6   Configuring logging, audit trail, and other guidelines.

Selecting an Interface

This step is the planning stage, from which the rest of the configuration for CBAC will work. You first must determine on what (an internal or external interface) you want to configure CBAC. You can also configure CBAC on both internal and external interfaces. Configure CBAC in two directions when the networks on both sides of the firewall require protection, such as with extranet or intranet configurations, and for protection against DoS attacks.

Configuring IP ACLs at the Interface

Configuring your ACL correctly is critical for CBAC to work properly. Follow these two general rules when evaluating your IP ACLs at the Cisco IOS Firewall:

- Permit CBAC traffic leaving the network through the Cisco IOS Firewall.

- Use extended ACLs to deny traffic entering the network (from the external interface) through the Cisco IOS Firewall.

All ACLs that evaluate traffic leaving the protected network should permit traffic that will be inspected by CBAC. If Telnet will be inspected by CBAC, for example, Telnet traffic should be permitted on all ACLs that apply to traffic leaving the network.

Configuring Global Timeouts and Thresholds

Global timeouts and thresholds help CBAC determine how long to manage state information for a session and when to drop sessions that do not become fully established. Table 15-3 lists all the available CBAC timeouts and thresholds along with the corresponding command and default value.

Table 15-3 *Default Timeout and Threshold Values for CBAC Inspections**

Timeout or Threshold Value to Change	Command	Default
The length of time the software waits for a TCP session to reach the established state before dropping the session	**ip inspect tcp synwait-time** *seconds*	30 seconds
The length of time a TCP session will still be managed after the firewall detects a FIN exchange	**ip inspect tcp finwait-time** *seconds*	5 seconds
The length of time a TCP session will still be managed after no activity (the TCP idle timeout)	**ip inspect tcp idle-time** *seconds*	3,600 seconds (1 hour)
The length of time a UDP session will still be managed after no activity (the UDP idle timeout)	**ip inspect udp idle-time** *seconds*	30 seconds
The length of time a Domain Name System (DNS) name lookup session will still be managed after no activity	**ip inspect dns-timeout** *seconds*	5 seconds
The number of existing half-open sessions that will cause the software to start deleting half-open sessions	**ip inspect max-incomplete high** *number*	500 existing half-open sessions
The number of existing half-open sessions that will cause the software to stop deleting half-open sessions	**ip inspect max-incomplete low** *number*	400 existing half-open sessions
The rate of new unestablished sessions that will cause the software to start deleting half-open sessions	**ip inspect one-minute high** *number*	500 half-open sessions per minute
The rate of new unestablished sessions that will cause the software to stop deleting half-open sessions	**ip inspect one-minute low** *number*	400 half-open sessions per minute
The number of existing half-open TCP sessions with the same destination host address that will cause the software to start dropping half-open sessions to the same destination host address	**ip inspect tcp max-incomplete host** *number* **block-time** *minutes*	50 existing half-open TCP sessions 0 minutes

* Source: Cisco.com: Context-Based Access Control. This table has been reproduced by Cisco Press with the permission of Cisco Systems Inc. Copyright © 2003 Cisco Systems, Inc. All Rights Reserved.

Port to Application Mapping

Port to Application Mapping (PAM) is a feature of the Cisco IOS Firewall set enabling the mapping of TCP or UDP port numbers to network or application services. Such mapping proves useful in environments that run services on ports that are different from the registered or well-known ports associated with the network or application service. An example of this is using port 8080 rather than the well-known port 80 for HTTP services that run via a proxy server. You can enable this mapping for a single host, a subnet, or the whole network.

PAM establishes a table of default port-to-application mapping information. During the firewall startup, system-defined mapping information populates the PAM table. CBAC uses the information in the table to inspect traffic on nonstandard ports.

PAM has three types of mapping entries:

- **System-defined port mapping**—System-defined entries are all the services using the well-known or registered port mapping. The mapping information cannot be deleted or changed. For example, SMTP services may not be moved to the H.323 port 1720 or vice versa.

- **User-defined port mapping**—For CBAC to inspect any application or network service that is using a nonstandard port, a user-defined entry must exist in the PAM table. It is possible to map a service to multiple ports. For example, if Telnet is running on port 5000 and 5001 rather than the default port 23, you can use PAM to map port 5000 and 5001 to Telnet services.

- **Host-specific port mapping**—Port mapping information may be mapped to specific hosts or subnets using host-specific port mapping. For example, hosts on subnet 192.168.10.0/24 may run Telnet services on port 5000, while other subnets on a network continue to use the default port 23.

PAM Configuration Task List

To configure PAM, you need to configure a standard access list prior to configuring PAM. The **ip port-map** command, in global configuration mode, enables you to associate a port with an application. Example 15-2 shows the mapping of Telnet services for host 192.168.10.1 to port 5000.

Example 15-2 *Host Telnet Service Mapping*

```
Router#configure terminal
Router(config)#access-list 1 permit 192.168.10.1
Router(config)#ip port-map telnet port 5000 list 1
```

Defining an Inspection Rule

The inspection rule defines the IP traffic monitored by CBAC. The **ip inspect name** command enables you to define a set of inspection rules. Table 15-4 shows the **ip inspect** command parameters:

```
ip inspect name inspection-name protocol [alert {on | off}]
   [audit-trail {on | off}] [timeout seconds]
   no ip inspect name [inspection-name protocol]
```

Table 15-4 **ip inspect name** *Command Parameters*

Parameter	Description
inspection-name	Names the set of inspection rules. If you want to add a protocol to an existing set of rules, use the same *inspection-name* as the existing set of rules.
protocol	A protocol keyword listed in (FTP, Java, SMTP, and so on).
**alert {on	off}**	(Optional) For each inspected protocol, the generation of alert messages can be set be on or off. If the **no** option is selected, alerts are generated based on the setting of the **ip inspect alert-off** command.
**audit-trail {on	off}**	(Optional) For each inspected protocol, the audit trail can be set on or off. If the **no** option is selected, audit trail messages are generated based on the setting of the **ip inspect audit-trail** command.
http	(Optional) Specifies the HTTP protocol for Java applet blocking.
timeout *seconds*	(Optional) To override the global TCP or UDP idle timeouts for the specified protocol, specify the number of seconds for a different idle timeout. This timeout overrides the global TCP and UPD timeouts but does not override the global DNS timeout.
java-list *access-list*	(Optional) Specifies the ACL (name or number) to use to determine "friendly" sites. This keyword is available only for the HTTP protocol for Java applet blocking. Java blocking only works with standard ACLs.
rpc program-number *number*	Specifies the program number to permit. This keyword is available only for the RPC protocol.
wait-time *minutes*	(Optional) Specifies the number of minutes to keep a small hole in the firewall to allow subsequent connections from the same source address and to the same destination address and port. The default wait time is 0 minutes. This keyword is available only for the RPC protocol.
fragment	Specifies fragment inspection for the named rule.
max *number*	(Optional) Specifies the maximum number of unassembled packets for which state information (structures) is allocated by Cisco IOS Software. Unassembled packets are packets that arrive at the Cisco IOS Firewall interface before the initial packet for a session. The acceptable range is from 50 to 10,000. The default is 256 state entries. Memory is allocated for the state structures, and setting this value to a larger number might cause memory resources to be exhausted.

Table 15-4 **ip inspect name** *Command Parameters*

timeout *seconds* (fragmentation)	(Optional) Configures the number of seconds that a packet state structure remains active. When the timeout value expires, the Cisco IOS Firewall drops the unassembled packet, freeing that structure for use by another packet. The default timeout value is 1 second. If this number is set to a value greater that 1 second, it is automatically adjusted by the Cisco IOS Software when the number of free state structures goes below certain thresholds: When the number of free states is less than 32, the timeout is divided by 2; when the number of free states is less than 16, the timeout is set to 1.

Usually, one inspection rule is defined per interface. Sometimes, however, you might want to configure an inspection rule in both directions on a single firewall interface. In these situations, you should configure two rules, one for each direction. The inspection rule includes a series of statements each listing a protocol and specifying the same inspection rule name.

Configuring Generic TCP and UDP Inspection

To configure CBAC inspection for TCP or UDP packets, use one or both of the following global configuration commands:

- **ip inspect name** *inspection-name* **tcp** [**timeout** *seconds*]

- **ip inspect name** *inspection-name* **udp** [**timeout** *seconds*]

With TCP and UDP inspection, packets entering the network must match the corresponding packet that previously exited the network. The entering packets must have the same source/destination addresses and source/destination port numbers as the exiting packet (but reversed); otherwise, the entering packets are blocked at the interface.

With UDP inspection configured, replies are only permitted back in through the firewall if they are received within a configurable time after the last request was sent out. (You configure this time with the **ip inspect udp idle-time** command.)

Configuring Java Inspection

To reduce the threats from malicious Java applets, you can configure CBAC to filter Java applets at the firewall. This filtering allows users to download only applets residing within the firewall and permitted applets from outside the firewall:

```
ip inspect name inspection-name http [java-list access-list] [alert {on | off}]
    [audit-trail {on | off}] [timeout seconds]
```

Example 15-3 shows an inspection named test1 for blocking Java applets.

Example 15-3 **Sample ip inspect** *Command Used for Java Inspection*

```
ip inspect name test1 http java-list 3 audit-trail on
```

> **NOTE** CBAC does not detect or block encapsulated Java applets. Java applets contained in ZIP or JAR format are not blocked at the firewall. CBAC also does not detect or block applets loaded from FTP, gopher, and HTTP on a nonstandard port.

> **NOTE** Java blocking forces a strict order on TCP packets. To properly verify that Java applets are not in the response, a firewall drops any TCP packet that is out of order. Because the network—not the firewall—determines how packets are routed, the firewall cannot control the order of the packets; the firewall can only drop and retransmit all TCP packets that are not in order.

Applying the Inspection Rule to an Interface

To apply an inspection rule to an interface, use the **ip inspect** *inspection-name* {**in** | **out**} command in interface configuration mode. In the following example, inspection test1 is applied to the Ethernet interface in an inward direction:

```
Router(config)#interface Ethernet0
Router(config-int)#ip inspect test1 in
```

Verifying and Debugging CBAC

You can check your CBAC configuration by using the **show** commands listed in Table 15-5. In addition to the **show** commands, the **debug** commands prove very useful in monitoring and troubleshooting your CBAC configuration.

Table 15-5 *Some of the Commands Used to Check Your CBAC Configuration*

Command	Purpose
show ip access-lists	Displays the contents of all current IP access lists
show ip inspect name *inspection-name*	Shows a particular configured inspection rule
show ip inspect config	Shows the complete CBAC inspection configuration
show ip inspect interfaces	Shows interface configuration with regard to applied inspection rules and access lists
show ip inspect session [detail]	Shows existing sessions currently being tracked and inspected by CBAC
show ip inspect all	Shows all CBAC configurations and all existing sessions currently being tracked and inspected by CBAC

Debugging CBAC

The following three types of **debug** commands are available for debugging CBAC:

- Generic **debug** commands

- Transport-level **debug** commands

- Application protocol **debug** commands

To assist CBAC debugging, you can turn on audit trail messages that will display on the console after each CBAC session closes. To turn on audit trail messages, use the following global configuration command:

```
ip inspect audit trail
```

Generic debug Commands

You can use the following generic **debug** commands listed in Table 15-6.

Table 15-6 *Generic* **debug** *Commands*

Command	Purpose
debug ip inspect function-trace	Displays messages about software functions called by CBAC.
debug ip inspect object-creation	Displays messages about software objects being created by CBAC. Object creation corresponds to the beginning of CBAC-inspected sessions.
debug ip inspect object-deletion	Displays messages about software objects being deleted by CBAC. Object deletion corresponds to the closing of CBAC-inspected sessions.
debug ip inspect events	Displays messages about CBAC software events, including information about CBAC packet processing.
debug ip inspect timers	Displays messages about CBAC timer events, such as when a CBAC idle timeout is reached.
debug ip inspect detailed	Enables the **detailed** option, which can be used in combination with other options to get additional information.

Transport-Level debug Commands

To monitor and troubleshoot CBAC TCP and UDP inspection configuration, use the transport-level **debug** commands:

- **debug ip inspect tcp**—Display messages about CBAC-inspected TCP events, including details about TCP packets

- **debug ip inspect udp**—Display messages about CBAC-inspected UDP events, including details about UDP packets

Configuring CBAC Example

For this example, CBAC is being configured to inspect inbound. As shown in Figure 15-3, interface Fa0/0 is the protected network, and interface Fa0/1 is the unprotected network. The security policy for the protected site uses ACLs to restrict inbound traffic on the unprotected interface to specific ICMP traffic, denying inbound access for TCP and UDP traffic. Inbound access for specific protocol traffic is provided through dynamic ACLs, which are generated according to CBAC inspection rules.

Figure 15-3 *Sample Application of CBAC*

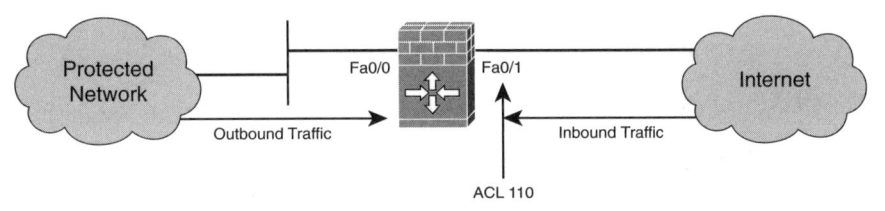

ACL 110 denies TCP and UDP traffic from any source or destination while permitting specific ICMP traffic. The final **deny** statement is not required but is included. The final entry in any ACL is an implicit denial of all IP traffic:

```
Firewall(config)#access-list 110 deny tcp any any
Firewall(config)#access-list 110 deny udp any any
Firewall(config)#access-list 110 permit icmp any any echo-reply
Firewall(config)#access-list 110 permit icmp any any time-exceeded
Firewall(config)#access-list 110 permit icmp any any traceroute
Firewall(config)#access-list 110 permit icmp any any unreachable
Firewall(config)#access-list 110 deny ip any any
```

ACL 110 is applied inbound at interface Fa0/1 to block all access from the unprotected network to the protected network:

```
Firewall(config)#interface Fa0/1
Firewall(config-if)#ip access-group 110 in
```

An inspection rule is created for Intranet users:

```
Firewall(config)#ip inspect name intranetusers ftp
Firewall(config)#ip inspect name intranetusers http
Firewall(config)#ip inspect name intranetusers rcmd
Firewall(config)#ip inspect name intranetusers realaudio
Firewall(config)#ip inspect name intranetusers smtp timeout 3600
Firewall(config)#ip inspect name intranetusers tftp timeout 30
Firewall(config)#ip inspect name intranetusers udp timeout 15
Firewall(config)#ip inspect name intranetusers tcp timeout 3600
```

The inspection rule is applied inbound at interface Ethernet0 to inspect traffic from users on the protected network:

```
Firewall(config)#interface Fa0/0
Firewall(config-if)#ip inspect intranetusers in
```

Foundation Summary

The "Foundation Summary" section of each chapter lists the most important facts from the chapter. Although this section does not list every fact from the chapter that will be on your SNRS exam, a well-prepared candidate should at a minimum know all the details in each "Foundation Summary" before going to take the exam.

- CBAC is one of the key features in Cisco IOS Firewall.

- The CBAC engine provides secure, per-application access control across network perimeters.

- CBAC intelligently filters TCP and UDP packets based on application layer protocol session information and can be used for intranets, extranets, and the Internet.

- Even though CBAC uses slightly less than 600 bytes of memory per connection, it can still impact the efficiency of your router. Because of this memory usage, you should use CBAC only when you need to.

- ACLs are created and deleted by CBAC dynamically at the Cisco IOS Firewall interfaces, according to the information maintained in the state tables.

- The three types of debugging command for CBAC are generic, transport-level, and application-level **debug** commands.

- To assist with debugging, you can enable audit trail messages that will display on the console after each CBAC session closes.

- CBAC is available only for IP traffic. Only TCP, UDP, and ICMP common type packets are inspected. Other IP traffic, such as routing protocols, cannot be inspected with CBAC and should be filtered with basic ACLs instead.

Q&A

As mentioned in the section "How to Use This Book" in the Introduction, you have two choices for review questions. The questions that follow present a bigger challenge than the exam itself because they use an open-ended question format. By reviewing using this more difficult format, you can exercise your memory better and prove your conceptual and factual knowledge of this chapter. You can find the answers to these questions in the appendix.

For more practice with exam-like question formats, including questions using a router simulator and multiple-choice questions, use the exam engine on the CD-ROM.

1. What are the steps in the CBAC configuration process?
2. Are inspection rules a requirement for CBAC configuration?
3. What three categories of **debug** commands are commonly used to debug CBAC configuration?
4. Can CBAC be configured to inspect all TCP, UDP, and ICMP packets?
5. Which command enables you to show a complete CBAC inspection configured on the Cisco IOS Firewall?
6. Which command do you use to turn on audit trail messages?
7. What indicators in half-open sessions does CBAC measure before it takes steps to prevent a DoS attack?
8. Does CBAC block malicious Java applets that are in JAR format?
9. Name two features of the CBAC.
10. Name one restriction with using CBAC.
11. What are half-open sessions, and how does CBAC mitigate to many half-open sessions?
12. RTSP may use many different data transport modes. What transport modes does CBAC support?
13. What performance improvements does the new release of Cisco IOS Software CBAC provide?

This chapter covers the following subjects:

- Understanding Authentication Proxy

- Authentication Proxy and the Cisco IOS Firewall

- Configuring Authentication Proxy on the Cisco IOS Firewall

- Using Authentication Proxy with TACACS+

- Using Authentication Proxy with RADIUS

- Limitations of Authentication Proxy

CHAPTER 16

Authentication Proxy and the Cisco IOS Firewall

Authentication proxy is a function that enables users to authenticate via the firewall when accessing specific resources. The Cisco IOS Firewall is designed to interface with authentication, authorization, and accounting (AAA) servers using standard authentication protocols to perform this function. This design enables administrators to create a granular and dynamic per-user security policy. This chapter discusses authentication proxy and how it is used to authenticate both inbound and outbound connections. The Cisco IOS Firewall supports TACACS+ and RADIUS AAA servers. The configuration steps for authentication proxy using TACACS+ or RADIUS are covered in this chapter along with the introduction to the Cisco Secure Access Control Server (Cisco Secure ACS), which can perform both TACACS+ and RADIUS functions. Because no solution can support the needs of every implementation, it is important to understand the limitations of authentication proxy.

"Do I Know This Already?" Quiz

The purpose of the "Do I Know This Already?" quiz is to help you decide whether you really need to read the entire chapter. If you already intend to read the entire chapter, you do not necessarily need to answer these questions now.

The 10-question quiz, derived from the major sections in the "Foundation Topics" portion of the chapter, helps you determine how to spend your limited study time.

Table 16-1 outlines the major topics discussed in this chapter and the "Do I Know This Already?" quiz questions that correspond to those topics.

Chapter 16: Authentication Proxy and the Cisco IOS Firewall

Table 16-1 *"Do I Know This Already?" Foundation Topics Section-to-Question Mapping*

Foundation Topics Section	Questions Covered in This Section
Understanding Authentication Proxy	7
Authentication Proxy and the Cisco IOS Firewall	3
Configuring Authentication Proxy on the Cisco IOS Firewall	1, 4
Using Authentication Proxy with TACACS+	8
Using Authentication Proxy with RADIUS	5
Limitations of Authentication Proxy	2, 6, 9, 10

> **CAUTION** The goal of self-assessment is to gauge your mastery of the topics in this chapter. If you do not know the answer to a question or are only partially sure of the answer, you should mark this question wrong for purposes of the self-assessment. Giving yourself credit for an answer you correctly guess skews your self-assessment results and might provide you with a false sense of security.

1. What four steps enable you to complete the connection from the source to the destination using the authentication proxy?

    a. Source initiates a connection through the firewall.

    b. Source is validated for current authentication rights.

    c. Send login prompt to the user.

    d. Username/password is validated against the AAA server.

    e. Source must reestablish connection.

    f. All of the above.

2. The authentication proxy is compatible with the following IOS security features except for which of the following?

    a. Context-Based Access Control

    b. IPSec

    c. VPN client

    d. Network Address Translation

    e. Cisco IOS Firewall Intrusion Prevention System (IPS)

    f. All of the above

3. What step is not a part of the process of configuring the authentication proxy on the Cisco IOS Firewall?

   a. Configure AAA.
   b. Configure the HTTP server.
   c. Configure the access list for the authentication proxy.
   d. Configure the authentication proxy.
   e. Verify the authentication proxy configuration.

4. What are the commands to configure the authentication proxy?

   a. **ip auth-proxy auth-cache-time 5**
   b. **ip auth-proxy auth-proxy-banner**
   c. **ip auth-proxy name myaccess http**
   d. **interface fe0**
   e. **ip auth-proxy myaccess**
   f. All of the above

5. Which three primary tasks are required to configure the Cisco Secure ACS as a TACACS+ or RADIUS server?

   a. Network configuration
   b. Privilege settings
   c. Interface configuration
   d. Group setup
   e. None of the above

6. Which is not a limitation of the authentication proxy?

   a. Triggers only on HTTP, HTTPS, and FTP connections.
   b. Client browsers must support JavaScript for authentication.
   c. Does not support concurrent usage to the same host at the same time.
   d. Load balancing using different AAA servers is not supported.
   e. Supports only traffic passing through the router.
   f. All of the above.

7. The authentication proxy supports access via which of the following ports? (Select four.)
   a. 443
   b. 8080
   c. 8081
   d. 23
   e. 80
   f. 21

8. Where is the Cisco IOS Firewall configured on the Cisco Secure ACS?
   a. Group Setup window
   b. Network Configuration window
   c. Interface Configuration window
   d. Command-line interface
   e. Client window
   f. None of the above

9. You can configure the Cisco IOS Firewall authentication proxy to utilize how many AAA servers?
   a. 32
   b. 255
   c. Unlimited
   d. 1
   e. None of the above

10. Authentication proxy support for user accounts prevents concurrent usage of user accounts.
    a. True
    b. False

The answers to the "Do I Know This Already?" quiz are found in the appendix. The suggested choices for your next step are as follows:

- **8 or less overall score**—Read the entire chapter. This includes the "Foundation Topics" and "Foundation Summary" sections and the "Q&A" section.

- **9 or 10 overall score**—If you want more review on these topics, skip to the "Foundation Summary" section and then go to the "Q&A" section. Otherwise, move on to Chapter 17, "Identity-Based Networking Services."

Foundation Topics

This chapter discusses authentication proxy and the Cisco IOS Firewall advanced security features. It explains how authentication proxy works and how it is used to increase the security of the Cisco IOS Firewall.

Understanding Authentication Proxy

Authentication proxy is one of the core components of the Cisco IOS Firewall feature set. Prior to the implementation of authentication proxy, access to a resource was normally restricted by the IP address of the requesting source, and a single policy was applied to that source or network using an access control list (ACL). There was no way to ensure that only authorized users had physical access to the workstation or that unauthorized users were not attempting to access a resource outside of their privilege level.

Authentication proxy enables administrators to restrict access to resources on a per-user basis and tailor the privileges of each individual instead of applying a generic policy to all users.

It is difficult to determine how authentication proxy will be addressed on the SNRS exam. At the time this writing, Cisco emphasized the importance of understanding "how" authentication proxy works. This does not mean that you should not be familiar with the commands used to configure authentication proxy, but you should certainly be very familiar with the mechanics of how authentication proxy functions and the steps required to implement it.

How Authentication Proxy Works

Unlike many Cisco IOS Firewall functions, authentication proxy is not a service that is transparent to the user. On the contrary, it requires user interaction. The authentication proxy is triggered when the user initiates an HTTP, HTTPS, FTP, or Telnet session through the Cisco IOS Firewall. The firewall checks to see whether the user has already been authenticated. If the user has previously authenticated, it allows the connection. If the user has not previously authenticated, the firewall prompts the user for a username and password and verifies the user input with a TACACS+ or RADIUS server.

The authentication proxy process entails three steps when a source initiates a connection to the destination on the other side of the Cisco IOS Firewall. "On the other side" in this context refers to internal sources initiating outbound connections and external hosts initiating inbound connections. Figure 16-1 depicts the steps required for an internal host to complete an outbound connection to an external destination.

Figure 16-1 *Internal Host Connection to External Destination*

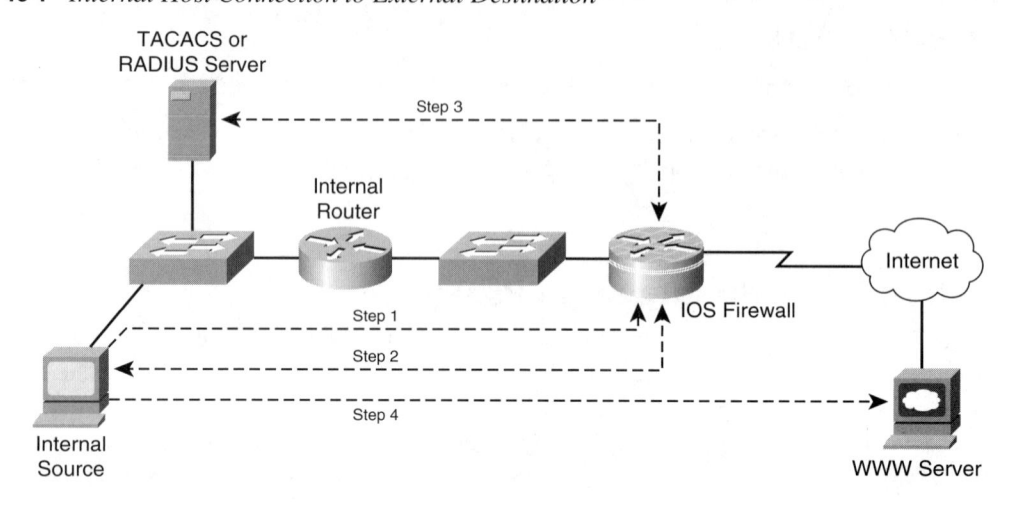

As Figure 16-1 shows, three steps are required to complete the connection from the source to the destination:

Step 1 The source host initiates an HTTP connection that is intended to pass through the Cisco IOS Firewall to reach its destination.

Step 2 The Cisco IOS Firewall checks to see whether the source has already been authenticated. If the source has not previously authenticated, the firewall sends a login prompt to the user.

Step 3 The user completes the username and password, and the Cisco IOS Firewall verifies the user account information with the AAA server.

Step 4 If the user provides the correct account information and is authenticated by the AAA server, the firewall allows the connection to complete.

Figure 16-2 shows that the same steps are required for communication in the opposite direction. It depicts a user on the internal network attempting to access a website on the Internet.

Figure 16-2 *External Host Connection to Internal Destination*

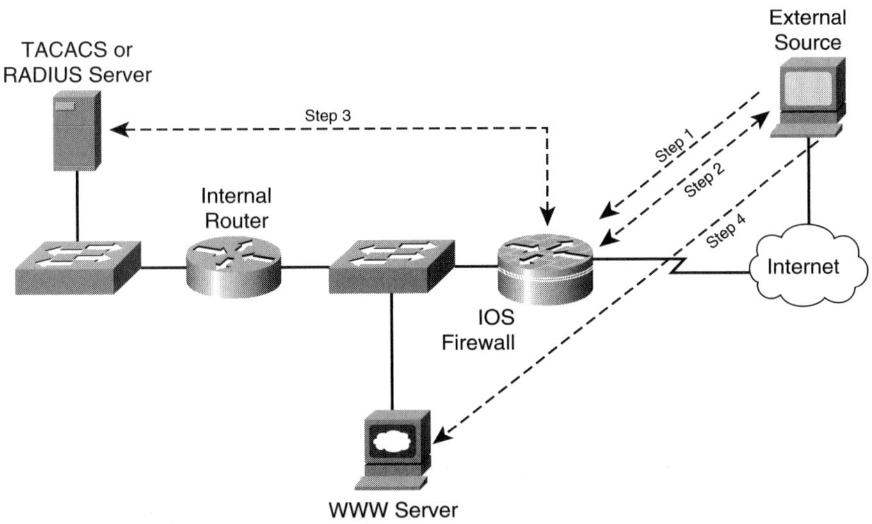

What Authentication Proxy Looks Like

When the user initiates the HTTP connection, the Cisco IOS Firewall checks to see whether the user has already been authenticated. If the user has not been previously authenticated, the firewall responds with an HTTP login page.

Figure 16-3 depicts the authentication proxy login page. The user must fill in the correct username and password to successfully authenticate and connect to the desired resource.

Figure 16-3 *Authentication Proxy Login Page*

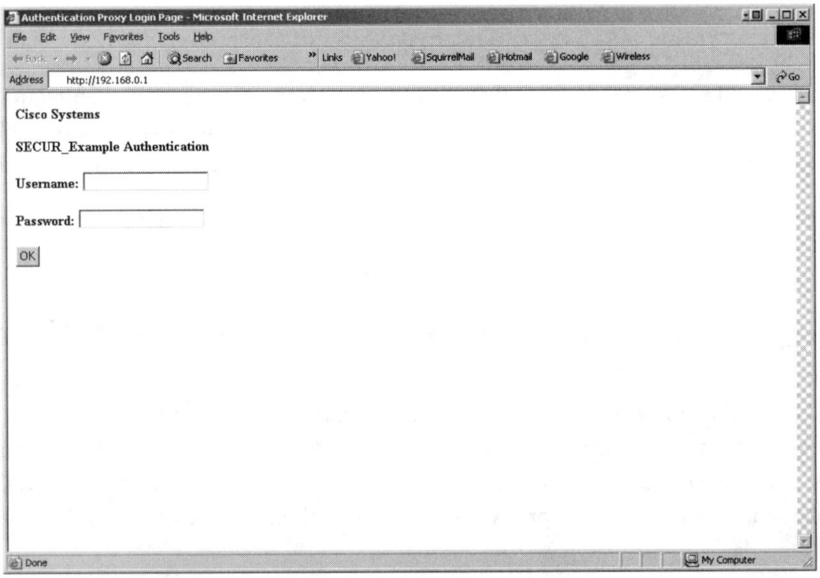

After successfully authenticating, the user sees a login success page similar to Figure 16-4.

Figure 16-4 *Successful Login Screen*

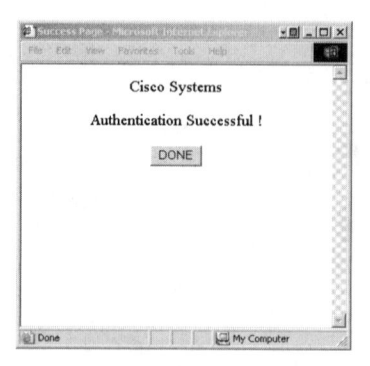

Authentication Proxy and the Cisco IOS Firewall

Authentication proxy is a feature that became available with Cisco IOS Software Release 12.0.5.T. Authentication proxy is compatible with the following Cisco IOS Software security features:

- **Context-Based Access Control (CBAC)**—CBAC was discussed in great detail in Chapter 15, "Context-Based Access Control." If you configure authentication proxy to work with CBAC, you can create dynamic access control entries. If you do not configure authentication proxy with CBAC, you need to reference static access lists on the Cisco IOS Firewall.

- **Network Address Translation (NAT)**—Enables you to translate internal addresses to external (normally public) addresses. If you are using authentication proxy on a firewall that is also performing NAT, you must also use CBAC to ensure that session translations do not conflict.

- **IPsec encryption**—Authentication proxy works transparently with IPsec encryption.

- **VPN client software**—Authentication proxy can be used for user authentication when creating a virtual private network (VPN) connection. This feature provides an additional level of security for administrators by authenticating the user before the encrypted connection is created.

- **Cisco IOS Firewall Intrusion Detection System (IDS)**—Authentication proxy works transparently with Cisco IOS Firewall IDS.

Configuring Authentication Proxy on the Cisco IOS Firewall

Authentication proxy enables users to connect through the firewall to a resource only after a AAA server has verified their credentials. After the authentication is complete, the Cisco IOS Firewall receives authorization information from the AAA server in the form of a dynamic access list. It is always a good idea to ensure that all traffic is properly flowing through the Cisco IOS Firewall prior to implementing authentication proxy. Access lists applied to the Cisco IOS Firewall determine the level of security (for example, what traffic requires authentication proxy). It is possible to require authentication proxy for all traffic or to limit the requirement only to specific sources or destinations. There are many different ways to configure authentication proxy, and each one is slightly different depending on the Cisco IOS Firewall services used and the direction the traffic is traveling in relation to the Cisco IOS Firewall. Cisco publishes specific configuration guides with examples for each type of configuration at http://www.cisco.com/en/US/products/sw/secursw/ ps1018/ prod_configuration_examples_list.html.

The authentication proxy configurations published by Cisco include the following:

- Authentication proxy inbound (no CBAC or NAT)
- Authentication proxy outbound (no CBAC or NAT)
- Authentication proxy inbound (with CBAC, but no NAT)
- Authentication proxy outbound (with CBAC, but no NAT)
- Authentication proxy inbound (with CBAC and NAT)
- Authentication proxy outbound (with CBAC and NAT)
- Authentication proxy inbound with IPsec and VPN client (no CBAC or NAT)
- Authentication proxy outbound with IPsec and VPN client (no CBAC or NAT)
- Authentication proxy inbound with IPsec and VPN client (with CBAC and NAT)
- Authentication proxy outbound with IPsec and VPN client (with CBAC and NAT)

This chapter focuses on configuring the Cisco IOS Firewall for inbound and outbound traffic without using CBAC, NAT, IPsec, or the VPN client.

Authentication Proxy Configuration Steps

You must complete a number of steps to configure authentication proxy on the Cisco IOS Firewall. Authentication proxy requires the firewall to communicate with many different systems, and each of these systems must be put into the firewall configuration. This section describes the configuration steps and individual commands used to configure the authentication proxy. There are examples of these configuration commands in the "Authentication Proxy Configuration Examples" section. It is important to understand the different steps, the commands within each step, and how they relate to each other. As with any other component configuration, you must understand how the commands relate to troubleshoot problems with the authentication proxy. For the purpose of the SNRS examination, it will most likely be critical to recognize which commands are missing or in the wrong syntax. Try to remember the four individual steps and consider how each step references the other.

This information will enable you to troubleshoot the configuration and determine which portion is not correctly configured. The configuration steps for the Cisco IOS Firewall are as follows:

Step 1 Configure AAA.

Step 2 Configure the HTTP server.

Step 3 Configure authentication proxy.

Step 4 Verify the authentication proxy configuration.

Step 1: Configure AAA

You must first configure the Cisco IOS Firewall to perform AAA functions. Doing so consists of seven individual Cisco IOS Software commands, listed in Table 16-2. You must enter each of these commands when in the global configuration mode.

Table 16-2 *IOS Commands Required to Configure AAA*

Command	Description
aaa new-model	Enables the AAA functionality on the Cisco IOS Firewall.
aaa authentication login default [*tacacs+* | *radius*]	Defines the authentication method to be utilized at login.
aaa authorization auth-proxy default [*method1* [*method2*...]]	The **auth-proxy** keyword in this command enables authentication proxy for that method of AAA authentication (for example, TACACS+ or RADIUS) and allows the router to download dynamic access control lists from the AAA server.
aaa accounting auth-proxy default start-stop group tacacs+	The **auth-proxy** keyword in this command activates authentication proxy accounting functions.
tacacs-server host *hostname* or **radius-server host** *hostname*	Identifies the AAA server by host name or IP address.
tacacs-server key *key* or **radius-server key** *key*	Configures the authentication and encryption key to ensure secure communication between the Cisco IOS Firewall and the AAA server.
access-list [*access-list-number*] **permit tcp host** *source* **eq tacacs** | **radius host** *destination*	Creates the access list to allow traffic from the AAA server to return to the Cisco IOS Firewall. The *source* address is the address of the AAA server, and the *destination* address is the address of the interface on the Cisco IOS Firewall that connects to the AAA server.

Step 2: Configure the HTTP Server

In Step 2, you configure the Cisco IOS Firewall to function as an HTTP server and set up the authentication method.

Table 16-3 lists and describes the commands required to configure the HTTP server. You must enter each of these commands when in the global configuration mode.

Table 16-3 *HTTP Server Configuration Commands*

Command	Description
ip http server	Enables the HTTP server on the Cisco IOS Firewall. The HTTP server is used by the Cisco IOS Firewall to send the login page to the client.
ip http secure-server	Enables the HTTP server on the Cisco IOS Firewall using Secure Sockets Layer (SSL). The Cisco IOS Firewall is accessed in the same manner as the HTTP server except it is accessed using HTTPS (SSL). Utilizing the secure server mitigates the vulnerabilities associated with allowing access to the Cisco IOS Firewall using the HTTP server.
ip http authentication aaa	Sets the HTTP server authentication method to AAA.
access-list *access-list-number* **deny any**	A standard access list must be created to deny any host.
ip http access-class *access-list-number*	Specifies the access list to be used by the HTTP server. The *access-list-number* that was created in the previous row is used to prevent any host from connecting directly to the HTTP server.

Step 3: Configure the Authentication Proxy

In Step 3, you configure authentication proxy on the Cisco IOS Firewall. Table 16-4 lists and describes commands and options used to configure authentication proxy on the Cisco IOS Firewall. You must enter each of these commands when in the global configuration mode.

Step 4: Verify the Authentication Proxy Configuration

After configuring authentication proxy, it is important that you verify the configuration. Doing so can greatly reduce troubleshooting if the users cannot connect because of an error in the configuration of the Cisco IOS Firewall. Use the **show ip auth-proxy configuration** command to display and verify the authentication proxy configuration.

Authentication Proxy Configuration Examples

The steps required to configure authentication proxy were listed and defined in the preceding section. In this section, authentication proxy is configured for both inbound and outbound connections through the Cisco IOS Firewall.

Table 16-4 *Authentication Proxy Configuration Commands*

Command	Description
ip auth-proxy auth-cache-time *min*	Sets the global authentication proxy idle timeout. All authentication entries and dynamic access lists are removed from the Cisco IOS Firewall when the idle timeout is exceeded. This value is in minutes, and the default value is 60 (minutes).
ip auth-proxy auth-proxy-banner	(Optional) Enables you to display the firewall name on the authentication proxy login page. This feature is disabled by default.
ip auth-proxy name *auth-proxy-name* **http** [**auth-cache-time** *min*] [**list** {**acl**	*acl-name*}]	Configures the individual authentication proxy rules. The [**auth-cache-time** *min*] portion of the command is optional and is used to specify the **auth-cache-time** for that specific rule instead of using the global configuration. The [**list** {**acl**	*acl-name*}] portion of the command is also optional and is used to specific access lists to the rule.
interface *type*	Specifies the interface type on which the authentication proxy is applied. This command also puts the Cisco IOS Firewall in the interface configuration mode.
ip auth-proxy *auth-proxy-name*	This command is run in interface configuration mode. It applies the authentication proxy rule (by the rule name) to the interface. It is important to ensure that you apply the authentication proxy rule to the correct interface. The rule must be applied to the first interface in the Cisco IOS Firewall that the request will hit. For example, the rule should be applied to the internal interface of the Cisco IOS Firewall for outbound traffic or the external interface for inbound traffic.

Figure 16-5 depicts the environment used for the configuration of authentication proxy on the 3640 Cisco IOS Firewall.

Figure 16-5 *Network Diagram of Authentication Proxy Source and Destination (External Host)*

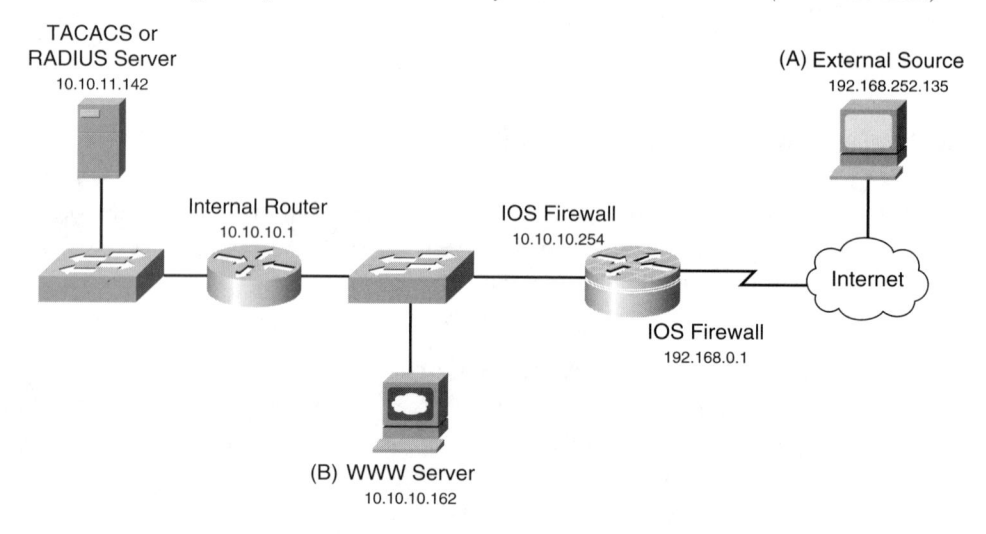

As Figure 16-5 illustrates, the source (A) is using the IP address 192.168.252.135, and the destination is the World Wide Web server (B) located on the internal network at 10.10.10.162. For the purpose of this exercise, NAT is not used for any address space. The Cisco IOS Firewall will require any external host attempting to access 10.10.10.162 to authenticate before allowing access.

> **NOTE** Remember that you are only using the 192.168.0.0/16 addresses to represent public Internet addresses. That address space is normally reserved per RFC 1918.

Example 16-1 depicts the configuration of the Cisco IOS Firewall to allow authentication proxy for sources requesting access to a resource on the internal network. Authentication proxy is not a bidirectional function; you must configure it for each direction that the user traffic should flow.

Example 16-1 *Configuring Inbound Authentication Proxy on the Cisco IOS Firewall*

```
Router1#configure terminal
! - - - Enable authentication on the Cisco IOS firewall
Router1(config)#aaa new-model
! - - - Define TACACS+ as the authentication method used for login
Router1(config)#aaa authentication login default group tacacs+
! - - - The auth-proxy keyword is used to enable authentication proxy for TACACS+
Router1(config)#aaa authorization auth-proxy default group tacacs+
! - - - Activate authentication proxy accounting
Router1(config)#aaa accounting auth-proxy default start-stop group tacacs+
! - - - Define the AAA server
Router1(config)#tacacs-server host 10.10.11.142
! - - - Define the key for encryption between the AAA server and the Cisco IOS firewall
Router1(config)#tacacs-server key abc123
! - - - Create an access list to allow traffic from the AAA server back to the router
Router1(config)#access-list 103 permit tcp host 10.10.11.142 eq tacacs host 10.10.10.254
! - - - Enable the HTTP server on the Cisco IOS firewall
Router1(config)#ip http server
! - - - Set the authentication to AAA
Router1(config)#ip http authentication aaa
! - - - Create a standard access list denying all traffic
Router1(config)#access-list 22 deny any
! - - - Define standard access list 22 for the HTTP server
Router1(config)#ip http access-class 22
! - - - Define the global authentication timeout to 30 minutes
Router1(config)#ip auth-proxy auth-cache-time 30
! - - - Display the firewall name on the login page
Router1(config)#ip auth-proxy auth-proxy-banner
! - - - Create the auth-proxy rules with the name allowed-inbound
Router1(config)#ip auth-proxy name allowed-inbound http
! - - - Enter the interface configuration mode
Router1(config)#interface s0
! - - - Configure the IP address of the interface
```

continues

Example 16-1 *Configuring Inbound Authentication Proxy on the Cisco IOS Firewall (Continued)*

```
Router1(config-if)#ip address 192.168.0.1 255.255.255.0
! - - - Apply the named auth-proxy rule to the interface
Router1(config-if)#ip auth-proxy allowed-inbound
! - - - Exit the interface configuration mode
Router1(config)#CTL-Z
Router1(config)#
```

Next, you configure the Cisco IOS Firewall for an internal source and an external destination. Figure 16-6 depicts the network with an internal source and external destination.

Figure 16-6 *Network Diagram of Authentication Proxy Source and Destination (Internal Host)*

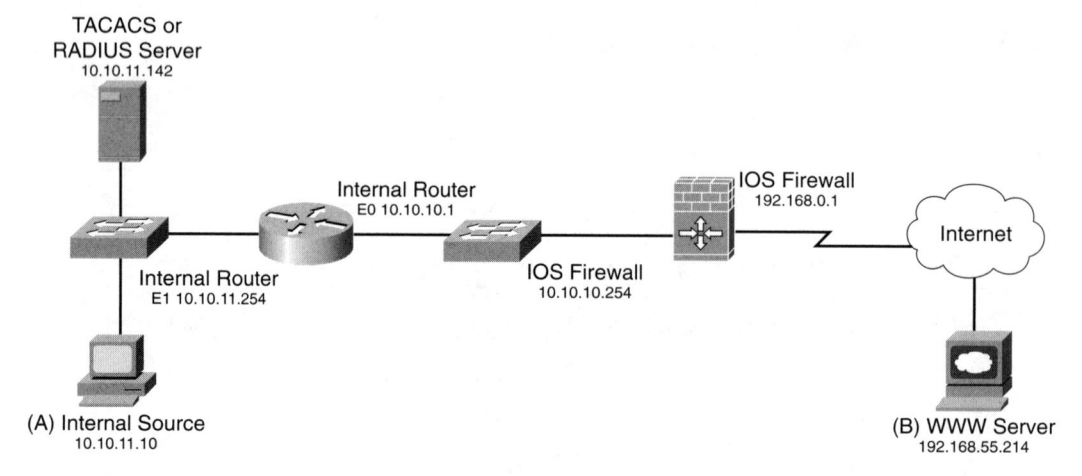

As Figure 16-6 illustrates, the source (A) is located on the internal network using the IP address 10.10.11.10, and the destination is a World Wide Web server (B) located on the Internet at 192.168.55.214. Again, this exercise does not use NAT for any address space. The Cisco IOS Firewall requires any internal host attempting to access the Internet to authenticate before allowing access.

> **NOTE** Remember that you are using only the 192.168.0.0/16 addresses to represent public Internet addresses. That address space is normally reserved per RFC 1918.

Example 16-2 depicts the configuration of the perimeter router to perform authentication proxy for sources on the internal network attempting to access resources on the Internet. Again, authentication proxy is not a bidirectional function; it must be configured for each direction that the user traffic should flow.

Example 16-2 *Configuring Outbound Authentication Proxy on the Perimeter Router*

```
Router1#configure terminal
! - - - Enable authentication on the Cisco IOS firewall
Router1(config)#aaa new-model
! - - - Define TACACS+ as the authentication method used for login
Router1(config)#aaa authentication login default group tacacs+
! - - - The auth-proxy keyword is used to enable authentication proxy for TACACS+
Router1(config)#aaa authorization auth-proxy default group tacacs+
! - - - Activate authentication proxy accounting
Router1(config)#aaa accounting auth-proxy default start-stop group tacacs+
! - - - Define the AAA server
Router1(config)#tacacs-server host 10.10.11.142
! - - - Define the key for encryption between the AAA server and the Cisco IOS firewall
Router1(config)#tacacs-server key abc123
! - - - Create an access list to allow traffic from the AAA server back to the router
Router1(config)#access-list 103 permit tcp host 10.10.11.142 eq tacacs host 10.10.10.254
! - - - Enable the HTTP server on the Cisco IOS firewall
Router1(config)#ip http server
! - - - Set the authentication to AAA
Router1(config)#ip http authentication aaa
! - - - Create a standard access list denying all traffic
Router1(config)#access-list 22 deny any
! - - - Define standard access list 22 for the HTTP server
Router1(config)#ip http access-class 22
- - - Define the global authentication timeout to 30 minutes
Router1(config)#ip auth-proxy auth-cache-time 30
! - - - Display the firewall name on the login page
Router1(config)#ip auth-proxy auth-proxy-banner
! - - - Create the auth-proxy rules with the name allowed-outbound
Router1(config)#ip auth-proxy name allowed-outbound http
! - - - Enter the interface configuration mode
Router1(config)#interface e0
! - - - Configure the IP address of the interface
Router1(config-if)#ip address 10.10.10.254 255.255.255.0
! - - - Apply the named auth-proxy rule to the interface
Router1(config-if)#ip auth-proxy allowed-outbound
! - - - Exit the interface configuration mode
Router1(config)#CTL-Z
Router1(config)#
```

Notice from Example 16-2 that the major difference in the configuration is where the access list is applied to the Cisco IOS Firewall. The access list must be applied on the interface facing the source to facilitate the communication between the source and the Cisco IOS Firewall.

Using Authentication Proxy with TACACS+

Cisco Secure ACS provides both TACACS+ and RADIUS functionality. Cisco Secure ACS was discussed in detail in Chapter 9, "Cisco Secure Access Control Server." This section discusses configuring TACACS+ using the CSACS. If the CSACS is already configured, you only need to make a few configuration changes to run TACACS+. You must complete three steps for this configuration:

Step 1   Complete the network configuration.

Step 2   Complete the interface configuration.

Step 3   Complete the group setup.

Step 1: Complete the Network Configuration

To complete the network configuration, connect to the Cisco Secure ACS using your browser and click the **Network Configuration** icon on the left border. Figure 16-7 depicts the Network Configuration page of the Cisco Secure ACS.

Ensure that the Cisco IOS Firewall is listed as a AAA client. The IP address should be the address of the interface that faces the AAA server, and the Authenticate Using field should match the authentication protocol being used—in this case, TACACS+ (Cisco IOS). To change any parameters for the AAA client, just click the client (link), and the Edit window will appear. Figure 16-8 depicts the Edit window for the AAA client.

Figure 16-7 *Cisco Secure ACS Network Configuration Page*

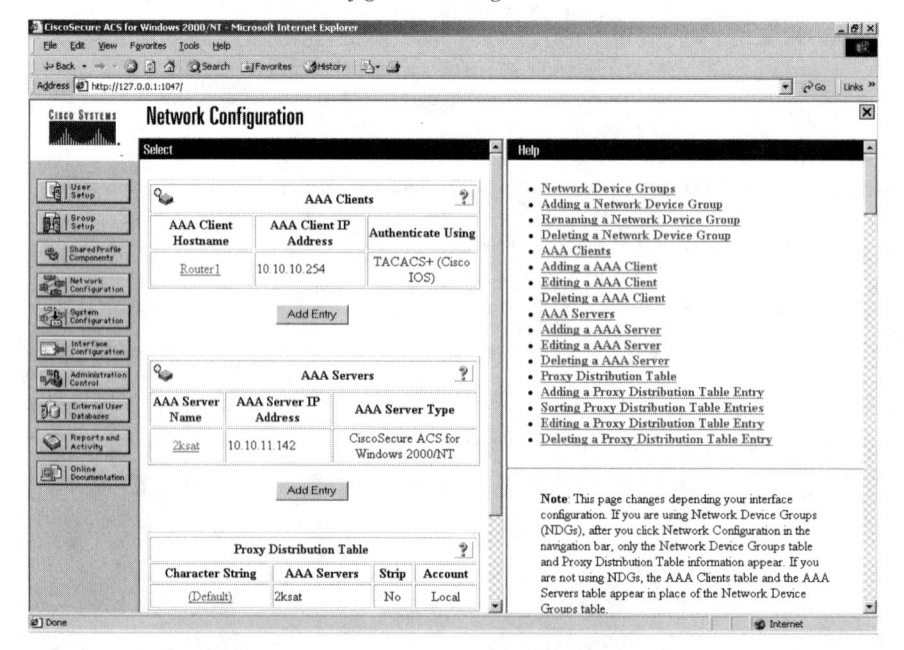

Figure 16-8 *AAA Client Edit Window*

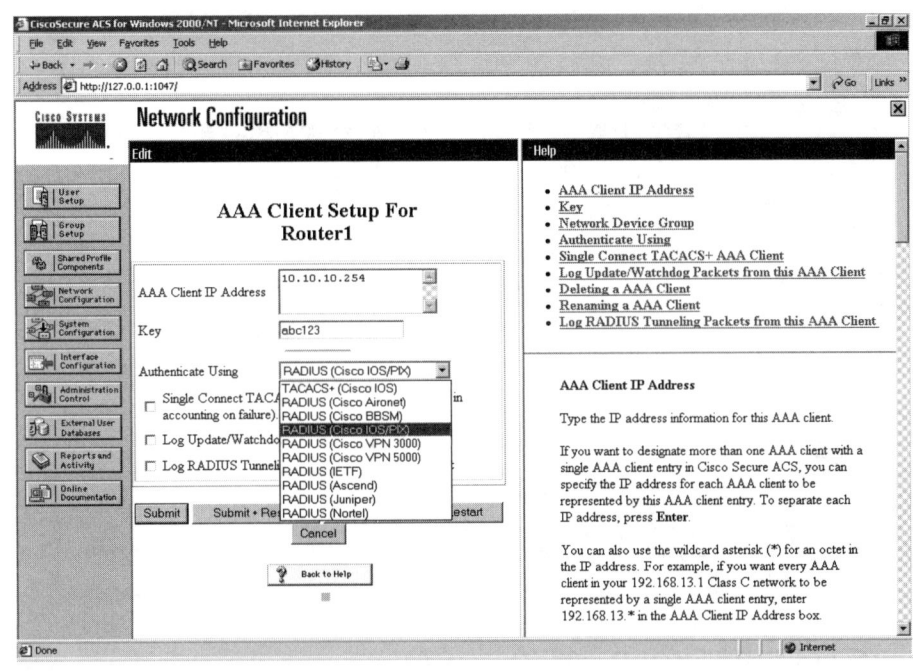

In Figure 16-8, you can see that it is possible to change the AAA client IP address and key. The authentication protocols are selected from a drop-down list.

> **NOTE** Ensure that you click **Submit-Restart** after making any changes to the AAA client configuration.

Step 2: Complete the Interface Configuration

The next step is to complete the interface configuration. Select the **Interface Configuration** icon on the left border and scroll down in the Edit window until you get to the TACACS+ Services configuration box. Figure 16-9 depicts this area.

In Figure 16-9, you can see that TACACS+ services can be assigned to either users or groups. In the New Services block, check the Group box and list the service as **auth-proxy**.

Figure 16-9 *Interface Configuration Window*

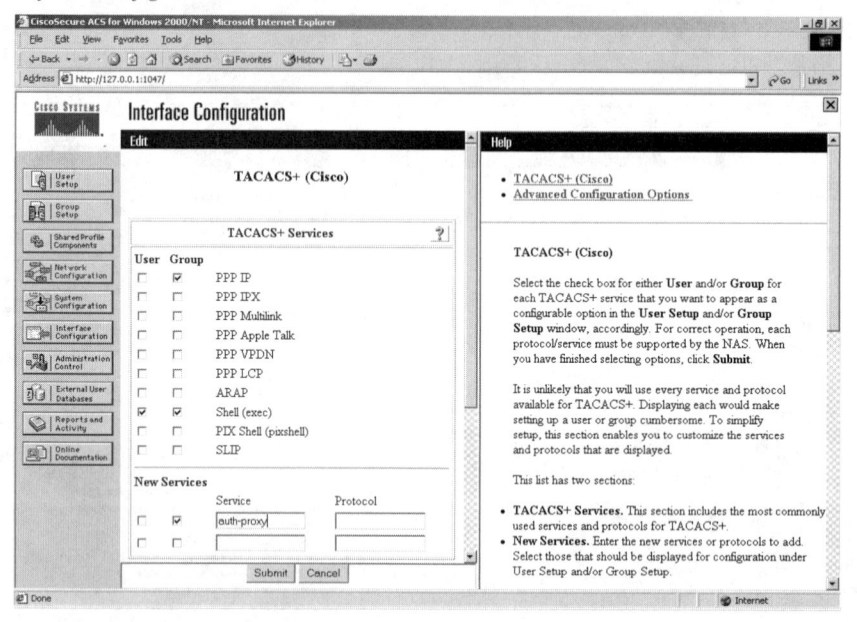

Step 3: Complete the Group Setup

The next step is to configure the parameters of the dynamic ACLs. You do this in the Group Setup window, which you access by clicking the **Group Setup** icon on the left border and scrolling down to the auth-proxy window. Figure 16-10 depicts the Group Setup window.

Figure 16-10 *Group Setup Configuration Window*

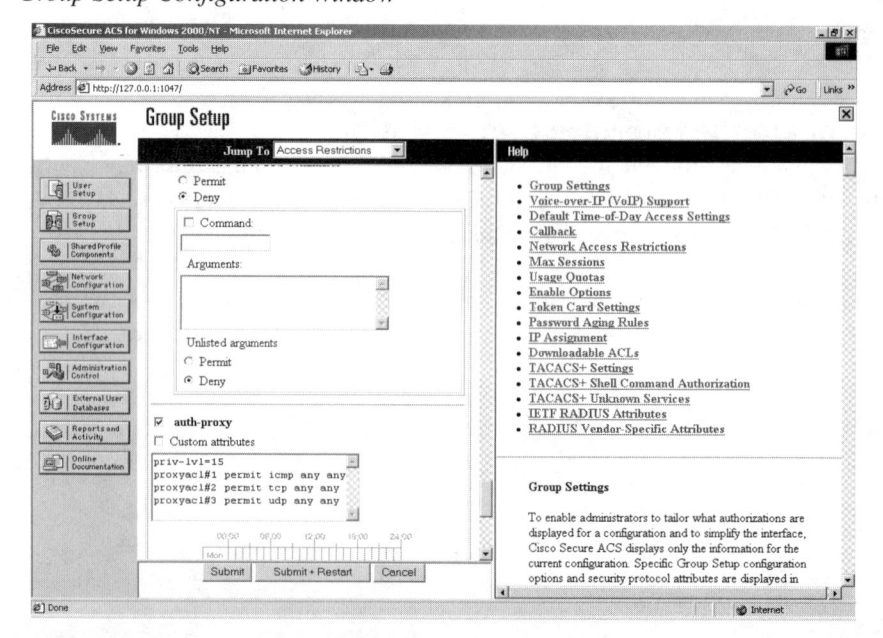

In Figure 16-10, four lines are added to the auth-proxy attributes. This is a very open policy and allows anyone who successfully authenticates to have open access to internal resources. Obviously, you want to use a more restrictive policy when configuring your authentication proxy in a production environment:

```
priv-lvl=15
proxyacl#1=permit icmp any any
proxyacl#2=permit tcp any any
proxyacl#3=permit udp any any
```

Using Authentication Proxy with RADIUS

Configuring the Cisco Secure ACS for RADIUS requires the same steps as TACACS+, although they need a slightly different configuration. Figure 16-11 depicts the Network Configuration window. This time you need to ensure that the AAA client is using the RADIUS authentication protocol.

In Figure 16-11, the Authenticate Using box now says RADIUS (Cisco IOS/PIX). The next step is to configure the RADIUS properties in the Interface Configuration window. Figure 16-12 depicts the Interface Configuration window.

In Figure 16-12, the boxes are checked for the different RADIUS configurations. In this case, you are only using the first item listed (cisco-av-pair). The final step is to configure the attributes for cisco-av-pair. You do this in the Group Setup window. Figure 16-13 depicts the Group Setup window.

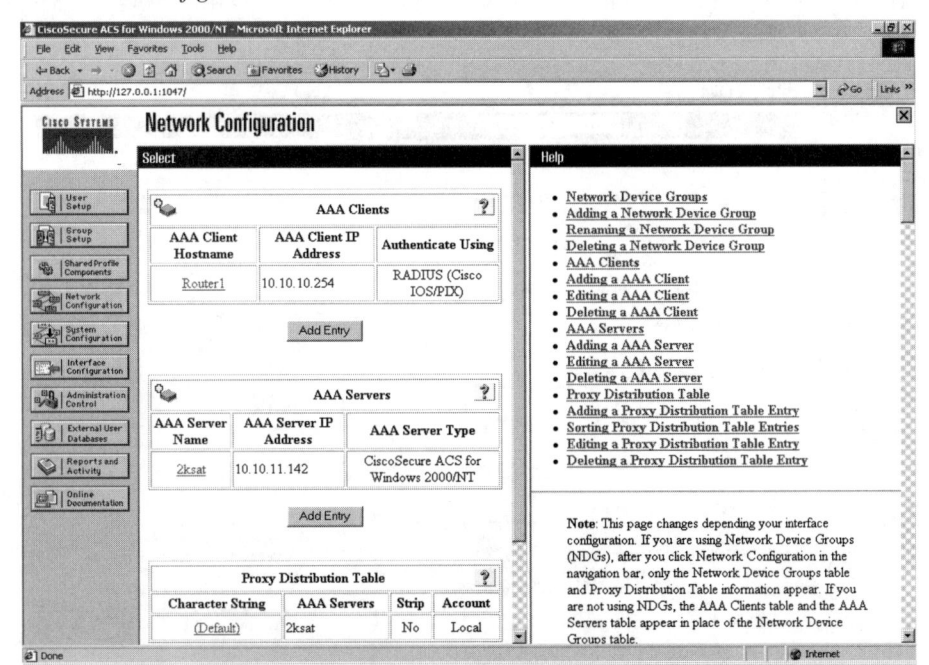

Figure 16-11 *Network Configuration Window*

Figure 16-12 *Interface Configuration Window*

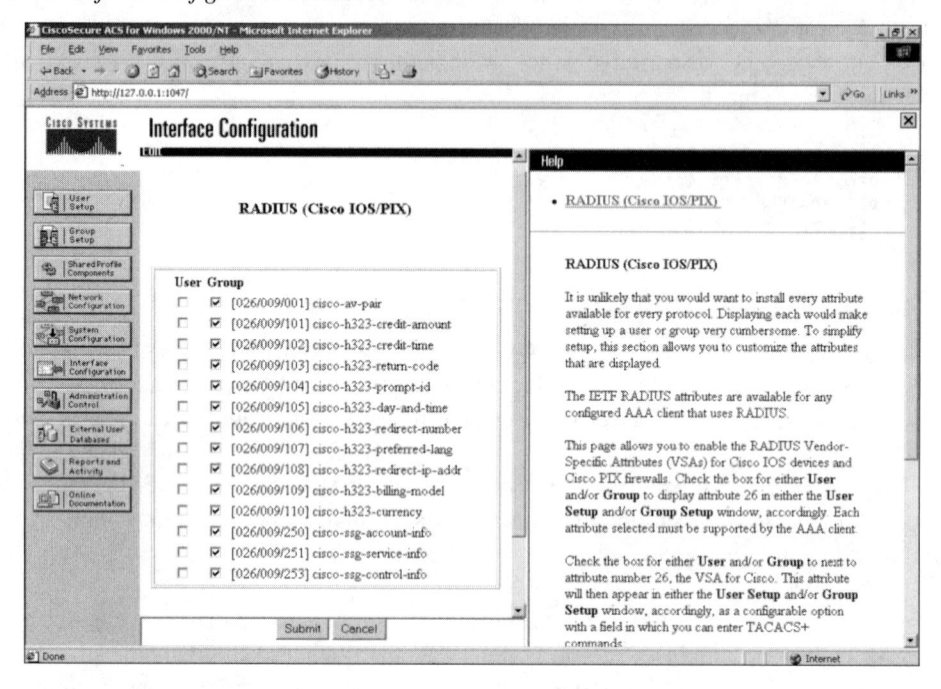

Figure 16-13 *Group Setup Configuration Window*

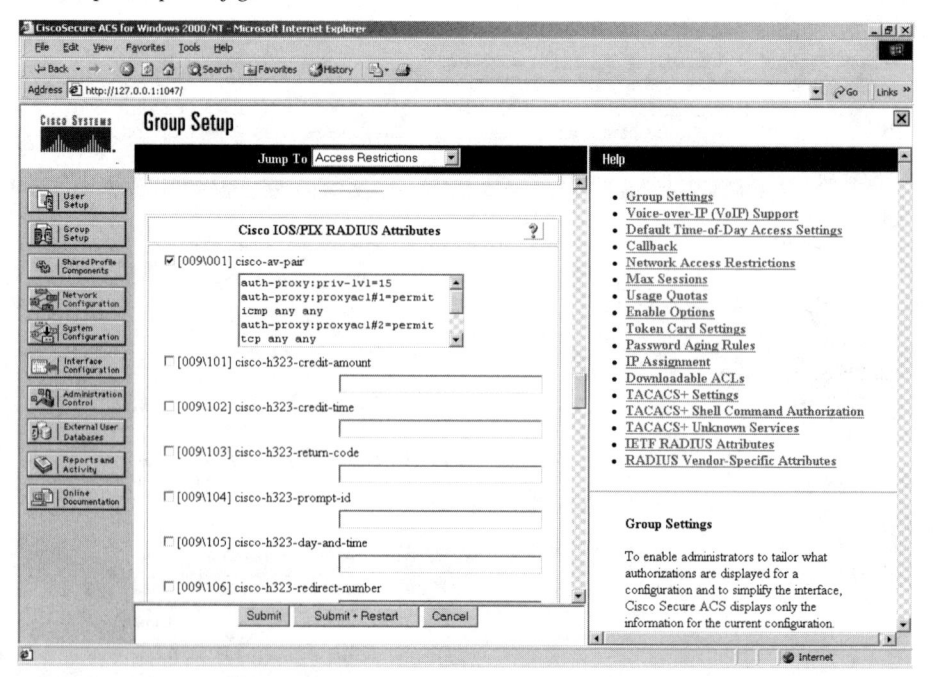

In Figure 16-13, it is only possible to see the first three of four configuration entries added to the Cisco IOS/PIX RADIUS Attributes box because of the limited size of the window. As with the TACACS+ configuration, this is a very liberal configuration that allows access to all resources after the user successfully authenticates. Four lines are added to the auth-proxy attributes. The four lines that are included in the cisco-av-pair window are as follows:

```
auth-proxy:priv-lvl=15
auth-proxy:proxyacl#1=permit icmp any any
auth-proxy:proxyacl#2=permit tcp any any
auth-proxy:proxyacl#3=permit udp any any
```

Limitations of Authentication Proxy

To properly design a solution using authentication proxy, it is important to understand the limitations. It is always best to design a solution that can completely fulfill the business requirement. The following are limitations to authentication proxy:

- The authentication proxy triggers only on HTTP connections.

- HTTP services must be running on the standard (well-known) port, which is port 80 for HTTP.

- Client browsers must enable JavaScript for secure authentication.

- The authentication proxy access lists apply to traffic passing through the router. Traffic destined to the router is authenticated by the existing Cisco IOS Software authentication methods.

- The authentication proxy does not support concurrent usage; that is, if two users try to log in from the same host at the same time, authentication and authorization applies only to the user who first submits a valid username and password.

- Load balancing using multiple or different AAA servers is not supported.

Foundation Summary

The "Foundation Summary" section of each chapter lists the most important facts from the chapter. Although this section does not list every fact from the chapter that will be on your SNRS exam, a well-prepared candidate should at a minimum know all the details in each "Foundation Summary" before going to take the exam.

Authentication proxy facilitates communication between the Cisco IOS Firewall and a AAA server. This communication enables administrators to restrict access to resources down to the individual "authenticated" user level. Authentication proxy requires you to configure both the Cisco IOS Firewall and the AAA server. Configuring the Cisco IOS Firewall requires four tasks:

1. Configure AAA.
2. Configure the HTTP server.
3. Configure authentication proxy.
4. Verify the authentication proxy configuration.

The following three primary tasks are required to configure the Cisco Secure ACS as a TACACS+ or RADIUS server:

- Network configuration
- Interface configuration
- Authentication proxy configuration

Authentication proxy is not a bidirectional service. You must configure authentication proxy to respond to requests from internal or external sources. If you need to configure authentication proxy to function in both directions, you must create an inbound configuration and an outbound configuration.

It is important to understand the limitations of authentication proxy to ensure that the correct solution is designed to fulfill the business requirement. Limitations of authentication proxy include the following:

- Authentication proxy only supports HTTP and HTTPS on standard ports (80 and 443).
- Authentication proxy requires that the client browser be configured to support JavaScript to perform secure authentication.
- Authentication proxy does not support access directly to the Cisco IOS Firewall.
- Only a single user account can be logged on at a time. Authentication proxy does not support concurrent usage.
- Authentication proxy can only be configured to a single AAA server or server type.

Q&A

As mentioned in the section "How to Use This Book" in the Introduction, you have two choices for review questions: the Q&A questions here or the exam simulation questions on the CD-ROM. The questions that follow present a bigger challenge than the exam itself because they use an open-ended question format. By using this more difficult format, you can exercise your memory better and prove your conceptual and factual knowledge of this chapter. You can find the answers to these questions in the appendix.

1. What is the job of the authentication proxy?
2. After the authentication is completed, the Cisco IOS Firewall receives authorization from the AAA server in what form?
3. What keyword must be added to the following AAA authentication command to enable the authentication proxy?

    aaa authorization xxxxxxxxx default

4. What does the following command do?
5. router1(config)

    #aaa accounting auth-proxy default start-stop group tacacs+

6. What command enables you to create an authentication proxy name called testauth?
7. Through the user of the Cisco Secure ACS, TACACS+ and RADIUS authentication are supported for authentication proxy. True or false?
8. Does the authentication proxy support concurrent user access to the same host?
9. What is the command for enabling authentication proxy using HTTPS?
10. Authentication proxy is configured as a new service in the Cisco Secure ACS in which window?
11. Can authentication proxy be trigger by connecting to the Cisco IOS Firewall?

This chapter covers the following subjects:

- IBNS Overview
- IEEE 802.1x
- How 802.1x Works
- Selecting EAP
- Cisco Secure ACS

CHAPTER 17

Identity-Based Networking Services

Cisco Identity-Based Networking Services (IBNS) is a technology framework for delivering logical and physical network access authentication. IBNS combines several Cisco products that offer authentication, user policies, and access control to provide a comprehensive solution for increasing network access security. An IBN incorporates capabilities defined in IEEE 802.1x standard. The 802.1x standard is a framework defined by the IEEE 802.1 working group to provide a standard link layer protocol for port-based access control and authentication.

This chapter discusses IBNS and 802.1x features and functionality.

"Do I Know This Already?" Quiz

The purpose of the "Do I Know This Already?" quiz is to help you decide whether you really need to read the entire chapter. If you already intend to read the entire chapter, you do not necessarily need to answer these questions now.

The 12-question quiz, derived from the major sections in "Foundation Topics" section of the chapter, helps you determine how to spend your limited study time.

Table 17-1 outlines the major topics discussed in this chapter and the "Do I Know This Already?" quiz questions that correspond to those topics.

Table 17-1 *"Do I Know This Already?" Foundation Topics Section-to-Question Mapping*

Foundation Topics Section	Questions Covered in This Section
IBNS Overview	11
IEEE 802.1x	1, 2
How 802.1x Works	3–5, 8, 9
Selecting EAP	6, 7, 10
Cisco Secure ACS	12

Chapter 17: Identity-Based Networking Services

> **CAUTION** The goal of self-assessment is to gauge your mastery of the topics in this chapter. If you do not know the answer to a question or are only partially sure of the answer, you should mark this question wrong for purposes of the self-assessment. Giving yourself credit for an answer you correctly guess skews your self-assessment results and might provide you with a false sense of security.

1. Which of the following is a framework defined by the IEEE 802.1 working group that provides a standard link layer protocol for port-based access control and authentication?

    a. 802.1q

    b. 802.11b

    c. 802.1x

    d. 802.1w

2. What are the three roles the IEEE 802.1x framework defines in the authentication process?

    a. Authentication server

    b. Administrator

    c. Authenticator

    d. Supplicant

    e. Client

    f. Object

3. Prior to the client authentication, which protocols are allowed to pass through a port? (Select three.)

    a. 802.1x

    b. RADIUS

    c. CDP

    d. TACACS+

    e. SNMP

    f. Spanning Tree Protocol

4. What transport mechanisms are used in 802.1x to authenticate supplicants against a back-end data store? (Select two.)

    a. RADIUS

    b. TACACS+

c. XML

d. PAP

e. EAP

5. With the encapsulation technique known as an EAPOL frame format, the value of the packet type field determines the type of packet being transmitted. If the value is set to 3, which type of packet is being transmitted?

   a. EAPOL-Encapsulated-ASF-Alert

   b. EAPOL-Start

   c. EAPOL-Logoff

   d. EAPOL-Key

   e. EAP-Packet

6. Match the EAP authentication type with the method used for authentication.

Type	Method
1. Cryptographic	A. EAP-MD5, LEAP, and EAP-MSCHAPv2
2. Challenge-response	B. EAP-TLS
3. Other	C. PEAP, EAP-TTLS, and EAP-FAST
4. Tunneling based	D. EAP-GTC

7. Cisco developed _____ to support customers that require strong password policy enforcement but do not want to deploy digital certificates.

   a. PEAP

   b. EAP-FAST

   c. EAP-TLS

   d. EAP-MD5

   e. EAP-GTC

8. How many clients per port does the IEEE 802.1x standard link layer protocol allow?

   a. 50

   b. 5

   c. 1

   d. 42

9. Which field in the EAP frame format is 1 octet and identifies the type of EAP packet.

   a. Code

   b. Identifier

   c. Length

   d. Data

10. Which of the following is an IETF draft RFC submitted by Cisco Systems, Microsoft, and RSA Security that does not have an internal method of client authentication?

    a. LEAP

    b. PEAP

    c. EAP-FAST

    d. EAP-TLS

11. Which of the following are IBNS features? (Select two.)

    a. Reduced operating cost and user productivity

    b. Reduced security restrictions on the LAN

    c. Increased network and resource connectivity

    d. Increased network availability

12. What ACS feature allows for EAP authentication capabilities?

    a. Administrator

    b. Authentication

    c. RADIUS

    d. TACACS

The answers to the "Do I Know This Already?" quiz are found in the appendix. The suggested choices for your next step are as follows:

- **10 or less overall score**—Read the entire chapter. This includes the "Foundation Topics" and "Foundation Summary" sections and the "Q&A" section.

- **11 or 12 overall score**—If you want more review on these topics, skip to the "Foundation Summary" section and then go to the "Q&A" section. Otherwise, move on to Chapter 18, "Configuring 802.1x Port-Based Authentication."

Foundation Topics

This chapter provides a general overview of IBNS and 802.1x technologies based on Cisco IOS Software Release 12.1.(19)EA1 for Catalyst switches.

IBNS Overview

In today's network environments, many enterprises have implemented virtual LANs (VLANs), wireless LANs (WLANs), DHCP, and other dynamic technologies to accommodate user mobility and flexibility. Although these technologies have increased benefits for the users in terms of network availability, they also render the corporate networks vulnerable to unauthorized access. In an unsecured environment, it is much easier for hackers and unauthorized entities to launch denial-of-service (DoS), hijack, or other types of attacks.

Cisco IBNS is an IEEE 802.1x-based technology solution that increases network security by authenticating users based on personal identity in addition to device MAC and IP address verification. IBNS controls who and what is on the network, keeping outsiders out through authentication and authorization, keeping the insiders honest via accountability, and increasing the overall network visibility.

IBNS provides the following benefits:

- **Improved user flexibility and mobility**—Associating user and group profiles with network resources allows the ease of authentication, authorization, and accounting (AAA) of users regardless of physical location. Hence, the users obtain the freedom to move to work from various locations.

- **Increased network and resource connectivity**—Because policies are associated with users and not physical ports, network administration is simplified. Ports may be provisioned dynamically, which provides greater scalability.

- **Reduced operating cost and user productivity**—Providing quicker secure access to new users and modifying existing user profiles allows for more efficient and timely access to resources.

IBNS spans a broad range of Cisco LAN infrastructure products, including Catalyst switches and Aironet access points, to identify users attempting to gain access to the wired or wireless LAN.

IEEE 802.1x

IEEE 802.1x is a standard set by IEEE 802.1 working group. It is a data link layer (Layer 2) protocol designed to provide port-based network access control using authentication unique to a device or user. This service is called port-level authentication. For security reasons, port authentication is limited to a single client per port.

Several enhancements to the IEEE 802.1x standard are offered by Cisco, including the following:

- **VLAN assignments**—In a standard 802.1x implementation, authenticated users are placed in a VLAN preconfigured for the connected port. Using Cisco IBNS VLAN assignments may be obtained via the user or device identity. This is accomplished using the username-to-VLAN and device-to-VLAN association maintained in the RADIUS server. When the user successfully authenticates, the RADIUS server sends the VLAN information to the switch, and the switch dynamically configures the attached port for the specific VLAN.

- **802.1x guest VLAN**—This is an important feature in environments where there is a mix of 802.1x clients and clients that do not support 802.1x. Using this feature, the non-802.1x-compatible users or devices gain access to network resources via the guest VLAN. A guest VLAN normally provides restricted access for basic network resources that would allow for limited functions (such as browsing or e-mail).

- **Port security**—802.1x provides the option to enable port security on a switch port. If this feature is enabled to allow only a single MAC address on the port, it will deny multiple MAC addresses to be used on that port. This eliminates the risk of users attaching hubs to a switch port to add more devices to the same port.

- **Voice VLAN ID**—This is a feature that incorporates the benefits of Cisco Architecture for Voice, Video, and Integrated Data (AVVID), VoIP, and dynamic port security mechanisms such as 802.1x. It enables administrators to configure an auxiliary VLAN for voice VLAN ID.

- **High availability**—In environments that have implemented high availability using active and standby supervisors in a modular switch, this feature provides the capability to synchronize the 802.1x port security information between the two supervisors. So in the case of a failover to the standby supervisor, the port security state is maintained.

- **Access control list (ACL) assignment**—ACL lists may be dynamically assigned using 802.1x authentication policy. This feature enables administrators to restrict the user to specific network segments or limit access to sensitive resources to specific users without compromising user mobility.

As described earlier in this section, 802.1x is a link layer technology, which means that the majority of the information described in this chapter focuses on the available features in Cisco Catalyst switches IOS Release 12.1(19)EA1 and later. However, there is a new feature in Cisco IOS Release 12.3(11)T for the router platforms in support of 802.1x and IBNS. It is called *Remote Site IEEE 802.1x Local Authentication Service*. The Remote Site IEEE 802.1X Local Authentication Service feature enables you to configure an access point or wireless-aware router to act as a local RADIUS server. This capability provides backup authentication in the event that the WAN link or the central RADIUS server fails. For more information on this feature refer to Cisco.com.

802.1x Components

The IEEE 802.1x framework defines three roles in the authentication process. The terminology for these roles is as follows:

Supplicant—The supplicant is the endpoint requesting access to the network. For example, this could be an end user device, a printer, or an IP phone.

Authentication server—It is the entity that validates the identity of the supplicant and notifies the authenticator to allow or deny the client request for access. For example, a RADIUS server such as ACS may provide authentication server services.

Authenticator—It is the device between the supplicant and the authentication server that facilitates authentication. The client is normally directly connected to the authenticator. For example, a switch or a wireless access point would provide authenticator services to clients attempting to access LAN.

Figure 17-1 shows the specific roles of the devices in the network during 802.1x port-based authentication.

Figure 17-1 *802.1x Device Roles*

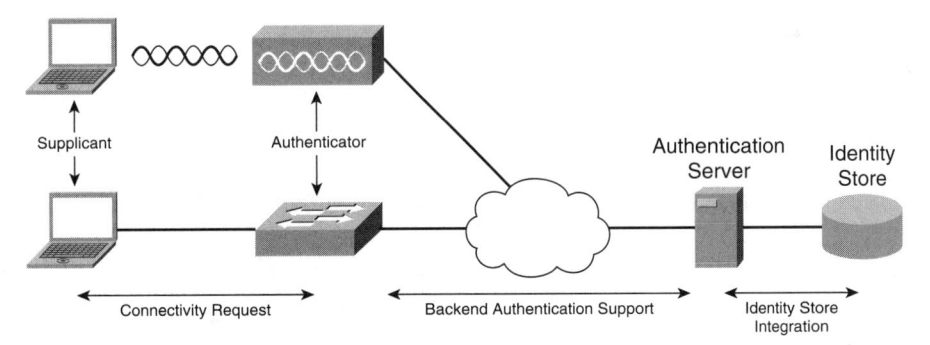

Prior to the client authentication, the port will only allow 802.1x protocol, Cisco Discovery Protocol (CDP), and Spanning Tree Protocol (STP) traffic through the connected port. After the authentication is successful, normal traffic may pass through the port.

How 802.1x Works

Extensible Authentication Protocol (EAP) is the transport mechanism used in 802.1x to authenticate supplicants against a back-end data store, typically a RADIUS server. EAP was initially defined in RFC 2284 as a general authentication framework running over Layer 2 PPP. In RFC 3748, the EAP definition has been updated to include IEEE 802 as a link layer. The IEEE 802 encapsulation of EAP does not involve PPP, and IEEE 802.1X does not include support for link or network layer negotiations. As a result, within IEEE 802.1X, it is not possible to negotiate non-EAP authentication mechanisms, such as Password Authentication Protocol (PAP) or Challenge Handshake Authentication Protocol (CHAP). EAP does not select a specific authentication mechanism during the link layer phase but rather postpones it until the authentication phase. Figure 17-2 depicts the EAP frame format.

Figure 17-2 *EAP Frame Format*

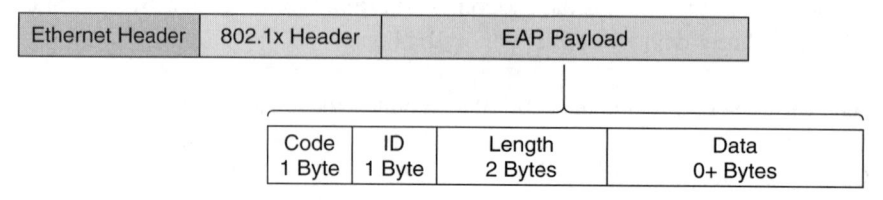

Ethernet Header	802.1x Header	EAP Payload

EAP Payload:

Code 1 Byte	ID 1 Byte	Length 2 Bytes	Data 0+ Bytes

The field descriptions are as follows:

- **Code**—The Code field is 1 octet and identifies the type of EAP packet. Table 17-2 shows the assigned EAP Codes.

Table 17-2 *Assigned EAP Codes*

Code	Description
1	Request
2	Response
3	Success
4	Failure

EAP only defines codes 1 through 4. Any EAP packets with other codes are silently discarded.

- **Identifier**—The Identifier field is 1 octet and aids in matching Responses with Requests.

- **Length**—The Length field is 2 octets and indicates the length, in octets, of the EAP packet including the Code, Identifier, Length, and Data fields. Octets outside the range of the Length field should be treated as data link layer padding and will be ignored upon reception. A message with the Length field set to a value larger than the number of received octets will be silently discarded.

- **Data**—The Data field is 0 or more octets. The Code field determines the format of the Data field.

On LAN media, the supplicant and the authenticator communicate using an encapsulation technique known as EAP over LANs (EAPOL). EAPOL supports various media types, such as Ethernet, Token Ring, Fiber Distributed Data Interface (FDDI), and WLANs. EAPOL is encapsulated EAP messages that can be handled directly by a LAN MAC service. Figure 17-3 shows the EAPOL frame format.

Figure 17-3 *EAPOL Frame Format*

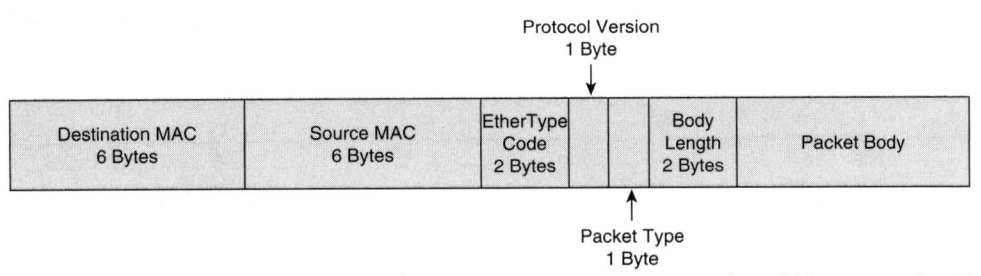

```
Destination MAC | Source MAC | EtherType Code | Body Length | Packet Body
6 Bytes         | 6 Bytes    | 2 Bytes        | 2 Bytes     |
```

Protocol Version: 1 Byte
Packet Type: 1 Byte

The Packet Type field is 1 octet in length, taken to represent an unsigned binary number. Its value determines the type of packet being transmitted. The following types are defined:

- **EAP-Packet**—A value of 0 indicates that the frame carries an EAP packet.

- **EAPOL-Start**—A value of 1 indicates that the frame is an EAPOL-Start frame.

- **EAPOL-Logoff**—A value of 2 indicates that the frame is an explicit EAPOL-Logoff request frame.

- **EAPOL-Key**—A value of 3 indicates that the frame is an EAPOL-Key frame.

- **EAPOL-Encapsulated-ASF-Alert**—A value of 4 indicates that the frame carries an EAPOL-Encapsulated-ASF-Alert.

There are a number of modes of operation for 802.1x authentication. Figure 17-4 depicts a generic message exchange between a client and the authentication server:

1. The authenticator sends an EAP-Request/Identity packet to the supplicant as soon as it detects that the link is active (for example, the client has connected to a switch port).

2. The supplicant sends an EAP-Response/Identity packet to the authenticator, which is then passed on to the authentication (RADIUS) server. Communications between the supplicant and authentication server also leverage the RADIUS protocol carried over standard User Datagram Protocol (UDP).

3. The authentication server sends back a challenge to the authenticator, such as with a token password system. The authenticator unpacks this from IP and repackages it into EAPOL and sends it to the supplicant. Different authentication methods will vary this message and the total number of messages. EAP supports client-only authentication and strong mutual authentication. Only strong mutual authentication is considered appropriate in a wireless environment.

4. The supplicant responds to the challenge via the authenticator and passes the response on to the authentication server.

5. If the supplicant provides proper identity, the authentication server responds with a success message, which is then passed on to the supplicant. The authenticator now allows access to the

LAN, possibly restricted based on attributes that came back from the authentication server. For example, the authenticator might switch the supplicant to a particular VLAN or install a set of access control rules.

Figure 17-4 *802.1x Message Exchange*

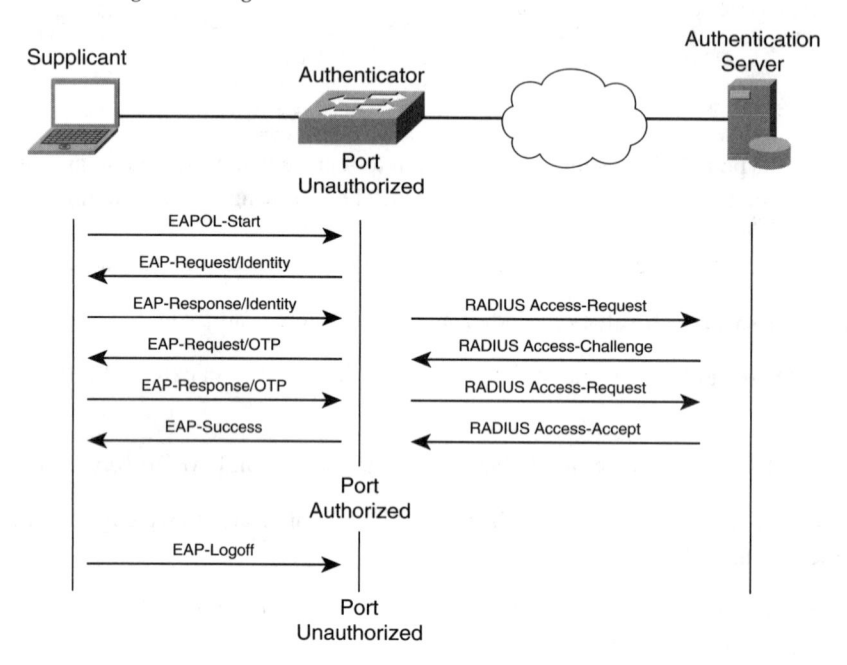

```
                                                    Authentication
   Supplicant         Authenticator                    Server
   [laptop]           [switch] ---- (cloud) ---- [server]
                        Port
                     Unauthorized

         EAPOL-Start →
         ← EAP-Request/Identity
         EAP-Response/Identity →         RADIUS Access-Request →
         ← EAP-Request/OTP               ← RADIUS Access-Challenge
         EAP-Response/OTP →              RADIUS Access-Request →
         ← EAP-Success                   ← RADIUS Access-Accept

                        Port
                     Authorized

         EAP-Logoff →

                        Port
                     Unauthorized
```

Port State

If IEEE 802.1x is configured on a switch port, the port starts in the *unauthorized* state. The port does not allow any traffic through except for IEEE 802.1x, STP, and CDP packets. When the supplicant successfully authenticates, the port changes to *authorized* state and allows through traffic from the client to network resources. If the client does not support IEEE 802.1x, the switch cannot authenticate the client unless a guest VLAN is preconfigured to provide some level of necessary network access.

In the scenario in which the switch port is an unauthorized port even though the client is 802.1x compatible, the client sends an authentication request to the switch using an EAPOL-Start frame. The client attempts the request a fixed number of times. Because no response is received from an unauthorized port, the client continues sending frames as if the port is in authorized state.

For more information about controlling the port state, visit Cisco.com.

Selecting EAP

Several types of EAP authentication are available for wired and wireless networks. Table 17-3 shows the types and methods for EAP authentication.

Table 17-3 *EAP Authentication Types*

Type	Methods
Challenge-response	EAP-MD5, LEAP, and EAP-MSCHAPv2
Cryptographic	EAP-TLS
Tunneling based	PEAP, EAP-TTLS, and EAP-FAST
Other	EAP-GTC

The following sections describe the most common methods of 802.1x authentication.

EAP-MD5

EAP-MD5 uses Message Digest 5 (MD5)-based challenge-response for authentication. Using this method, the client identity is transmitted over the network, but the password is not sent. The server generates a random string and sends it to the user as a challenge. The client MD5 hashes the challenge using its password as the key. The server then authenticates the subscriber by verifying the user's MD5 hash password.

This type of authentication is well supported and provides a simple mechanism for authentication using username and passwords. It also does not burden the server or the client because of its lightweight processing requirements.

The cons of using MD5 are the security weaknesses inherit in this authentication method. MD5 requires the storage of plain-text or reversible passwords on the authentication server. Microsoft is also phasing it out.

Figure 17-5 shows the MD5 authentication process is an 802.1x environment.

Cisco Lightweight EAP

Cisco Lightweight Extensible Authentication Protocol (LEAP) uses the concept of mutual authentication to validate a user. Mutual authentication relies on a shared secret and the user's password, which is known by the client and the network. The authentication server sends a challenge to the client. The client uses a one-way hash of the user password to send a response to the challenge. The server creates its own response based on the user database information and compares it to the response received from the client. When the server authenticates the client, the same process is repeated in reverse so the client can authenticate the server. When this process is completed, an EAP success message is sent to the client, and both the client and the authentication server derive the dynamic Wired Equivalent Privacy (WEP) key.

This type of mutual authentication reduces the risk of access point vulnerability to man-in-the-middle attack. However, as with MD5 authentication, the user password may be vulnerable to attackers.

Figure 17-5 *MD5 Authentication Process*

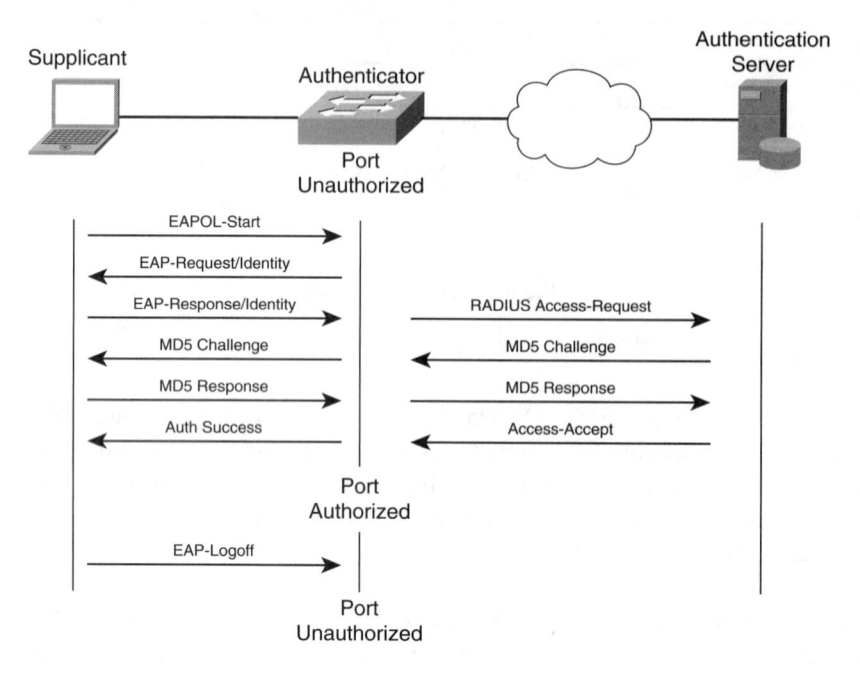

EAP Transport Layer Security

EAP-TLS is a standard developed by Microsoft and accepted by the Internet Engineering Task Force (IETF) as RFC 2716. It is based on the Transport Layer Security (TLS) protocol as described in another standard (RFC 2246).

Similarly to Cisco LEAP method, EAP-TLS mutually authenticates the client and the server, but in this case, passwords are not used at all. Instead, Public Key cryptography based on the Rivest, Shamir, and Adelman (RSA) handshake is used. EAP-TLS uses digital certificates or smart cards to validate both the user and the server's identity.

Figure 17-6 shows the process of authentication via EAP-TLS.

As shown in Figure 17-6, the RADIUS server sends its certificate to the client in Phase 1 of the authentication sequence (server-side TLS). The client validates the RADIUS server certificate by verifying the issuer of the certificate, a certificate authority (CA) server entity, and the contents of the digital certificate. When this is complete, the client sends its certificate to the RADIUS server in Phase 2 of the authentication sequence (client-side TLS). The RADIUS server validates the client's certificate by verifying the issuer of the certificate (CA server entity) and the contents of the digital certificate. When this is complete, an EAP-Success message is sent to the client, and both the client and the RADIUS server derive the dynamic WEP key.

Figure 17-6 *EAP-TLS Authentication Process*

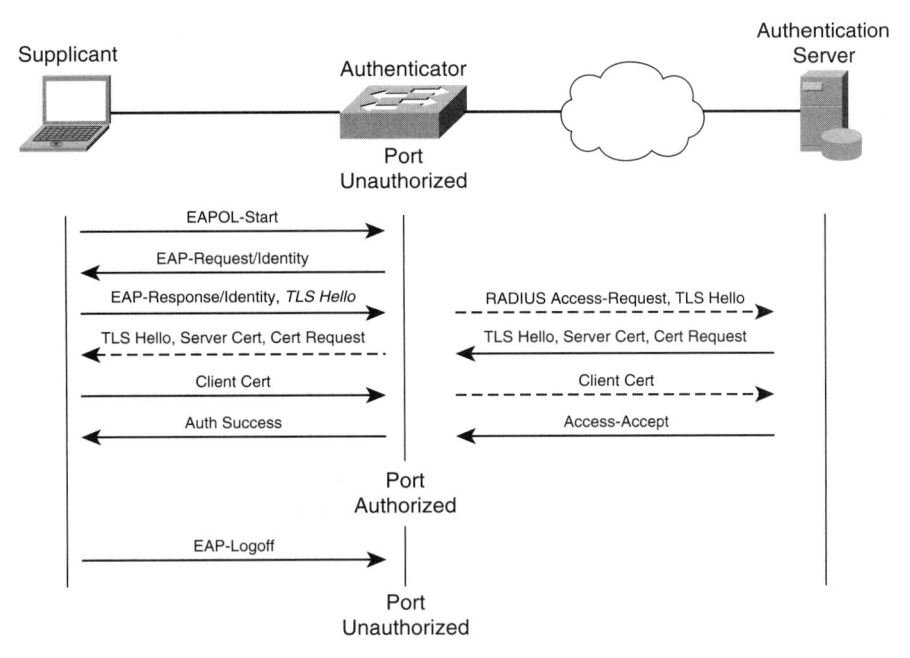

The advantage of EAP-TLS is that it may be configured for a two-factor authentication systems; it is one of the strongest forms of authentication today.

The cons of EAP-TLS are that it may be more complex to deploy because of the various components, such as a CA. It is also more computationally intensive on both the client and the server.

EAP-TLS is suited for large enterprise environments that only use Microsoft Windows XP/2000/2003 with deployed certificates.

Protected EAP

Protected EAP (PEAP) is an IETF draft RFC submitted by Cisco Systems, Microsoft, and RSA Security. PEAP does not have an internal method of client authentication. It supports various EAP-encapsulated methods within a protected TLS tunnel.

PEAP supports an extensible set of user authentication methods, such as one-time token authentication and password change or aging. It uses server-side digital certificate authentication based on the PKI standard. In environments where certificates are not issued to every client, PEAP can use a Microsoft Windows username and password instead by querying the Windows domain controller, Active Directory, or other existing user database (for example Lightweight Directory Access Protocol [LDAP], Novell NetWare Directory Services [NDS], and one-time password databases).

PEAP uses the same TLS mechanisms as EAP-TLS, but adds the record protocol for encryption.

Figure 17-7 shows the PEAP authentication process.

Figure 17-7 *PEAP Authentication Process*

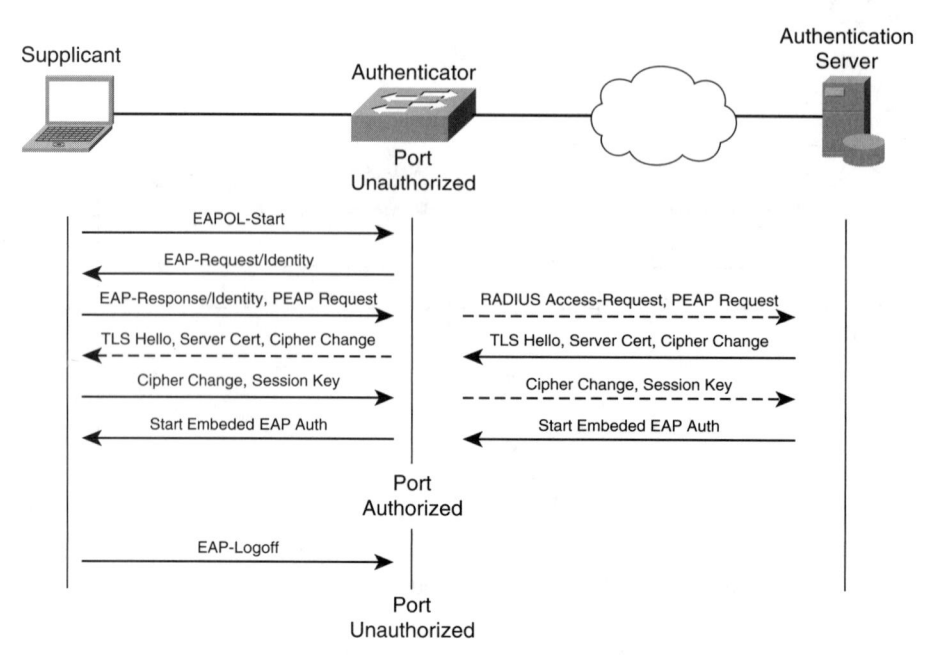

As shown in Figure 17-7, Phase 1 of the authentication sequence is the same as that for EAP-TLS (server-side TLS). A server-side TLS authentication is performed to create an encrypted tunnel and complete server-side authentication. At the end of Phase 1, an encrypted TLS tunnel is created between the user and the RADIUS server for transporting EAP authentication messages.

In Phase 2, the RADIUS server authenticates the client through the encrypted TLS tunnel via another EAP type. As an example, a user can be authenticated using a one-time password (OTP) using the EAP-GTC subtype or MS-CHAP version 2 (as defined by the PEAP draft). The RADIUS server will relay the OTP credentials (user ID and OTP) to an OTP server to validate the user login. When this is complete, an EAP-Success message is sent to the client, and both the client and the RADIUS server derive the dynamic WEP key.

EAP Flexible Authentication via Secure Tunneling

EAP Flexible Authentication via Secure Tunneling (EAP-FAST) was developed by Cisco and is available in an IETF informational draft. Cisco developed EAP-FAST to support customers that require strong password policy enforcement but do not want to deploy digital certificates. EAP-FAST provides protection against a variety of network attacks, including man-in-the-middle, replay, and dictionary attacks.

EAP-FAST uses symmetric key algorithms to achieve a tunneled authentication process. The tunnel establishment relies on a Protected Access Credential (PAC) that can be provisioned and managed dynamically by EAP-FAST through the authentication server, such as the Cisco Secure Access Control Server.

Figure 17-8 shows the EAP-FAST authentication process.

Figure 17-8 *EAP-FAST Authentication Process*

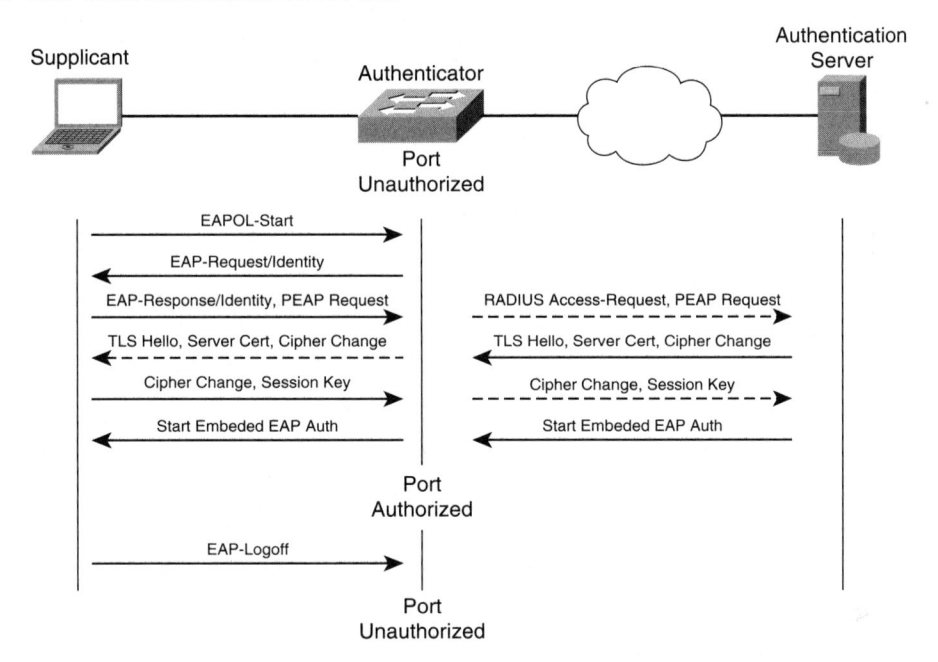

Phase 1 establishes a mutually authenticated tunnel. The client and server use PAC to authenticate each other and establish a secure tunnel.

Phase 2 performs client authentication in the established tunnel. The client sends username and password to authenticate and establish client authorization policy.

Optionally, Phase 0 is used infrequently to enable the client to be dynamically provisioned with a PAC. During this phase, a per-user access credential is generated securely between the user and the network. This per-user credential, known as the PAC, is used in Phase 1 of EAP-FAST authentication.

EAP-FAST supports Microsoft Windows 2000, Windows XP, and Windows CE (Power PC 2002, Power PC 2003, and CE.Net 4.2) operating systems.

EAP Methods Comparison

Table 17-4 compares the various types of EAP as to network requirements.

Table 17-4 *EAP Methods Comparison*

	PEAP with Generic Token Card (GTC)	PEAP with MS-CHAP Version 2	EAP-FAST	Cisco LEAP	EAP-TLS
User Authentication Database and Server	OTP, LDAP, Novell NDS, Windows NT domains, Active Directory	Microsoft Windows NT domains, Active Directory	Windows NT domains, Active Directory, LDAP (limited)	Windows NT domains, Active Directory	OTP, LDAP, Novell NDS, Windows NT domains, Active Directory
Requires Server Certificates	Yes	Yes	No	No	Yes
Requires Client Certificates	No	No	No	No	Yes
Operating System Support	Driver: Windows XP, Windows 2000, Windows CE With third-party utility: Other OS	Driver: Windows XP, Windows 2000, Windows CE With third-party utility: Other OS	Driver: Windows XP, Windows 2000, Windows CE With third-party utility: Other OS	Driver: Microsoft Windows 98, Windows 2000, Windows NT, Microsoft Windows Me, Windows XP, Mac OS, Linux, Windows CE, DOS	Driver: Windows XP, Windows 2000, Windows CE With third-party utility: Other OS
Application-Specific Device (ASD) Support	No	No	Yes	Yes	No
Credentials Used	Client: Windows, Novell NDS, LDAP password; OTP or token Server: Digital certificate	Windows password	Windows password, LDAP user ID/password (manual provisioning required for PAC provisioning)	Windows password	Digital certificate
Single Sign-On Using Windows Login	No	Yes	Yes	Yes	Yes
Password Expiration and Change	No	Yes	Yes	No	N/A
Works with Fast Secure Roaming	No	No	Yes	Yes	No
Works with WPA and WPA2	Yes	Yes	Yes	Yes	Yes

Cisco Secure ACS

Traditionally, access control was managed at the edge of the network. The growth of wired and wireless LANs has prompted concerns with the overall network security. As the infrastructure expands, the edge is not the only vulnerable point in the enterprise. A combination of asynchronous communications server (ACS) and IEEE 802.1x provides the capability of using RADIUS-based AAA for LAN access. The following benefits may be realized as a result of combining IEEE 802.1x and ACS:

- User accounting and auditing with the ability to track and monitor users on the LAN
- Strong authentication using a variety of authentication methods such as smart cards, tokens, PKI, and external user databases
- Flexible policy assignments, such as per user VLANs
- Granular control of supplicants on the LAN via centralized means

ACS, using its RADIUS features, could act as the transport for EAP communications from the authenticator. It also carries policy instructions to the authenticator in the form of attribute-value (AV) pairs.

Figure 17-9 shows wired and wireless client authentication to the ACS server providing RADIUS services to gain access to network resources.

Figure 17-9 *ACS Client Authentication*

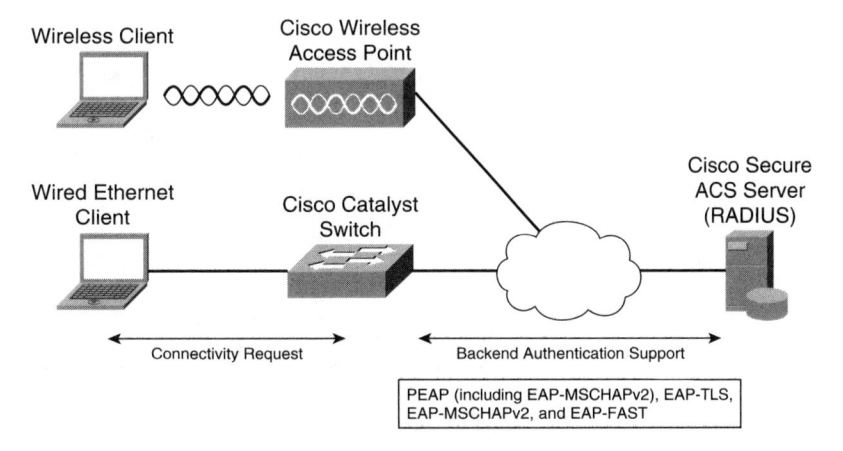

ACS support PEAP (including EAP-MSCHAPv2), EAP-TLS, EAP-MSCHAPv2, and EAP-FAST authentication protocols and a variety of external databases. For more information about ACS, see Chapter 9, "Cisco Secure Access Control Server," and refer to Cisco.com.

Foundation Summary

The "Foundation Summary" section of each chapter lists the most important facts from the chapter. Although this section does not list every fact from the chapter that will be on your SNRS exam, a well-prepared candidate should at a minimum know all the details in each "Foundation Summary" before going to take the exam.

Cisco IBNS is an IEEE 802.1x-based technology solution that increases network security by authenticating users based on personal identity rather than the traditional MAC or IP address verification methods.

IBNS provides improved user flexibility and mobility, increased network and resource connectivity, and reduced operating cost and user productivity.

IEEE 802.1x is a standard set by IEEE 802.1 working group. Several enhancements to the IEEE 802.1x standard are offered by Cisco, including VLAN assignments, 802.1x guest VLAN, port security, voice VLAN ID, high availability, and ACL assignment.

The IEEE 802.1x framework defines three roles in the authentication process: supplicant, authentication server, and authenticator.

On LAN media, the supplicant and the authenticator communicate using an encapsulation technique known EAPOL. The types of EAPOL packets are EAP-Packet, EAPOL-Start, EAPOL-Logoff, EAPOL-Key, and EAPOL-Encapsulated-ASF-Alert.

Several types of EAP authentication are available for wired and wireless networks.

Type	Method
Challenge-response	EAP-MD5, LEAP, and EAP-MSCHAPv2
Cryptographic	EAP-TLS
Tunneling based	PEAP, EAP-TTLS, and EAP-FAST
Other	EAP-GTC

ACS, using its RADIUS features, can act as the transport for EAP communications from the authenticator. It also carries policy instructions to the authenticator in the form of AV pairs.

Q&A

As mentioned in the section "How to Use This Book" in the Introduction, you have two choices for review questions: the Q&A questions here or the exam simulation questions on the CD-ROM. The questions that follow present a bigger challenge than the exam itself because they use an open-ended question format. By using this more difficult format, you can exercise your memory better and prove your conceptual and factual knowledge of this chapter. You can find the answers to these questions in the appendix.

1. What benefits does IBNS provide?
2. What enhancements to the IEEE 802.1x standard are offered by Cisco?
3. Explain the parts of the EAP frame format.
4. Describe the method EAP-MD5 uses for challenge-response authentication.
5. What benefits may be realized as a result of combining IEEE 802.1x and ACS?
6. What is the purpose of the guest VLAN?
7. Describe the concept of mutual authentication used by Cisco LEAP.

This chapter covers the following subjects:

- 802.1x Port-Based Authentication Configuration Tasks

- 802.1x Mandatory Configuration

- 802.1x Optional Configurations

- Displaying 802.1x Statistics and Status

CHAPTER 18

Configuring 802.1x Port-Based Authentication

This chapter describes how to configure 802.1x port-based authentication on a Catalyst switch to prevent unauthorized clients (supplicants) from gaining access to the network.

"Do I Know This Already?" Quiz

The purpose of the "Do I Know This Already?" quiz is to help you decide whether you really need to read the entire chapter. If you already intend to read the entire chapter, you do not necessarily need to answer these questions now.

The 10-question quiz, derived from the major sections in the "Foundation Topics" portion of the chapter, helps you determine how to spend your limited study time.

Table 18-1 outlines the major topics discussed in this chapter and the "Do I Know This Already?" quiz questions that correspond to those topics.

Table 18-1 *"Do I Know This Already?" Foundation Topics Section-to-Question Mapping*

Foundation Topics Section	Questions Covered in This Section
802.1x Port-Based Authentication Configuration Tasks	1
802.1x Mandatory Configuration	2–5
802.1x Optional Configurations	6–8, 10
Displaying 802.1x Statistics and Status	9

The goal of self-assessment is to gauge your mastery of the topics in this chapter. If you do not know the answer to a question or are only partially sure of the answer, you should mark this question wrong for purposes of the self-assessment. Giving yourself credit for an answer you correctly guess skews your self-assessment results and might provide you with a false sense of security.

1. What port type supports the 802.1x protocol?

   a. Trunk port
   b. Dynamic and dynamic-access ports
   c. SPAN ports
   d. Layer 2 static access ports
   e. RSPAN ports
   f. All of the above

2. Which one of the tasks is a required task for configuring 802.1x port-based authentication on a Catalyst switch?

   a. Configuring the host mode
   b. Changing the quiet period
   c. Configuring a guest VLAN
   d. Changing the switch-to-client retransmission time
   e. Configuring the switch-to-RADIUS-server communication
   f. Debugging the 802.1x configuration

3. If the **dotx system-auth-control** command is issued in global configuration mode, it enables authentication on _____.

   a. the port it is specified on
   b. the network related services on the switch
   c. the interface types specified in the method list
   d. the global configuration mode
   e. the entire switch

4. Which of the following are attributes used for RADIUS VLAN assignments? (Select three.)

   a. Tunnel-Type=VLAN
   b. Tunnel-Medium-Type=802
   c. Tunnel-Auth-Type=radius
   d. Tunnel-Private-Group-ID=VLANID
   e. None of the above

5. What are the required command parameters for specifying RADIUS servers in global configuration mode?

   a. *hostname* or *ip-address*, *auth-port*, *logging-port*, *key*

   b. *hostname* or *ip-address*, *logging-port*, *key*

   c. *hostname* or *ip-address*, *auth-port*, *key*

   d. *hostname* or *ip-address*, *admin-port*, *key*

   e. *hostname* or *ip-address*, *key*

   f. *hostname* or *ip-address*

6. On a per-port basis, periodic re-authentication may be enabled using either the _____ or the _____ commands.

   a. **dot1x reauth-period** *seconds*, **dot1x timeout** *seconds*

   b. **dot1x reauthentication, dot1x timeout reauth-period** *seconds*

   c. **dot1x reauthentication, dot1x reauth-period** *minutes*

   d. **dot1x reaccredidation, dot1x timeout reacc-period** *seconds*

   e. **dot1x reauthentication retry-period seconds, dot1x timeout period** *seconds*

   f. **dot1x timeout** *seconds*, **dot1x timeout reauth-period** *minutes*

7. To set the switch-to-client retransmission time to 60 seconds and the switch-to-client frame-retransmission number to 4, use which of the following commands?

   a. **dot1x timeout tx-period** *60* and **dot1x max-req** *4*

   b. **dot1x timeout period** *60* and **dot1x req** *4*

   c. **dot1x retransmission-timeout tx-period** *60* and **dot1x frame-max-req** *4*

   d. **dot1x client retransmission timeout tx-period** *60* and **dot1x frame-max-req** *4*

   e. **dot1x client retransmission timeout tx-period** *60* and **dot1x max-req** *4*

   f. None of the above

8. What optional display expressions help narrow the display of 802.1x statistics and status?

   a. **include** and **exclude**

   b. **start**, **contain**, and **end**

   c. **grep** and **egrep**

   d. **begin**, **include**, and **exclude**

   e. None of the above

9. When running the **show dot1x all** command, the output includes all of the following except what?

   a. Interface

   b. MAC

   c. EAPOL packet count

   d. Quiet period

   e. None of the above

10. What kind of clients could benefit from a guest VLAN within an 802.1x-based environment? (Select two.)

    a. Printers

    b. Microsoft Windows 98 workstations and servers

    c. PBXs

    d. Users with forgotten passwords

    e. Wireless access points

    f. All of the above

The answers to the "Do I Know This Already?" quiz are found in the appendix. The suggested choices for your next step are as follows:

- **8 or less overall score**—Read the entire chapter. This includes the "Foundation Topics" and "Foundation Summary" sections and the "Q&A" section.

- **9 or 10 overall score**—If you want more review on these topics, skip to the "Foundation Summary" section and then go to the "Q&A" section. Otherwise, move on to Chapter 19, "Building a VPN Using IPsec."

Foundation Topics

Prior to configuring 802.1x port-based authentication, you need to consider several limitations. The 802.1x protocol support is available on Layer 2 static access ports, voice virtual LAN (VLAN) ports, and multi-VLAN access ports(MVAP).

Trunk, dynamic, dynamic-access, EtherChannel, Switched Port Analyzer (SPAN), and Remote SPAN (RSPAN) destination ports do not support the 802.1x protocol. You must remove any existing EtherChannel ports from the EtherChannel group prior to configuring 802.1x. RSPAN and voice VLANs may not be used as an 802.1x guest VLAN. The guest VLAN feature is not supported on routed or trunk ports.

Support for Cisco Secure Access Control Server (ACS) application for 802.1x authentication on switches running Cisco IOS Software Release 12.1(14)EA1 is available in ACS version 3.2.1 and later.

In environments upgrading to Cisco IOS Software Release 12.1(14) or later, the 802.1x commands have either changed or new commands have been introduced. To enable 802.1x, refer to the Cisco.com online documentation.

The 802.1x protocol allows the switch to facilitate communications between the client and the AAA server. The switch is actually using 802.1x to communicate with the client and is communicating with the authentication, authorization, and accounting (AAA) server via RADIUS. These communications are used to authenticate the user and computer and apply the necessary policies to allow access to the network. During this communication, the switch is aware of the transaction, and upon successful authentication, it dynamically configures the policies and enables the connected switch interface. Figure 18-1 depicts the typical communications scenario for a system attempting to gain access to the network.

Figure 18-1 *802.1x Communications*

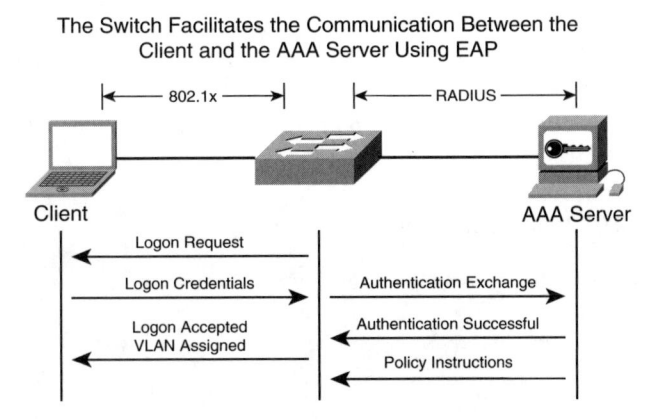

Several transactions occur between the client and the AAA server. In each transaction, the switch is facilitating the transaction. The transaction steps are as follows:

Step 1   The client physically connects to the switch and the switch responds by sending the client a logon request.

Step 2   The client sends logon credentials.

Step 3   The switch forwards the logon credentials to the AAA server.

Step 4   The AAA server verifies the credentials and sends the policy instructions to the switch.

Step 5   The switch applies the policies and enables the port.

Step 6   The client now has access to the network via the assigned VLAN.

Figure 18-2 depicts the steps in the negotiation that results in the client gaining the correct access.

Figure 18-2   *802.1x Client Negotiation*

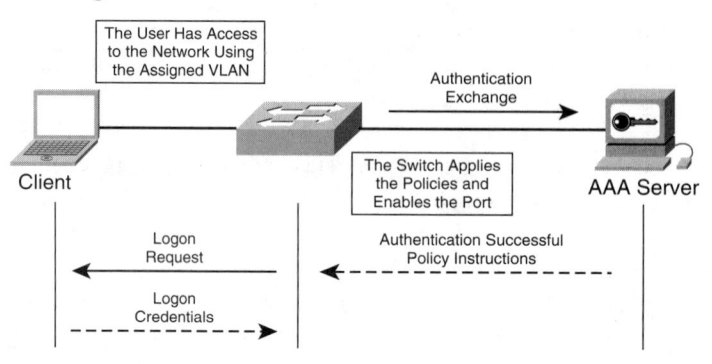

802.1x Port-Based Authentication Configuration Tasks

During 802.1x port authentication, the client only receives Extensible Authentication Protocol over LAN (EAPOL), Cisco Discovery Protocol (CDP), and Spanning Tree Protocol (STP) traffic through the port. When the authentication is successful, normal traffic is passed through.

The tasks involved in configuring 802.1x port-based authentication on a Catalyst switch are as follows:

- Configuring 802.1x authentication
- Configuring the switch-to-RADIUS server communication

The following are several optional tasks that might prove helpful to further customize the 802.1x-related features:

- Configuring periodic re-authentication
- Manually re-authenticating
- Changing the quiet period
- Changing the switch-to-client retransmission time
- Setting the switch-to-client frame-retransmission number
- Configuring the host mode
- Configuring a guest VLAN
- Resetting the 802.1x configuration to the default values
- Debugging 802.1x configuration

802.1x Mandatory Configuration

The following section provides the mandatory command description and examples to activate 802.1x port-based authentication.

Enabling 802.1x Authentication

To enable 802.1x, port-based AAA must be enabled. To enable AAA, use the **aaa new-model** command in global configuration mode.

After AAA is enabled, you must specify the authentication list for 802.1x. The **aaa authentication dot1x** command specifies the authentication method list. A method list is a sequential list that describes the authentication methods used to authenticate a user or device and the security protocols to be used for authentication. If a named list is not specified, a default list is used. Use the **default** keyword followed by the method for this type of situation. The default method list is automatically applied to all ports. The following is the syntax for this command:

```
aaa authentication dot1x {default} method1
```

For *method1*, enter the **group radius** keyword to use the list of RADIUS servers for authentication. This command must be executed in global configuration mode.

> **NOTE** The current version of the Catalyst IOS 12.1(22)EA3 only supports the **default group radius** keyword. This is even though other keywords are available in the command-line help.

After the AAA commands have been configured, use the **dot1x system-auth-control** command to enable 802.1x authentication for the whole switch. This command must be used in global configuration mode.

To activate 802.1x port-based authentication on a specific port, use the interface configuration command **dot1x port-control**. The syntax used is as follows:

```
dot1x port-control {auto | force-authorized | force-unauthorized}
```

Table 18-2 shows the command parameters and the description of this command.

Table 18-2 *The Parameters for the* **dot1x port-control** *Command*

Parameter	Description
auto	Enables 802.1x authentication on the interface and causes the port to transition to the authorized or unauthorized state based on the 802.1x authentication exchange between the switch and the client.
force-authorization	Disables the 802.1x authentication on the interface and causes the port to transition to the authorized state without any authentication exchange required. The port sends and receives normal traffic without 802.1x-based authentication of the client.
force-unauthorized	Denies all access through the interface by ignoring all client authentication attempts and forcing the port to the unauthorized state. The switch does not provide authentication services to the client through the interface.

The **aaa authorization network default group radius** command is typically used on the routers for RADIUS user authorization of all network-related service requests, such as PPP, SLIP, and ARAP. However, in the context of the Catalyst switch, it allows RADIUS server authorization of VLAN assignment or per-user ACLs. The RADIUS server tunnel and 802.1x attributes, which are defined in Internet Engineering Task Force (IETF) RFCs 2868 and 3580, provide fields that can be populated with user VLAN information. Three tunnel attributes are used for RADIUS VLAN assignment:

- **Attribute 64**—Tunnel-Type=VLAN (type 13)

- **Attribute 65**—Tunnel-Medium-Type=802 (type 6)

- **Attribute 81**—Tunnel-Private-Group-ID=VLANID

The first two attributes have an integer value, and the last one is a text string identifying the VLAN *name*. If a VLAN number is assigned to this attribute, the 802.1x authorization process will place the port to an authorized state after successful authentication, but the port will remain in the default VLAN.

> **NOTE** For per-user ACLs, single-host mode must be configured. This is the default setting.

Example 18-1 is the basic configuration to enable 802.1x and AAA on a switch port.

Example 18-1 *Enabling 802.1x and AAA on a Switch Port*

```
Switch#configure terminal
Switch(config)#aaa new-model
Switch(config)#aaa authentication dot1x default group radius
Switch(config)#dot1x system-auth-control
Switch(config)#interface fastethernet0/1
Switch(config-if)#switchport mode access
Switch(config-if)#dot1x port-control auto
Switch(config-if)#end
```

Configuring the Switch-to-RADIUS Server Communication

The 802.1x port-based authentication scheme relies on RADIUS for authentication of supplicants (that is, the desktop, wireless LAN [WLAN], and so on). The authenticator (which is the switch, router, and so on) must be configured for RADIUS to communicate with the authentication server. It is possible to specify multiple RADIUS servers on a switch. RADIUS servers are identified by host name or IP address and specific port numbers. The combination of these two parameters creates a unique identifier providing redundancy and availability. The host entries are tried sequentially in the order configured. Use the following command in global configuration mode to specify a RADIUS server host, User Datagram Protocol (UDP) port, and shared secret text string:

```
radius-server host {hostname | ip-address} [auth-port port-number] key string
```

The **auth-port** *port-number* is an optional parameter to specify a different UDP destination port for authentication requests. The default UDP port for RADIUS on Cisco router running Cisco IOS Software Release 12.3(x) is 1645 for authentication and 1646 for accounting. The default UDP ports on a Cisco Catalyst IOS-based switch version 12.1 or higher is 1812 for authentication and 1813 for accounting.

The **key** *string* specifies the authentication and encryption key used between the authenticator and the RADIUS daemon on the authentication server. The key value is a text string that must match the key used on the RADIUS server.

> **NOTE** The key string must be configured as the last item in the preceding command syntax. This is because any leading spaces in between and after the key are used as part of the key string. This could cause mismatched key strings between the RADIUS server and authenticator device.

The **radius-server host** command may be reentered for each RADIUS server and UDP port combinations that are available.

Chapter 18: Configuring 802.1x Port-Based Authentication

Example 18-2 shows how to specify the server with IP address 10.200.200.1 as the RADIUS server, to use port 1633 as the authorization port (leaving the authentication port as 1812), and to set the encryption key to *radkey*, matching the key on the RADIUS server.

Example 18-2 *Configure RADIUS on a Switch*

```
Switch#configure terminal
Switch(config)#radius-server host 10.200.200.1 auth-port 1633 key radkey
```

The timeout, retransmission, and encryption key values can be set for all RADIUS servers specified on the switch by using the **radius-server timeout**, **radius-server retransmit**, and the **radius-server key** global commands.

802.1x Optional Configurations

The following section provides the optional command description and examples to activate 802.1x port-based authentication.

Enabling Periodic Re-Authentication

When a client is authenticated to a switch port using 802.1x, the port remains set with the supplicant attributes until a reboot occurs. To validate the user on an ongoing basis, periodic re-authentication may be enabled on a per-port basis. You can also specify the time period between authentications. Table 18-3 describes the command syntax for periodic re-authentication.

Table 18-3 *Periodic Re-Authentication Commands*

Command	Description
dot1x reauthentication	Enables periodic re-authentication of the client. This feature is disabled by default.
dot1x timeout reauth-period *seconds*	Sets the number of seconds between re-authentication attempts. The range from 1 to 65,535 with a default setting of 3600 seconds.

Both of the commands listed in Table 18-3 must be used in interface configuration mode. In Cisco IOS Software versions prior to 12.1.14EA1, the 802.1x re-authentication was configured in global configuration mode via **dot1x re-authentication** command.

> **NOTE** Re-authentication does not disturb the client connectivity of already authorized clients.

Example 18-3 shows the configuration of periodic re-authentication on port 3 of a Catalyst switch and specifies 2 hours between each re-authentication.

Example 18-3 *Configuring Port Re-Authentication Parameters*

```
Switch#configure terminal
Switch(config)#interface fa0/3
Switch(config-if)#dot1x reauthentication
Switch(config-if)#dot1x timeout reauth-period 7200
```

Manually Re-Authenticating a Client Connected to a Port

To force an immediate re-authentication on a specific port, use the following command in privileged EXEC mode:

   `dot1x re-authenticate interface` *interface-id*

The *interface-id* is the interface port to re-authenticate.

Example 18-4 shows how to manually re-authenticate the client connected to port 3 of a Catalyst switch.

Example 18-4 *Manual Re-Authentication*

```
Switch#dot1x re-authenticate interface fa0/3
```

This command does not disturb the client connectivity of already authorized clients. If the objective is to force a new authentication session, use the **dot1x initialize interface interface-id** command in privileged EXEC mode. Example 18-5 provides an example of forcing a new authentication session for the client on port 3 of a Catalyst switch.

Example 18-5 *Re-Initialization Interface fa0/3*

```
Switch#dot1x initialize interface fa0/3
```

Changing the Quiet Period

The quiet period is the idle time between failed client authentications. There are many reasons for an authentication failure, such as invalid username or password entry. During the quiet period, the switch does not accept or initiate any authentication requests. Setting the quiet period to a smaller number provides faster response time to the client. The following syntax in interface configuration mode enables you to change the quiet period:

   `dot1x timeout quiet-period` *seconds*

The *seconds* parameter is the time the switch remains in the quiet state following a failed authentication. The range is from 1 to 65,535. The default value is 60.

Example 18-6 shows a sample configuration of the quiet period set to 45 seconds on port 3 of a Cisco Catalyst switch.

Example 18-6 *Configuring the Quite Period*

```
Switch#configure terminal
Switch(config)#interface fa0/3
Switch(config-if)#dot1x timeout quiet-period 45
```

Changing the Switch-to-Client Retransmission Time

During the 802.1x authentication, the switch sends an EAP-Request/Identity frame to the client. The client responds with an EAP-Response/Identity frame. If the switch does not receive this response, it waits a defined period of time before resending the request frame. This wait time is known as the switch-to-client retransmission time. To change the value of the transmission time, use **dot1x timeout tx-period** *seconds*. The *seconds* parameter is an integer value from 15 and 65,535. The default value is 30 seconds.

Example 18-7 sets the retransmission time to 90 seconds on port 3 of a Cisco Catalyst switch.

Example 18-7 *Configuring the Switch-to-Client Retransmission Time*

```
Switch#configure terminal
Switch(config)#interface fa0/3
Switch(config-if)#dot1x timeout tx-period 90
```

NOTE It is recommended that this command only be used in special circumstances, such as the existence of unreliable links or other problematic symptoms with clients and authentication servers.

Setting the Switch-to-Client Frame-Retransmission Number

If during the 802.1x authentication process the switch does not receive a response from the client, it resends the EAP-Request/Identity frame twice before restarting the authentication process. To change the number of times the EAP-Request/Identity frame is sent, use the following interface configuration command:

   dot1x max-req *count*

The *count* value is an integer from 1 and 10 designating the maximum number of requests that are sent to the client.

Example 18-8 shows the command used to set the maximum frame retransmissions on port 3 of a Catalyst switch to 4.

Example 18-8 *Configuring the Frame-Retransmission Number*

```
Switch#configure terminal
Switch(config)#interface fa0/3
Switch(config-if)#dot1x max-req count 4
```

> **NOTE** It is recommended that this command only be used in special circumstances, such as the existence of unreliable links or other problematic symptoms with clients and authentication servers.

Enabling Multiple Hosts

You can limit the number of hosts attached on an 802.1x-enabled port. You can configure the port to allow one client connection using the **dot1x host-mode single-host**. This is the default state for the port. To allow multiple clients on an 802.1x-authorized port, use the **dot1x host-mode multi-host** interface configuration command. In multi-host mode, only one of the attached clients needs to be successfully authorized for all clients to be granted network access. If the port becomes unauthorized (in case of re-authentication failure or an EAPOL-Logoff message is received), all attached clients are denied access to the network.

To use this command, you must set the **dot1x port-control** interface configuration command to **auto**.

Example 18-9 shows how to enable multi-host mode on port 3 of a Catalyst switch.

Example 18-9 *Enabling Multiple Hosts on an 802.1x Port*

```
Switch#configure terminal
Switch(config)#interface fastethernet0/3
Switch(config-if)#dot1x port-control auto
Switch(config-if)#dot1x host-mode multi-host
```

Configuring a Guest VLAN

In an 802.1x-based environment, clients that fail the 802.1x port authentication will not be granted access to the network resources. This could be because these clients are not 802.1x capable, such as a printer or Microsoft Windows 98 operating system, or that they did not provide valid credentials, such as forgotten password, visitors, and so on.

In these types of scenarios, it is possible to assign a VLAN as guest VLAN with limited services to the clients. The switch assigns clients to a guest VLAN when the authentication server does not receive a response to its EAP-Request/Identity frame or when the client does not send EAP packets.

Prior to release 12.2(22)EA2, both the 802.1x and non-802.1x-capable clients could join the guest VLAN. This is because the switch did not keep a history of EAPOL packets, therefore granting access to the guest VLAN for all clients that failed to authenticate. In release 12.2(22)EA2, the behavior of the guest VLAN changed so that it keeps a history of the EAPOL packet and denying access to a non-802.1x-based client.

In addition, if an 802.1x-capable client attempts to join a guest VLAN port (multiport) on which non-802.1x clients are configured, the port is put into the unauthorized state, and authentication is restarted. This affects network access for all clients on that 802.1x port.

The **dot1x guest-vlan supplicant** global configuration command on the switch is used to allow 802.1x clients that fail authentication into the guest VLAN.

> **NOTE** If a Dynamic Host Configuration Protocol (DHCP) client is connected to a guest VLAN, the DHCP process on the client might time out during the 802.1x authentication process. In this case, the authentication process might need to be restarted. Also, the settings for quiet period and switch-to-client transmission time may be decreased to expedite the re-authentication.

To assign a guest VLAN on a port, use the **dot1x guest-vlan** *vlanid* command. The *vlanid* parameter is the active VLAN intended to be used as the guest VLAN.

> **NOTE** The guest VLANs are supported locally on the switch and are not propagated through Virtual Terminal Protocol (VTP).

Example 18-10 shows how to enable the guest VLAN behavior and specifies VLAN 2 as a guest VLAN for port 3 of a Catalyst.

Example 18-10 *Configuring Guest VLAN Parameters*

```
Switch#configure terminal
Switch(config)#dot1x guest-vlan supplicant
Switch(config)#interface fa0/3
Switch(config-if)#dot1x guest-vlan 2
```

Resetting the 802.1X Configuration to the Default Values

To reset the 802.1x configuration on a switch port to the default values, use the following interface configuration command:

   dot1x default

Example 18-11 shows how to reset the configurable 802.1x parameters to the default values on port 3 of a Catalyst switch.

Example 18-11 *Resetting Port 802.1x Values to Default*

```
Switch#configure terminal
Switch(config)#interface fa0/3
Switch(config-if)#dot1x default
```

Displaying 802.1x Statistics and Status

To verify successful 802.1x configuration, use the following command in privileged EXEC mode:

```
show dot1x [all] | [interface interface-id] | [statistics interface interface-id] [{ |
begin | exclude | include} expression]
```

This command displays 802.1x administrative and operational status for the switch or a specified interface. It also provides statistics information. Table 18-4 describes the command syntax and parameters.

Table 18-4 show 802.1x *Syntax Description*

all	This optional parameter displays the 802.1x status for all interfaces.
interface *interface-id*	This optional parameter displays the 802.1x status for a specified interface.
statistics [interface *interface-id*]	This optional parameter displays 802.1x statistics for the specified interface.
| begin	This optional parameter displays an expression that begins with a specific value.
| include	This optional parameter includes lines that match the expression.
| exclude	This optional parameter excludes lines that match the specified expression.
expression	Expression in the output to use as a reference point. Expressions are case-sensitive.

Example 18-12 displays the output the **show dot1x** command without any of the optional parameters.

Example 18-12 **show dot1x** *Command Output*

```
Switch# show dot1x
Sysauthcontrol                         = Enabled
Supplicant Allowed In Guest Vlan       = Disabled
Dot1x Protocol Version                 = 1
Dot1x Oper Controlled Directions       = Both
Dot1x Admin Controlled Directions      = Both
```

The output is useful for obtaining general summary information on 802.1x status. The *sysauthcontrol* output verifies 802.1x has been enabled globally.

Example 18-13 displays the output of **show dot1x all** command. This command shows 802.1x status on all commands.

Example 18-13 **show dot1x all** *Command Output*

```
Switch#sho dot1x all
Dot1x Info for interface FastEthernet0/3
----------------------------------------------------
Supplicant MAC 0040.4513.075b
```

continues

388 Chapter 18: Configuring 802.1x Port-Based Authentication

Example 18-13 show dot1x all *Command Output (Continued)*

```
AuthSM State        = AUTHENTICATED
BendSM State        = IDLE
PortStatus          = AUTHORIZED
MaxReq              = 2
HostMode            = Single
Port Control        = Auto
QuietPeriod         = 60 Seconds
Re-authentication   = Enabled
ReAuthPeriod        = 120 Seconds
ServerTimeout       = 30 Seconds
SuppTimeout         = 30 Seconds
TxPeriod            = 30 Seconds
Guest-Vlan          = 0
```

The output provides information on all the 802.1x-enabled ports. Key information that displays includes the MAC address of the client attached to the interface and status of the port and authentication state. If the client has not been authenticated or is in the process of being authenticated, the *AuthSM State* would be *CONNECTING*, and the port status would be *UNAUTHORIZED*. The following shows the output:

```
Switch# show dot1x all
Dot1x Info for interface FastEthernet0/3
-------------------------------------------------
Supplicant MAC <Not Applicable>
    AuthSM State        = CONNECTING
    BendSM State        = IDLE
PortStatus          = UNAUTHORIZED
MaxReq              = 2
HostMode            = Single
Port Control        = Auto
QuietPeriod         = 60 Seconds
Re-authentication   = Enabled
ReAuthPeriod        = 30 Seconds
ServerTimeout       = 30 Seconds
SuppTimeout         = 30 Seconds
TxPeriod            = 30 Seconds
Guest-Vlan          = 0
```

To display 802.1x information on a specific interface, use the **interface** *interface-id* parameter. Example 18-14 provides the output of this parameter on port 10 of a Catalyst switch.

Example 18-14 show dot1x interface fa0/10 *Command Output*

```
Switch#show dot1x interface fa0/10
Supplicant MAC 0040.4513.075b
AuthSM State        = AUTHENTICATED
BendSM State        = IDLE
PortStatus          = AUTHORIZED
MaxReq              = 2
```

Example 18-14 show dot1x interface fa0/10 *Command Output (Continued)*

```
HostMode            = Single
Port Control        = Auto
QuietPeriod         = 60 Seconds
Re-authentication   = Enabled
ReAuthPeriod        = 120 Seconds
ServerTimeout       = 30 Seconds
SuppTimeout         = 30 Seconds
TxPeriod            = 30 Seconds
Guest-Vlan          = 0
```

The output of this command is similar to the **show dot1x all** command. The difference is that this command shows the status for only the specified interface versus all interfaces.

Example 18-15 shows a sample output of the **show dot1x statistics interface** command.

Example 18-15 show dot1x statistics interface *Command Output*

```
Switch#show dot1x statistics interface fa0/3
PortStatistics Parameters for Dot1x
-------------------------------------------
TxReqId = 140 TxReq = 273     TxTotal = 410
RxStart = 0 RxLogoff = 0    RxRespId = 133  RxResp = 266
RxInvalid = 0 RxLenErr = 0    RxTotal= 267
RxVersion = 1 LastRxSrcMac 0040.4513.075b
```

Table 18-5 provides the field descriptions for the **show dot1x statistics interface** command.

Table 18-5 show dot1x statistics interface *Field Descriptions*

Field	Description
TxReqId	The number of EAPOL-Request/Identity frames that have been sent
TxReq	The number of EAPOL-Request frames that have been sent that are not Request/Identity
TxTotal	Total number of EAPOL frames of any type that have been sent
RxStart	The number of valid EAPOL frames that have been received
RxLogoff	The number of EAPOL-Logoff frames that have been received
RxRespId	The number of EAPOL-Response/Identity frames that have been received
RxResp	The number of valid EAPOL-Response frames that have been received that are not responses
RxInvalid	The number of EAPOL frames that have been received and have an unrecognized frame type
RxLenErr	The number of EAPOL frames that have been received in which the packet body length field is invalid

continues

Table 18-5 show dot1x statistics interface *Field Descriptions (Continued)*

Field	Description
RxTotal	The number of valid EAPOL frames of any type that have been received
RxVersion	Packets received with 802.1x version 1 format
LastRxSrcMac	The source MAC address frame received in the last EAPOL packet

In addition to the **show** commands, the **debug** commands are useful in monitoring and troubleshooting the 802.1x configuration.

To enable the debugging for the 802.1x feature, use the **debug dot1x** command in privilege EXEC mode.

```
debug dot1x {errors | events | packets | registry | state-machine | all}
```

Table 18-6 Debug dot1x *Command Parameters*

Parameter	Description
errors	Activates the debugging of 802.1x error messages
events	Activates the debugging of 802.1x events
packets	Activates the debugging of 802.1x packets
registry	Activates the debugging of 802.1x registry invocations messages
state-machine	Activates the debugging of state machine related events
all	Activates the debugging of all 802.1x conditions

Example 18-16 displays a partial output of the **debug dot1x all** command. The useful information is emphasized with shading.

Example 18-16 debug dot1x all Partial *Command Output*

```
04:34:41: dot1x-ev:dot1x_port_cleanup_author: cleanup author on interface FastEthernet0/3
04:34:41: dot1x-ev:dot1x_update_port_status: Called with host_mode=0 state UNAUTHORIZED
04:34:41: dot1x-ev:dot1x_update_port_status: using mac 0040.4513.075b to send port to
  unauthorized on vlan 0
04:34:41: dot1x-ev:Found a supplicant block for mac 0040.4513.075b 80CC20B4
04:34:41: dot1x-ev:dot1x_port_unauthorized: Host-mode=0 radius/guest vlan=0 on
  FastEthernet0/3
04:34:41: dot1x-ev:     GuestVlan configured=0
04:34:41: dot1x-ev:supplicant 0040.4513.075b is last
04:34:41: dot1x-ev:Found a supplicant block for mac 0040.4513.075b 80CC20B4
  . . .
  . . .
  . . .
04:34:41: dot1x-ev:Delaying initial EAP-Request/Identity packet
```

Example 18-16 debug dot1x all Partial *Command Output (Continued)*

```
04:34:51: dot1x-ev:dot1x_tx_eap: EAP Ptk
04:34:51: dot1x-ev:EAP-code=REQUEST
04:34:51: dot1x-ev:EAP Type= IDENTITY
04:34:51: dot1x-ev:ID=0
04:34:51: dot1x-packet:Received an EAPOL frame on interface FastEthernet0/3
04:34:51: dot1x-ev:Received pkt saddr =0040.4513.075b , daddr = 0180.c200.0003,pae-ether-
   type = 34958
04:34:51: dot1x-ev:Found a supplicant block for mac 0040.4513.075b 80CC20B4
04:34:51: dot1x-packet:Received an EAP packet on interface FastEthernet0/3
04:34:51: dot1x-ev:sending AUTH_START to BEND for supp_info=80CC20B4
04:34:51: dot1x-ev:Received AuthStart from Authenticator for supp_info=80CC20B4
04:34:51: dot1x-ev:Managed Timer in sub-block attached as leaf to master
04:34:51: dot1x-ev:Going to Send Request to AAA Client on RP for id = 0 and length = 9
04:34:51: dot1x-ev:The Interface on which we got this AAA Request is FastEthernet0/3
04:34:51: dot1x-ev:Found a supplicant block for mac 0040.4513.075b 80CC20B4
04:34:51: dot1x-ev:Username is test
04:34:51: dot1x-ev:MAC Address is 0040.4513.075b
04:34:51: dot1x-ev:RemAddr is 00-40-45-13-07-5B/00-12-01-5E-4D-C3
04:34:51: dot1x-ev:Found a supplicant block for mac 0040.4513.075b 80CC20B4
04:34:51: dot1x-ev:going to send to backend on SP, length = 29
04:34:51: dot1x-ev:Received VLAN is No Vlan
04:34:51: dot1x-ev:Enqueued the response to BackEnd
04:34:51: dot1x-ev:Going to Send Request to AAA Client on RP for id = 1 and length = 26
04:34:52: dot1x-ev:Dot1x matching request-response id 249 found
04:34:52: dot1x-ev:Length of recv eap packet from radius = 4
04:34:52: dot1x-ev:Received VLAN Id -1
04:34:52: dot1x-ev:dot1x_bend_success_enter:0040.4513.075b: Current ID=1
04:34:52: dot1x-ev:dot1x_update_port_status: using mac 0040.4513.075b to send port to
   authorized
04:34:52: dot1x-ev:dot1x_port_authorized:supplicant 0040.4513.075b is first, old vlan 1, new
   vlan 0
04:34:52: dot1x-ev:dot1x_port_authorized: Host-mode=0 radius/guest vlan=0
04:34:52: dot1x-ev:    GuestVlan configured=0
04:34:52: dot1x-ev:vlan 1 vp is added on the interface FastEthernet0/3
04:34:52: dot1x-ev:dot1x_port_authorized: clearing HA table from vlan 1
04:34:52: dot1x-ev:dot1x_port_authorized: Added 0040.4513.075b to HA table on vlan 1
04:34:52: dot1x-ev:dot1x_update_port_status:0040.4513.075b: Current ID=1
04:34:52: dot1x-ev:dot1x_tx_eap: EAP Ptk
04:34:52: dot1x-ev:EAP-code=SUCCESS
04:34:52: dot1x-ev:EAP Type= MD5 CHALLENGE
04:34:52: dot1x-e
```

Foundation Summary

The "Foundation Summary" section of each chapter lists the most important facts from the chapter. Although this section does not list every fact from the chapter that will be on the SNRS exam, a well-prepared candidate should at the minimum know all the details in each "Foundation Summary" before going to take the exam.

The following are the required steps to configure 802.1x port-based authentication on a Cisco Catalyst switch:

Step 1 Enable AAA. Use the **aaa new-model** global configuration to enable AAA.

Step 2 Configure AAA services. Use the **aaa authentication** global configuration command to define method lists that use RADIUS for authentication.

Step 3 Enable 802.1x globally on the switch using **dot1x system-auth-control**.

Step 4 Activate 802.1x port-based authentication on a specific port using the **dot1x port-control auto** interface mode command.

Step 5 Specify the RADIUS server host, UDP port, and shared secret text string using the **radius-server host** command.

You can use several optional parameters to tweak the 802.1x configuration. These parameters include configuring periodic re-authentication, manual re-authentication, changing the quiet period, changing the switch-to-client retransmission time, setting the switch-to-client frame-retransmission number, configuring the host mode, configuring a guest VLAN, resetting the 802.1x configuration to the default values, and debugging 802.1x configuration.

The following troubleshooting commands enable testing and verification of 802.1x port-based authentication:

- **show dot1x**
- **show dot1x all**
- **show dot1x interface**
- **show dot1x statistics interface**
- **debug dot1x**

Q&A

As mentioned in the section "How to Use This Book" in the Introduction, you have two choices for review questions: the Q&A questions here or the exam simulation questions on the CD-ROM. The questions that follow present a bigger challenge than the exam itself because they use an open-ended question format. By using this more difficult format, you can exercise your memory better and prove your conceptual and factual knowledge of this chapter. You can find the answers to these questions in the appendix.

1. The 802.1x protocol support is available on which types of ports?

2. During 802.1x port authentication, the client only receives EAPOL, CDP, and _____ traffic through the port.

3. To enable 802.1x port-based AAA, use the _____ command in global configuration mode.

4. The three parameters of the **dot1x port-control** command are _____, **force-authentication**, and **force-unauthorized**.

5. The combination of which two parameters creates a unique identifier for RADIUS servers providing redundancy and availability?

6. The **radius-server timeout**, **radius-server retransmit**, and **radius-server key** commands set what keys for the RADIUS servers specified on the switch?

7. What is meant by the quiet period when dealing with authentication on the Catalyst switch?

8. If one of the attached clients fails authentication when multiple clients are connecting on an 8021.x-enabled port, then _____.

9. Because the switch keeps a history of _____ packets for clients that failed to authenticate, only non-802.1x-capable clients can join a guest VLAN.

10. If a client has not been authenticated, or is in the process of being authenticated when the **show dot1x all** command is run, the port status would be _____.

Part V: VPN

Chapter 19 Building a VPN Using IPsec

Chapter 20 Scaling a VPN Using IPsec with a Certificate Authority

Chapter 21 Troubleshooting the VPN Configuration on a Cisco Router

Chapter 22 Configuring Remote Access Using Easy VPN

This chapter covers the following subjects:

- Configuring a Cisco Router for IPsec VPNs Using Preshared Keys

- Configuring Manual IPsec

- Configuring IPsec Using RSA-Encrypted Nonces

CHAPTER 19

Building a VPN Using IPsec

Prior to the development of virtual private networking (VPN) technology, the only way to secure communications between two locations was to purchase a "dedicated circuit." To secure communications across an enterprise was tremendously expensive, and securing communications with remote users was cost-prohibitive. VPN technology enables you to secure communications that travel across the public infrastructure such as the Internet. VPN technology allows organizations to interconnect their different locations without having to purchase dedicated connections and significantly reduces the cost of the enterprise network infrastructure.

"Do I Know This Already?" Quiz

The purpose of the "Do I Know This Already?" quiz is to help you decide whether you really need to read the entire chapter. If you already intend to read the entire chapter, you do not necessarily need to answer these questions now.

The 10-question quiz, derived from the major sections in "Foundation Topics" section of the chapter helps you determine how to spend your limited study time.

Table 19-1 outlines the major topics discussed in this chapter and the "Do I Know This Already?" quiz questions that correspond to those topics.

Table 19-1 *"Do I Know This Already?" Foundation Topics Section-to-Question Mapping*

Foundation Topics Section	Questions Covered in This Section
Configuring a Cisco Router for IPsec Using Preshared Keys	1–8
Configuring Manual IPsec	9
Configuring IPsec Using RSA-Encrypted Nonces	10

> **CAUTION** The goal of self-assessment is to gauge your mastery of the topics in this chapter. If you do not know the answer to a question or are only partially sure of the answer, you should mark this question wrong for purposes of the self-assessment. Giving yourself credit for an answer you correctly guess skews your self-assessment results and might provide you with a false sense of security.

1. Using a client software package for connecting to corporate resources over a secure VPN in known as what?

    a. Access VPN
    b. Remote-access VPN
    c. Dialup VPN
    d. Network resource VPN
    e. None of the above

2. What is not a step required to create an IPsec VPN tunnel?

    a. User initiates network access.
    b. Endpoint devices authenticate each other via IKE.c.
    c. IPsec SA between the peers is established.
    d. User credential information is validated on the devices.
    e. All of the above.

3. Which of the following are components of defining the IKE (Phase 1) policy?

    a. Select a key distribution method.
    b. Select an authentication method.
    c. Identify the ISAKMP peer.
    d. Select the ISAKMP polices.
    e. Select the IKE SA settings.
    f. All of the above.

4. What is not an ISAKMP policy for connection within the Cisco IOS Firewall feature set?

    a. Message encryption algorithm
    b. DES
    c. Diffie-Hellman
    d. Triple DES
    e. SHA-1
    f. All of the above

5. All of the following are features of ESP except for what?

   a. Provides origin authentication
   b. Implements data integrity
   c. Provides antireplay protection
   d. Encrypts only the payload
   e. Support tunnel and transport mode
   f. All of the above

6. What valid IPsec transforms can be used in an IPsec tunnel configuration? (Select four.)

   a. esp-md5-hmac
   b. ah-sha-hmac
   c. esp-aes (aes 128,192,256)
   d. AH, DES
   e. AH, AES-256
   f. esp-sha-hmac

7. What command is missing from the creation of this IKE policy on the router?

   a. router(config-isakmp)#**authentication pre-share**
   b. router(config-isakmp)#**encryption aes 256**
   c. router(config-isakmp)#**group 5**
   d. router(config-isakmp)#**lifetime 86400**
   e. #**hash sha**
   f. #**acl 300**
   g. #**pre-share key test123**
   h. #**hashtype md5**
   i. #**address 192.100.1.0 mask 0.0.0.255**
   j. None of the above

8. Which of the following is not a feature of the crypto map?

   a. Identify what traffic is to be encrypted
   b. Where data should be sent (SA peer)
   c. What security should be applied to the traffic
   d. How granular the logging of traffic should be
   e. How the IPsec SA should be established
   f. All of the above

9. When you manually configure the IPsec connection from the crypto map configuration, it removes this functionality.

   a. Ability to use preshared secrets.
   b. Multipoint tunnels.
   c. Strong authentication and encryption support.
   d. Peers can renegotiate and constantly change connection parameters.
   e. Tracking and monitoring.
   f. You cannot manually configure the IPsec connection.

10. What are the drawbacks of using RSA nonces for an IPsec connection? (Select two.)

    a. Initial key exchange
    b. Limited logging
    c. Difficult management
    d. Special privilege requirements
    e. Additional firmware
    f. All of the above

The answers to the "Do I Know This Already?" quiz are found in the appendix. The suggested choices for your next step are as follows:

- **8 or less overall score**—Read the entire chapter. This includes the "Foundation Topics" and "Foundation Summary" sections and the "Q&A" section.

- **9 or 10 overall score**—If you want more review on these topics, skip to the "Foundation Summary" section and then go to the "Q&A" section. Otherwise, move on to Chapter 20, "Scaling a VPN Using IPsec with a Certificate Authority."

Foundation Topics

Before beginning to configure the Cisco router as an endpoint for a VPN tunnel, it is important to understand the different types of VPNs and what part the Cisco router plays in each of these connections. The two types of VPNs are site-to-site VPNs and access VPNs:

- **Site-to-site VPNs**—A site-to-site VPN usually consists of two or more endpoints configured as a VPN peer with the other end. The term *endpoint* refers to the point where the VPN terminates. In other words, this is where the encrypted connection begins and ends. Designated traffic that travels from one end to the other is encrypted as it passes through the first endpoint and decrypted as it passes through the other endpoint. The encryption is manually configured on both ends or negotiated by each VPN peer. The endpoints for site-to-site VPNs are normally routers, VPN-enabled firewalls, or VPN hardware appliances. For the purposes of this chapter, the focus is on creating VPN connections using Cisco routers. Figure 19-1 depicts a site-to-site VPN connection between New York and San Francisco.

Figure 19-1 *Site-to-Site VPN Between New York and San Francisco*

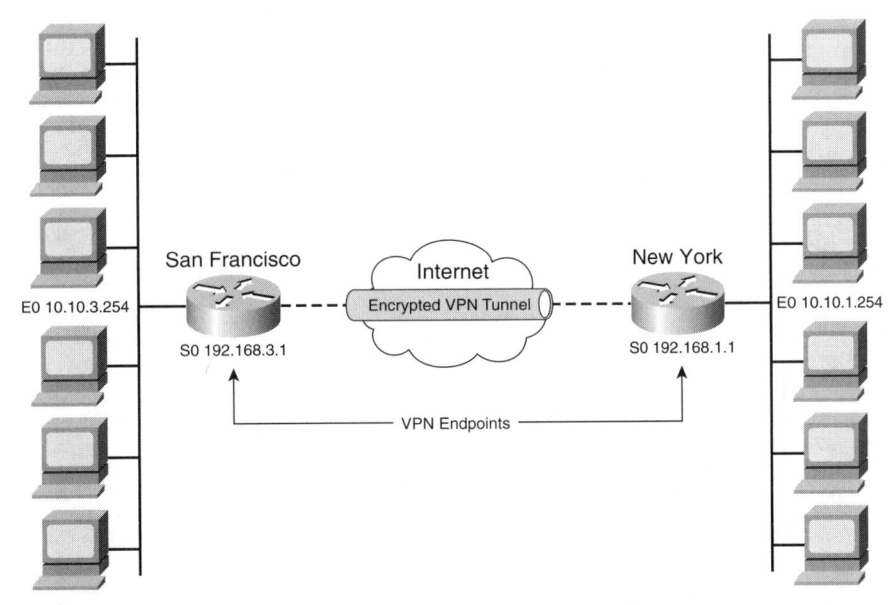

As Figure 19-1 illustrates, all traffic between the New York and San Francisco locations travels across the Internet but through an encrypted VPN tunnel to maintain data confidentiality and integrity. Not all data is required to travel through the VPN tunnel. You can configure the endpoints to encrypt only specific traffic, referred to as *interesting traffic*. It is also important to note that traffic on the internal network is not encrypted. This traffic is only encrypted as it enters the router on its way to its destination. Site-to-site VPNs are divided into two categories:

— **Intranet VPNs**—Intranet VPNs are used to securely interconnect the different locations of an organization. Intranet VPNs allow an organization that might be spread across multiple locations globally to function as a single secure enterprise network.

— **Extranet VPNs**—Extranet VPNs provide a secure channel for communications between an organization and its business partners (for example, suppliers, customers, and so on).

- **Access VPNs**—Access VPNs, also known as *remote-access VPNs*, normally incorporate a VPN client software package installed on the remote user's computer. The remote user can connect to the Internet via dialup, cable modem, DSL, or even from a different organization's LAN connection. When a connection is made to the user's network, the VPN client software creates an encrypted connection from that workstation to the VPN endpoint. The VPN endpoints for remote access VPNs are routers, VPN-enabled firewalls, or VPN concentrators. This chapter focuses on VPN routers because they are commonly used to create both types of VPNs and will most likely be emphasized on the exam. Figure 19-2 depicts remote access VPN connections to the office in New York and illustrates that all traffic from the remote users to New York is encrypted at the user's computer and remains encrypted until it reaches the router in New York.

Figure 19-2 *Remote-Access VPNs to the New York Office*

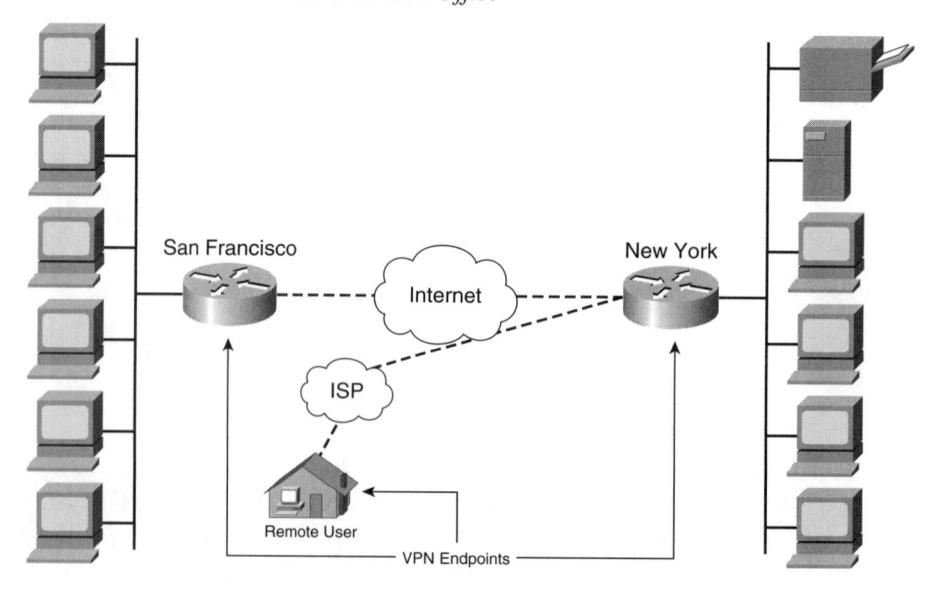

The method used to access the Internet does not normally affect the use of VPN client software so long as it is not blocked by an access list or firewall rule. The standard design for a large enterprise network (such as the SAFE Blueprint) is to install the VPN endpoint inside of the perimeter router or firewall.

Configuring a Cisco Router for IPsec Using Preshared Keys

IPsec is not a protocol but a framework of open-standard protocol suites that provides origin authentication, data integrity, data confidentiality, and antireplay protection. IPsec runs over IP and uses Internet Key Exchange (IKE) to negotiate the security association (SA) between the peers. Parameters must be configured for both IKE and IPsec SAs.

How IPsec Works

Five specific steps are required to create and terminate an IPsec VPN tunnel. The endpoints perform different functions to establish the encrypted connection at each step. Figure 19-3 describes the steps required to create and terminate the IPsec tunnel.

Figure 19-3 *Creating an IPsec VPN Tunnel*

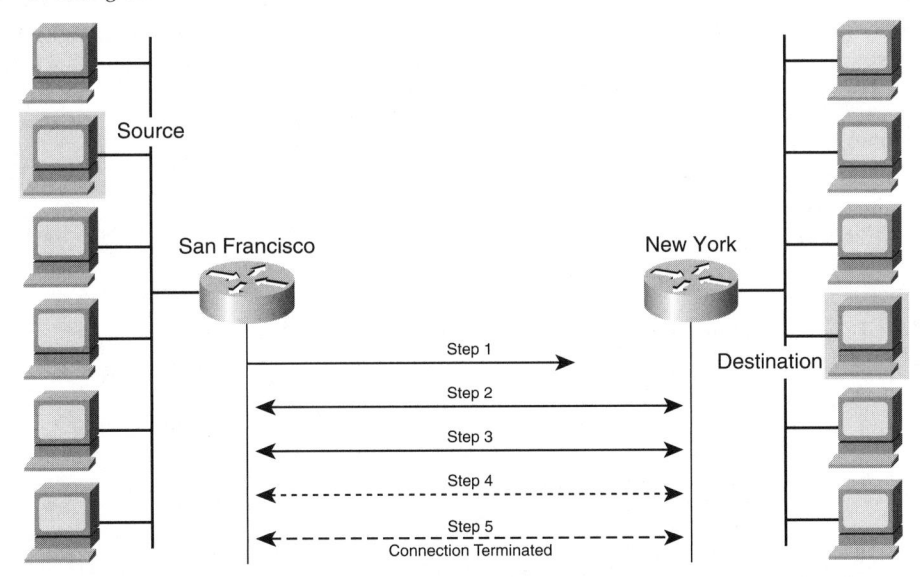

Step 1   The user at the source computer in San Francisco initiates a connection to the destination system in New York. The router in San Francisco recognizes the traffic as "interesting traffic" because of the configured ACL and initiates the IKE process with the router in New York.

404 Chapter 19: Building a VPN Using IPsec

Step 2 The endpoint routers use IKE to authenticate each other as IKE peers and negotiate the IKE SA. At this time, a secure channel is established allowing for negotiation of the IPsec SA. This is referred to as *IKE Phase 1*.

Step 3 IKE is again used to negotiate the IPsec SA between the peers. When the negotiation is completed, the IPsec peers have an established SA and are prepared to transfer data. This is referred to as *IKE Phase 2*.

The tunnel is established and the IPsec SA information is stored in the SA database on both SA peers. Further key negotiations take place per the parameters negotiated during Phase 2.

Step 4 The connection terminates when it times out or is deleted from either peer.

This section describes configuring VPN routers with preshared keys for site-to-site connectivity with VPNs.

> **NOTE** Internet Key Exchange (IKE) is a protocol based on ISAKMP/Oakley, which stands for *Internet Security Association and Key Management Protocol* (with Oakley distribution), and supports IPsec by providing a private, authenticated key management channel through which the peers can communicate and negotiate session keys for AH and ESP and to negotiate encryption, authentication, and compression algorithms for the session. IKE is used to perform authentication for IPsec peers, negotiation of IKE and IPsec SAs, and establishment of keys for the encryption algorithms used by IPsec. The terms *IKE* and *ISAKMP* are used interchangeably throughout this chapter.

To configure IPsec encryption on the Cisco router, you must complete four steps. Each task includes specific subtasks:

Step 1 Select the IKE and IPsec parameters.

Step 2 Configure IKE.

Step 3 Configure IPsec.

Step 4 Test and verify the IPsec configuration.

Step 1: Select the IKE and IPsec Parameters

The process for configuring a router for an IPsec VPN is not a difficult one. It is, however, a complex process with multiple tasks and subtasks and requires significant attention to detail. The first task involves selecting the initial configuration parameters for the VPN connection and determining which configuration is most appropriate. If you make all the configuration decisions prior to configuring either device, you can greatly reduce the risk of a configuration error. This task is divided into five subtasks:

Step 1   Define the IKE (Phase 1) policy.

Step 2   Define the IPsec policies.

Step 3   Verify the current router configuration.

Step 4   Verify connectivity.

Step 5   Ensure compatible access lists.

Define the IKE (Phase 1) Policy

The importance of IKE Phase 1 is that it provides the negotiation to create a secure channel through which the Phase 2 negotiation can take place. You must consider several items when determining the IKE Phase 1 policies. Following is a list of each item with its specific purpose:

- **Select a key distribution method**—This item is usually determined by the expected size of the network. For networks that only require a few VPN peers, you can manually distribute the keys (configure each peer manually). For large networks, it is recommended to use a certificate authority (CA) server. This method allows for significant growth because a trusted CA identifies each IPsec peer. If you are not manually distributing the keys, you need to implement the ISAKMP to support the method of key distribution you have selected.

- **Select an authentication method**—There are several ways to configure the routers to authenticate themselves during Phase 1 of the IKE negotiation when establishing the SA. The configuration to be used is usually determined by the number of VPNs connected to the router and how dynamic the network environment will be. Three different configuration types are used:

    — **Preshared keys**—If your organization only requires VPN connectivity with few locations, you might want to use a static configuration on the router. This static configuration is referred to as *preshared keys* because the keys are manually configured on both peers. Preshared keys are alphanumeric keys (similar to passwords) configured on each router and must match exactly for the routers to negotiate the connection. Management of multiple VPN connections using preshared keys can become cumbersome as the number of connections grows.

    — **RSA signatures**—RSA is a public key cryptography system using digital certificates authenticated by RSA signatures.

    — **RSA-encrypted nonces**—An RSA nonce is a random value generated by the peer that is encrypted using RSA encryption. This method requires you to configure the

RSA public key and designate the peer. This method is more secure because a different nonce is created with every negotiation.

- **Identify the ISAKMP peer**—The ISAKMP peer is the router at the other end of the VPN connection that is functioning as the termination point. It is the device that you negotiate with to create the VPN tunnel. The IPsec peer is identified either by IP address or host name.

- **Select the ISAKMP policies for the connection**—It is important that the ISAKMP policies for both peers match. If the configurations differ on each peer, they cannot negotiate the VPN connection. You can configure multiple policies on each router, however, because each router will search for a matching policy. The following items must be determined when selecting the ISAKMP policies:

  — **Message-encryption algorithm**—Cisco IOS VPN routers support two encryption algorithms:

  Data Encryption Standard (DES)—DES is a 56-bit symmetric encryption algorithm. It uses a 64-bit block of plain text and converts it into cipher text of the same size, encrypting it with a secret key. The key length is also 64 bits, but 8 bits are used for parity, leaving the effective key length at 56 bits. Although still widely used, DES is a somewhat outdated algorithm and should not be used if your data is highly sensitive. It is commonly used for VPN connections to locations outside the United States that cannot purchase higher levels of encryption because of U.S. technology export policies.

  Triple Data Encryption Standard (3DES)—3DES is a 168-bit symmetric encryption algorithm. 3DES just applies three different phases of DES, effectively tripling the key length to 168 bits. The data is encrypted using three stages of DES using a different 56-bit key for each stage.

  — **Message-integrity (hash) algorithm**—The hash algorithm converts message input into a fixed-length output called the *message digest*. The message digest is then put into a digital signature algorithm, and the output becomes a digital signature for the message. Because the message digest is usually much smaller than the actual message, it is more efficient to sign the digest rather than the message itself. Keyed-Hashing for Message Authentication (HMAC) is a variant that provides an additional layer of security by performing additional cryptographic keying and a secret key for calculation and verification of the message authentication values. HMAC is a variant that can be added to the supported hash algorithms. Cisco IOS VPN routers support two hash algorithms:

  Secure Hash Algorithm 1 (SHA-1)—The output of SHA-1 is 160 bit. Because the output is larger than MD5, SHA-1 is considered to be more secure; however, it requires more CPU cycles to process.

Message Digest 5 (MD5)—The output of MD5 is 128 bit. MD5 is slightly faster to process because of its smaller message digest.

- **Peer-authentication method**—This is the method that each peer uses to authenticate itself to the other peer. The three methods are explained in the previous section (preshared keys, RSA signatures, and RSA-encrypted nonces).

- **Diffie-Hellman key exchange**—Diffie-Hellman is a public key cryptography protocol used between two IPsec peers to derive a shared secret over an unsecured channel without transmitting it to each other. There are seven Diffie-Hellman groups with varying key lengths. This chapter focuses on the first two Diffie-Hellman groups because they are currently the most commonly used on VPN enabled routers: group 1 is 768 bits, and group 2 is 1024 bits.

- **IKE SA lifetime**—The SA lifetime is the time that each system waits before initiating another key exchange. This allows the systems to constantly renegotiate the connection, greatly reducing any chance of an unauthorized listener being able to decrypt the connection.

Table 19-2 lists the IKE policy parameters that can be used when determining the IKE Phase 1 policies.

Table 19-2 *IKE Policy Parameters (Phase 1)*

Parameter	Strong	Stronger
Encryption algorithm	DES	3DES, AES, AES-192, AES-265
Hash algorithm	MD5	SHA-1
Authentication method	Preshare	RSA signature RSA-encrypted nonce
Diffie-Hellman key exchange	Group 1	Group 2 Group 5
SA lifetime	86,400 seconds	<86,400 seconds

The peers cannot negotiate the connection if the policies on both peers do not match. Figure 19-4 depicts the peer configurations for the VPN connection between New York and San Francisco.

As Figure 19-4 illustrates, the IKE configuration parameters on both routers must match for the negotiation to complete successfully.

Figure 19-4 *VPN Peer Configuration (New York and San Francisco)*

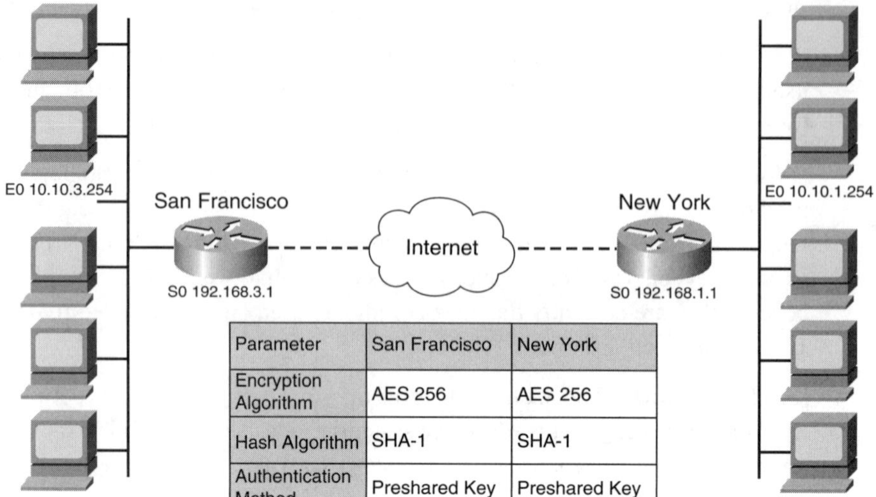

Parameter	San Francisco	New York
Encryption Algorithm	AES 256	AES 256
Hash Algorithm	SHA-1	SHA-1
Authentication Method	Preshared Key	Preshared Key
Diffie-Hellman Key Exchange	2	2
SA Lifetime	86,400 Seconds	86,400 Seconds
Peer Address	192.168.1.1	192.168.3.1

Define the IPsec Policies

IPsec uses two different modes for VPNs. Each of these modes has a specific purpose, and it is important to select the correct mode when configuring your VPN. The two VPN modes are as follows:

- **Tunnel mode**—The tunnel mode is commonly used for both site-to-site VPNs and access VPNs where the destination is not the VPN endpoint. Both the source and destination (in the original IP header) information are encrypted when using the tunnel mode and are not decrypted until they reach the destination endpoint.

- **Transport mode**—The transport mode is most commonly used when the destination is the VPN endpoint. The original IP header is not encrypted when using the transport mode.

The IPsec policies are often referred to as the *IKE Phase 2* policies because they occur during Phase 2 of the IKE negotiation. IKE Phase 1 establishes a bidirectional secure tunnel known as the *IKE SA*, which is used to complete the negotiation of the IPsec SA. The routers must negotiate two separate unidirectional IPsec SAs to facilitate bidirectional traffic between the peers.

Many configuration options are available when configuring IPsec. It is important to select the best possible configuration for the VPN connection. Follow these steps when defining the IPsec parameters:

Step 1 **Identify the IPsec protocol necessary for the type of traffic.** Two different IPsec protocols perform specific functions:

- **Encapsulating Security Payload (ESP)**—ESP provides data authentication, encryption, and antireplay services. ESP is protocol number 50 assigned by the Internet Assigned Numbers Authority (IANA). ESP is primarily responsible for getting the data from the source to the destination in a secure manner, verifying that the data has not been altered, and ensuring that the session cannot be hijacked. ESP provides origin authentication, data integrity, and antireplay protection. ESP can also be used to authenticate the sender either by itself or in conjunction with AH. ESP can be configured to encrypt the entire data packet or only the payload of the packet. Figure 19-5 shows how ESP encapsulates the normal IPv4 packet in the transport mode and in the tunnel mode.

  Figure 19-5 illustrates the normal IPv4 packet before and after encapsulation by ESP.

Figure 19-5 *ESP and the IPv4 Packet*

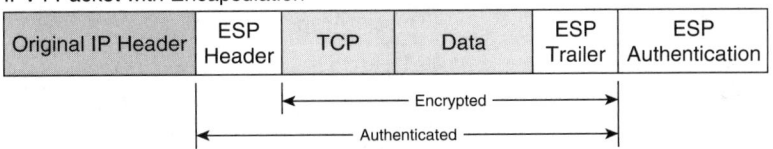

IPV4 Packet Without ESP Encapsulation

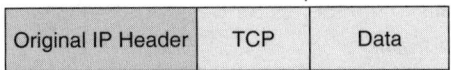

Original IP Header	TCP	Data

IPV4 Packet with Encapsulation

Original IP Header	ESP Header	TCP	Data	ESP Trailer	ESP Authentication

Encrypted: ESP Header to ESP Trailer
Authenticated: ESP Header to ESP Authentication

- **Authentication Header (AH)**—AH provides data authentication and antireplay services. AH is protocol number 51 assigned by the IANA. The primary function of AH is origin authentication. AH does not provide any data encryption; it provides only origin authentication or verification that the data is from the sender. This functionality also prevents session hijacking. It is important to note that ESP only authenticates the payload, whereas AH authenticates the IP header. AH is not compatible with Network Address Translation (NAT) or Port Address Translation (PAT) because they change the source IP address, making it different from the source address in the authentication header. The IPsec peer then rejects the traffic because the source address in the IP header and the authentication header do not match. Figure 19-6 shows how AH effects the normal IPv4 packet in both the transport and tunnel modes.

Figure 19-6 *AH and the IPv4 Packet*

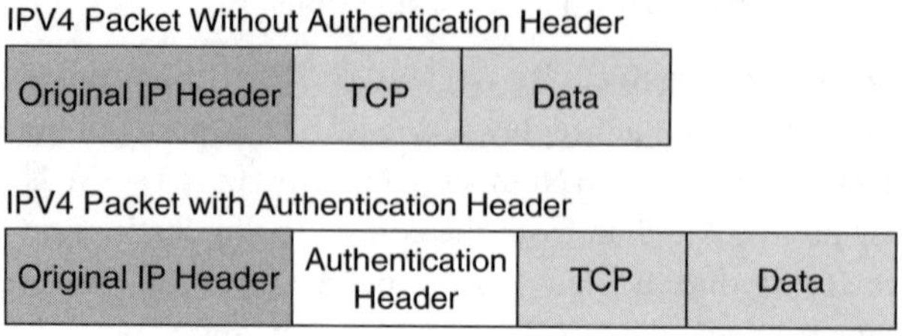

IPV4 Packet Without Authentication Header

| Original IP Header | TCP | Data |

IPV4 Packet with Authentication Header

| Original IP Header | Authentication Header | TCP | Data |

Step 2 **Select the appropriate IPsec transforms.** Transforms and transform sets are the defined combination of IPsec algorithm and encryption algorithm. The combination you select can focus more on authentication, encryption, or combine to cover both. The following protocols, algorithms, and so on, are combined to create your transforms:

- **IPSec protocol**—AH and ESP
- **Encryption algorithm**—DES, 3DES, or AES
- **Hash algorithm**—SHA-1 and MD5 (with or without HMAC)

Table 19-3 lists the possible combinations for transforms. When combined, the transforms make a *transform set*.

Table 19-3 *IPsec Transforms*

AH Transform	Description
ah-md5-hmac	AH, MD5 hash, HMAC variant (authentication)
ah-sha-hmac	AH, SHA-1 hash, HMAC variant (authentication)
ESP Encryption Transform	**Description**
esp-null	ESP, Null encryption
esp-des	ESP, DES (56 bit encryption)
esp-aes	ESP, AES (128 bit encryption)
esp-3des	ESP, 3DES (168 bit encryption)
esp-aes 192	ESP, AES-192 (192 bit encryption)
esp-aes 256	ESP, AES-256 (256 bit encryption)
ESP Authentication Transform	**Description**
esp-md5-hmac	ESP, MD5 hash (HMAC variant)
esp-sha-hmac	ESP, SHA-1 hash (HMAC variant)

Step 3 **Define the IPsec peer.** You must define the router at the other end of the VPN connection by either host name or IP address.

Step 4 **Define the local hosts or networks.** Identify which local hosts or networks are allowed to send traffic through the VPN connection. This is completed by configuring ACLs on the router to define "interesting traffic," which will initiate the encryption.

Step 5 **Select the type of SA initiation.** Determine whether the IPsec SA should be negotiated by IKE or by using manual IPsec.

Verify the Current Router Configuration

You must verify that the current configuration of the router will not conflict with the new items that you want to add. You can use three commands to display the current router configuration:

- **show running-configuration**—This command displays the current configuration that is running on the router. The **show configuration** command shows the last configuration that was saved to memory but does not display any changes that took place but were not saved. This command is the same as the **show startup configuration** because it displays the configuration the router will have when it starts—any unsaved configuration changes are lost during a reboot.

- **show crypto isakmp policy**—This command displays the current ISAKMP policy that is configured on the router. You can then verify that your planned configuration will not conflict with the current configuration.

- **show crypto map**—This command can include the interface or the map name and displays any crypto map entries configured on the router. The crypto map includes the name, interface, and local address of the router; peer address; crypto access list; SA lifetime; and transform set name. The **show crypto map** command also tells you whether Perfect Forward Secrecy has been enabled for that connection. *Perfect Forward Secrecy* is a key-establishment protocol that generates a new public/private key pair with each session. The result is a dynamic key exchange that prevents an eavesdropper from decrypting messages using keys derived from previously captured data.

Verify Connectivity

Remember that encryption adds complexity to any network connection. You should always ensure connectivity between the SA peers before attempting to create a VPN connection. The best way to verify your connectivity is to attempt the type of connection that you intend to use after the encrypted connection is established. By doing so, you can ensure that you do not have any filters preventing the connection. If there is no connectivity between the peers, it will be impossible for them to negotiate the IKE SA and build the encrypted tunnel.

Ensure Compatible Access Lists

You must ensure that the access lists on the perimeter routers allow IPsec traffic. IKE runs over UDP port 500. If your perimeter routers are blocking UPD 500, you cannot negotiate the IKE SA. In addition, ESP and AH are protocol numbers 50 and 51, respectively. It is important that you verify that your current access lists do not block these protocols.

Step 2: Configure IKE

Now that you understand the importance of planning the configuration beforehand, refer to Figure 19-7 for the configuration settings for task 2.

Figure 19-7 *IKE Configuration Settings*

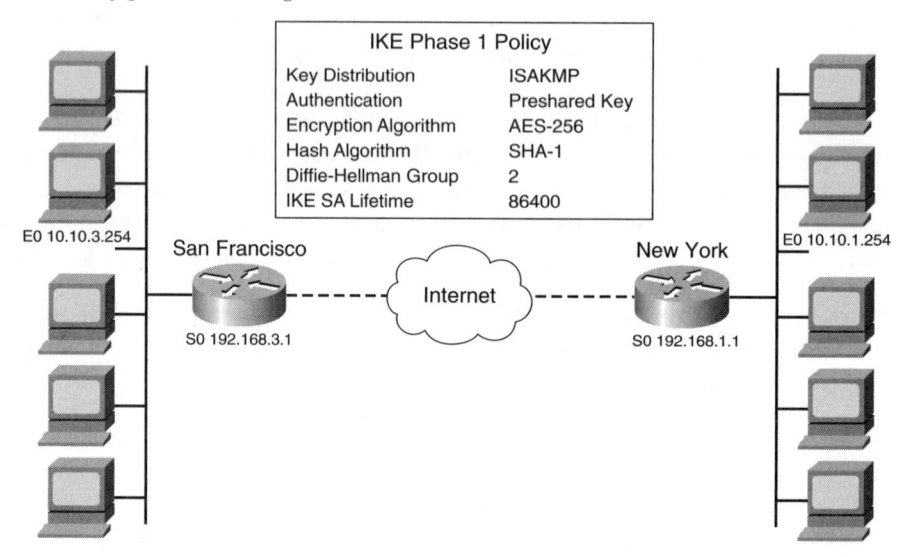

```
IKE Phase 1 Policy
Key Distribution      ISAKMP
Authentication        Preshared Key
Encryption Algorithm  AES-256
Hash Algorithm        SHA-1
Diffie-Hellman Group  2
IKE SA Lifetime       86400
```

E0 10.10.3.254 — San Francisco — S0 192.168.3.1 — Internet — New York — S0 192.168.1.1 — E0 10.10.1.254

Figure 19-7 shows the standards that you plan to use when configuring IKE on both peers. This section shows the configuration of the router in New York for the VPN between New York and San Francisco. Configuration steps and commands are common for the Cisco certification exams. This exercise is designed to familiarize you with the different commands for configuring IKE, the purpose for each command, and how the commands interact. To configure IKE on the router, follow these four steps:

Step 1   Enable IKE.

Step 2   Create the IKE policy.

Step 3   Configure the preshared key.

Step 4   Verify the IKE configuration.

Each of these steps is discussed in detail in the following sections.

Enable IKE

IKE is enabled by default. You should verify that it is enabled when you check the current configuration. IKE is enabled globally (on all interfaces) and must be enabled before you can use it on the router. Example 19-1 shows the command for enabling IKE on the router.

Example 19-1 *Enabling IKE*

```
NewYork#configure terminal
NewYork (config)#crypto isakmp enable
```

Create the IKE Policy

You can configure multiple IKE policies on a single router. The router runs through each policy in order based on the *policy priority*. The *policy priority* is a number between 1 and 10,000 assigned when the policy is created: the lower the number, the higher the priority. Because the router checks each policy in order, it is a good idea to give the most secure policy the lower number. The **crypto isakmp policy** command is input from the global configuration mode and will put you in the config-isakmp mode. Example 19-2 shows the command for creating the IKE policy.

Example 19-2 *Creating IKE Policy*

```
NewYork#configure terminal
NewYork (config)#crypto isakmp policy 100
NewYork (config-isakmp)#authentication pre-share
NewYork (config-isakmp)#encryption aes 256
NewYork (config-isakmp)#hash sha
NewYork (config-isakmp)#group 5
NewYork (config-isakmp)#lifetime 86400
```

Configure Preshared Key

Preshared keys (also known as *preshared secrets*) are alphanumeric keys manually configured on each peer. You must identify the peer and the key when performing the configuration. The peer is identified either by host name or IP address. The command for configuring the preshared key is as follows:

```
crypto isakmp key keystring address  peer-address [mask]
```

or

```
crypto isakmp key keystring hostname peer-hostname
```

Example 19-3 shows the command for defining the preshared key and the peer.

Example 19-3 *Defining the Preshared Key and Peer*

```
NewYork#configure terminal
NewYork (config)#crypto isakmp policy 100
NewYork (config-isakmp)#authentication pre-share
NewYork (config-isakmp)#encryption aes 256
```

continues

Example 19-3 *Defining the Preshared Key and Peer (Continued)*

```
NewYork (config-isakmp)#hash sha
NewYork (config-isakmp)#group 5
NewYork (config-isakmp)#lifetime 3600
NewYork (config)#crypto isakmp key abc123 address 192.168.3.1 255.255.255.255
```

Verify the IKE Configuration

Before moving on to the next task, it is a good idea to verify the changes you have just completed. It is much easier to correctly configure each step rather than troubleshoot the configuration if the VPN does not work. The **show crypto isakmp policy** command enables you to view the current configuration, and the output should match the combined examples listed previously. Example 19-4 shows the output from **show crypto isakmp policy**.

Example 19-4 *Verifying the IKE Policy*

```
NewYork#configure terminal
NewYork (config)#crypto isakmp key abc123 address 192.168.3.1 255.255.255.255
NewYork (config)#crypto isakmp policy 100
NewYork (config-isakmp)#authentication pre-share
NewYork (config-isakmp)#encryption aes 256
NewYork (config-isakmp)#hash sha
NewYork (config-isakmp)#group 5
NewYork (config-isakmp)#lifetime 86400
NewYork (config-isakmp)#exit

NewYork#sh crypto isakmp policy

Global IKE policy
Protection suite of priority 100
        encryption algorithm:   AES - Advanced Encryption Standard (256 bit keys).
        hash algorithm:         Secure Hash Standard
        authentication method:  Pre-Shared Key
        Diffie-Hellman group:   #5 (1536 bit)
        lifetime:               86400 seconds, no volume limit
Default protection suite
        encryption algorithm:   DES - Data Encryption Standard (56 bit keys).
        hash algorithm:         Secure Hash Standard
```

Step 3: Configure IPsec

Just like the IKE configuration, it is important that the IPsec configuration matches on both peers for them to negotiate the IPsec SA. Figure 19-8 shows the configuration settings for Step 3.

Figure 19-8 *IPsec Configuration Settings*

IKE Phase 1 Policy

Key Distribution	ISAKMP
Authentication	Preshared Key
Encryption Algorithm	AES-256
Hash Algorithm	SHA-1
Diffie-Hellman Group	2
IKE SA Lifetime	86400

San Francisco — Internet — New York

E0 10.10.3.254
S0 192.168.3.1
S0 192.168.1.1
E0 10.10.1.254

IPsec Policy

Authentication Transform	esp-sha-hmac
Encryption Transform	esp-aes-256

IPsec Peer

Local/Host/Network

IPsec SA Lifetime   3600 Seconds

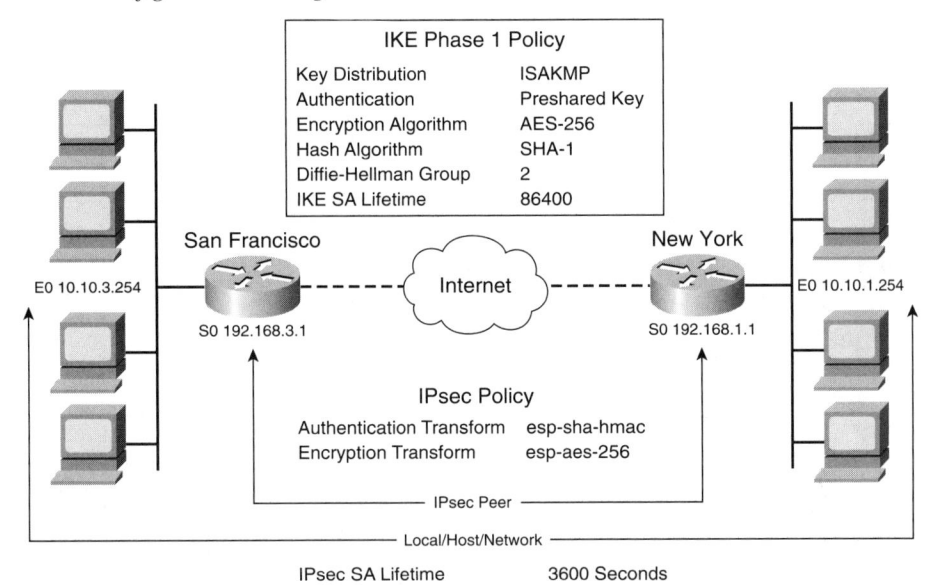

Figure 19-8 shows all the information needed to configure IPsec on both peers. This exercise configures only the router in New York. To configure IPsec on the router, follow these five steps:

Step 1 Create the IPsec transform set.

Step 2 Configure IPsec SA lifetimes.

Step 3 Create the crypto access lists.

Step 4 Create the crypto map.

Step 5 Apply the crypto map.

Each of these steps is discussed in detail in the following sections.

Create the IPsec Transform Set

The IPsec transform set defines the parameters that each peer uses when negotiating the VPN connection. It is possible to configure multiple transform sets on the router. When negotiating the connection, each router compares transform sets until finding a match. (This negotiation takes place during IKE Phase 2.) This is not to say that the routers negotiate the transform sets; they just compare the transform sets until they find a match. If the routers do not find transform sets that match, they cannot create the VPN tunnel. Figure 19-9 depicts how the routers compare the transform sets when negotiating the VPN connection.

Figure 19-9 shows the routers comparing transform sets until they find a match at **esp-aes 256 esp-sha-hmac**.

Figure 19-9 *Transform Set Comparison*

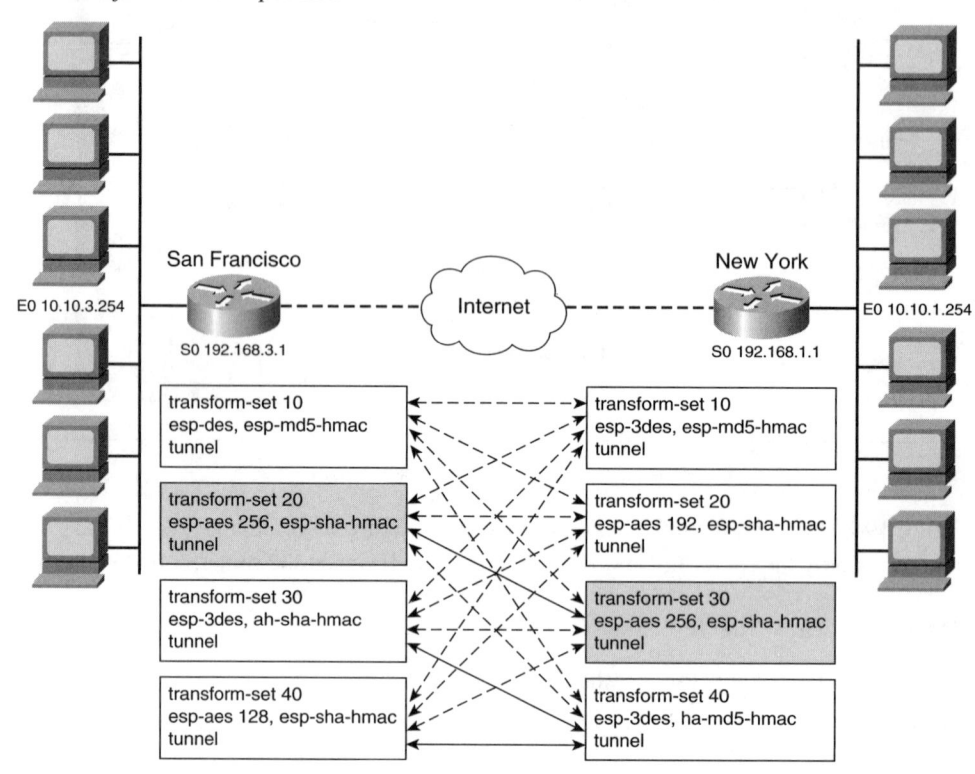

The command syntax for creating the IPsec transform set is as follows:

```
crypto ipsec transform-set transform set name [transform1][transform2][transform3]
```

Example 19-5 shows the command for creating the IPsec transform set.

Example 19-5 *Creating the IPsec Transform Set*

```
NewYork#configure terminal
NewYork (config)#crypto ipsec transform-set 20 esp-aes 256 esp-sha-hmac
```

Configure IPsec SA Lifetimes

The IPsec SA lifetime determines at what interval the routers renegotiate the connection. A constantly changing connection is extremely difficult to decrypt because the data sample continues to change, leaving no consistent data available for a cryptographer to decrypt. If a cryptographer were able to discover the keys for a connection, it would take more than the hour that those keys are valid. The global IPsec SA lifetime is configured in either seconds or kilobytes (amount of data that pass through the tunnel). The command syntax for defining the global IPsec SA lifetime is as follows:

```
crypto ipsec security-association lifetime { seconds #### | kilobytes ####}
```

The default value is 3600 seconds (1 hour) or 4,608,000 kilobytes. The global IPsec SA lifetime is overridden by the SA lifetime that is added to the crypto map, and Cisco recommends that you use the default values. Example 19-6 shows the command for configuring the global IPsec SA lifetime to 30 minutes, or 4,400,000 kilobytes.

Example 19-6 *Configuring the Global IPsec SA Lifetime*

```
NewYork#configure terminal
NewYork (config)#crypto ipsec security-association lifetime seconds 1800
or
NewYork (config)#crypto ipsec security-association lifetime kilobytes 4400000
```

Create the Crypto ACLs

The crypto ACL defines interesting traffic for the router. *Interesting traffic* is the traffic protected by the VPN connection. Crypto ACLs determine which outbound traffic is encrypted and which goes out as clear text. Inbound traffic is also compared to the crypto ACL. If traffic comes in as clear text and should be encrypted, the router drops the traffic. When creating the crypto ACL, the terms **permit** and **deny** refer to *encrypt* and *do not encrypt,* respectively. The syntax of the command is as follows:

```
access-list ACL-number permit | deny protocol
    source-ip source wildcard destination-ip destination wildcard
```

Example 19-7 shows the command for configuring the crypto ACL for the connection from New York to San Francisco to encrypt all TCP traffic between those networks.

Example 19-7 *Crypto ACL on the New York Router*

```
NewYork#configure terminal
NewYork (config)#access-list 105 permit ip 10.10.1.0 0.0.0.255 10.10.3.0 0.0.0.255
```

With the preceding configuration, the router encrypts all traffic from the internal network at New York (10.10.1.0/24) destined for the San Francisco network (10.10.3.0/24) and expects that all traffic from San Francisco be encrypted. It is important that the crypto ACLs match at both ends of the connection to ensure that the traffic is able to flow.

> **NOTE** Because the IPsec SA is a unidirectional connection, you can configure the peers to only allow encrypted traffic to travel in one direction.

Create the Crypto Map

The crypto map is the component that consolidates all the IPsec configuration pieces. The following items are defined in the crypto map:

- Which traffic is to be encrypted (reference to the access list)
- How granular the protected data flow should be
- Where the encrypted data should be sent (the SA peer)
- The local address used for encrypted data (local router interface address)
- What security should be applied to the traffic (reference to transform sets)
- How the IPsec SA should be established (manual or IKE)
- Any other parameters for the IPsec SA

This command requires multiple lines. Each line of the command addresses a different portion of the configuration. The syntax of the **crypto map** command is as follows:

```
crypto map map-name seq-number connection
```

The command **crypto map** is entered from the global configuration mode and identifies the crypto map by name and sequence number. It also configures how the IPsec SA should be established. Table 19-4 lists the possible commands for configuring the IPsec SA.

Table 19-4 *Crypto Map IPsec SA Commands*

Command	Description
cisco	This is the default value and indicates that IPsec will not be used but will be replaced with CET. This transform is being phased out.
ipsec-manual	This value indicates that IKE will not be used to establish the IPsec SA.
ipsec-isakmp	This value indicates that IKE will be used to establish the IPsec SA.
dynamic	This optional command specifies that a pre-existing static crypto map be referenced for the correct configuration. This option is only available after the **ipsec-isakmp** parameter.

Example 19-8 shows the command for configuring the crypto map for the connection from New York to San Francisco.

Example 19-8 *Crypto Map on the New York Router*

```
NewYork#configure terminal
NewYork(config)#crypto map SanFran 120 ipsec-isakmp
NewYork(config-crypto-map)#match address 105
```

Example 19-8 *Crypto Map on the New York Router (Continued)*

```
NewYork(config-crypto-map)#set peer 192.168.3.1
NewYork(config-crypto-map)#set pfs group5
NewYork(config-crypto-map)#set transform-set 20
NewYork(config-crypto-map)#set security-association lifetime seconds 1800
```

If you want to create the same type of VPN connection to another IPsec peer, you can add another **set peer** line to the crypto map if the traffic matches that same access list.

Apply the Crypto Map to the Correct Interface

For the crypto map to take effect, it must be applied to the interface facing the peer. You must enter the interface configuration mode to apply the crypto map. Example 19-9 shows the command applying the crypto map to the Serial 0 interface on the router in New York.

Example 19-9 *Applying the Crypto Map to Interface S0*

```
NewYork#configure terminal
NewYork(config)#interface fa0/1
NewYork(config-if)#crypto map SanFran
```

Step 4: Test and Verify the IPsec Configuration

It is best to verify your configuration beforehand instead of having to troubleshoot the connection if it is not working. Various **show** and **debug** commands enable you to check the current configuration, including the following:

- **show crypto isakmp policy**—Displays the configured IKE policies.
- **show crypto ipsec transform-set**—Displays the configured transform sets.
- **show crypto ipsec sa**—Displays the current state of your IPsec SAs.
- **show crypto map**—Displays your current crypto maps.
- **show crypto dynamic-map**—Displays your dynamic crypto map set.
- **debug crypto isakmp**—Enables debugging of IKE events. This command generates a tremendous amount of output and should be used only when traffic is low.
- **debug crypto ipsec**—Enables debugging of IPsec events. This command generates a tremendous amount of output and should be used only when traffic is low.

Remembering that the configurations must match on both peers, it is best to compare the configuration from each router if possible. Figure 19-10 shows the configuration settings for this task.

Chapter 19: Building a VPN Using IPsec

Figure 19-10 *Configuration Settings Between New York and San Francisco*

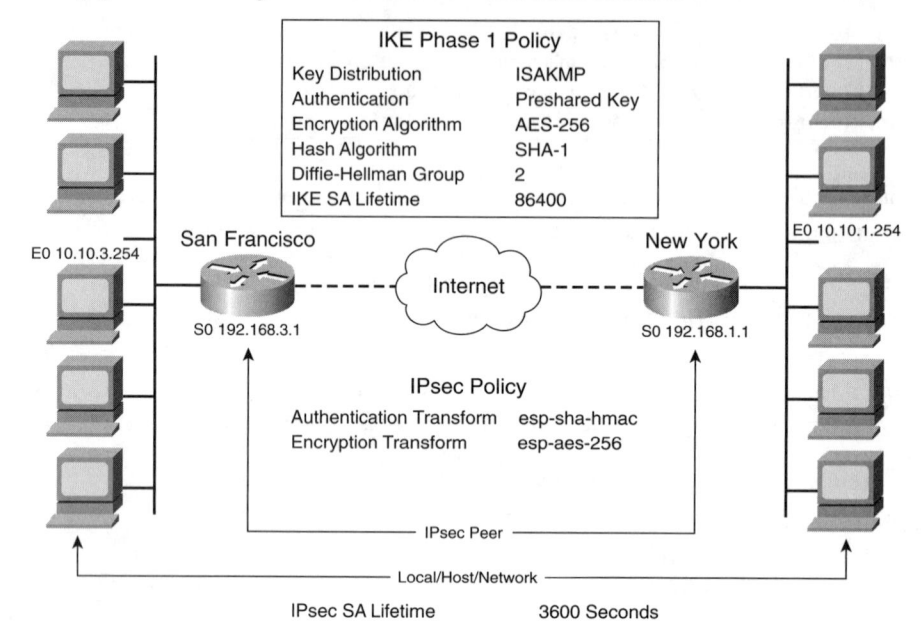

```
IKE Phase 1 Policy
Key Distribution        ISAKMP
Authentication          Preshared Key
Encryption Algorithm    AES-256
Hash Algorithm          SHA-1
Diffie-Hellman Group    2
IKE SA Lifetime         86400
```

San Francisco — E0 10.10.3.254, S0 192.168.3.1
New York — E0 10.10.1.254, S0 192.168.1.1

```
IPsec Policy
Authentication Transform    esp-sha-hmac
Encryption Transform        esp-aes-256
```

IPsec Peer
Local/Host/Network
IPsec SA Lifetime   3600 Seconds

Example 19-10 shows the output from each command. You must verify that the configurations are sufficient for the VPN to function between New York and San Francisco, as shown in Example 19-10.

Example 19-10 show *Command Output from Peers*

```
NewYork#show crypto isakmp policy
Global IKE policy
Protection suite of priority 100
 encryption algorithm:   AES - Advanced Encryption Standard (256 bit keys).
 hash algorithm:         Secure Hash Standard
 authentication method:  Pre-Shared Key
 Diffie-Hellman group:   #5 (1536 bit)
 lifetime:               86400 seconds, no volume limit
Default protection suite
 encryption algorithm:   DES - Data Encryption Standard (56 bit keys).
 hash algorithm:         Secure Hash Standard
 authentication method:  Rivest-Shamir-Adleman Signature
 Diffie-Hellman group:   #1 (768 bit)
 lifetime:               86400 seconds, no volume limit
NewYork#

San Francisco#show crypto isakmp policy
Global IKE policy
Protection suite of priority 100
 encryption algorithm:   AES - Advanced Encryption Standard (256 bit keys).
 hash algorithm:         Secure Hash Standard
```

Example 19-10 show *Command Output from Peers (Continued)*

```
    authentication method:   Pre-Shared Key
    Diffie-Hellman group:    #5 (1536 bit)
    lifetime:                86400 seconds, no volume limit
 Default protection suite
    encryption algorithm:    DES - Data Encryption Standard (56 bit keys).
    hash algorithm:          Secure Hash Standard
    authentication method:   Rivest-Shamir-Adleman Signature
    Diffie-Hellman group:    #1 (768 bit)
    lifetime:                86400 seconds, no volume limit
 SanFrancisco#
```

```
NewYork#show crypto ipsec transform-set
Transform set 20: {esp-256-aes esp-sha-hmac}
   will negotiate = { Tunnel, },



NewYork#
```

```
San Francisco#show crypto ipsec transform-set
Transform set 10: (esp-3des)
    will negotiate = (Tunnel, ),


Transform set 20: {esp-256-aes esp-sha-hmac}
   will negotiate = { Tunnel, },
Transform set 30: (esp-3des esp-md5-hmac)
   Will negotiate = (Tunnel, ),


Transform set 40 (esp-3des ha-md5-hmac)
   Will negotiate = (Tunnel, ),
```

```
NewYork#show crypto ipsec sa
interface: FastEthernet0/1
    Crypto map tag: SanFran, local addr. 192.168.1.1
    protected vrf:
    local  ident (addr/mask/prot/port): (10.10.1.0/255.255.255.0/0/0)
    remote ident (addr/mask/prot/port): (10.10.3.0/255.255.255.0/0/0)
    current_peer: 192.168.3.1:500
      PERMIT, flags={origin_is_acl,}
    #pkts encaps: 321, #pkts encrypt: 321, #pkts digest 321
    #pkts decaps: 321, #pkts decrypt: 321, #pkts verify 321
    #pkts compressed: 0, #pkts decompressed: 0
    #pkts not compressed: 0, #pkts compr. failed: 0
    #pkts not decompressed: 0, #pkts decompress failed: 0
    #send errors 4, #recv errors 0

     local crypto endpt.: 192.168.1.1, remote crypto endpt.: 192.168.3.1
     path mtu 1500, ip mtu 1500, ip mtu idb FastEthernet0/1
     current outbound spi: AE399F1B

     inbound esp sas:
```

continues

Example 19-10 show *Command Output from Peers (Continued)*

```
          spi: 0xFE4F2F8(266662648)
            transform: esp-256-aes esp-sha-hmac ,
            in use settings ={Tunnel, }
            slot: 0, conn id: 2000, flow_id: 1, crypto map: SanFran
            sa timing: remaining key lifetime (k/sec): (4570617/1455)
            IV size: 16 bytes
            replay detection support: Y

       inbound ah sas:

       inbound pcp sas:

       outbound esp sas:
          spi: 0xAE399F1B(2923011867)
            transform: esp-256-aes esp-sha-hmac ,
            in use settings ={Tunnel, }
            slot: 0, conn id: 2001, flow_id: 2, crypto map: SanFran
            sa timing: remaining key lifetime (k/sec): (4570617/1453)
            IV size: 16 bytes
            replay detection support: Y

       outbound ah sas:

       outbound pcp sas:
NewYork#
```

```
NewYork#show crypto map
Crypto Map "SanFran" 120 ipsec-isakmp
        Peer = 192.168.3.1
        Extended IP access list 105
access-list 105 permit ip 10.10.1.0 0.0.0.255 10.10.3.0 0.0.0.255
        Current peer: 192.168.3.1
        Security association lifetime: 4608000 kilobytes/1800 seconds
        PFS (Y/N): Y
        DH group:  group5
        Transform sets={20, }
        Interfaces using crypto map SanFran:
               FastEthernet0/1

NewYork#
```

```
San Francisco#show crypto map
Crypto Map "NY" 120 ipsec-isakmp
        Peer = 192.168.1.1
        Extended IP access list 105
access-list 105 permit ip 10.10.3.0 0.0.0.255 10.10.1.0 0.0.0.255
        Current peer: 192.168.1.1
        Security association lifetime: 4608000 kilobytes/1800 seconds
        PFS (Y/N): Y
```

Example 19-10 show *Command Output from Peers (Continued)*

```
        DH group:   group5
        Transform sets={20, }
        Interfaces using crypto map NY:
                FastEthernet0/1

SanFrancisco#
```

Configuring Manual IPsec

You can manually configure your IPsec connection from the crypto-map configuration mode. When you manually configure the IPsec parameters, you manually input all the keys necessary to create the connection. This configuration removes the functionality that allows the peers to renegotiate and constantly change the connection parameters and greatly reduces the security of the connection. The commands for configuring manual IPsec are as follows:

```
set session-key inbound | outbound ah/esp spi hex-key-string
set session-key inbound | outbound ah/esp spi authentication cipher hex-key-string
```

NOTE Cisco recommends against the use of manual IPsec because it is difficult to manage and relatively insecure as compared to configurations using IKE.

Configuring IPsec Using RSA-Encrypted Nonces

As discussed earlier in this chapter, an RSA nonce is a random value generated by the peer that is encrypted using RSA encryption. It provides a strong method of authentication using Diffie-Hellman key exchange. RSA nonces require that peers possess each other's public key without the use of a CA. It is important that the encrypted nonces are initially exchanged via a secure source. There are two drawbacks to using RSA-encrypted nonces:

- **Initial key exchange**—If you are using RSA-encrypted nonces, you must either manually configure and exchange RSA keys or use RSA signatures from a previously successful ISAKMP negotiation with the peer.
- **Management**—RSA-encrypted nonces can be difficult (and complex) to configure and exchange and are, therefore, more difficult to manage for enterprise networks.

The tasks required to configure IPsec using RSA-encrypted nonces are similar to the normal IPsec configuration with the exception of the second task (configuring RSA keys):

1. Select the IKE and IPsec parameters.
2. Configure the RSA keys.
3. Configure IKE.

4. Configure IPsec.

5. Test and verify the IPsec configuration.

This section focuses on the second task because all the other tasks have been discussed earlier in this chapter.

Configure the RSA Keys

As with any VPN configuration, management of RSA keys is not a difficult task, but it can be a complex undertaking. It is important to completely plan your implementation before you begin to configure the peers. To configure and generate your public keys and enter the public keys of your peer, follow these six steps:

Step 1 Plan the implementation using RSA keys.

Step 2 Configure the router host name and domain name.

Step 3 Generate the RSA keys.

Step 4 Enter your peer RSA public keys.

Step 5 Verify the key configuration.

Step 6 Manage the RSA keys.

Each of these steps is discussed in detail in the following sections.

Plan the Implementation Using RSA Keys

Planning a VPN implementation using RSA keys follows the same process as other IPsec configurations; however, you should ensure that you have carefully planned the key generation and exchange. As with any VPN configuration, little margin for error exists when configuring a VPN using RSA keys. It is best to define every portion of the configuration before you begin the implementation.

Configure the Router Host Name and Domain Name

An important part of authentication is that the system must be able to correctly identify itself. For this reason, you must configure the host name and domain name of the router. By configuring the host name and domain name on the router prior to generating the RSA keys, you can be sure that the router keys properly identify the router. To configure the host name of the router, use the **hostname** command while in the global configuration mode. To configure the domain name of the router, use the **ip domain-name** command with the correct domain name for the router. Example 19-11 shows the commands required to configure the host name and domain name for the router in New York.

Example 19-11 *Configuring the Host Name and Domain Name*

```
NewYork#configure terminal
NewYork(config)#hostname NewYork
NewYork(config)#ip domain-name NY.com
```

Generate the RSA Keys

By default, RSA key pairs do not exist on the Cisco router. You need to add the optional command **usage-keys** to the command to generate an encryption key pair and an authentication key pair. The command for generating RSA key pairs is **crypto key generate rsa usage-keys**. This command generates a key pair (one public and one private key). When generating RSA keys, you must select a "modulus length." RSA keys can be generated in four lengths: 360 bits, 512 bits, 1024 bits, and 2048 bits. The longer the modulus length, the more secure the key, and the more time required to generate the key. Cisco recommends a minimum modulus length of 1024 bits. This is a simple process for generating RSA keys. Example 19-12 depicts the generation of RSA keys for NewYork.NY.com with a modulus length of 1024 bits. Notice that the router tells you that it will take longer to generate keys with a longer key length.

Example 19-12 *Generating RSA Keys*

```
NewYork(config)#crypto key generate rsa usage-keys
The name for the keys will be: NewYork.NY.com
Choose the size of the key modulus in the range of 360 to 2048 for your
  Signature Keys. Choosing a key modulus greater than 512 may take
  a few minutes.

How many bits in the modulus [512]: 1024
Choose the size of the key modulus in the range of 360 to 2048 for your
  Encryption Keys. Choosing a key modulus greater than 512 may take
  a few minutes.

How many bits in the modulus [512]: 1024
% Generating 1024 bit RSA keys ...[OK]
% Generating 1024 bit RSA keys ...[OK]

NewYork(config)#
```

Verifying the RSA Keys

After you have generated your RSA key pairs, you can verify them with the command **show crypto key mypubkey rsa**. You can copy the output from this command and transmit it in a secure manner to a peer that you intend to utilize public key encryption with. Example 19-13 depicts the output from this command on the router NewYork.NY.com.

Example 19-13 *Verifying RSA Keys*

```
NewYork#sh crypto key mypubkey rsa
% Key pair was generated at: 00:22:09 UTC Mar 1 1993
Key name: NewYork.NY.com
 Usage: Signature Key
 Key is not exportable.
 Key Data:
  30819F30 0D06092A 864886F7 0D010101 05000381 8D003081 89028181 00D7518F D76BD206 DBFC4635
 DEBAD90A 749CF811 60D9AB5C 3E7EBA99 6201B691 43762A63 BFA033EB 9FD2EED9 C2B96512 06A40169
 0B989DBF 5F61A4FE 06736EE7 A3484B23 4B9B112E A0167D3C 6C47DBE4 3C8E17BD 316E0F10 744315D9
 A7A0D802 63FE54F4 6BA0FEA2 177C6074 159DFB9A 2C786C59 015CCA35 6A0D253C C8513B6D 29020301 0001
% Key pair was generated at: 00:22:21 UTC Mar 1 1993
Key name: NewYork.NY.com
 Usage: Encryption Key
 Key is not exportable.
 Key Data:
  30819F30 0D06092A 864886F7 0D010101 05000381 8D003081 89028181 00E83666 1BC9A809 97E4CDC1
 A99B8451 F4E1E9E1 50C8E3B2 3256B8E4 5831E1C9 C4CE1E9D 67548A8E EE59D86E 92CC694E 068952F5
 38EE6528 17FA6B7F C814D391 F14625E1 2C876FF4 FF178ADD 1C5DAE33 C333E0D1 CD954EC0 D36C4AB3
 A39EAE42 C7EF1A2F F998BA94 0A231099 7350651F 243BEE09 BFA32644 71D5C9E0 ABE8437A AF020301
 0001
% Key pair was generated at: 00:22:38 UTC Mar 1 1993
Key name: NewYork.NY.com.server
 Usage: Encryption Key
 Key is not exportable.
 Key Data:
  307C300D 06092A86 4886F70D 01010105 00036B00 30680261 00B2A10F D2AE3292
  26B04421 A0719F09 2B8F0A32 052283F8 4B47F93D BAFD355F 23C13F34 694D5EC8
  C2E1A79F 67E28F5D 8A31B1BC AE15C890 AC36F3DA D7D3DBEF 8810D69F 4AD006B8
  F5597E35 265B1896 B5684824 887BDA4F B5FCEFCF FF7A5FFA 21020301 0001
NewYork#
```

Enter Your Peer RSA Public Keys

After receiving the public key from your peer (in a secure manner), you need to enter the public key into the router. Several commands are required to complete this process:

```
crypto key pubkey-chain rsa
addressed-key key-address
named-key key-name
key-string
```

Example 19-14 shows the commands required to install the public key from San Francisco into the peer router in New York.

Example 19-14 *Installing the Public Key in the New York Router*

```
NewYork(config)#crypto key pubkey-chain rsa
NewYork(config-pubkey-chain)#addressed-key 192.168.3.1
NewYork(config-pubkey-key)#key-string
NewYork(config-pubkey)#00302017 4A7D385B 1234EF29 335FC973
NewYork(config-pubkey)#2DD50A37 C4F4B0FD 9DADE748 429618D5
```

Example 19-14 *Installing the Public Key in the New York Router (Continued)*

```
NewYork(config-pubkey)#18242BA3 2EDFBDD3 4296142A DDF7D3D8
NewYork(config-pubkey)#08407685 2F2190A0 0B43F1BD 9A8A26DB
NewYork(config-pubkey)#07953829 791FCDE9 A98420F0 6A82045B
NewYork(config-pubkey)#90288A26 DBC64468 7789F76E EE21
NewYork(config-pubkey)#quit
NewYork(config-pubkey-key)#exit
```

Verify the Key Configuration

Two commands are used to show the current key configurations on the router. The first command (**show crypto key mypubkey rsa**) displays the public keys installed on the router, and the second (**show crypto key pubkey-chain rsa**) displays all peer keys installed. Example 19-15 shows the output from the **show crypto key mypubkey rsa** command.

Example 19-15 *Viewing RSA Public Keys on the New York Router*

```
NewYork#show crypto key mypubkey rsa

% Key pair was generated at: 18:13:49 UTC Mar 23 2003
Key name: NewYork.newyork.com
Usage: Signature Key
Key Data:
005C300D 06092A86 4886F70D 01010105 00034B00 30480241 00C5E23B 55D6AB22
04AEF1BA A54028A6 9ACC01C5 129D99E4 64CAB820 847EDAD9 DF0B4E4C 73A05DD2
BD62A8A9 FA603DD2 E2A8A6F8 98F76E28 D58AD221 B583D7A4 71020301 0001
% Key pair was generated at: 18:13:49 UTC Mar 23 2003
Key name: NewYork.newyork.com
Usage: Encryption Key
Key Data:
00302017 4A7D385B 1234EF29 335FC973 2DD50A37 C4F4B0FD 9DADE748 429618D5
18242BA3 2EDFBDD3 4296142A DDF7D3D8 08407685 2F2190A0 0B43F1BD 9A8A26DB
07953829 791FCDE9 A98420F0 6A82045B 90288A26 DBC64468 7789F76E EE21
```

Example 19-16 shows the output from the **show crypto key pubkey-chain rsa** command.

Example 19-16 *Viewing RSA Public Keys on the New York Router*

```
NewYork#show crypto key pubkey-chain rsa
Codes: M - Manually Configured, C - Extracted from certificate
Code  Usage       IP-address    Name
M     Signature   192.168.3.1   San Francisco.San Francisco.com
M     Encryption  192.168.3.1   San Francisco.San Francisco.com
```

Manage the RSA Keys

When generated and installed, the only management of RSA keys that is required is to remove old unused keys. The **crypto key zeroize rsa** command enables you to remove old keys.

Foundation Summary

The "Foundation Summary" section of each chapter lists the most important facts from the chapter. Although this section does not list every fact from the chapter that will be on your SNRS exam, a well-prepared candidate should at a minimum know all the details in each "Foundation Summary" section before going to take the exam.

Configure a Cisco Router for IPsec Using Preshared Keys

Several tasks and subtasks are required to configure the router for an IPsec VPN using preshared keys:

1. Select the IKE and IPsec parameters.
   a. Define the IKE (Phase 1) policy.
   b. Define the key distribution method.
      i. Manual key distribution
      ii. ISAKMP
   c. Define the authentication method.
      i. Preshared secret
      ii. RSA signatures
      iii. RSA nonces
   d. Identify the IKE SA peer by IP address or host name.
   e. Define the IKE Phase 1 policy.
      i. Encryption algorithm (DES, 3DES)
      ii. Hash algorithm (SHA-1, MD5)
      iii. IKE SA lifetime
2. Define the IPsec policies.
   a. Select the IPsec protocol (AH, ESP).
   b. Configure transforms and transform sets.
   c. Define the IPsec peer by host name or IP address.
   d. Define local hosts/networks.

        e. Select SA initiation type (manual, IKE).
   3. Verify the current router configuration (**show running-configuration**).
        a. Verify connectivity.
        b. Ping through to the peer.
        c. Ensure compatible access lists.
        d. Verify you are not blocking protocol 50/51 or UDP 500.
   4. Configure IKE.
        a. Enable IKE.
        b. Create policies (per plan listed previously).
        c. Validate the configuration.
   5. Configure IPsec.
        a. Define transforms.
        b. Create the crypto ACLs.
        c. Create the crypto maps.
        d. Apply the crypto maps.

Verifying the IKE and IPsec Configuration

The best way to verify your VPN configuration is to attempt to connect to a destination at the other end of the VPN. If you cannot successfully connect, the following commands enable you to verify your VPN configuration:

- **show** commands:

    - **show crypto isakmp policy** displays the current IKE policies.
    - **show crypto ipsec transform-set** displays the current transform sets.
    - **show crypto ipsec sa** displays the current state of your IPsec SAs.
    - **show crypto map** displays your current crypto maps.
    - **show crypto dynamic-map** displays your dynamic crypto map set.

- **debug** commands:

    - **debug crypto isakmp** enables debugging of IKE events.
    - **debug crypto ipsec** enables debugging of IPsec events.

— Both **debug** commands generate a tremendous amount of output and should be used only when traffic is low.

Explain the Issues Regarding Configuring IPsec Manually and Using RSA-Encrypted Nonces

The only additional process required for implementing IPsec using RSA nonces is the key-generation and -exchange process. The following steps are required to generate and exchange RSA keys:

Step 1   Plan the implementation using RSA keys.

Step 2   Configure the router host name and domain name.

Step 3   Generate the RSA keys.

Step 4   Enter your peer RSA public keys.

Step 5   Verify the key configuration.

Step 6   Manage the RSA keys.

Q&A

As mentioned in the section "How to Use This Book" in the Introduction, you have two choices for review questions: the Q&A questions here or the exam simulation questions on the CD-ROM. The questions that follow present a bigger challenge than the exam itself because they use an open-ended question format. By using this more difficult format, you can exercise your memory better and prove your conceptual and factual knowledge of this chapter. You can find the answers to these questions in the appendix.

1. When an organization wants to connect two or more offices to one another over VPNs, this is called _____.

2. What kind of authentication method is used if the endpoints agree on the key of "VPNK3y" on their routers?

3. To access information on the far end of a VPN that is destined *behind* the VPN endpoint, it is best to use what IPsec mode?

4. What is the primary function of the AH (authentication header)?

5. What command enables you to display the current ISAKMP policy configured on the router?

6. Why is it important to verify that the ACLs on the perimeter router are correct?

7. What values need to be placed in the following IKE Phase 1 policy?

    - Key distribution:ISAKMP
    - Authentication:Preshared key
    - Encryption algorithm:?
    - Hash algorithm:?
    - Diffie-Hellman group:2
    - IKE SA lifetime:86400

8. What is the following command used for when established an IPsec policy?
    ```
    router(config)#crypto isakmp key abc123 address 200.10.41.254 255.255.255.255
    ```

9. Why is the IPsec SA lifetime so important to an IPsec policy?

10. Is it possible to establish multiple transform sets on a router? If so, why is this beneficial?

11. For a crypto map to take effect, it must be applied to _____.

12. What are the tasks required to configure IPsec using RSA-encrypted nonces?

13. After you receive your peer's RSA public key, what command do you use to enter this key into the crypto map?

This chapter covers the following subject:

- Advanced IPsec VPNs Using Cisco Routers and CAs

CHAPTER 20

Scaling a VPN Using IPsec with a Certificate Authority

Cisco IOS devices are designed with a feature called *CA interoperability support* that allows them to interact with certificate authorities (CA) when deploying IPsec. This functionality allows for a scalable and manageable enterprise virtual private network (VPN) solution.

In this chapter, you'll learn to describe how Cisco IOS Software supports open CA standards and CA interoperability, and how SCEP manages the certificate life cycle. You'll also learn each of the five tasks in configuring IPsec encryption using digital certificates on a Cisco router.

"Do I Know This Already?" Quiz

The purpose of the "Do I Know This Already?" quiz is to help you decide whether you really need to read the entire chapter. If you already intend to read the entire chapter, you do not necessarily need to answer these questions now.

The 10-question quiz, derived from the major sections in "Foundation Topics" section of the chapter helps you determine how to spend your limited study time.

Table 20-1 outlines the major topics discussed in this chapter and the "Do I Know This Already?" quiz questions that correspond to those topics.

Table 20-1 *"Do I Know This Already?" Foundation Topics Section-to-Question Mapping*

Foundation Topics Section	Questions Covered in This Section
Advanced IPsec VPNs Using Cisco Routers and CAs.	1–10

> **CAUTION** The goal of self-assessment is to gauge your mastery of the topics in this chapter. If you do not know the answer to a question or are only partially sure of the answer, you should mark this question wrong for purposes of the self-assessment. Giving yourself credit for an answer you correctly guess skews your self-assessment results and might provide you with a false sense of security.

1. Which of the following items are not components of a digital certificate?

    a. Name
    b. Organization
    c. IP address
    d. Serial number
    e. Private key
    f. All of the above

2. A CA is responsible for all of the following except what?

    a. Receives certificate requests
    b. Issues certificates
    c. Maintains certificates
    d. Revokes certificates
    e. Validates certificates
    f. All of the above

3. Which of the following is the advantage of using CA support on peer routers for an IPsec tunnel?

    a. Multiple peers can be managed and configured.
    b. Peers no longer need to manually exchange preshared keys or nonces.
    c. Peer encryption/decryption is done more efficiently.
    d. Peers can be centrally monitored.
    e. Each peer can report detailed IPsec latency information.
    f. Peers can utilize stronger encryption methods.

4. Cisco IOS Software supports all of the following CA standards except for _____ and _____.

    a. IPsec
    b. RSA keys
    c. IKE
    d. X.508 certificates
    e. PKCS#7
    f. PKCS#11

5. Which of the following CA services use X.509v3 certificates and support SCEP? (Select four.)

  a. VeriSign Onsite 4.5
  b. Entrust Technologies
  c. Baltimore Technologies
  d. Twarte 4.6
  e. OpenSSL 3.0
  f. Windows 2000 Certificate Server 5.0

6. What is not a step in the process of configuring a router for CA support?

  a. Configure the router host name and domain name.
  b. Set the router date, time, and time zone.
  c. Declare the CA.
  d. Generate the RSA key pair.
  e. Request your certificate.
  f. All of the above.

7. Why is it important to add the CA server to the router host table?

  a. It can be contacted by IP address only.
  b. It protects the security of the CA.
  c. It provides more detailed logging information for the CA.
  d. It is required for generating private/public key pairs.
  e. It increases performance of the router.
  f. None of the above.

8. What command is required to identify the trusted CA Chicago_CA for a Cisco router?

  a. **crypto ca trustpoint Chicago_CA**
  b. **ca server trust Chicago_CA**
  c. **crypto ca primary Chicago_CA**
  d. **ca server primary Chicago_CA**
  e. **crypto ca primary trustpoint Chicago_CA**
  f. None of the above

9. When a certificate is requested from a server using the **cypto ca enroll** command, which of the following occurs? (Select four.)

   a. Router sends the key pairs to the CA server.
   b. Server generates and signs the identity certificates.
   c. Server validates key pair with peer.
   d. CA servers send the certificates back to the router.
   e. CA server sends certificate to the peer.
   f. CA posts a copy of certificate in its public repository.

10. What commands can you use to validate the CA configuration? (Select at least three.)

   a. **show crypto ca certificates**
   b. **show crypto key mypubkey rsa**
   c. **show crypto map**
   d. **show crypto key rsa**
   e. **show crypto key pubkey-chain rsa**
   f. **show crypto ca**

The answers to the "Do I Know This Already?" quiz are found in the appendix. The suggested choices for your next step are as follows:

- **8 or less overall score**—Read the entire chapter. This includes the "Foundation Topics" and "Foundation Summary" sections and the "Q&A" section.

- **9 or 10 overall score**—If you want more review on these topics, skip to the "Foundation Summary" section and then go to the "Q&A" section. Otherwise, move on to Chapter 21, "Troubleshooting the VPN Configuration on a Cisco Router."

Foundation Topics

Building VPNs using IPsec was discussed at length in Chapter 19, "Building a VPN Using IPsec." This chapter builds on the information from the preceding chapter and includes what is needed to build more scalable enterprise VPNs using CAs.

Advanced IPsec VPNs Using Cisco Routers and CAs

This section is dedicated to configuring the Cisco router for advanced scalable IPsec VPNs using CAs. It contains an overview of the CA support and configuration steps required to deploy IPsec VPNs using CA support. Clearly, the use of CAs is not a requirement for building IPsec VPNs (as noted in the previous chapter), but the interoperability between Cisco IOS devices and CAs normally results in a more scalable and manageable IPsec solution.

Digital Signatures, Certificates, and Certificate Authorities

Digital signatures are a Public Key Infrastructure (PKI) component used to digitally identify and authenticate a device or user. In public key cryptography, each user has a key pair that contains the public and private keys. Anything that is encrypted using the private key can only be decrypted using the public key and the digital signature is applied when the data is encrypted using the users' private key. The receiver is able to authenticate the sender because only the sender's public key will decrypt the message. A digital certificate provides the information to identify the sender, such as name, organization, IP address, and serial number, and will include a copy of the sender's public key. The CA is a system that is responsible for receiving certificate requests, and issuing, maintaining, and revoking certificates. The CA is a trusted third party that validates the identities of each entity.

The use of CAs provides an automated method for key distribution and allows for more dynamic network configurations. Consider the network depicted in Figure 20-1. Each router in this network must be manually configured with the key pairs for each VPN segment. This requires that you configure the keys for each router connection to every other router. As you can see, each additional connection requires a significant amount of effort, and network management will become extremely complex and difficult as the network continues to grow.

The CA provides a management system that enables you to scale the network because each router can get certificates for each new device that enters the network. Figure 20-2 depicts how each device communicates with the CA to acquire certificates for every other potential peer.

Figure 20-1 *Multipoint Network Without a CA*

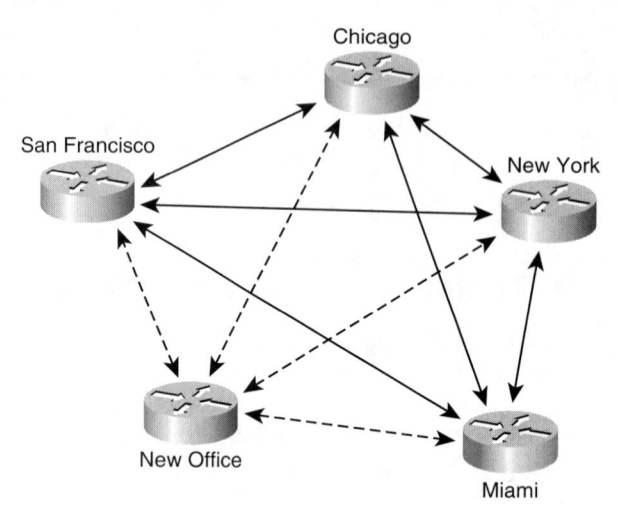

Figure 20-2 *Communication Between the Network Peers and the CA*

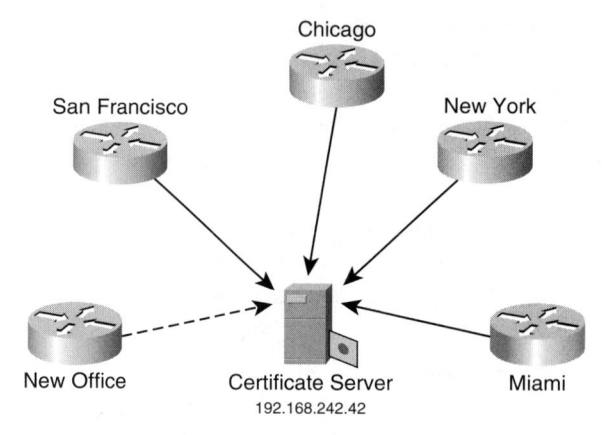

Overview of Cisco Router CA Support

When two peer routers initiate an exchange of IPsec-protected traffic, they must first authenticate each other. The authentication is performed with Internet Key Exchange (IKE) and the preselected key-exchange method. The advantage of using CA support is that peers no longer need to manually exchange preshared keys or nonces. When two peers begin the IKE negotiation, they just exchange public keys, which are then authenticated by the CA. This process greatly improves manageability because there is no requirement to track keys. As a result, this solution is easy to scale.

Cisco IOS Software supports the following CA standards:

- **IPsec**—IPsec is a framework of open standards that provide authentication, integrity, and confidentiality of data between peers.

- **RSA keys**—Rivest, Shamir, and Adelman (RSA) is an asymmetric public key cryptography system. RSA keys come in key pairs (one public and one private). When generating RSA keys, it is also possible to generate an authentication pair and an encryption pair.

- **IKE**—IKE is a combination of ISAKMP and the Oakley key exchange protocols. Also referred to as *ISAKMP/Oakley*, it provides a method of authentication and negotiation to create a secure environment for the IPsec negotiation.

- **CA interoperability**—CA interoperability is the component that provides communication functionality between Cisco devices and CA servers. A main component of CA interoperability is the Simple Certificate Enrollment Protocol (SCEP):

    — SCEP is a lightweight transaction-oriented protocol that uses Public Key Cryptography Standard #7 (PKCS#7) and PKCS#10. SCEP requires either manual authentication or a preshared key during enrollment.

- **X.509v3 certificates**—X.509 is a digital certificate standard that was created using the X.500 standard as its foundation. It allows peers to exchange digital certificates to authenticate their identity. This solution removes the requirement for manually exchanging public keys between peers. The following CA services use X.509v3 certificates and also support SCEP:

    — VeriSign OnSite 4.5

    — Entrust Technologies

    — Baltimore Technologies

    — Windows 2000 Certificate Server 5.0

- **PKCS#7**—A standard from RSA used to encrypt, sign, and package certificate enrollment messages.

- **PKCS#10**—A standard from RSA that defines the syntax for certificate requests.

Several things take place when sending traffic from the source to the destination via a VPN connection. Figure 20-3 shows how traffic gets from the source system on the New York network to the destination system on the Boston network and how the peers communicate with each other and the CA server for peer authentication. Source A generates traffic destined for destination B, and the traffic is passed to the router. The router (New York) compares the traffic to its current security policy and determines that the traffic must be encrypted and forwarded to the VPN endpoint router (Boston). The New York network checks for an existing IPsec security association (SA) with the Boston router, and if no IPsec SA exists, the router initiates the negotiation of an IKE SA. As part of the IKE SA negotiation, both routers exchange digital certificates that have been signed by a CA that is trusted by both peers. Upon receiving the certificate from the peer, each router downloads a certificate revocation list (CRL) from either the CA or a CRL distribution point and verifies that the certificate from the peer has not been revoked. After verifying the certificates, the peers complete the negotiation of the IKE SA, followed by the negotiation of the IPsec SA. The data is transferred between the source and destination as soon as the VPN tunnel is created.

As Figure 20-3 illustrates, the CA server is accessible to both peers. The peers exchange digital certificates, and the CA server validates the certificate. The CA maintains a list of certificates that are no longer valid called a CRL. The peers ensure that a certificate is valid by checking it against the CRL.

Figure 20-3 *Peer Authentication Using a CA*

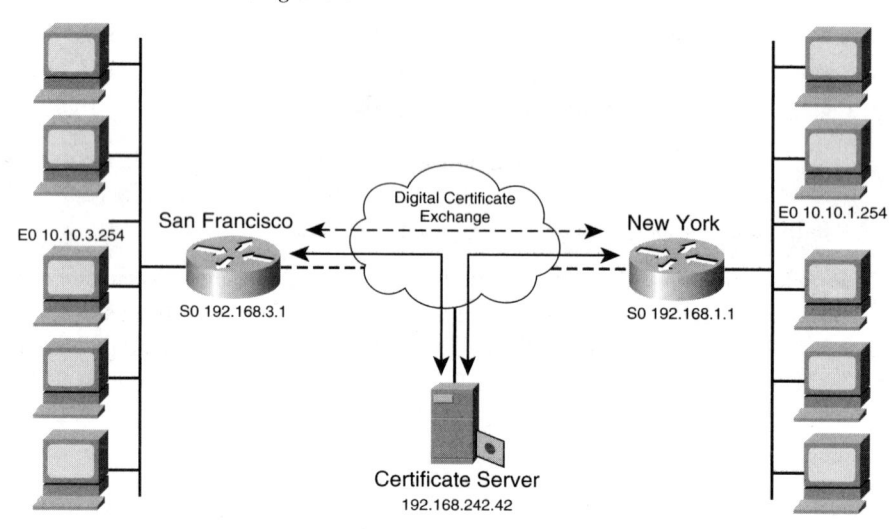

SCEP

The SCEP was introduced by Cisco in late 1999 as one of the first protocols designed to facilitate the transfer of digital certificates between CAs. The key to SCEP is that it introduced a common method for requesting and receiving digital certificates in a scalable, open, web-based solution.

Configuring the Cisco Router for IPsec VPNs Using CA Support

To configure the router for IPsec VPNs using CA support, you must complete five tasks. Each task contains several subtasks. As always, the most important component is thorough planning and meticulous implementation. Because of the complexity of this process, any error can prevent the VPN from functioning properly. The five steps are as follows:

Step 1 Select the IKE and IPsec parameters.

Step 2 Configure the router CA support.

Step 3   Configure IKE using RSA signatures.

Step 4   Configure IPsec using RSA signatures.

Step 5   Test and verify the configuration.

Step 1: Select the IKE and IPsec Parameters

Selecting the IKE and IPsec parameters is just the process of predetermining which settings will be used on both peers to ensure successful negotiation of the connection. Many of these items were covered in Chapter 19. This task is divided into six subtasks, as follows:

1. Plan for CA support.

   Get the details of the CA server to include the server type, IP address, host name, Uniform Resource Locator (URL), and server administrator contact information. Coordinate with the CA server administrator to ensure that your certificates are properly validated.

2. Define the IKE (Phase 1) policy.

   See Chapter 19.

3. Define the IPsec policies.

   See Chapter 19.

4. Verify the current router configuration.

5. Verify connectivity.

   See Chapter 19.

6. Ensure compatible access lists.

   See Chapter 19.

Step 2: Configure the Router CA Support

To configure a router for CA support and verify that configuration, you must complete 11 different steps. These steps include configuring the router, generating keys, and communicating with the CA server:

Step 1   Configure the router host name and domain name. The host name and domain name are written to the key pairs. It is important that you have the correct identity information configured on the router before you generate the key pair. The syntax for these global configuration commands is as follows:

- **hostname** *name* sets the host name for the router.

- **ip domain-name** *name* sets the default domain name on the router, which is used to convert unqualified host names to fully qualified domain names (FQDN).

Example 20-1 shows the correct syntax for configuring the host name and domain name on the router in New York.

Example 20-1 *Configuring the Host Name and Domain Name*

```
router#configure terminal
router(config)#hostname NewYork
router(config)#ip domain-name NY.com
```

Step 2   Set the router date, time, and time zone. The time on the router must be accurate to enroll with the CA server. The **clock set** command is entered with the router in privileged EXEC mode, and the **clock timezone** command is entered in global configuration mode. The syntax for these commands is as follows:

- **clock timezone** sets the correct time zone on the router.
- **clock set** configures the date and time on the router.

Example 20-2 shows the correct syntax for entering the date, time, and time zone on the router in New York.

Example 20-2 *Configuring the Date, Time, and Time Zone*

```
NewYork#configure terminal
NewYork(config)#clock timezone est -5
NewYork(config)#exit
NewYork#clock set 19:00:00 30 May 2005
```

The most accurate way to ensure the correct time on the router is to configure it to synchronize time with a network time server using the Network Time Protocol (NTP). Cisco products support both NTP and the Simple Network Time Protocol (SNTP). NTP maintains communication between the router and the NTP server via User Datagram Protocol (UDP) port 123, and the router can then relay time to other systems on the network. SNTP is used mainly by low-end routers and acts as a client-only mode. SNTP maintains communication between the client and server using UDP port 580. To activate NTP, you must enable NTP while in the interface configuration mode. You should limit NTP to a specific interface and only allow the router to access time updates from specific NTP peers. The command for configuring NTP on the Cisco router is **ntp access-group [query-only | serve-only | server | peer]** *access-list-number*. Table 20-2 lists the configuration options for the **ntp access-group** command.

Table 20-2 **ntp access-group** *Command Options*

Term	Definition
query-only	Allows NTP control queries only
serve-only	Allows and serves NTP time requests only
serve	Allows NTP time requests, but does not synchronize with the remote system
peer	Allows NTP control queries and time requests; also allows the router to synchronize time with the remote system

Step 3 Generate the RSA key pair. The RSA keys are used to authenticate the router to its SA peer. The command syntax for key generation is **crypto key generate rsa usage keys**. The **usage keys** option enables you to generate two special-purpose key pairs (authentication pair and encryption pair for RSA-encrypted nonces). If you do not use the optional command, you will generate a single "general-purpose" public/private key pair. It is also possible to select the modulus length when generating keys. General-purpose keys are sufficient for standard authentication using RSA signatures. The available modulus lengths are 360, 512, 1024, and 2048 bits. The longer the modulus length, the longer it will take the router to generate the keys.

To delete all configured RSA keys from the router, use the **crypto key zeroize rsa** command.

Step 4 Add the CA server to the router host table. By adding the CA server IP address to the router host table, you define a static host name-to-IP address mapping and remove the requirement for using Domain Name System (DNS). Removing the requirement for DNS increases the performance of the router because it is no longer affected by any delay of the DNS server:

- **ip host** *name address1* [*address2*]

Example 20-3 shows the correct syntax for adding the CA server to the host table on the router in New York.

Example 20-3 *Adding the CA Server to the Host Table*

```
NewYork#configure terminal
NewYork(config)#ip host CA-Server 192.168.242.42
```

Step 5 Declare the CA. Configure the trusted CA on the router in the global configuration mode with the **crypto ca trustpoint** *name* command. This identifies the trusted CA for the router and puts you in the **ca-identity** configuration mode.

> **NOTE** The **crypto ca trustpoint** command was introduced with Cisco IOS Software Release 12.2(8)T and replaced the **crypto ca identity** command in Cisco IOS Software Release 12.3.

A variety of subcommands are available in the ca-trustpoint configuration mode:

- **enrollment**—This optional subcommand specifies the enrollment parameters. The **enrollment mode ra** command enrolls the CA as a Registration Authority (RA). The IPsec peers complete transactions with the RA, which then forwards the requests to the CA. Both peers must be configured with the public keys for the CA and RA.
- **enrollment http proxy**—Configures access to the CA via using http via an HTTP proxy server.
- **root**—Defines the TFTP to get the CA Certificate and defines the server name and filename for the Certificate.
- **match certificate**—This optional command specifies a certificate-based existing access list (ACL) that is associated with a specific **crypto ca certificate map** command.
- **primary**—Defines the primary CA trustpoint for the router.
- **crl**—The CRL command queries the CRLs to ensure the certificate for its peer is valid and has not been revoked.
- **default (ca-trustpoint)**—This command resets the CA trustpoint configuration to its default values.

Example 20-4 shows the correct syntax declaring the CA server as the CA on the router in New York.

Example 20-4 *Declaring the CA*

```
NewYork(config)#crypto ca trustpoint CA_Server
NewYork(ca-trustpoint)#enrollment url http://CS-Server/certserv/mscep/mscep.dll
NewYork(ca-trustpoint)#enrollment mode ra
```

Step 6 Authenticate the CA. The router authenticates the CA by retrieving the CA self-signed certificate and the CA's public key. The command for this action is **crypto ca authenticate**. This command initiates the authentication process with the CA by sending the CA/RA request to the CA. The CA generates the CA/RA certificate and returns it to the router. The router authenticates the CA/RA certificate using the CA/RA fingerprint. Example 20-5 shows the correct syntax for authenticating the CA on the router in New York.

Example 20-5 *Authenticating the CA*

```
NewYork#configure terminal
NewYork(config)#crypto ca authenticate CA-Server
```

Step 7 Request your certificate. The router must request a certificate from the CA server with the **crypto ca enroll** command. This command requests certificates from the CA for all the router RSA key pairs. The router sends the key pairs to the CA server, which generates and signs the identity certificates. Finally, the CA server sends the identity certificates back to the router and posts a copy in its public repository. Example 20-6 shows the correct syntax for requesting a certificate from the CA.

Example 20-6 *Requesting a Certificate from the CA*

```
NewYork#configure terminal
NewYork(config)#crypto ca enroll CA-Server
% Start certificate enrollment…
% Create a challenge password. You need to verbally provide this
password to the CA administrator to revoke your certificate.
For security reasons, your password will not be saved in the configuration.
Please make a note of it.

Password: <password>
Re-enter password: <password>

% The subject name in the certificate will be: NewYork.NY.com
% Include the router serial number in the subject name? (yes/no): no
% Include the IP address in the subject name? (yes/no): no
Request certificate from CA? (yes/no) yes
% Certificate request sent to certificate authority
% The certificate request fingerprint will be displayed.
% The show crypto ca certificate command will also show the fingerprint.

NewYork(config)#
      Signing Certificate Request Fingerprint:
      1D017C1F 9AE457BD 501BA5DF CF472D21

      Encryption Certificate Request Fingerprint:
      2FF054AB 01DC2A22 AB147620 05C5AB5F
```

To delete the current enrollment, use the **no crypto ca enroll** command.

Step 8 Save the configuration to the router. Ensure that the current configuration is saved. Write the configuration to memory using the **copy running-config startup-config** command.

Chapter 20: Scaling a VPN Using IPsec with a Certificate Authority

> **NOTE** It is also a good idea to save the configuration to prevent certificate loss because of a system reboot and to back up the configuration if hardware fails.

Figure 20-4 depicts the communications between the router and the CA server that are required to complete the authentication, enrollment, and certificate-generation process.

Figure 20-4 *Communication Between the Router and CA*

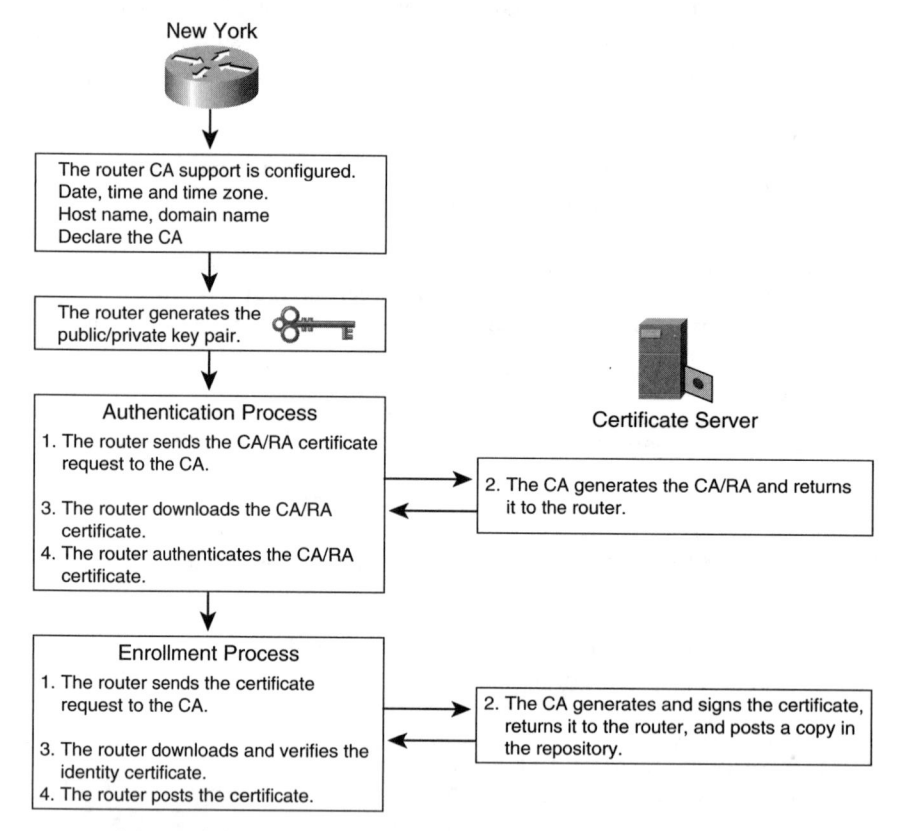

```
                            New York
                            (router)
                               │
                               ▼
        ┌──────────────────────────────────────┐
        │ The router CA support is configured. │
        │ Date, time and time zone.            │
        │ Host name, domain name               │
        │ Declare the CA                       │
        └──────────────────────────────────────┘
                               │
                               ▼
        ┌──────────────────────────────────────┐
        │ The router generates the             │
        │ public/private key pair.             │
        └──────────────────────────────────────┘
                               │
                               ▼
        ┌──────────────────────────────────┐        ┌──────────────────┐
        │ Authentication Process           │        │ Certificate Server│
        │ 1. The router sends the CA/RA    │        └──────────────────┘
        │    certificate request to the CA.│──────▶ 2. The CA generates the CA/RA and returns
        │                                  │           it to the router.
        │ 3. The router downloads the CA/RA│◀──────
        │    certificate.                  │
        │ 4. The router authenticates the  │
        │    CA/RA certificate.            │
        └──────────────────────────────────┘
                               │
                               ▼
        ┌──────────────────────────────────┐
        │ Enrollment Process               │
        │ 1. The router sends the          │──────▶ 2. The CA generates and signs the certificate,
        │    certificate request to the CA.│           returns it to the router, and posts a copy in
        │                                  │           the repository.
        │ 3. The router downloads and      │◀──────
        │    verifies the identity         │
        │    certificate.                  │
        │ 4. The router posts the          │
        │    certificate.                  │
        └──────────────────────────────────┘
```

Many of the steps shown in Figure 20-4 are completed automatically by SCEP.

Step 9   Manage key storage in NVRAM. Memory management is an option available to prevent the number of stored certificates and CRLs from occupying memory space.

Step 10 Manage the keys on the router. Key management is an option that enables you to delete keys and certificates from the router and to request a CRL from the CA.

Step 11 Verify the CA configuration. Three commands enable you to view the status of certificates and keys on the router, as follows:

- **show crypto ca certificates** displays certificates currently on the router. Example 20-7 shows the output from the **show crypto ca certificates** command.

Example 20-7 show crypto ca certificates *Command Output*

```
NewYork#show crypto ca certificates
Certificate
Subject Name
Name: NewYork.NY.com
IP Address: 192.168.1.1
Status: Available
Certificate Serial Number: 428125BDA34196003F6C78316CD8FA95
Key Usage: Signature
Certificate
Subject Name
Name: NewYork.NY.com
IP Address: 192.168.1.1
Status: Available
Certificate Serial Number: AB352356AFCD0395E333CCFD7CD33897
Key Usage: Encryption
CA Certificate
Status: Available
Certificate Serial Number: 3051DF7123BEE31B8341DFE4B3A338E5F
Key Usage: Not Set
```

- **show crypto key mypubkey rsa** displays public keys for the router. Example 20-8 shows the output from the **show crypto key mypubkey rsa** command.

Example 20-8 show crypto key mypubkey rsa *Command Output*

```
NewYork#show crypto key mypubkey rsa
% Key pair was generated at: 19:07:49 UTC May 30 2005
Key name: NewYork.NY.com
Usage: Signature Key
Key Data:
005C300D 06092A86 4886F70D 01010105 00034B00 30480241 00C5E23B 55D6AB22
04AEF1BA A54028A6 9ACC01C5 129D99E4 64CAB820 847EDAD9 DF0B4E4C 73A05DD2
BD62A8A9 FA603DD2 E2A8A6F8 98F76E28 D58AD221 B583D7A4 71020301 0001
% Key pair was generated at: 19:07:50 UTC May 30 2005
Key name: NewYork.NY.com
Usage: Encryption Key
Key Data:
00302017 4A7D385B 1234EF29 335FC973 2DD50A37 C4F4B0FD 9DADE748 429620D5
18242BA3 2EDFBDD3 4296142A DDF7D3D8 08407685 2F2190A0 0B43F1BD 9A8A26DB
07953829 791FCDE9 A98420F0 6A82045B 90288A26 DBC64468 7789F76E EE21
```

- **show crypto key pubkey-chain rsa** displays the peer public keys on the router. Example 20-9 shows the output from the **show crypto key pubkey-chain rsa** command.

Example 20-9 show crypto key pubkey-chain rsa *Command Output*

```
NewYork#show crypto key pubkey-chain rsa
Codes: M - Manually Configured, C - Extracted from certificate
Code   Usage        IP-address      Name
C      Signature    192.168.2.1     Chicago.CH.com
C      Encryption   192.168.20.1    Chicago.CH.com
M      Signature    192.168.3.1     SanFran.SF.com
M      Encryption   192.168.3.1     SanFran.SF.com
C      General      192.168.4.1     Maimi.MI.com
```

Step 3: Configure IKE Using RSA Signatures

Chapter 19 covered these configuration steps in great detail. For the purpose of this exercise, Figure 20-5 provides the configuration parameters for the VPN connection between San Francisco and New York.

Figure 20-5 *IKE Configuration Parameters*

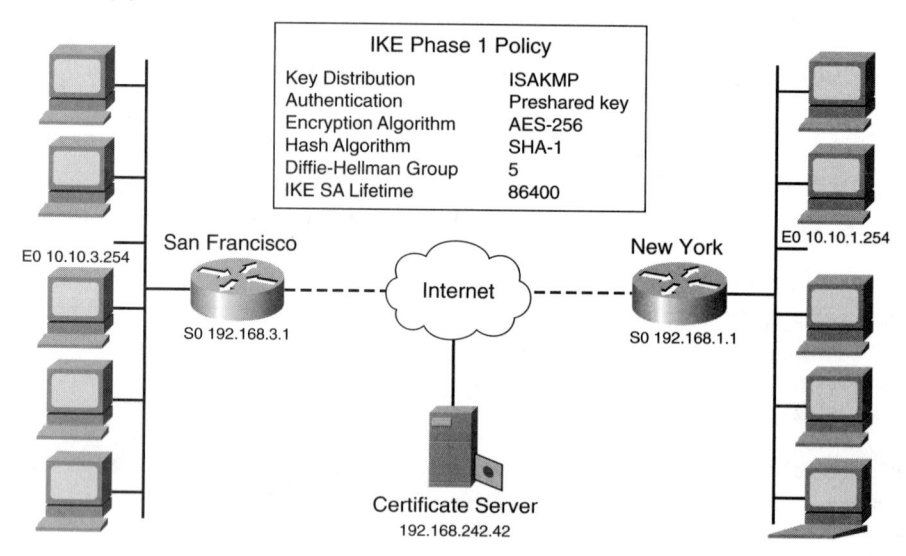

IKE Phase 1 Policy
- Key Distribution: ISAKMP
- Authentication: Preshared key
- Encryption Algorithm: AES-256
- Hash Algorithm: SHA-1
- Diffie-Hellman Group: 5
- IKE SA Lifetime: 86400

San Francisco — E0 10.10.3.254, S0 192.168.3.1
New York — E0 10.10.1.254, S0 192.168.1.1
Certificate Server 192.168.242.42

Example 20-10 shows the commands used to configure IKE using RSA signatures on the router in New York.

Example 20-10 *Configuring IKE Using RSA Signatures*

```
NewYork#configure terminal
NewYork (config)#crypto isakmp policy 120
NewYork (config-isakmp)#authentication rsa-sig
NewYork (config-isakmp)#encryption aes-256
NewYork (config-isakmp)#hash sha
NewYork (config-isakmp)#group 5
NewYork (config-isakmp)#lifetime 86400
```

Step 4: Configure IPsec

Chapter 19 also covered these configuration steps in great detail. For the purpose of this exercise, Figure 20-6 provides the configuration parameters for the VPN connection between San Francisco and New York.

Figure 20-6 *IPsec Configuration Parameters*

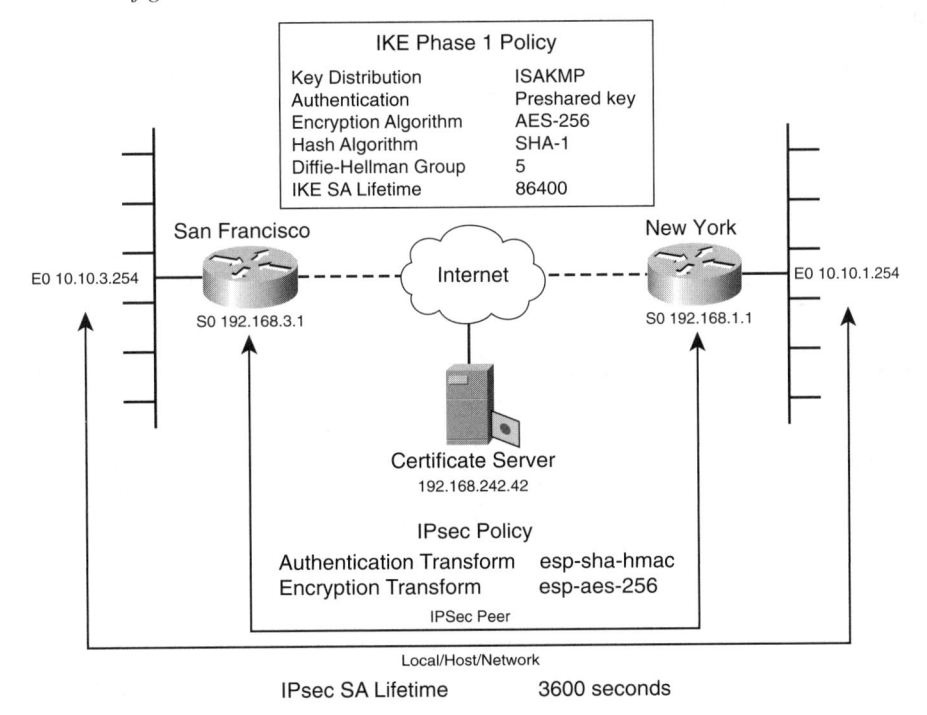

IKE Phase 1 Policy

Key Distribution	ISAKMP
Authentication	Preshared key
Encryption Algorithm	AES-256
Hash Algorithm	SHA-1
Diffie-Hellman Group	5
IKE SA Lifetime	86400

San Francisco — E0 10.10.3.254, S0 192.168.3.1
New York — E0 10.10.1.254, S0 192.168.1.1
Certificate Server 192.168.242.42

IPsec Policy

Authentication Transform	esp-sha-hmac
Encryption Transform	esp-aes-256

IPSec Peer
Local/Host/Network
IPsec SA Lifetime   3600 seconds

Example 20-11 shows the commands used to configure the following IPsec parameters on the San Francisco router:

1. Create the IPsec transform set.

2. Configure IPsec SA lifetimes.

3. Create the crypto access lists.
4. Create the crypto map.
5. Apply the crypto maps.

Example 20-11 *Configuring IPsec Parameters*

```
NewYork#configure terminal
NewYork (config)#crypto ipsec transform-set 20 esp-256-aes esp-sha-hmac
NewYork (cfg-crypto-trans)#exit

NewYork (config)#crypto ipsec security-association lifetime seconds 3600
NewYork (config)#access-list 105 permit ip 10.10.1.0 0.0.0.255 10.10.3.0 0.0.0.255

NewYork(config)#crypto map SanFran 120 ipsec-isakmp
NewYork(config-crypto-map)#match address 105
NewYork(config-crypto-map)#set peer 192.168.3.1
NewYork(config-crypto-map)#set pfs group 5
NewYork(config-crypto-map)#set transform-set 20
NewYork(config-crypto-map)#set security-association lifetime seconds 86400
NewYork(config-crypto-map)#interface Fa0/1
NewYork(config-if)#crypto map SanFran
```

Step 5: Test and Verify the Configuration

The following three commands enable you to verify your configuration when working with CAs:

- **crypto ca identity**—Displays the CA that your router is configured to use

- **debug crypto pki**{*callbacks*, *messages*, *transactions*}—Enables you to display the callbacks, transactions, or messages that occur between the router and the CA

- **show crypto ca certificates**—Displays information about the certificate of your CA and any RA

Foundation Summary

The "Foundation Summary" section of each chapter lists the most important facts from the chapter. Although this section does not list every fact from the chapter that will be on your SNRS exam, a well-prepared candidate should at a minimum know all the details in each "Foundation Summary" section before going to take the exam.

Advanced IPsec VPNs Using Cisco Routers and CAs

Although configuring the connection to a CA server is complex, when correctly configured the functionality is scalable and easy to manage. The main focus of this chapter has been the configuration and enrollment process. Cisco IOS Software supports the following CA products using CA interoperability:

- VeriSign OnSite 4.5
- Entrust Technologies
- Baltimore Technologies
- Microsoft Windows 2000 Certificate Server 5.0

Multiple tasks are required to configure the router for CA support:

- Configure the router host name and domain name.
- Set the router date, time, time zone, and configure for NTP.
- Add the CA server to the router host table.
- Generate the RSA key pair.
- Declare the CA.
- Authenticate the CA.
- Request your certificate.

- Save the configuration to the router.
- Manage key storage in NVRAM.
- Manage the keys on the router.
- Verify the CA configuration.

Q&A

As mentioned in the section "How to Use This Book" in the Introduction, you have two choices for review questions: the Q&A questions here or the exam simulation questions on the CD-ROM. The questions that follow present a bigger challenge than the exam itself because they use an open-ended question format. By using this more difficult format, you can exercise your memory better and prove your conceptual and factual knowledge of this chapter. You can find the answers to these questions in the appendix.

1. What is a digital signature?
2. What keys are exchanged between peers that utilize a certificate authority?
3. The CA maintains a list of certificates that are no longer valid, which is called _____.
4. The _____ facilitates the transfer of digital certificates between certificate authorities on Cisco devices.
5. What is the X.509v3 certificate?
6. Why is the host name, domain name, router date, time, and time zone so important when configuring the Cisco device for CA support?
7. What command enables you to authenticate the CA server Chicago_CA for the router?
8. What does the **no crypto ca enroll** command do when implemented on a router?
9. The option on the router that enables you to delete keys and certificates from the router is called what?
10. What commands enable you to debug and validate the CA configuration on the router?

This chapter covers the following subjects:

- **show** Commands
- **debug** Commands
- **clear** Commands

CHAPTER 21

Troubleshooting the VPN Configuration on a Cisco Router

Much of the configuration of VPNs took place in Chapters 19, "Building a VPN Using IPsec," and 20, "Scaling a VPN Using IPsec with a Certificate Authority." This chapter covers primarily verifying the configuration of the router for Internet Key Exchange (IKE), IPsec, and interoperability with certification authorities (CA).

"Do I Know This Already?" Quiz

The purpose of the "Do I Know This Already?" quiz is to help you decide whether you really need to read the entire chapter. If you already intend to read the entire chapter, you do not necessarily need to answer these questions now.

The 10-question quiz, derived from the major sections in the "Foundation Topics" portion of the chapter, helps you determine how to spend your limited study time.

Table 21-1 outlines the major topics discussed in this chapter and the "Do I Know This Already?" quiz questions that correspond to those topics.

Table 21-1 *"Do I Know This Already?" Foundation Topics Section-to-Question Mapping*

Foundation Topics Section	Questions Covered in This Section
show Commands	1, 3–6
debug Commands	7, 8
clear Commands	9, 10

Chapter 21: Troubleshooting the VPN Configuration on a Cisco Router

> **CAUTION** The goal of self-assessment is to gauge your mastery of the topics in this chapter. If you do not know the answer to a question or are only partially sure of the answer, you should mark this question wrong for purposes of the self-assessment. Giving yourself credit for an answer you correctly guess skews your self-assessment results and might provide you with a false sense of security.

1. Which of the following are not **show crypto** commands?

   a. **show crypto ca certificates**

   b. **show crypto isakmp policy**

   c. **show crypto engine connections**

   d. **show crypto ipsec map**

   e. **show crypto ipsec security-association lifetime**

   f. **show crypto key mypubkey rsa**

2. What does the **clear** command do when utilized for IPsec?

   a. Clears out the IPsec counters

   b. Resets the VPN parameters

   c. Clears the IOS screen

   d. Resets bad configuration and negotiation

   e. Resets the IPsec settings

   f. None of the above

3. To show the global IPsec policy used, which command do you need to use?

   a. **show crypto isakmp policy**

   b. **show crypto policy**

   c. **show crypto isakmp map policy**

   d. **show crypto policy map**

   e. **show crypto policy isakmp**

   f. None of the above

4. What optional commands for the **show crypto ipsec sa** commands provide more granular details? (Select four.)

   a. **address**

   b. **tunnel**

c. interface

d. map

e. encryption

f. peer

5. Which command enables you to get information on the configured crypto maps?

   a. show ipsec

   b. show crypto

   c. show ipsec encryption

   d. show crypto map

   e. show map

   f. None of the above

6. What is the rest of the command that you need to display the public keys configured on the router? **show crypto key** _____

   a. mypubkey

   b. string rsa

   c. pubkey rsa

   d. pubkey

   e. mypubkey rsa

   f. None of the above

7. Which command enables you to get a dump of events that shows the source, destination, ESP, and lifetime information of an IPsec connection?

   a. debug crypto sa

   b. debug crypto log

   c. debug crypto ipsec

   d. debug crypto log

   e. debug crypto

   f. None of the above

8. Use the **debug crypto pki transactions** command to show which kind of communication?

   a. Communication between peer and CA
   b. Interaction between CA and router
   c. Peer-to-peer communication
   d. Host-to-peer communication
   e. None of the above

9. Which specific optional commands can you use with the **clear crypto sa** command? (Select four.)

   a. **peer**
   b. **map**
   c. **key**
   d. **counters**
   e. **entry**
   f. **type**

10. What does the **clear crypto isakmp** command do?

    a. Clears IKE connections
    b. Resets the IKE peers
    c. Resets the IKE crypto map
    d. Clears the IKE crypto map
    e. Clears IKE security associations
    f. None of the above

The answers to the "Do I Know This Already?" quiz are found in the appendix. The suggested choices for your next step are as follows:

- **8 or less overall score**—Read the entire chapter. This includes the "Foundation Topics" and "Foundation Summary" sections and the "Q&A" section.

- **9 or 10 overall score**—If you want more review on these topics, skip to the "Foundation Summary" section and then go to the "Q&A" section. Otherwise, move on to Chapter 22, "Configuring Remote Access Using Easy VPN."

Foundation Topics

Chapters 19 and 20 covered the configuration of IPsec virtual private networks (VPN) on Cisco IOS devices. This chapter discusses the various commands used to verify and troubleshoot your VPN. There are several ways to verify your configuration, but it is best to break up the configuration into pieces and look at each piece individually. Because the VPN configuration is built in pieces (IKE parameters, IPsec parameters, crypto maps, and so on), and an error in any single component will most likely prevent the VPN from building, it is best to verify the configuration or each component separately. Some of the items to look at include the following:

- Display the configured IKE policies.
- Display the configured transform sets.
- Display the current state of the IPsec security associations (SAs).
- View configured crypto maps.
- Debug IKE and IPsec traffic.
- Debug CA events.

Any time you configure any communication between devices, especially encrypted communication, it is critical that the configuration of both devices match. With the complexity of today's encryption, it is common for a single mistake to prevent the connection from completing successfully. Notice the word *complexity*. This does not mean the configuring VPN connectivity is difficult, but it can be complex and requires attention to detail. The best way to troubleshoot any VPN is to verify the configuration of both peers. Doing so is the only way to ensure that the configurations match on both ends of the connection. In many cases, the most difficult challenge is to get the configuration to a single location in a secure manner. This chapter addresses three simple commands most commonly used to troubleshoot VPN connectivity:

1. **show**—Used to view how the device is configured, what actions have been completed, or what actions have failed. The **show** command tends to be static because it displays what has happened rather than how the device is reacting to the environment. The **show** command is most commonly used to compare the configuration of the peers of a connection. This chapter covers the following **show** commands:

    - **show crypto ca certificates**
    - **show crypto isakmp policy**
    - **show crypto ipsec sa**
    - **show crypto ipsec security-association lifetime**

- **show crypto ipsec transform-set**
- **show crypto map** (IPsec)
- **show crypto engine connections**
- **show crypto key pubkey-chain rsa**
- **show crypto key mypubkey rsa**

2. **debug**—Shows what is happening when the devices are attempting to complete a function, such as negotiating the connection. The **debug** command is a more dynamic command because it shows how the device handles the traffic that it is receiving in real time rather than after the fact. This chapter covers the following **debug** commands:

   - **debug crypto isakmp**
   - **debug crypto engine**
   - **debug crypto ipsec**
   - **debug crypto pki messages**
   - **debug crypto pki transactions**

3. **clear**—Used to reset something on the device. The **clear** command is most commonly used to clear out a configuration or negotiation that has caused a problem in the device. For example, the **clear** command can be used to reset a SA if there is a problem with the negotiation of that SA. This chapter covers the following clear commands:

   - **clear crypto sa**
   - **clear crypto sa peer** (peer address or name)
   - **clear crypto isakmp**
   - **clear crypto ipsec sa**
   - **clear crypto sa counters**

show Commands

The **show** commands listed in this section will produce much of the information necessary to determine how a device in configured. The **show** commands are a key component in troubleshooting.

show crypto ca certificates Command

This command is executed in the privileged EXEC mode and is used to gather information about the certificate, the CA, and any registration authorities. Example 21-1 depicts the output on the New York router after it has been configured with special usage Rivest, Shamir, and Adelman (RSA) key pairs (usage keys) with the keys and certificates requested and received.

Example 21-1 *Output from the* **show crypto ca certificates** *Command*

```
NewYork#show crypto ca certificates
Certificate
Subject Name
Name: NewYork.NY.com
IP Address: 192.168.1.1
Status: Available
Certificate Serial Number: 428125BDA34196003F6C78316CD8FA95
Key Usage: Signature
Certificate
Subject Name
Name: NewYork.NY.com
IP Address: 192.168.1.1
Status: Available
Certificate Serial Number: AB352356AFCD0395E333CCFD7CD33897
Key Usage: Encryption
CA Certificate
Status: Available
Certificate Serial Number: 3051DF7123BEE31B8341DFE4B3A338E5F
Key Usage: Not Set

NewYork#
```

show crypto isakmp policy Command

This command is executed from the privileged EXEC mode used to view the current Internet Security Association and Key Management Protocol (ISAKMP) (IKE) policy. The IKE policy determines the encryption algorithm, hash algorithm, Diffie-Hellman group, and ISAKMP SA lifetime for the IKE Phase 1 connection. Example 21-2 depicts the output from this command on the New York router. Notice that the configuration includes not only the configured parameters (AES-256, SHA-1, HD group 5, and 86,400 seconds), but it also includes a default ISAKMP policy. It is possible to create VPN connections using the default policy, but it is not recommended.

Example 21-2 *Output from the* **show crypto isakmp policy** *Command*

```
NewYork#show crypto isakmp policy

Global IKE policy
Protection suite of priority 100
        encryption algorithm:    AES - Advanced Encryption Standard (256 bit keys).
        hash algorithm:          Secure Hash Standard
        authentication method:   Pre-Shared Key
        Diffie-Hellman group:    #5 (1536 bit)
        lifetime:                86400 seconds, no volume limit
```

continues

Example 21-2 *Output from the* **show crypto isakmp policy** *Command (Continued)*

```
Default protection suite
    encryption algorithm:   DES - Data Encryption Standard (56 bit keys).
    hash algorithm:         Secure Hash Standard
    authentication method:  Rivest-Shamir-Adleman Signature
    Diffie-Hellman group:   #1 (768 bit)
    lifetime:               86400 seconds, no volume limit
NewYork#R-614
```

show crypto ipsec sa Command

This command is executed from the privileged EXEC mode to view established IPsec SAs. To remove the existing SAs, use the **clear crypto sa** command. This command shows you all established SAs, and the output can become difficult to sort through if there are several established VPNs on the router. Several optional commands enable you to specify exactly what you are looking for on the router, as follows:

- **address**—Enables you to define the destination peer IP address

- **identity**—Displays flow information only

- **interface**—Enables you to select connections terminating to a specific interface on the router

- **map**—Enables you to select SAs that were created for a specific crypto map

- **active**—Displays active security associations

- **standby**—Displays security associations that are in a standby state

- **Virtual routing and forwarding (VRF):**

    — **peer**—Enables you to identify all VRF SAs based on address or VRF name

    — **detail**—Provides detailed error counters

Example 21-3 depicts the output from the New York router when it has an established SA with the router in San Francisco (with no optional commands used).

Example 21-3 *Output from the* **show crypto ipsec sa** *Command*

```
NewYork#sh crypto ipsec sa

interface: FastEthernet0/1
    Crypto map tag: SanFran, local addr. 192.168.1.1

    protected vrf:
    local   ident (addr/mask/prot/port): (10.10.1.0/255.255.255.0/0/0)
```

Example 21-3 *Output from the* **show crypto ipsec sa** *Command (Continued)*

```
      remote ident (addr/mask/prot/port): (10.10.3.0/255.255.255.0/0/0)
      current_peer: 192.168.3.1:500
        PERMIT, flags={origin_is_acl,}
       #pkts encaps: 321, #pkts encrypt: 321, #pkts digest 321
       #pkts decaps: 321, #pkts decrypt: 321, #pkts verify 321
       #pkts compressed: 0, #pkts decompressed: 0
       #pkts not compressed: 0, #pkts compr. failed: 0
       #pkts not decompressed: 0, #pkts decompress failed: 0
       #send errors 4, #recv errors 0

        local crypto endpt.: 192.168.1.1, remote crypto endpt.: 192.168.3.1
        path mtu 1500, ip mtu 1500, ip mtu idb FastEthernet0/1
        current outbound spi: AE399F1B

        inbound esp sas:
         spi: 0xFE4F2F8(266662648)
           transform: esp-256-aes esp-sha-hmac ,
           in use settings ={Tunnel, }
           slot: 0, conn id: 2000, flow_id: 1, crypto map: SanFran
           sa timing: remaining key lifetime (k/sec): (4570617/1455)
           IV size: 16 bytes
           replay detection support: Y

        inbound ah sas:

        inbound pcp sas:

        outbound esp sas:
         spi: 0xAE399F1B(2923011867)
           transform: esp-256-aes esp-sha-hmac ,
           in use settings ={Tunnel, }
           slot: 0, conn id: 2001, flow_id: 2, crypto map: SanFran
           sa timing: remaining key lifetime (k/sec): (4570617/1453)
           IV size: 16 bytes
           replay detection support: Y

        outbound ah sas:

        outbound pcp sas:
NewYork#
```

show crypto ipsec security-association lifetime Command

This command is executed from the privileged EXEC mode and displays the SA lifetime value configured for a particular crypto map entry. Example 21-4 depicts the output for the New York router.

Example 21-4 *Output from the* **show crypto ipsec security-association lifetime** *Command*

```
NewYork#show crypto ipsec security-association lifetime

crypto ipsec security-association lifetime seconds 1800

NewYork#
```

show crypto ipsec transform-set Command

This command is executed from the privileged EXEC mode to display the configured transform sets on the router. To display a specific transform set, you can add the optional **tag** command variable by listing the *transform-set name*. Example 21-5 depicts the output of this command on the New York router.

Example 21-5 *Output from the* **show crypto ipsec transform-set** *Command*

```
NewYork#sh crypto ipsec transform-set
Transform set 20: { esp-256-aes esp-sha-hmac  }
   will negotiate = { Tunnel,  },
NewYork#
```

show crypto isakmp key Command

This command is issued from the privileged EXEC mode and displays the key rings, which include the ISAKMP peer address and any preshared key. Example 21-6 depicts the output from this command on the New York router.

Example 21-6 *Output from the* **show crypto isakmp key** *Command*

```
NewYork#show crypto isakmp key
Hostname/Address  Preshared Key
192.168.3.1       abc123

NewYork#
```

show crypto map Command (IPsec)

This command is executed from the privileged EXEC mode and displays the configured crypto maps. As with the **show crypto ipsec transform-set** command, you can request specific information by using the **tag** and/or **interface** command options. Example 21-7 depicts the output of this command on the New York router.

Example 21-7 *Output from the* **show crypto map** *Command*

```
NewYork#sh crypto map
Crypto Map "SanFran" 120 ipsec-isakmp
        Peer = 192.168.3.1
        Extended IP access list 105
            access-list 105 permit ip 10.10.1.0 0.0.0.255 10.10.3.0 0.0.0.255
        Current peer: 192.168.3.1
        Security association lifetime: 4608000 kilobytes/1800 seconds
        PFS (Y/N): Y
        DH group: group2
        Transform sets={ 20,  }
        Interfaces using crypto map SanFran:
            FastEthernet0/1

NewYork#
```

show crypto key pubkey-chain rsa Command

This command displays the RSA public keys for other devices that are stored on the router. You can request information for specific keys by using the optional command values **name** *key*-name and/or **address** *key*-address. Example 21-8 depicts the output from this command on the New York router.

Example 21-8 *Output from the* **show crypto key pubkey-chain rsa** *Command*

```
NewYork#show crypto key pubkey-chain rsa
Codes: M - Manually Configured, C - Extracted from certificate
Code Usage IP-address Name
M Signature 192.168.3.1 FanFrancisco.SF.com
M Encryption 192.168.3.1 FanFrancisco.SF.com
```

Example 21-9 depicts the command using the optional *name* value.

Example 21-9 *Output from the* **show crypto key pubkey-chain rsa** *Command with the Optional name Value Added*

```
NewYork#show crypto key pubkey rsa name SanFrancisco.SF.com
Key name: SanFrancisco.SF.com
Key address: 192.168.3.1
Usage: Signature Key
Source: Manual
Data:
305C300D 06092A86 4886F70D 01010105 00034B00 30480241 00C5E23B 55D6AB22
04AEF1BA A54028A6 9ACC01C5 129D99E4 64CAB820 847EDAD9 DF0B4E4C 73A05DD2
BD62A8A9 FA603DD2 E2A8A6F8 98F76E28 D58AD221 B583D7A4 71020301 0001
```

continues

Example 21-9 *Output from the* **show crypto key pubkey-chain rsa** *Command with the Optional name Value Added (Continued)*

```
Key name: SanFrancisco.SF.com
Key address: 192.168.3.1
 Usage: Encryption Key
 Source: Manual
 Data:
  00302017 4A7D385B 1234EF29 335FC973 2DD50A37 C4F4B0FD 9DADE748 429618D5
  18242BA3 2EDFBDD3 4296142A DDF7D3D8 08407685 2F2190A0 0B43F1BD 9A8A26DB
  07953829 791FCDE9 A98420F0 6A82045B 90288A26 DBC64468 7789F76E EE21
```

show crypto key mypubkey rsa Command

This command is executed from the privileged EXEC mode and displays the RSA public keys that are configured on the router. Example 21-10 depicts the output from the command on the New York router.

Example 21-10 *Output from the* **show crypto key mypubkey rsa** *Command*

```
NewYork#sh crypto key mypubkey rsa
% Key pair was generated at: 00:22:09 UTC Mar 1 1993
Key name: NewYork.NY.com
 Usage: Signature Key
 Key is not exportable.
 Key Data:
  30819F30 0D06092A 864886F7 0D010101 05000381 8D003081 89028181 00D7518F D76BD206 DBFC4635
  DEBAD90A 749CF811 60D9AB5C 3E7EBA99 6201B691 43762A63 BFA033EB 9FD2EED9 C2B96512 06A40169
  0B989DBF 5F61A4FE 06736EE7 A3484B23 4B9B112E A0167D3C 6C47DBE4 3C8E17BD 316E0F10 744315D9
  A7A0D802 63FE54F4 6BA0FEA2 177C6074 159DFB9A 2C786C59 015CCA35 6A0D253C C8513B6D 29020301
  0001
% Key pair was generated at: 00:22:21 UTC Mar 1 1993
Key name: NewYork.NY.com
 Usage: Encryption Key
 Key is not exportable.
 Key Data:
  30819F30 0D06092A 864886F7 0D010101 05000381 8D003081 89028181 00E83666 1BC9A809 97E4CDC1
  A99B8451 F4E1E9E1 50C8E3B2 3256B8E4 5831E1C9 C4CE1E9D 67548A8E EE59D86E 92CC694E 068952F5
  38EE6528 17FA6B7F C814D391 F14625E1 2C876FF4 FF178ADD 1C5DAE33 C333E0D1 CD954EC0 D36C4AB3
  A39EAE42 C7EF1A2F F998BA94 0A231099 7350651F 243BEE09 BFA32644 71D5C9E0 ABE8437A AF020301
  0001
% Key pair was generated at: 00:22:38 UTC Mar 1 1993
Key name: NewYork.NY.com.server
 Usage: Encryption Key
 Key is not exportable.
 Key Data:
  307C300D 06092A86 4886F70D 01010105 00036B00 30680261 00B2A10F D2AE3292
  26B04421 A0719F09 2B8F0A32 052283F8 4B47F93D BAFD355F 23C13F34 694D5EC8
  C2E1A79F 67E28F5D 8A31B1BC AE15C890 AC36F3DA D7D3DBEF 8810D69F 4AD006B8
  F5597E35 265B1896 B5684824 887BDA4F B5FCEFCF FF7A5FFA 21020301 0001
NewYork#
```

debug Commands

The **debug** commands enable you to determine what is actually happening in the communications between peers. The **debug** commands are used to view the actual negotiations between peers and are commonly used to troubleshoot connectivity issues.

debug crypto isakmp Command

This command is executed from the privileged EXEC mode and is used to display the IKE events as they occur. To terminate this output, use the **no debug crypto isakmp** command. Example 21-11 depicts the output from the command on the New York router as it negotiates a connection with the router in San Francisco.

Example 21-11 *Output from the* **debug crypto isakmp** *Command*

```
NewYork#debug crypto isakmp
09:19:27: ISAKMP (8): beginning Main Mode exchange
09:19:27: ISAKMP (8): processing SA payload. message ID = 0
09:19:27: ISAKMP (8): Checking ISAKMP transform 1 against priority 100 policy
09:19:27: ISAKMP:      encryption AES-256
09:19:27: ISAKMP:      hash SHA
09:19:27: ISAKMP:      default group 5
09:19:27: ISAKMP:      auth pre-share
09:19:27: ISAKMP (8): atts are acceptable. Next payload is 0
09:19:27: ISAKMP (8): SA is doing pre-shared key authentication
09:19:28: ISAKMP (8): processing KE payload. message ID = 0
09:19:28: ISAKMP (8): processing NONCE payload. message ID = 0
09:19:28: ISAKMP (8): SKEYID state generated
09:19:28: ISAKMP (8): processing ID payload. message ID = 0
09:19:28: ISAKMP (8): processing HASH payload. message ID = 0
09:19:28: ISAKMP (8): SA has been authenticated
09:19:28: ISAKMP (8): beginning Quick Mode exchange, M-ID of 192036475
09:19:28: ISAKMP (8): processing SA payload. message ID = 192036475
09:19:28: ISAKMP (8): Checking IPsec proposal 1
09:19:28: ISAKMP:      transform 1, AES-256
09:19:28: ISAKMP:      attributes in transform:
09:19:28: ISAKMP:      encaps is 1
09:19:28: ISAKMP:      SA life type in seconds
09:19:28: ISAKMP:      SA life duration (basic) of 1800
```

debug crypto key-exchange Command

The command is executed from the privileged EXEC mode and displays the exchange of public keys between peers. To terminate this command, use the **no debug crypto key-exchange** command. Example 21-12 depicts the output from the command on the New York router as it negotiates a connection with the router in San Francisco.

Example 21-12 *Output from the* **debug crypto key-exchange** *Command*

```
NewYork#debug crypto key-exchange
CRYPTO-KE: Sent 4 bytes.
CRYPTO-KE: Sent 2 bytes.
CRYPTO-KE: Sent 2 bytes.
CRYPTO-KE: Sent 2 bytes.
CRYPTO-KE: Sent 64 bytes.
NewYork# debug crypto key-exchange
CRYPTO-KE: Received 4 bytes.
CRYPTO-KE: Received 2 bytes.
CRYPTO-KE: Received 2 bytes.
CRYPTO-KE: Received 2 bytes.
CRYPTO-KE: Received 49 bytes.
CRYPTO-KE: Received 15 bytes.
```

debug crypto engine Command

All encryption and decryption are performed in the crypto engine. The **debug crypto engine** command enables you to display the message that occurs and the crypto engine functions. To terminate this command, use the **no debug crypto engine** command. Example 21-13 depicts the output from the command on the New York router as it negotiates a connection with the router in San Francisco.

Example 21-13 *Output from the* **debug crypto engine** *Command*

```
NewYork#debug crypto engine
09.22.47:%SYS-5-CONFIG_I:Configured from console by console
09.22.50:CryptoEngine0:generate alg parameter
09.22.50:CRYPTO_ENGINE:Dh phase 1 status:0
09.22.50:CRYPTO_ENGINE:Dh phase 1 status:0
09.22.50:CryptoEngine0:generate alg parameter
09.22.51:CryptoEngine0:calculate pkey hmac for conn id 0
09.22.51:CryptoEngine0:create ISAKMP SKEYID for conn id 1
09.22.51:Crypto engine 0:RSA decrypt with public key
09.22.51:CryptoEngine0:CRYPTO_RSA_PUB_DECRYPT
09.22.51:CryptoEngine0:generate hmac context for conn id 1
09.22.51:CryptoEngine0:generate hmac context for conn id 1
```

debug crypto ipsec Command

This command is executed from the privileged EXEC mode and displays IPsec events. To terminate this command, use the **no debug crypto ipsec** command. Example 21-14 depicts the output from the command on the New York router as it negotiates a connection with the router in San Francisco.

Example 21-14 *Output from the* **debug crypto ipsec** *Command*

```
NewYork#debug crypto ipsec
09.22.55: IPSEC(sa_request):              <<<< IPSec requests and negotiations begin
(key eng. msg.) src= 192.168.1.1, dest= 192.168.3.1,
src_proxy= 192.168.1.1/255.255.255.255/0/0 (type=1),
dest_proxy= 192.168.3.1/255.255.255.255/0/0 (type=1),
protocol= ESP, transform= esp-256-aes  esp-sha-hmac ,
lifedur= 1800s and 4608000kb,
spi= 0x0(0), conn_id= 0, keysize= 0, flags= 0x4
09.22.55: IPSEC(sa_request): ,
(key eng. msg.) src= 192.168.1.1, dest= 192.168.3.1,
src_proxy= 192.168.1.1/255.255.255.255/0/0 (type=1),
dest_proxy= 192.168.3.1/255.255.255.255/0/0 (type=1).,
protocol= AH, transform= ah-sha-hmac ,
lifedur= 1800s and 4608000kb,
spi= 0x0(0), conn_id= 0, keysize= 0, flags= 0x0.    <<<< IKE initiates SPI Exchange
00:24:34: IPSEC(key_engine): got a queue event...
00:24:34: IPSEC(spi_response): getting spi 302974012ld for SA
from 192.168.3.1 to 192.168.1.1 for prot 3
00:24:34: IPSEC(spi_response): getting spi 525075940ld for SA
from 192.168.3.1 to 192.168.1.1 for prot 2
00:24:34: IPSEC(validate_proposal_request): proposal part #1,   <<<< IKE Initiates SA
  proposal
(key eng. msg.) dest= 192.168.3.1, src= 192.168.1.1,
dest_proxy= 192.168.3.1/255.255.255.255/0/0 (type=1),
src_proxy= 192.168.1.1/255.255.255.255/0/0 (type=1),
protocol= ESP, transform= esp-256-aes esp-sha-hmac ,
lifedur= 0s and 0kb,
spi= 0x0(0), conn_id= 0, keysize= 0, flags= 0x4
09.23.00: IPSEC(initialize_sas): ,       <<<< IKE Negotiation complete INBOUND SA established
(key eng. msg.) dest= 192.168.1.1, src= 192.168.3.1,
dest_proxy= 192.168.1.1/255.255.255.255/0/0 (type=1),
src_proxy= 192.168.3.1/255.255.255.255/0/0 (type=1),
protocol= ESP, transform= esp-256-aes esp-sha-hmac ,
lifedur= 120s and 4608000 kb,
spi= 0x120F043C(302974012), conn_id= 29, keysize= 0, flags= 0x4
09.23.00: IPSEC(initialize_sas): ,
(key eng. msg.) src= 192.168.1.1, dest= 192.168.3.1,
src_proxy= 192.168.1.1/255.255.255.255/0/0 (type=1),
dest_proxy= 192.168.3.1/255.255.255.255/0/0 (type=1),
protocol= ESP, transform= esp-256-aes esp-sha-hmac ,
lifedur= 120s and 4608000kb,
spi= 0x38914A4(59315364), conn_id= 30, keysize= 0, flags= 0x4
```

continues

Example 21-14 *Output from the* **debug crypto ipsec** *Command (Continued)*

```
09.23:00: IPSEC(create_sa): sa created,
(sa) sa_dest= 192.168.1.1, sa_prot= 50,
sa_spi= 0x120F043C(302974012),
sa_trans= esp-256-aes esp-sha-hmac , sa_conn_id= 29
09.23:00: IPSEC(create_sa): sa created,
(sa) sa_dest= 192.168.3.1, sa_prot= 50,
sa_spi= 0x38914A4(59315364),
sa_trans= esp-256-aes esp-sha-hmac , sa_conn_id= 30
```

debug crypto pki messages Command

This command is executed from the privileged EXEC mode and displays debugging messages for the details of the communication between the router and the CA. To terminate this command, use the **no debug crypto pki messages** command. Example 21-15 depicts the output from the command on the New York router as it communicates with the CA at 192.168.242.42.

Example 21-15 *Output from the* **debug crypto pki messages** *Command*

```
NewYork#debug crypto pki messages
Fingerprint: 2CFC6265 77BA6496 3AEFCB50 29BC2BF2
10:04:12:Write out pkcs#10 content:274
10:04:12:30 82 01 0E 30 81 B9 02 01 00 30 22 31 20 30 1E 06 09 2A 86
10:04:12:48 86 F7 0D 01 09 02 16 11 70 6B 69 2D 33 36 61 2E 63 69 73
10:04:12:63 6F 2E 63 6F 6D 30 5C 30 0D 06 09 2A 86 48 86 F7 0D 01 01
10:04:12:01 05 00 03 4B 00 30 48 02 41 00 DD 2C C6 35 A5 3F 0F 97 6C
10:04:12:11 E2 81 95 01 6A 80 34 25 10 C4 5F 3D 8B 33 1C 19 50 FD 91
10:04:12:6C 2D 65 4C B6 A6 B0 02 1C B2 84 C1 C8 AC A4 28 6E EF 9D 3B
10:04:12:30 98 CB 36 A2 47 4E 7E 6F C9 3E B8 26 BE 15 02 03 01 00 01
10:04:12:A0 32 30 10 06 09 2A 86 48 86 F7 0D 01 09 07 31 03 13 01 63
10:04:12:30 1E 06 09 2A 86 48 86 F7 0D 01 09 0E 31 11 14 0F 30 0D 30
10:04:12:0B 06 03 55 1D 0F 04 04 03 02 05 A0 30 0D 06 09 2A 86 48 86
10:04:12:F7 0D 01 01 04 05 00 03 41 00 2C FD 88 2C 8A 13 B6 81 88 EA
10:04:12:5C FD AE 52 8F 2C 13 95 9E 9D 8B A4 C9 48 32 84 BF 05 03 49
10:04:12:63 27 A3 AC 6D 74 EB 69 E3 06 E9 E4 9F 0A A8 FB 20 F0 02 03
10:04:12:BE 90 57 02 F2 75 8E 0F 16 60 10 6F BE 2B
10:04:12:Enveloped Data ...
10:04:12:30 80 06 09 2A 86 48 86 F7 0D 01 07 03 A0 80 30 80 02 01 00
10:04:12:31 80 30 82 01 0F 02 01 00 30 78 30 6A 31 0B 30 09 06 03 55
10:04:12:04 06 13 02 55 53 31 0B 30 09 06 03 55 04 08 13 02 43 41 31
10:04:12:13 30 11 06 03 55 04 07 13 0A 53 61 6E 74 61 20 43 72 75 7A
10:04:12:31 15 30 13 06 03 55 04 0A 13 0C 43 69 73 63 6F 20 53 79 73
10:04:12:74 65 6D 31 0E 30 0C 06 03 55 04 0B 13 05 49 50 49 53 55 31
10:04:12:Signed Data 1382 bytes
10:04:12:30 80 06 09 2A 86 48 86 F7 0D 01 07 02 A0 80 30 80 02 01 01
10:04:12:31 0E 30 0C 06 08 2A 86 48 86 F7 0D 02 05 05 00 30 80 06 09
```

Example 21-15 *Output from the* **debug crypto pki messages** *Command (Continued)*

```
10:04:12:2A 86 48 86 F7 0D 01 07 01 A0 80 24 80 04 82 02 75 30 80 06
10:04:12:02 55 53 31 0B 30 09 06 03 55 04 08 13 02 43 41 31 13 30 11
10:04:12:33 34 5A 17 0D 31 30 31 31 31 35 31 38 35 34 33 34 5A 30 22
10:04:12:31 20 30 1E 06 09 2A 86 48 86 F7 0D 01 09 02 16 11 70 6B 69
10:04:12:2D 33 36 61 2E 63 69 73 63 6F 2E 63 6F 6D 30 5C 30 0D 06 09
10:04:12:2A 86 48 86 F7 0D 01 01 01 05 00 03 4B 00 30 48 02 41 00 DD
10:04:12:2C C6 35 A5 3F 0F 97 6C 11 E2 81 95 01 6A 80 34 25 10 C4 5F
10:04:12:3D 8B 33 1C 19 50 FD 91 6C 2D 65 4C B6 A6 B0 02 1C B2 84 C1
10:04:12:86 F7 0D 01 01 01 05 00 04 40 C6 24 36 D6 D5 A6 92 80 5D E5
10:04:12:15 F7 3E 15 6D 71 E1 D0 13 2B 14 64 1B 0C 0F 96 BF F9 2E 05
10:04:12:EF C2 D6 CB 91 39 19 F8 44 68 0E C5 B5 84 18 8B 2D A4 B1 CD
10:04:12:3F EC C6 04 A5 D9 7C B1 56 47 3F 5B D4 93 00 00 00 00 00 00
10:04:12:00 00
10:04:13:Received pki message:1778 types
```

debug crypto pki transactions Command

This command is executed from the privileged EXEC mode and displays debugging messages pertaining to the interaction between the router and the CA. To terminate this command, use the **no debug pki transactions** command. Example 21-16 depicts the output from the command on the New York router as it communicates with the CA at 192.168.242.42.

Example 21-16 *Output from the* **debug crypto pki transactions** *Command*

```
NewYork#debug crypto pki transactions
00:44:00:CRYPTO_PKI:Sending CA Certificate Request:
GET /certsrv/mscep/mscep.dll/pkiclient.exe?operation=GetCACert&message=msca HTTP/1.0
10:04:22:CRYPTO_PKI:http connection opened
10:04:23:CRYPTO_PKI:HTTP response header:
HTTP/1.1 200 OK
Server:Microsoft-IIS/5.0
Date:Fri, 17 Nov 2000 18:50:59 GMT
Content-Length:2693
Content-Type:application/x-x509-ca-ra-cert
Content-Type indicates we have received CA and RA certificates.
10:04:23:CRYPTO_PKI:WARNING:A certificate chain could not be constructed while selecting
certificate status
10:04:23:CRYPTO_PKI:WARNING:A certificate chain could not be constructed while selecting
certificate status
10:04:23:CRYPTO_PKI:Name:CN = msca-rootRA, O = Cisco System, C = US
10:04:23:CRYPTO_PKI:Name:CN = msca-rootRA, O = Cisco System, C = US
10:04:23:CRYPTO_PKI:transaction GetCACert completed
10:04:23:CRYPTO_PKI:CA certificate received.
10:04:23:CRYPTO_PKI:CA certificate received.
```

continues

Example 21-16 *Output from the* **debug crypto pki transactions** *Command (Continued)*

```
Router(config)# crypto ca enroll msca
%
% Start certificate enrollment ..
% Create a challenge password. You will need to verbally provide this
password to the CA Administrator in order to revoke your certificate.
For security reasons your password will not be saved in the configuration.
Please make a note of it.
Password:
Re-enter password:
% The subject name in the certificate will be:Router.cisco.com
% Include the router serial number in the subject name? [yes/no]:n
% Include an IP address in the subject name? [yes/no]:n
Request certificate from CA? [yes/no]:y
% Certificate request sent to Certificate Authority
% The certificate request fingerprint will be displayed.
% The 'show crypto ca certificate' command will also show the fingerprint.
Router(config)# Fingerprint: 2CFC6265 77BA6496 3AEFCB50 29BC2BF2
10:04:41:CRYPTO_PKI:transaction PKCSReq completed
10:04:41:CRYPTO_PKI:status:
10:04:41:CRYPTO_PKI:http connection opened
10:04:41:CRYPTO_PKI: received msg of 1924 bytes
10:04:41:CRYPTO_PKI:HTTP response header:
HTTP/1.1 200 OK
Server:Microsoft-IIS/5.0
Date:Fri, 17 Nov 2000 18:51:28 GMT
Content-Length:1778
Content-Type:application/x-pki-message
10:04:41:CRYPTO_PKI:signed attr:pki-message-type:
10:04:41:13 01 33
10:04:41:CRYPTO_PKI:signed attr:pki-status:
10:04:41:13 01 30
10:04:41:CRYPTO_PKI:signed attr:pki-recipient-nonce:
10:04:41:04 10 B4 C8 2A 12 9C 8A 2A 4A E1 E5 15 DE 22 C2 B4 FD
10:04:41:CRYPTO_PKI:signed attr:pki-transaction-id:
10:04:41:13 20 34 45 45 41 44 42 36 33 38 43 33 42 42 45 44 45 39 46
10:04:41:34 38 44 33 45 36 39 33 45 33 43 37 45 39
10:04:41:CRYPTO_PKI:status = 100:certificate is granted
10:04:41:CRYPTO__PKI:All enrollment requests completed.
10:04:41:%CRYPTO-6-CERTRET:Certificate received from Certificate Authority
```

clear Commands

If you have identified connectivity issues using the **debug** command, you might determine that negotiations have stalled because of a partly established state between peers. The **clear** command can flush the databases of the peers and force a renegotiation of the connection.

clear crypto sa Command

This command is executed from the privileged EXEC mode to delete established IPsec SAs. This command forces the renegotiation of SAs. The **clear crypto sa** command deletes all established SAs, but it is possible to delete specific SAs by using the optional command values, as follows:

- **entry**—Enables you to select a specific host name, IP address, protocol, or security parameter index (SPI)
- **peer**—Enables you to select a specific IPsec SA by the SA peer address
- **map**—Enables you to select a SA that was established by a specific crypto map entry
- **counters**—Does not remove the SA but merely resets the system counters
- **vrf**—Clears IPsec SAs within specific virtual routing and forwarding tables by name

clear crypto isakmp Command

This command is executed from the privileged EXEC mode and deletes the IKE SAs. It is common to view the established SAs using the **show crypto isakmp sa** command to identify the connection ID of the SA that you want to remove. If you do not specify a connection ID, this command removes all established SAs. As with the other **clear** commands, this command forces a renegotiation of the IKE SAs. Example 21-17 depicts the output from the **show crypto isakmp sa** command on the New York router and identifies established SAs with San Francisco and Chicago. The **clear crypto isakmp 1** command removes the SA established with San Francisco.

Example 21-17 *Output from the* **clear crypto isakmp sa** *Command*

```
NewYork#show crypto isakmp sa
dst src state conn-id slot
192.168.1.1 192.168.3.1 QM_IDLE 1 0
192.168.1.1 192.168.2.1 QM_IDLE 8 0
Router# clear crypto isakmp 1
Router# show crypto isakmp sa
dst src state conn-id slot
192.168.1.1 192.168.2.1 QM_IDLE 8 0
NewYork#
```

clear crypto sa counters Command

This command is executed from the privileged EXEC mode and resets the SA counters. It is commonly used when attempting to watch a specific SA built on a router that supports a large number of VPN connections. This command enables you to watch the counters increment as the additional SAs are established.

Foundation Summary

The "Foundation Summary" section of each chapter lists the most important facts from the chapter. Although this section does not list every fact from the chapter that will be on your SNRS exam, a well-prepared candidate should at a minimum know all the details in each "Foundation Summary" section before going to take the exam.

It is important to remember that encryption can be an extremely complex undertaking, and attention to detail is required when implementing secured communication paths. The most effective way to troubleshoot any VPN connectivity issue is to verify connectivity without the encryption and then verify that the configurations match on the VPN peers.

Three categories of commands are commonly used to troubleshoot VPN connectivity. Each category of command has a specific type of information that they provide and action that is taken on the router. The three categories of commands are

- **show commands**—Display the current configuration of the router. The **show** commands tend to be more static because they depict how the device is configured rather than which functions it may be undergoing. The **show** commands are run from the privileged EXEC mode, and the command terminates automatically after it produces its output.

- **debug commands**—Display actions as they are taking place and tend to be more dynamic. These commands display communications between devices as they take place. The **debug** commands run from the privileged EXEC mode and are normally terminated by typing the **no debug** command. It is possible to turn off all **debug** commands using the **no debug all** command.

- **clear commands**—Used to remove an existing negotiated element. The **clear** commands are normally used to force a renegotiation of a connection.

This chapter covered the following commands:

- **show** commands

    — **show crypto ca certificates**

    — **show crypto isakmp policy**

    — **show crypto isakmp key**

    — **show crypto ipsec sa**

    — **show crypto ipsec security-association lifetime**

- **show crypto ipsec transform-set**
- **show crypto map** (IPsec)
- **show crypto key pubkey-chain rsa**
- **show crypto key mypubkey rsa**

- **debug** commands
  - **debug crypto isakmp**
  - **debug crypto key-exchange**
  - **debug crypto engine**
  - **debug crypto ipsec**
  - **debug crypto pki messages**
  - **debug crypto pki transactions**

- **clear** commands
  - **clear crypto sa**
  - **clear crypto sa peer** (peer address or name)
  - **clear crypto isakmp**
  - **clear crypto sa counters**

Q&A

As mentioned in the section "How to Use This Book" in the Introduction, you have two choices for review questions: the Q&A questions here or the exam simulation questions on the CD-ROM. The questions that follow present a bigger challenge than the exam itself because they use an open-ended question format. By using this more difficult format, you can exercise your memory better and prove your conceptual and factual knowledge of this chapter. You can find the answers to these questions in the appendix.

1. When you configure any type of encrypted communication between devices, it is critical that the configuration of both devices _____.

2. If you want to debug IPsec, what are the various **debug** commands you can utilize?

3. What does the **show crypto ca certificates** command do on the router?

4. What does the **peer** option on the **show crypto ipsec sa** command show?

5. To show the SA lifetime value, which command do you use?

6. Which command generates the following output?

   ```
   Codes: M - Manually Configured, C - Extracted from certificate
   Code Usage IP-address Name
   M Signature 192.168.3.1 FanFrancisco.SF.com
   M Encryption 192.168.3.1 FanFrancisco.SF.com
   ```

7. Which **debug** command enables you to display all IKE events?

8. The **debug crypto key-exchange** command displays the exchange of keys between the CA and peers. True or False?

9. Which command enables you to display information about the communication between the router and the CA?

10. Which command enables you to clear the SA counters?

This chapter covers the following subjects:

- Describe the Easy VPN Server
- Describe the Easy VPN Remote
- Easy VPN Server Functionality

CHAPTER 22

Configuring Remote Access Using Easy VPN

Cisco Easy VPN is a client/server application that allows for virtual private network (VPN) security parameters to be "pushed out" to the remote locations that connect using Cisco SOHO/ROHO products. The server portion is a component of Cisco IOS Software Release 12.2(8)T available in numerous router platforms and Cisco PIX security appliances and all the Cisco VPN 3000 concentrators. The client portion, which as of Cisco IOS Software Release 12.2(15)T is referred to as Easy VPN Remote, is available for the 800 to 1700 series routers, PIX 501 and 506E Firewall, 3002 VPN hardware client, and Easy Remote VPN software client 3.x.

> **NOTE** For a complete list of products that support Cisco Easy VPN, see the products listing at Cisco.com.

"Do I Know This Already?" Quiz

The purpose of the "Do I Know This Already?" quiz is to help you decide whether you really need to read the entire chapter. If you already intend to read the entire chapter, you do not necessarily need to answer these questions now.

The 10-question quiz, derived from the major sections in the "Foundation Topics" portion of the chapter, helps you determine how to spend your limited study time.

Table 22-1 outlines the major topics discussed in this chapter and the "Do I Know This Already?" quiz questions that correspond to those topics.

Table 22-1 *"Do I Know This Already?" Foundation Topics Section-to-Question Mapping*

Foundation Topics Section	Questions Covered in This Section
Describe the Easy VPN Server	1, 2
Describe the Easy VPN Remote	4, 6, 9, 10
Easy VPN Server Functionality	3, 5, 7, 8

> **CAUTION** The goal of self-assessment is to gauge your mastery of the topics in this chapter. If you do not know the answer to a question or are only partially sure of the answer, you should mark this question wrong for purposes of the self-assessment. Giving yourself credit for an answer you correctly guess skews your self-assessment results and might provide you with a false sense of security.

1. Which version of Cisco IOS Software introduced Easy VPN Server?

    a. 12.1(13)

    b. 12.2(8)T

    c. 12.5

    d. 12.0(8)J

    e. None of the above

2. Which device does not support Easy VPN client?

    a. Cisco 800 series router

    b. Cisco 3002 hardware VPN client

    c. Cisco 2500 series router

    d. Cisco PIX 501 Firewall

    e. Cisco 1700 series router

3. What is "group-based policy control"?

    a. Group-based policy control enables you to apply policies on a per-user or per-group basis.

    b. Group-based policy control enables you to apply policies if you are a member of the administrators group.

    c. Group-based policy control enables you to apply policies to users only.

    d. Group-based policy control enables you to apply policies to groups only.

    e. None of the above.

4. Which Diffie-Hellman groups are supported by Easy VPN Server?

    a. 1, 2, 3, 4, and 5

    b. 2 only

    c. 2 and 4 only

    d. 1 and 4 only

    e. 2 and 5 only

5. Which configuration mode must you be in to configure the IP address pool?

    a. Pool-configuration mode
    b. Global configuration mode
    c. Privileged EXEC mode
    d. Interface configuration mode
    e. Enable mode

6. What is not configured when creating the ISAKMP policy for the remote VPN clients?

    a. Peer-authentication method
    b. Policy priority
    c. Encryption algorithm
    d. Hash algorithm
    e. Diffie-Hellman group

7. Which configuration mode must you be in to configure RRI?

    a. Crypto-map configuration mode
    b. Global configuration mode
    c. Privileged EXEC mode
    d. Interface configuration mode
    e. Enable mode

8. What is the time range (in seconds) for DPD keepalive "retries"?

    a. 10 to 3600
    b. 60 to 3600
    c. 2 to 3600
    d. 2 to 60
    e. 2 to 1800

9. What are the three modes of operation for Easy VPN?

    a. Server mode
    b. Client mode
    c. Encryptor mode
    d. Network extension plus
    e. Network extension mode

10. How does the command **crypto isakmp client configuration group default** effect the application of the ISAKMP policy?

   a. It blocks all VPN traffic.

   b. It allows all traffic.

   c. It applies the VPN policy to all traffic.

   d. It applies the VPN policy to all remote-access users.

   e. None of the above

The answers to the "Do I Know This Already?" quiz are found in the appendix. The suggested choices for your next step are as follows:

- **8 or less overall score**—Read the entire chapter. This includes the "Foundation Topics" and "Foundation Summary" sections and the "Q&A" section.

- **9 or 10 overall score**—If you want more review on these topics, skip to the "Foundation Summary" section and then go to the "Q&A" section. Otherwise, move on to Chapter 23, "Security Device Manager."

Foundation Topics

Cisco Easy VPN consists of two components: Cisco Easy VPN Remote and Cisco Easy VPN Server. The following sections describe each of these components.

Describe the Easy VPN Server

As mentioned in the beginning of this chapter, Easy VPN is a client/server product that allows for simplified VPN connectivity with branch offices, remote offices, and remote users. The server portion of this product is called Cisco Easy VPN Server and is a component first introduced in Cisco IOS Software Release 12.2(8)T. The client component installs on Cisco routers designed for remote office/home office (ROHO) use; Cisco PIX 500 series Firewall; the 3000 series VPN devices; the 800, 1700, 2600, 3620, 3640, 7100, 7200, and 7500 series routers; and the Cisco VPN client software (version 3.x). The functionality that makes Easy VPN "easy" is that it allows the client to connect with the server and download its VPN configuration. This precludes the requirement of configuring each client endpoint for the VPN. Figure 22-1 depicts how Easy VPN would be used by the New York headquarters to provide secure connectivity to its branch offices, remote offices, and remote users.

Figure 22-1 *Easy VPN Deployment for New York*

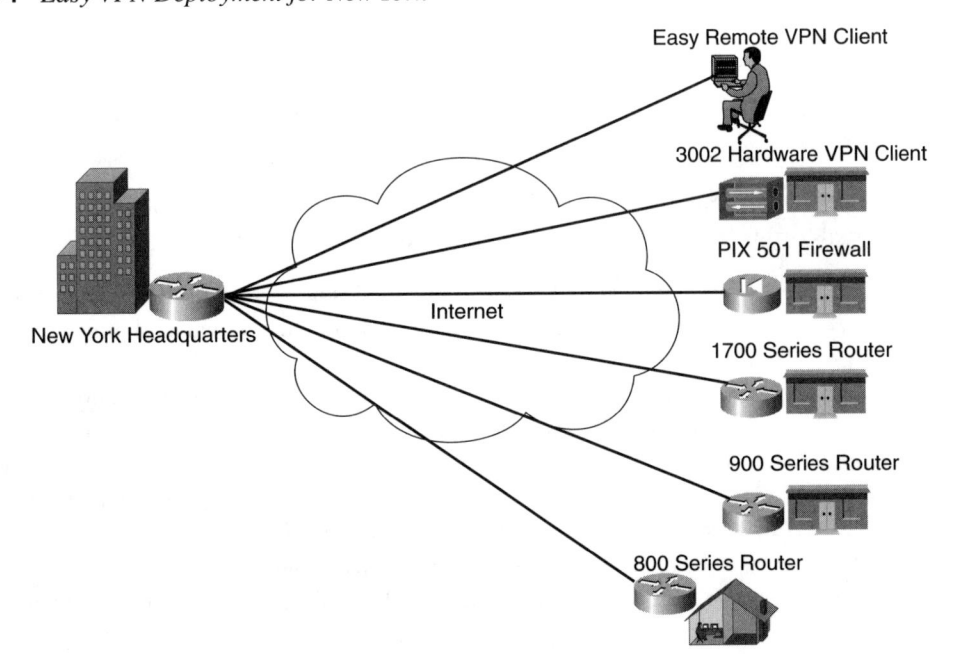

As Figure 22-1 shows, the VPN connection for the remote location terminates at the Cisco IOS router in the New York headquarters.

Describe the Easy VPN Remote

The Cisco Easy VPN Remote feature implements the Cisco Unity Client Protocol, which allows Cisco VPN clients to receive security policies from a Cisco Easy VPN Server. The server can be a dedicated VPN device or a Cisco IOS router that supports the Cisco Unity Client Protocol. This solution minimizes the amount of configuration necessary at a remote site, which leads to a reduction in IT support requirements, lower operational costs, and increased security through implementing a policy change from a single location.

The benefits of the Cisco Easy VPN Remote features follow:

- Allows the provider to change or replace equipment without impacting the end-user equipment
- Eliminates the requirement for Easy VPN client software installation on the end user PCs
- Provides centralized management of security policies
- Enables expediency in large-scale deployments
- Provides flexibility for dynamic configuration of end-user policy, which requires less manual configuration by end users
- Eliminates the need for end users to purchase and configure external VPN devices
- Creates and maintains of the VPN connections are performed on the router rather than the PC
- Reduces interoperability problems between the different VPN client software on the PC, external hardware-based VPN solutions, and other VPN applications

The Cisco Easy VPN Remote feature supports three modes of operation: client, network extension, and network extension plus. All modes of operation support the optional split tunneling.

In the client mode, Network Address Translation (NAT) or Port Address Translation (PAT) is required so that the PCs do not use any IP addresses of the destination server subnet. In this mode, the Cisco Easy VPN Remote automatically configures NAT or PAT translation and the access lists that would be needed to implement the VPN tunnel. The configurations are automatically created when the IPsec VPN connection is initiated. When the session is terminated and the tunnel is torn down, the NAT or PAT and access list configurations are automatically deleted. This mode is often used for remote-access clients working from home or a kiosk. Figure 22-2 shows a typical depiction for a client mode Cisco Easy VPN Remote.

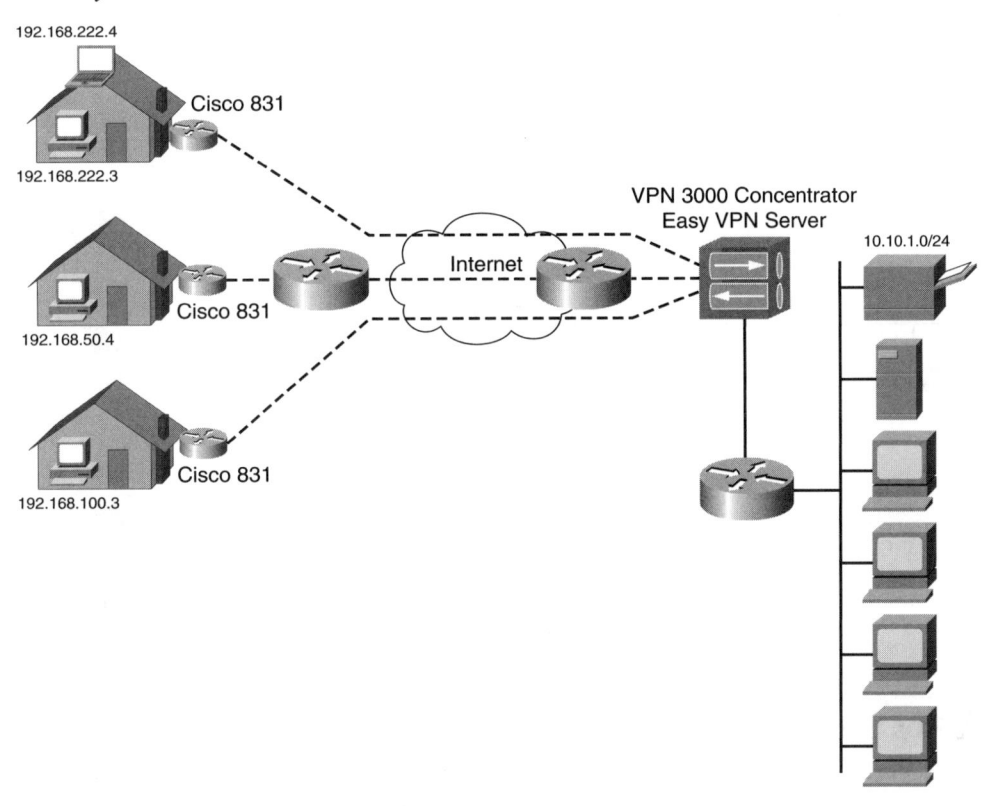

Figure 22-2 *Easy VPN Remote Client Mode*

In the network extension mode, the PCs and other hosts at the client end of the VPN tunnel are given IP addresses that are fully routable and reachable by the destination network over the tunneled network. This is so that they form one logical network. PAT is not used, which allows the client PCs and hosts to have direct access to the PCs and hosts at the destination network. This type of scenario is used for a small remote office. Figure 22-3 shows the network extension mode for Easy VPN Remote.

> **NOTE** The Unity Protocol supports only Internet Security Association Key Management Protocol (ISAKMP) policies that use group 2 (1024-bit Diffie-Hellman) Internet Key Exchange (IKE) negotiation, so the Easy VPN Server being used with the Cisco Easy VPN Remote feature must be configured for a group 2 ISAKMP policy. The Easy VPN Server cannot be configured for ISAKMP group 1 or group 5 when being used with a Cisco Easy VPN client.

The network extension plus mode (network plus) has the same functionality as the network extension mode but includes the additional functionality that the IP address is requested via the mode configuration and assigned to the loopback interface.

For more information on Easy VPN Remote, refer to Cisco.com.

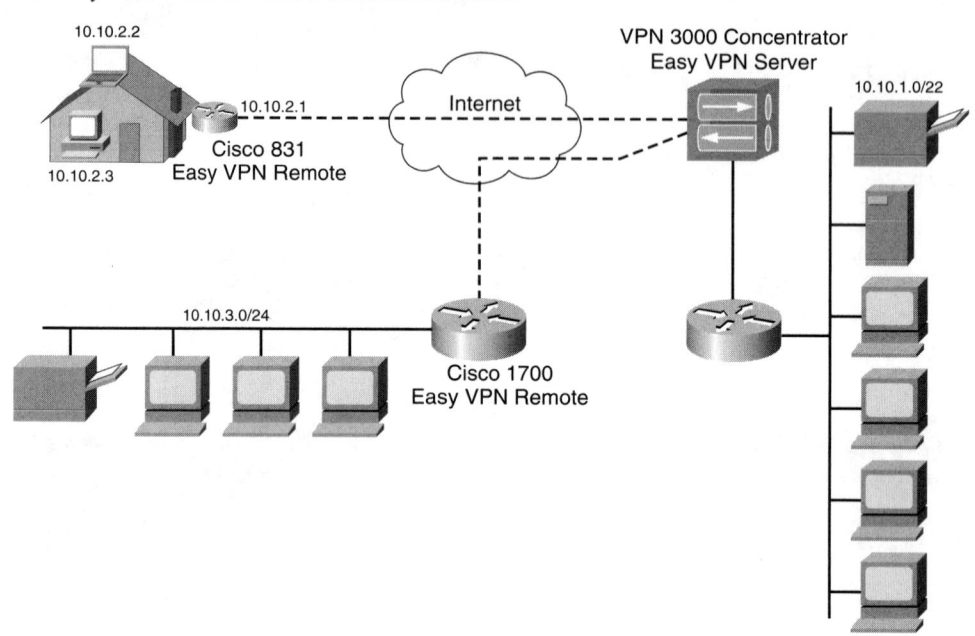

Figure 22-3 *Easy VPN Remote Network Extension Mode*

Easy VPN Server Functionality

Easy VPN Server was introduced with Cisco IOS Software Release 12.2(8)T. It is the first Cisco IOS Software version to provide server support for Cisco VPN client 3.x and the Cisco VPN 3002 hardware clients. The Easy VPN Server manages all IPsec policies centrally and pushes the policy out to the client. This design minimizes the configuration required on the client end.

The following functionality is integrated into the Cisco IOS Software 12.3(11)T with Easy VPN Server:

- **Split tunneling control**—Split tunneling occurs when the remote user is able to connect to the corporate intranet using the VPN, while being able to access the Internet and other resources local to the client VPN user. Many security experts consider split tunneling to be a significant security risk because it allows somewhat of a backdoor connection to the intranet via the remote user system. Split tunneling control allows the VPN administrator to force all traffic from the client to go through the VPN tunnel to include traffic destined for the Internet. Split tunneling is often mitigated by using a firewall at the remote end. The type of firewall deployed at the remote end depends on the connection at that end and could range from a firewall appliance, such as the Cisco PIX Firewall, on the LAN to a personal firewall that is installed on the workstation that is using the Easy Remote VPN client. The Cisco VPN client also prevents split tunneling.

- **IKE dead peer detection (DPD)**—DPD incorporates a series of keepalive messages between the IPsec peers when there is no other traffic passing through the VPN tunnel. DPD is automatically configured on the client end and must be configured on the server to determine the health of the client connection. If either end determines the connection is lost, it notifies the user and redistributes resources that were used by that connection.

- **Initial contact**—If a client is inadvertently disconnected, his connection is not removed from the IKE and IPsec security association (SA) tables until it exceeds the timeout values. If the user attempts to reconnect, the connection is refused because the server thinks there is currently a valid connection running. The initial contact feature integrates an initial contact flag into the message that tells the peer to remove any previous IKE and IPsec information for that SA from the connection tables. This feature resolves connection problems associated with invalid security parameter index (SPI) messages resulting from SA synchronization problems.

- **Extended authentication (Xauth) version 6 support**—IKE Xauth enables you to configure an authentication list using the **crypto map** command. Additional features have been integrated into Xauth version 6.

- **Mode configuration version 6 support**—Mode configuration is the method by which a VPN client receives the configuration settings necessary to successfully create the VPN tunnel.

- **Group-based policy control**—The Easy VPN Server can configure policy parameters on a per-group or per-user basis.

Table 22-2 lists the IPSec attributes supported by Cisco IOS Software 12.2(8)T and Cisco Easy VPN Server.

Table 22-2 *Supported IPSec Attributes*

Option	Supported Attribute
Authentication type	Preshared keys
	RSA signatures
Hash algorithm	MD5-HMAC
	SHA1-HMAC
Diffie-Hellman group	Group 2 (768 bit)
IKE encryption algorithm	DES (56 bit)
	3DES (168 bit)
IPsec encryption algorithm	DES (56 bit)
	3DES (168 bit)
	Null

continues

Table 22-2 *Supported IPSec Attributes (Continued)*

Option	Supported Attribute
IPsec protocols	ESP
	IPCOMP-LZS
IPsec mode	Tunnel mode

How Cisco Easy VPN Works?

The communication between the client and the Easy VPN Server begins with the client requesting a connection. The following is an overview of the connection steps:

Step 1 **Preshared key**—The client initiates IKE Phase 1 via aggressive mode (AM). The accompanying group name entered in the configuration GUI (ID_KEY_ID) is used to identify the group profile associated with this client.

Step 2 **Digital certificates**—The client initiates IKE Phase 1 via main mode (MM). The organizational unit (OU) field of a distinguished name (DN) is used to identify the group profile.

Because the client may be configured for preshared key authentication, which initiates IKE AM, it is recommended that the administrator change the identity of the Cisco IOS VPN device via the **crypto isakmp identity** *hostname* command. This will not affect certificate authentication via IKE MM:

- The client attempts to establish an IKE SA between its public IP address and the public IP address of the Cisco IOS VPN device. To reduce the amount of manual configuration on the client, every combination of encryption and hash algorithms, in addition to authentication methods and Diffie-Hellman group sizes, is proposed.

- Depending on its IKE policy configuration, the Cisco IOS VPN device will determine which proposal is acceptable to continue negotiating Phase 1. IKE policy is global for the Cisco IOS VPN device and can consist of several proposals. In the case of multiple proposals, the Cisco IOS VPN device will use the first match, so you should always list your most secure policies first.

NOTE Device authentication ends and user authentication begins at this point.

- After the IKE SA is successfully established, and if the Cisco IOS VPN device is configured for xauth, the client waits for a "username/password" challenge and then responds to the challenge of the peer. The information that is entered is checked against authentication entities using authentication, authorization, and accounting (AAA) protocols such as RADIUS and

TACACS+. Token cards may also be used via AAA proxy. During xauth, it is also possible for a user-specific attribute to be retrieved if the credentials of that user are validated via RADIUS.

> **NOTE** VPN devices that are configured to handle remote clients should always be configured to enforce user authentication.

- If the Cisco IOS VPN device indicates that authentication was successful, the client requests further configuration parameters from the peer. The remaining system parameters (for example, IP address, Domain Name System [DNS], and split tunnel attributes) are pushed to the client at this time using mode configuration.

> **NOTE** The IP address pool and group preshared key (if Rivest, Shamir, and Adelman [RSA] signatures are not being used) are the only required parameters in a group profile, all other parameters are optional.

- After each client is assigned an internal IP address via mode configuration, it is important that the Cisco IOS VPN device knows how to route packets through the appropriate VPN tunnel. Reverse route injection (RRI) will ensure that a static route is created on the Cisco IOS VPN device for each client internal IP address.

> **NOTE** It is recommended that you enable RRI on the crypto map (static or dynamic) for the support of VPN clients unless the crypto map is being applied to a Generic Routing Encapsulation (GRE) tunnel that is already being used to distribute routing information.

- After the configuration parameters have been successfully received by the client, IKE quick mode is initiated to negotiate IPsec SA establishment. After IPsec SAs are created, the connection is complete.

Configuring the Easy VPN Server

Remember the Easy VPN Server configuration is important because it is the central location where the other VPN client connections terminate. To configure Easy VPN Server on your Cisco IOS 12.2(8)T or later router, follow these steps:

Step 1   Create the IP address pool (remote router).

Step 2   Prepare the router for Easy VPN Server.

Step 3   Configure the group policy lookup.

Chapter 22: Configuring Remote Access Using Easy VPN

Step 4   Create the ISAKMP policy for the remote VPN clients.

Step 5   Define a group policy for a mode configuration push.

Step 6   Create the transform set.

Step 7   Create the dynamic crypto maps with RRI.

Step 8   Apply the mode configuration to the dynamic crypto map.

Step 9   Apply the dynamic crypto map to the interface.

Step 10   Enable IKE DPD.

Step 11   Configure Xauth.

For the purpose of this exercise, see Figure 22-4. This figure depicts the address space used between the headquarters and the remote office. The remote office is located in the resort town of Windham, New York, and is connected to the Internet via a 1700 series router.

Figure 22-4   *VPN Connection Between New York Headquarters and Remote Office*

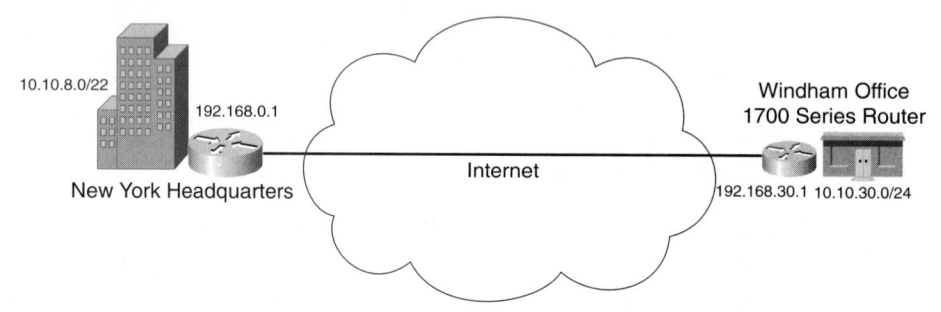

[Diagram: New York Headquarters with 10.10.8.0/22 network and 192.168.0.1 router connected via Internet to Windham Office 1700 Series Router with 192.168.30.1 and 10.10.30.0/24]

Create IP Address Pool

The configuration of IP DHCP address pool is required for Easy VPN Remote in client mode. The local router uses DHCP to assign IP addresses to the PCs connected to the router's LAN interface. The router then uses NAT or PAT to translate these IP addresses into a single of group of IP addresses that is transmitted across the VPN tunnel connection. Use the following steps to configure DHCP on the remote router:

Step 1   Create a name for the DHCP server address pool using the **ip dhcp pool pool-name** global configuration command. This command will put you in the DHCP pool configuration mode.

Step 2   Specify the IP network number and subnet mask of the DHCP address pool that is to be used by using the network command in dhcp configuration mode.

Step 3   Specify the IP address of the default router for DHCP client using the **default-router address** command in the DHCP configuration mode.

Step 4 Import the domain name, DNS server, and NetBIOS Windows Internet Name Service (WINS) server from a central DHCP server into the router's local DHCP database using **import all** command in DHCP configuration mode. These items are also defined in the mode configuration push.

Step 5 Exclude the specified IP addresses, such as the router's IP address, from the DHCP server pool using the **ip dhcp excluded-address lan-ip-address** command in global configuration mode.

Example 22-1 shows an Easy VPN Remote router using 10.1.1.0 for the LAN subnet pool called NYCRemote. In this case, the router's interface IP address of 10.1.1.1 is being excluded from the DHCP pool.

Example 22-1 *IP DHCP Pool Setup*

```
Router1#configure terminal
Router1(config)#ip dhcp pool NYCRemote
Router1(dhcp-config)#network 10.1.1.0 255.255.255.0
Router1(dhcp-config)#default-router 10.1.1.1
Router1(dhcp-config)#import all
Router1(dhcp-config)#exit
Router1(config)#ip dhcp excluded-address 10.1.1.1
```

Prepare the Router for Easy VPN Server

When preparing for the Easy VPN Server, the first configuration task is to enable AAA on the router. The command is entered in the global configuration mode. The next step is to configure a local address pool that will be used for assigning addresses to remote users. Use the following commands:

```
aaa new-model
ip local pool pool-name low-address high-address
```

Example 22-2 shows what the configuration would look like on the router at the New York headquarters.

Example 22-2 *Preparing the Router for Easy VPN Server*

```
NewYork#configure terminal
NewYork(config)#aaa new-model
NewYork(config)#ip local pool windham-office 10.10.8.1 10.10.8.50
```

Configure the Group Policy Lookup

The group policy lookup is the method used to authenticate the remote users attempting to gain access. You can use a RADIUS server as well as the local group. The servers will be tried in the order listed. The command for this configuration is as follows:

```
aaa authorization network group-name local [radius]
```

Example 22-3 shows the command for configuring the group policy lookup for the local group.

492 Chapter 22: Configuring Remote Access Using Easy VPN

Example 22-3 *Configuring the Policy Lookup*

```
NewYork#configure terminal
NewYork(config)#aaa authorization network windham-vpn-users local group radius
```

Create the ISAKMP Policy for the Remote VPN Clients

Configuring the ISAKMP policy for the VPN users is no different from the configuration required for any other VPN connection. ISAKMP is enabled by default on the Cisco router; however, you need to select the following IKE parameters:

- Peer-authentication method
- Encryption algorithm
- Diffie-Hellman group

See Table 22-2 for a list of supported options for each category. Example 22-4 lists the commands used to configure the ISAKMP policy.

Example 22-4 *Configuring the ISAKMP Policy*

```
crypto isakmp enable
crypto isakmp policy priority_number
authentication peer_authentication_method
encryption encryption_algorithm
group diffie-hellman_group
exit
```

Example 22-5 shows the configuration of ISAKMP policy for the remote users on the New York headquarters router.

Example 22-5 *Defining the ISAKMP Policy*

```
NewYork#configure terminal
NewYork(config)#crypto isakmp enable
NewYork(config)#crypto isakmp policy 10
NewYork(config-isakmp)#authentication pre-share
NewYork(config-isakmp)#encryption 3des
NewYork(config-isakmp)#group 2
NewYork(config-isakmp)#exit
```

Define a Group Policy for a Mode Configuration Push

The *mode configuration push* is the policy configuration pushed out to the remote users when they connect to the Easy VPN Server. To configure this group policy, follow these steps:

Step 1 Create the group being defined.

Easy VPN Server Functionality

Step 2 Configure the preshared key. This is the password that the user enters when using the VPN client software.

Step 3 Designate the DNS servers–that is, designate the DNS servers to be used via the VPN connection.

Step 4 Define the DNS domain, which identifies the fully qualified domain name (FQDN) for the network the Easy VPN Server is allowing authorized protected access to.

Step 5 Define the WINS servers—that is, designate the WINS servers to be used via the VPN connection.

Step 6 Define the local IP address pool, which identifies the IP address scope assigned to VPN clients.

The commands required for each step of this configuration are as follows:

```
crypto isakmp client configuration group {group-name | default}
```

NOTE If you use the **default** group name, the policy will apply to all remote-access users.

```
key preshared_key
dns primary_server secondary_server
domain domain_name
wins primary_server secondary_server
pool name
```

Example 22-6 displays the group policy for the configuration push to the Windham remote office.

Example 22-6 *Defining the Group Policy Configuration Mode*

```
NewYork#configure terminal
NewYork(config)#crypto isakmp client configuration group windham-vpn-users
NewYork(config-isakmp-group)#key abc123
NewYork(config-isakmp-group)#dns 10.10.8.252 10.10.10.252
NewYork(config-isakmp-group)#domain newyork.com
NewYork(config-isakmp-group)#wins 10.10.8.251 10.10.10.251
NewYork(config-isakmp-group)#pool windham-office
NewYork(config-isakmp-group)#exit
```

Create the Transform Set

This transform set is used by the remote clients that attempt to establish the IPsec tunnel to this endpoint. The steps required are the same as for any other transform set. If you refer back to Table 22-2, you will notice that AH is not a supported IPsec protocol. The command for this configuration is as follows:

```
crypto ipsec transform-set name [transform1] [transform2] [transform 3]
```

Example 22-7 shows the transform set configured for 3DES and MD5-HMAC.

Example 22-7 *Creating the Transform Set*

```
NewYork#configure terminal
NewYork(config)#crypto ipsec transform-set windhamtransform esp-3des esp-md5-hmac
NewYork(cfg-crypto-trans)#exit
```

Create the Dynamic Crypto Maps with RRI

This three-step process creates and assigns the crypto maps that the remote client connections will use. RRI is enabled to ensure that returning data destined for a specific tunnel can find that tunnel. The commands used for this configuration step are as follows:

```
crypto dynamic-map dynamic_map_name sequence_number
set transform-set transform-set_name
reverse route
```

It is possible to configure multiple transform sets.

Example 22-8 depicts the configuration of the Easy VPN Server for the connection to the Windham office.

Example 22-8 *Configuring Crypto Maps with RRI*

```
NewYork#configure terminal
NewYork(config)#crypto dynamic-map windham-map 1
NewYork(config-crypto-map)#set transform-set windhamtransform
NewYork(config-crypto-map)#reverse-route
NewYork(config-crypto=map)#exit
```

Apply the Mode Configuration to the Dynamic Crypto Map

This three-step process configures the Easy VPN Server to respond to mode configuration requests and begin sending the information required to create the VPN connection with the remote client. The following three steps are required for this task:

Step 1 Configure the router to respond to requests.

Step 2 Enable IKE queries for group policy lookup.

Step 3 Apply the changes to the dynamic crypto map.

The commands for these configuration steps are as follows:

```
crypto map map_name client configuration address respond
crypto map map_name isakmp authorization list list_name
crypto map map_name sequence_number ipsec-isakmp dynamic dynamic_map_name
```

Example 22-9 shows this configuration being completed on the Easy VPN Server for the connection to the Windham office.

Example 22-9 *Applying Mode Configuration*

```
NewYork#configure terminal
NewYork(config)#crypto map windham-map client configuration address respond
NewYork(config)#crypto map windham-map isakmp authorization list windham-vpn-users
NewYork(config)#crypto map windham-map 10 ipsec-isakmp dynamic windham-map
NewYork(config)#exit
```

Apply the Dynamic Crypto Map to the Interface

This command applies the dynamic crypto map to the interface. It is the same process as applying any other function to an interface:

```
interface interface_name
crypto map map_name
```

Example 22-10 shows the crypto map being applied to the router at the New York headquarters.

Example 22-10 *Applying Dynamic Crypto Maps to the Interface*

```
NewYork#configure terminal
NewYork(config)#interface serial 0/0
NewYork(config-if)#crypto map windham-map
NewYork(config-if)#exit
```

Enable IKE DPD

As discussed previously in this chapter, IKE DPD monitors the status of the connection by sending keepalives when there is no traffic passing over the connection. This monitoring allows the system to ensure the connection is functioning and removes any resources that are not required when the connection drops. When configuring IKE DPD, you just need to tell the router how often to send the keepalive message and how long to wait between retries if it does not get a response. The range for the keepalive messages is between 10 and 3600 seconds, and the range for the retries is between 2 and 60 seconds. The command for enabling IKE DPD is as follows:

```
crypto isakmp keepalive seconds retries
```

Example 22-11 depicts this configuration on the router, enabling a keepalive packet every 60 seconds and specifying to retry every 20 seconds if it does not get a response.

Example 22-11 *Enabling IKE DPD*

```
NewYork#configure terminal
NewYork (config)#crypto isakmp keepalive 60 20
```

Configure Xauth

IKE Xauth is a process for using AAA authentication for VPN connections. The following three steps are required to configure Xauth on the Easy VPN Server:

Step 1   Enable AAA login authentication.

Step 2   Configure the Xauth timeout value. (This is the time that the user will have to input the user ID and password.)

Step 3   Configure the Xauth dynamic crypto map.

The commands for configuring Xauth are as follows:

```
aaa authentication login list_name method1 [method 2]
crypto isakmp xauth timeout seconds
crypto map map_name client authentication list list_name
```

Example 22-12 shows how Xauth is configured on the Easy VPN Server for the connection to the Windham remote office.

Example 22-12   *Configuring xauth*

```
NewYork#configure terminal
NewYork(config)#aaa authentication login windham-vpn-users pre-share
NewYork(config)#crypto isakmp xauth timeout 30
NewYork(config-if)#crypto map windham-map client authentication list windham-vpn-users local
```

Easy VPN Modes of Operation

The Easy VPN can use three different remote Phase 2 modes for VPN connectivity, which mainly affect how the remote user is addressed when connected to the destination network. Both configurations support split tunneling. The three modes are as follows:

- **Client mode**—Allows whatever changes necessary to connect the client to the destination network via the VPN connection. In the client mode, the client is automatically configured with NAT or PAT and the access lists needed to create the VPN connection.

- **Network extension mode**—Treats the VPN client systems as components of the original network. The client systems must have fully routable IP addresses and cannot use NAT or PAT.

- **Network extension plus mode**—Acts much the same as the network extension mode except that it pulls an IP address and assigns it to the loopback interface.

> **NOTE**   The term *fully routable* only refers to address space that does not conflict on either end of the connection. This is not a reference to the use of RFC 1918 addressing.

Foundation Summary

The "Foundation Summary" section of each chapter lists the most important facts from the chapter. Although this section does not list every fact from the chapter that will be on your SNRP exam, a well-prepared SECUR candidate should at a minimum know all the details in each "Foundation Summary" before going to take the exam.

Describe the Easy VPN Server

The Easy VPN Server is a product introduced in Cisco IOS Software Release 12.2(8)T. It enables administrators to consolidate IPsec and user policies at a single manageable location that is the endpoint for multiple VPN connections. Each client that connects to this endpoint will download its policy during the VPN negotiation. This centralized management reduces management overhead and increases security.

Easy VPN Server Functionality

The Easy VPN Server provides the following functionality:

- Split tunneling control
- IKE DPD
- Initial contact
- IKE xauth version 6 support
- Mode configuration version 6 support
- Group-based policy control

Configuring the Easy VPN Server

To configure Easy VPN Server on your Cisco IOS router, you must complete the following tasks using the listed commands:

1. Prepare the router for Easy VPN Server.

    Enable AAA on the router:

    aaa new-model

Define an address pool:

ip local pool pool-name low-address high-address

2. Configure the group policy lookup:

aaa authorization network *group-name* **local [local][radius]**

3. Create the ISAKMP policy for the remote VPN clients.

Enable ISAKMP:

crypto isakmp enable

Define the IKE priority:

crypto isakmp policy priority_number

Define the peer-authentication method:

authen peer_authentication_method

Define the encryption algorithm:

encryption encryption_algorithm

Diffie-Hellman group (group 2 supported):

group diffie-hellman_group

4. Define a group policy for a mode configuration push.

Create the group that is being defined:

crypto isakmp client configuration group group-name

Configure the preshared key. This is the password that the user enters when using the VPN client software:

key preshared_key

Define the DNS servers. By doing so, you designate the DNS servers to be used via the VPN connection:

dns primary_server secondary_server

Define the DNS domain. By doing so, you identify the FQDN for the network the Easy VPN Server is protecting:

domain domain_name

Define WINS servers. By doing so, you designate the WINS servers to be used via the VPN connection:

wins primary_server secondary_server

Define the local IP address pool. By doing so, you identify the IP address scope to be assigned to remote VPN users that connect via the Easy VPN Server:

pool *name*

5. Create the transform set:

    crypto ipsec transform-set *name* [*transform1*] [*transform2*] [*transform 3*]

6. Create the dynamic crypto maps with RRI.

    Create the dynamic crypto map:

    crypto dynamic-map dynamic_map_name sequence_number

    Define the transform set:

    set transform-set transform-set_name

    Enable RRI:

    reverse-route

7. Apply the mode configuration to the dynamic crypto map.

    Configure the router to respond to requests:

    crypto map map_name client configuration address respond

    Enable IKE queries for group policy lookup:

    crypto map map_name isakmp authorization list list_name

    Apply the changes to the dynamic crypto map:

    crypto map map_name sequence_number ipsec-isakmp dynamic dynamic_map_name

8. Apply the dynamic crypto map to the interface.

    Enter the interface configuration mode:

    interface interface_name

    Apply the crypto map:

    crypto map map_name

9. Enable IKE DPD:

    crypto isakmp keepalive seconds retries

10. Configure Xauth.

    Enable AAA login authentication:

    aaa authentication login list_name method1 [method 2]

    Configure the Xauth timeout value:

    crypto isakmp xauth timeout seconds

    Configure the Xauth dynamic crypto map:

    crypto map map_name client authentication list list_name

Easy VPN Modes of Operation

- **Client mode**—Supports and requires NAT or PAT
- **Network extension mode**—Does not support NAT or PAT
- **Network extension plus**—Like network extension, but assigns IP to loopback interface

Q&A

As mentioned in the section "How to Use This Book" in the Introduction, you have two choices for review questions: the Q&A questions here or the exam simulation questions on the CD-ROM. The questions that follow present a bigger challenge than the exam itself because they use an open-ended question format. By using this more difficult format, you can exercise your memory better and prove your conceptual and factual knowledge of this chapter. You can find the answers to these questions in the appendix.

1. How does the Easy VPN Server control VPN policies for remote clients?
2. What is DPD?
3. How does the **aaa new model** command prepare the router for Easy VPN Server?
4. What must you do before selecting your IKE parameters for remote VPN clients?
5. What servers should you designate when defining the group policy for mode configuration push?
6. What must you do to make a dynamic crypto map function?
7. What is the difference between **crypto isakmp keepalive seconds** and **retries**?
8. What is Xauth?
9. How many different remote Phase 2 modes does Easy VPN Server support?
10. Which remote Phase 2 modes do not support NAT or PAT?

Part VI: Enterprise Network Management

Chapter 23 Security Device Manager

This chapter covers the following subjects:

- Security Device Manager Overview
- Installing SDM Software
- SDM User Interface
- SDM Wizards
- Using SDM Advanced Options
- Using SDM Monitor Mode

CHAPTER 23

Security Device Manager

The Cisco Security Device Manager (SDM) is a Java-based web management tool used for configuration and monitoring of Cisco IOS Software-based routers. Cisco SDM features, such as smart wizards and security audit, simplify router security. Cisco SDM is supported on a wide range of Cisco routers and Cisco IOS Software releases.

This chapter provides general installation and configuration guidance for SDM.

"Do I Know This Already?" Quiz

The purpose of the "Do I Know This Already?" quiz is to help you decide whether you really need to read the entire chapter. If you already intend to read the entire chapter, you do not necessarily need to answer these questions now.

The 11-question quiz, derived from the major sections in "Foundation Topics" section of the chapter, helps you determine how to spend your limited study time.

Table 23-1 outlines the major topics discussed in this chapter and the "Do I Know This Already?" quiz questions that correspond to those topics.

Table 23-1 *"Do I Know This Already?" Foundation Topics Section-to-Question Mapping*

Foundation Topics Section	Questions Covered in This Section
Security Device Manager Overview	1, 2
Installing SDM Software	9
SDM User Interface	3
SDM Wizards	4–8
Using SDM Advanced Options	10
Using SDM Monitor Mode	11

Chapter 23: Security Device Manager

> **CAUTION** The goal of self-assessment is to gauge your mastery of the topics in this chapter. If you do not know the answer to a question or are only partially sure of the answer, you should mark this question wrong for purposes of the self-assessment. Giving yourself credit for an answer you correctly guess skews your self-assessment results and might provide you with a false sense of security.

1. What features does the SDM include? (Select three.)

    a. Smart wizards

    b. Security audit

    c. Logical workflow

    d. Logic audit

    e. Security wizards

2. What is the minimum amount of free Flash memory that the router needs to store the SDM files?

    a. 2 MB

    b. 6 MB

    c. 8 MB

    d. 4 MB

    e. 3 MB

3. What are the preference options configuration available under the Edit menu?

    a. Preview commands before delivering to router

    b. Confirm before exiting SDM

    c. Write to startup config

    d. Reset to factory defaults

    e. Continue monitoring interface status when switching mode/task

4. Which two types of guided configurations does the Firewall Wizard provide?

    a. Stateful firewall

    b. Basic firewall

    c. Proxy firewall

    d. VPN firewall

    e. Advanced firewall

5. What are the two types of ways to configure site-to-site VPNs in SDM?

   a. Basic site-to-site VPN
   b. Advanced site-to-site VPN
   c. Site-to-site VPN using secure GRE tunnel
   d. Site-to-site VPN using secure IP over IP tunnel

6. What are the two modes in which Easy VPN Remote may be configured in SDM?

   a. Client
   b. Server
   c. Transparent
   d. Remote
   e. Network extension

7. What are the two methods of running the SDM Security Audit Wizard?

   a. Security Audit Wizard or one-step lockdown
   b. One-step Audit Wizard or security lockdown
   c. Basic Audit Wizard or advanced lockdown
   d. Client Audit Wizard or network lockdown

8. What are the types of hubs in a DMVPN configuration?

   a. Primary hub
   b. First hub
   c. Spoke hub
   d. Backup hub
   e. Meshed hub

9. Which prerequisite services are required on a router prior to support SDM? (Select two.)

   a. NTP
   b. Username/password
   c. TFTP
   d. HTTPS
   e. XML

Chapter 23: Security Device Manager

10. Which type of queuing policy does the QoS module in SDM create for handling traffic?

    a. CBWFQ

    b. PQ

    c. IP RTP priority

    d. Low-latency queuing (LLQ)

11. Which of the following are screen options under SDM monitor mode? (Select three.)

    a. Logging

    b. Interface Status

    c. Caching Options

    d. VPN Status

    e. Router Performance

The answers to the "Do I Know This Already?" quiz are found in the appendix. The suggested choices for your next step are as follows:

- **9 or less overall score**—Read the entire chapter. This includes the "Foundation Topics" and "Foundation Summary" sections and the "Q&A" section.

- **10 or 11 overall score**—If you want more review on these topics, skip to the "Foundation Summary" section and then go to the "Q&A" section. Otherwise, move on to Chapter 24, "Final Scenarios."

Foundation Topics

This chapter provides basic installation and configuration instruction for SDM version 2.1.1 features.

Security Device Manager Overview

As previously mentioned, Cisco SDM provides a Java-based web management tool for easy deployment and monitoring of Cisco IOS routers. The features of SDM include the following:

- **Smart wizards**—Smart wizards intelligently detect incorrect configurations and propose fixes. They also provide industry best practices configuration parameters for IOS services, such as firewall and virtual private network (VPN) features. Smart wizards represent a helpful way for users to deploy a basic implementation of a service without having expert knowledge of the technology. An interactive step-by-step configuration menu gathers the parameters necessary to configure and enable features on the Cisco IOS routers.

- **Security audit**—Cisco SDM also offers security auditing capability to check and provide advice on changes to router configuration, based on International Computer Security Association (ICSA) Labs and Cisco Technical Assistance Center (TAC) recommendations.

- **Logical workflow**—SDM provides a step-by-step menu that guides users through router and security setup of the LAN, WAN, firewall, VPN, and intrusion prevention systems (IPS). It also manages an integrated a suite of router services such as routing, switching, and wireless.

This section provides information about SDM software installation requirements.

Hardware Requirements

The computer running SDM must be a Pentium III or later series processor.

If the router is running the SDM files, it must have a minimum of 6 MB of free Flash memory on the router to store the SDM files. A minimum of 2 MB of free Flash memory is required on the router for Cisco SDM Express, and wireless management files requires 1.7 MB.

Operating System Requirements

SDM is designed to run on a personal computer with any of the following operating systems:

- Microsoft Windows XP Professional
- Microsoft Windows 2003 Server (Standard Edition)

- Microsoft Windows 2000 Professional
- Microsoft Windows NT 4.0 Workstation (Service Pack 4)
- Microsoft Windows Me
- Microsoft Windows 98 (second edition)
- Japanese, Simplified Chinese, French, German, Spanish, and Italian language OS support:
    — Microsoft Windows XP Professional
    — Microsoft Windows 2000 Professional

Browser Compatibility

The Cisco SDM must have a compatible browser and Java software installed. It supports the following browsers on Microsoft Windows operating systems:

- Microsoft Internet Explorer 5.5 or later
- Netscape Navigator 7.1 and 7.2
- Firefox 1.0.2

The following Java software are supported:

- Java Virtual Machine (JVM) built-in browsers required
- Java plug-in (Java Runtime Environment version 1.4.2_05 or later)

Installing SDM Software

SDM software may be configured to run from a PC local hard drive or from the router's Flash memory.

The Cisco IOS Software running on a router must support SDM, which is generally supported in Cisco IOS Software Release 12.3 or higher. However, the support is heavily dependent on a combination of hardware platform and software. For up-to-date information about SDM support in Cisco IOS Software and to obtain a copy of Cisco IOS Software, refer to the Cisco.com. The software may be downloaded by providing a valid login account with the appropriate Cisco maintenance contract.

The SDM files are compressed in a zip file. A decompression utility must be installed on the PC to decompress these files.

For the router to support SDM, Hypertext Transfer Protocol (HTTP) or Hypertext Transfer Protocol Secure (HTTPS), services must be enabled on the router and have authentication, authorization, and accounting (AAA) or local router login account. For enhanced security, use the HTTPS service and disable HTTP on the router. HTTPS is supported in all images that support the Crypto/IPsec feature set, starting in Cisco IOS Software Release 12.2(25)T. Example 23-1 shows an example of the configuration parameters for this configuration.

Example 23-1 *HTTP Services Configuration*

```
Router#configure terminal
Router(config)#no ip http server
Router(config)#ip http secure-server
Router(config)#username NYCAdmin password 0 Admin4NYC
```

To install SDM software, decompress the SDM software on the designated PC. In the extracted directory, locate the setup.exe file, and then run it. The setup screen will appear. During the installation, the setup prompts for the location of the SDM files. The options for installation are the PC, the router, or both. Installing the SDM on the PC saves router memory and allows for management of various routers from the same location.

When the installation is complete, SDM is ready for use. If SDM was installed on the router, use the web browser and the router IP address or Domain Name System (DNS) name to connect to it via HTTP or HTTPS protocol.

If SDM was installed on the PC, use the SDM launcher icon to start it up.

SDM User Interface

When SDM initially starts up, it prompts for a username and password. After the user has successfully authenticated, the SDM home page displays. The home page supplies basic information about the router's hardware, software, and configuration. Figure 23-1 shows the home page for a Cisco router 2621XM running Cisco IOS Software Release 12.3(11)T.

The File menu commands provides options to do the following:

- Save running configuration to PC
- SDM deliver to router
- Write to startup config
- Reset to factory defaults
- Exit

Figure 23-1 *SDM Home Page*

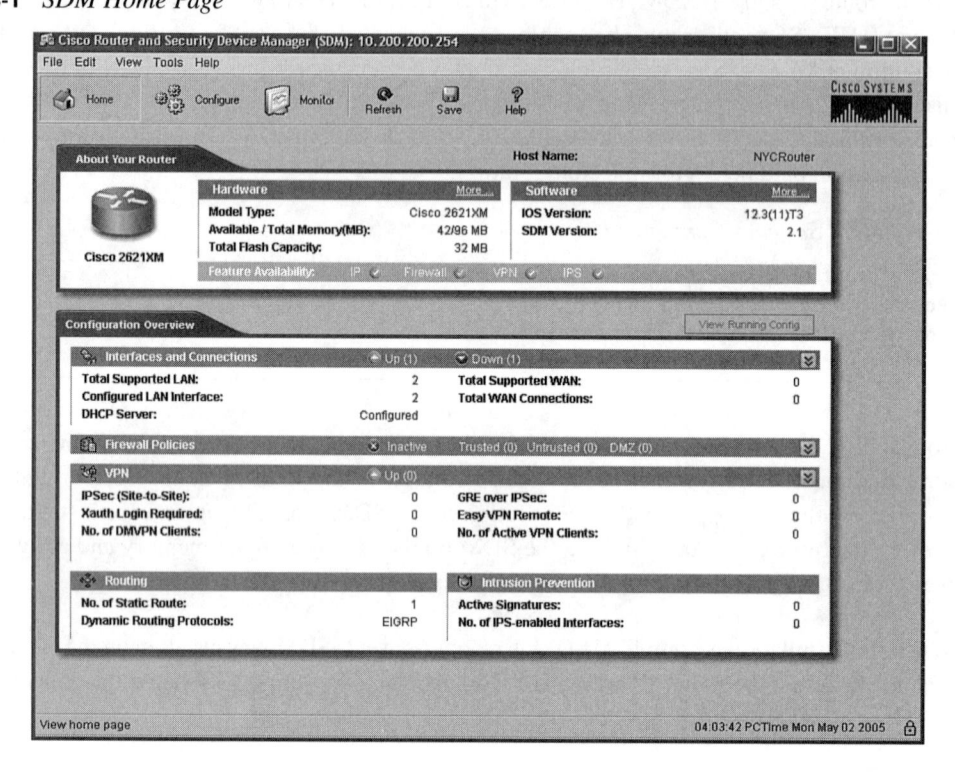

The Edit menu commands allow the configuration of preferences options for the following:

- Preview commands before delivering to router
- Confirm before exiting SDM
- Continue monitoring interface status when switching mode/task

The View menu commands provide the option to view the following:

- Home, configure, and monitor windows
- Running configuration on the router
- **show** commands on the router
- SDM default routes
- Refresh the window

The Tools menu commands provide ping, Telnet, security audit, and SDM update options.

The Help menu commands provide getting started, help topic, and other help resources.

SDM Wizards

The SDM wizards provide a step-by-step guide to configure different features on a router. SDM version 2.1.1 provides six wizards:

- **LAN Wizard**—Provides step-by-step configuration for a LAN interface on a router
- **WAN Wizard**—Provides step-by-step configuration for a WAN interface on a router
- **Firewall Wizard**—Provides step-by-step configuration for the Firewall feature set on a router
- **VPN Wizard**—Provides step-by-step configuration for the VPN feature set on a router
- **Security Audit Wizard**—Provides step-by-step examination of the router configuration and update of the router's security posture
- **Reset Wizard**—Resets the configuration of the router to factory default

All of these modules, except for the WAN Wizard, are discussed in the following sections.

SDM LAN Wizard

The SDM LAN Wizard guides the user through the configuration of a LAN interface. The LAN interfaces that are not configured on the router are listed on the Create Connection screen. An interface may be configured using the Create New Connection button on this page. Figure 23-2 shows this screen.

After an IP address and subnet mask have been assigned to the interface, the option of activating the Dynamic Host Configuration Protocol (DHCP) server service is given. If this is a service that will be required by the private LAN users, choose **Yes** from the menu and provide the IP address range that hosts may use.

Review the configuration parameters in the Summary screen and click **Finish** to deliver the settings to the router's running configuration. Figure 23-3 shows the Summary window.

To edit the LAN configuration, use the **Edit Interface/Connection** tab. Figure 23-4 shows the LAN Wizard Edit Window screen.

514 Chapter 23: Security Device Manager

Figure 23-2 *LAN Wizard Screen*

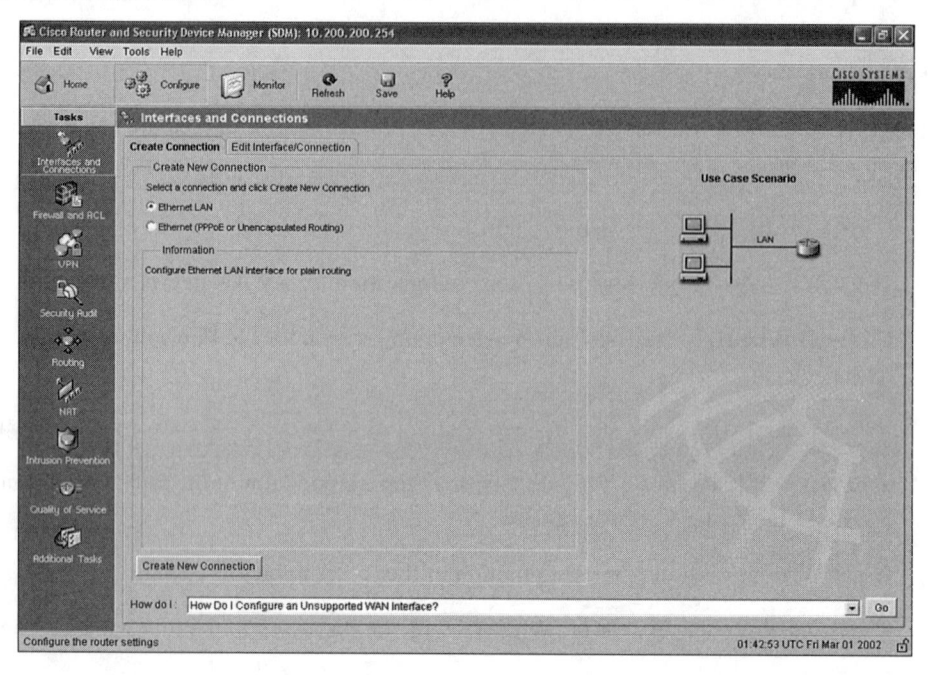

Figure 23-3 *LAN Wizard Summary Window*

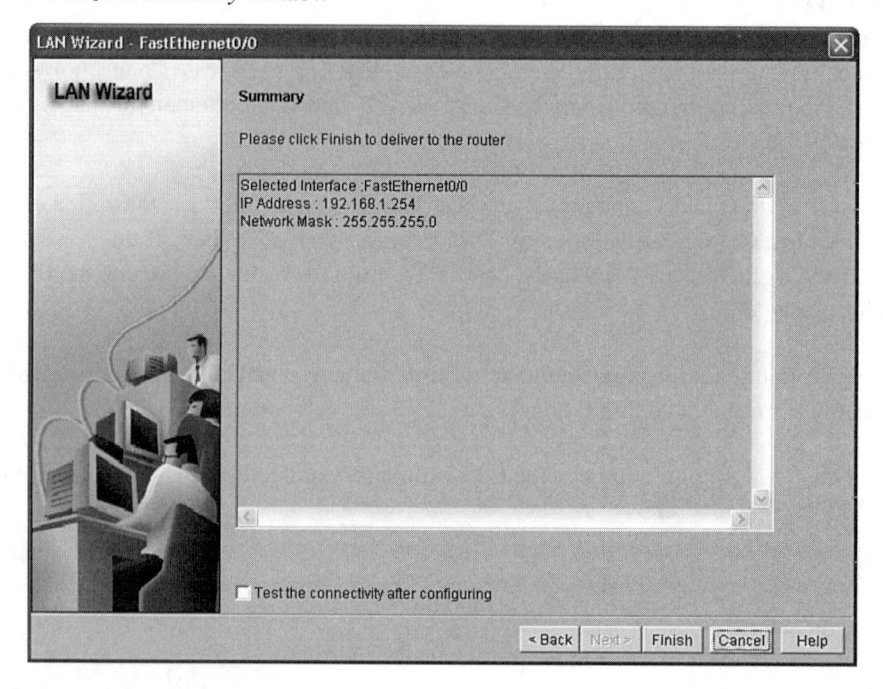

Figure 23-4 *LAN Wizard Edit Window*

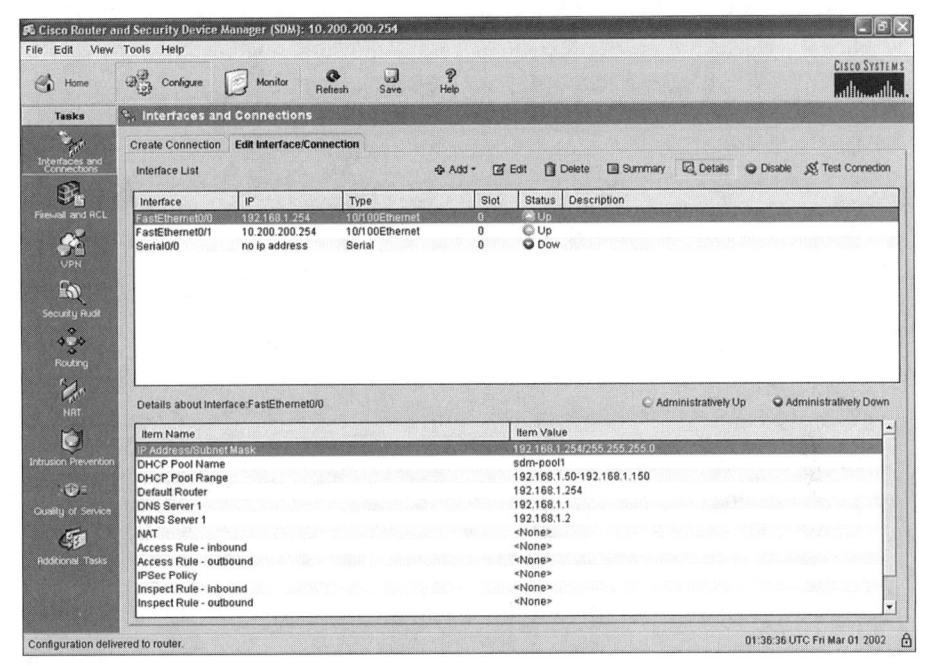

Using SDM to Configure a Firewall

Cisco SDM provides a wizard to assist in configuring the firewall parameters on a router by answering prompts in a set of screens. Prior to using this SDM feature, the router must be running a Cisco IOS image that supports the Firewall feature set, and the LAN/WAN configuration must be complete. The main tasks required to configure the IOS Firewall feature set are as follows:

- Selecting the type of firewall configuration
- Configuring the firewall interface
- Applying the firewall configuration to the router
- Advanced firewall configuration options

Selecting the Type of Firewall Configuration

The Firewall Wizard in SDM provides two types of guided configurations:

- **Basic Firewall**—This option is used if there is only one outside interface and no requirements for a demilitarized zone (DMZ) exist. SDM applies its default rule set to the trusted (inside) and untrusted (outside) interfaces. It also the applies default inspection rule and unicast reverse-path forwarding (uRPF) to the untrusted interface. The basic firewall configuration would be applicable to a small branch office connected to the corporate headquarters via the Internet. Figure 23-5 shows the initial SDM basic firewall screen.

Figure 23-5 *SDM Basic Firewall Screen*

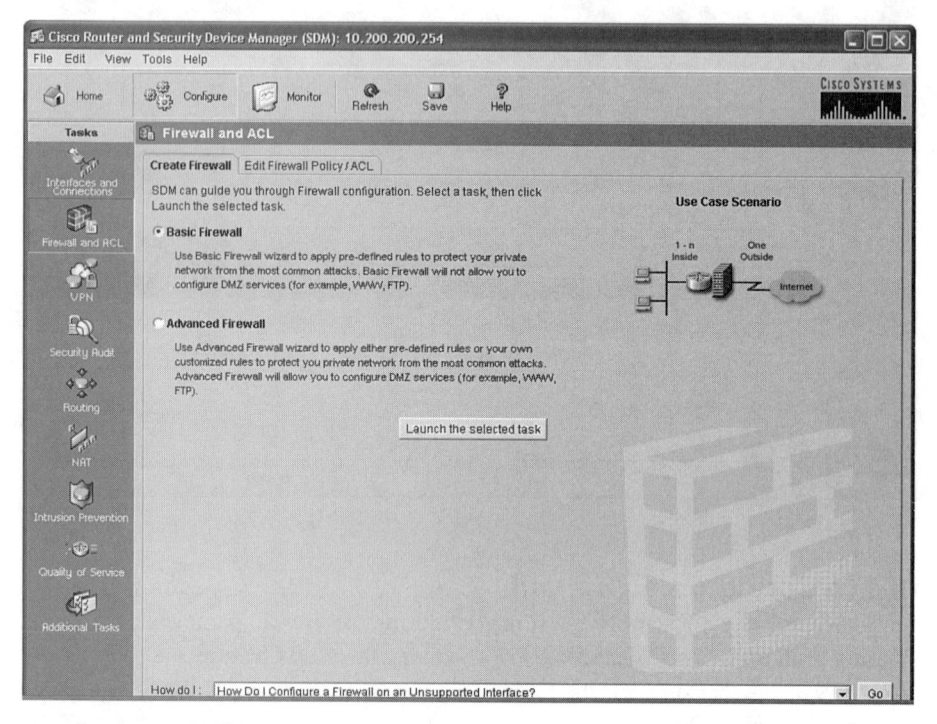

Advanced Firewall—This option is used if there are several outside interfaces. During the configuration, SDM provides the option to create a DMZ and specify inspection rules. SDM applies user-defined access rules to the trusted, untrusted, and DMZ interfaces and enables IP uRPF to the untrusted interface. An example of this would be a corporate headquarters with publicly accessible web services. Figure 23-6 shows the advanced firewall screen in SDM.

Configuring the Firewall Interfaces

In the basic configuration option, you are prompted to select the inside and outside interfaces from the wizard. Figure 23-7 shows the basic configuration wizard screen used to select the trusted and untrusted interfaces.

From the drop-down box, you select the outside untrusted interface, and select the inside interface via the check box menu.

SDM Wizards **517**

Figure 23-6 *SDM Advanced Firewall Screen*

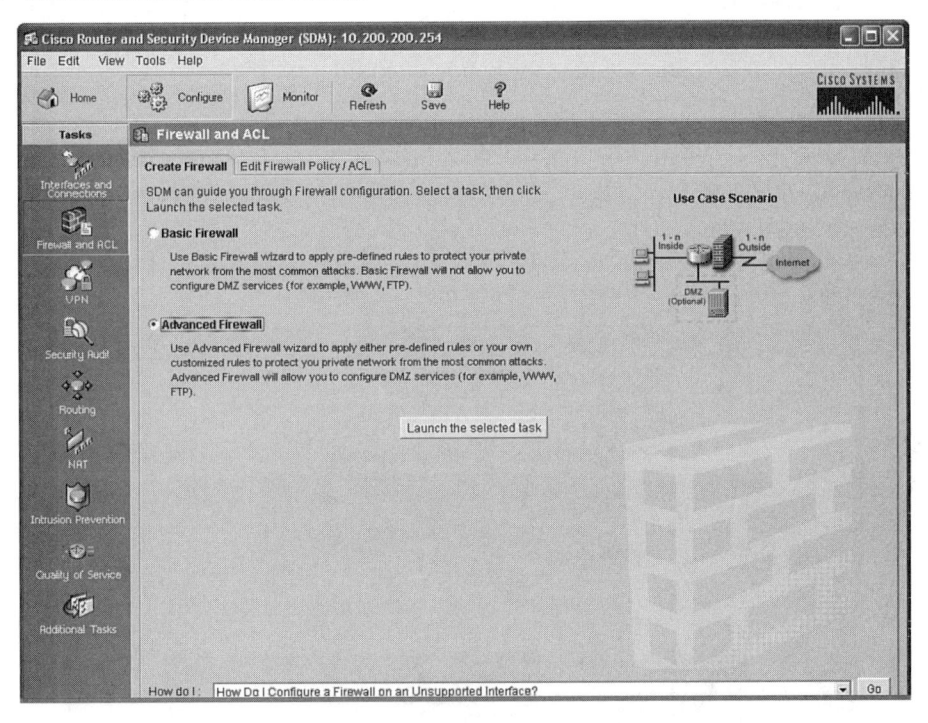

Figure 23-7 *Basic Firewall Wizard Interface Selection*

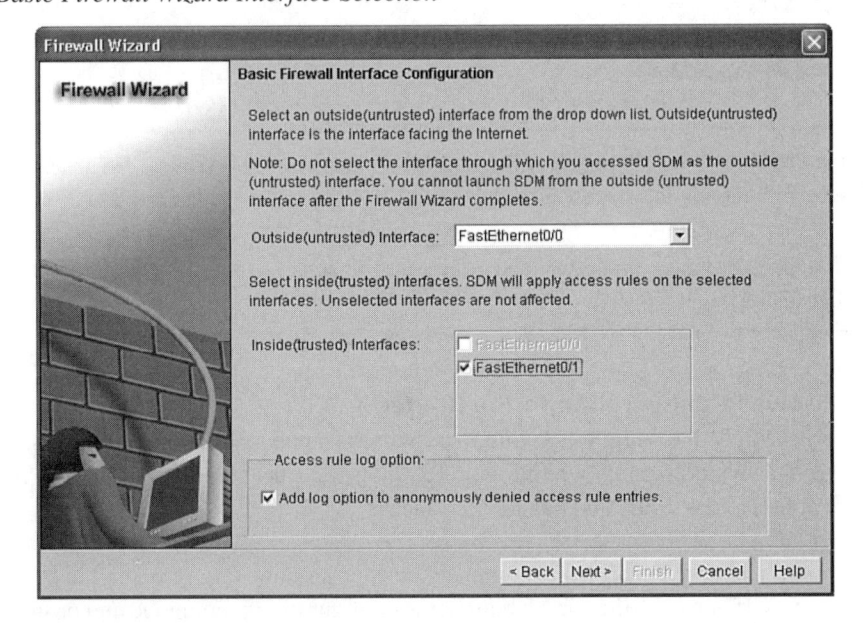

518 Chapter 23: Security Device Manager

> **NOTE** Make sure that SDM is launched from the inside trusted interface. If the outside interface is used to access SDM, the connectivity to SDM will be lost after the firewall configuration wizard completes and the policy is loaded.

In the advanced configuration option, the SDM Firewall Wizard assists in selecting the inside, outside, and DMZ interfaces. Figure 23-8 shows the advanced configuration wizard screen used to select these interfaces.

Figure 23-8 *Advanced Firewall Wizard Interface Selection*

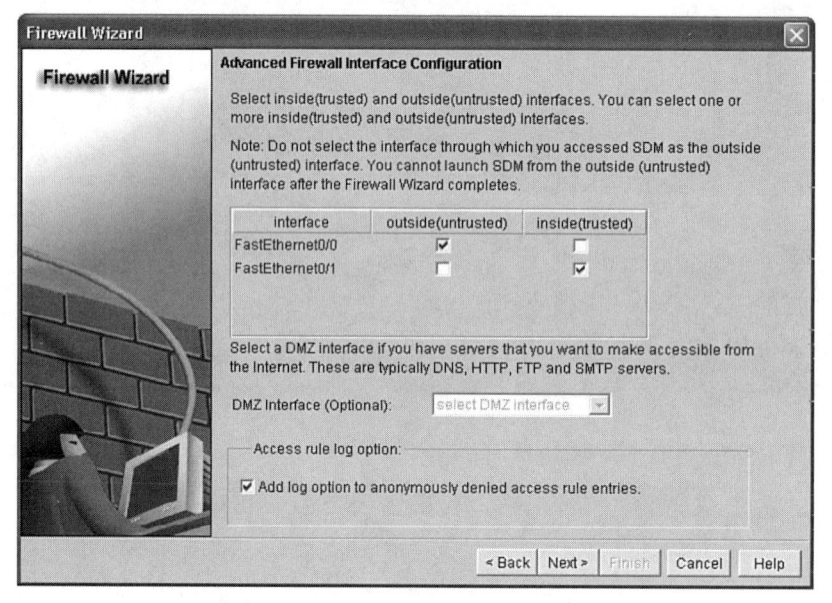

From the drop-down box, you select the interface used for the DMZ. Use the check boxes to select the inside and outside interfaces of the router.

> **NOTE** Both the basic and advanced configuration screens have a check box option to log any failed access attempts.

Applying the Firewall Configuration to the Router

The Internet Firewall Configuration Summary window appears after the interface selection is complete. This window, as shown in Figure 23-9, displays the summary of the configuration that will be applied to the inside and outside interfaces.

At this point, you have the option to complete the configuration or go back to the previous screen to make changes. If you click the **Finish** button, the configuration is compiled and pushed to the router. Figure 23-10 shows the screen displayed during this process.

Figure 23-9 *Summary Firewall Configuration*

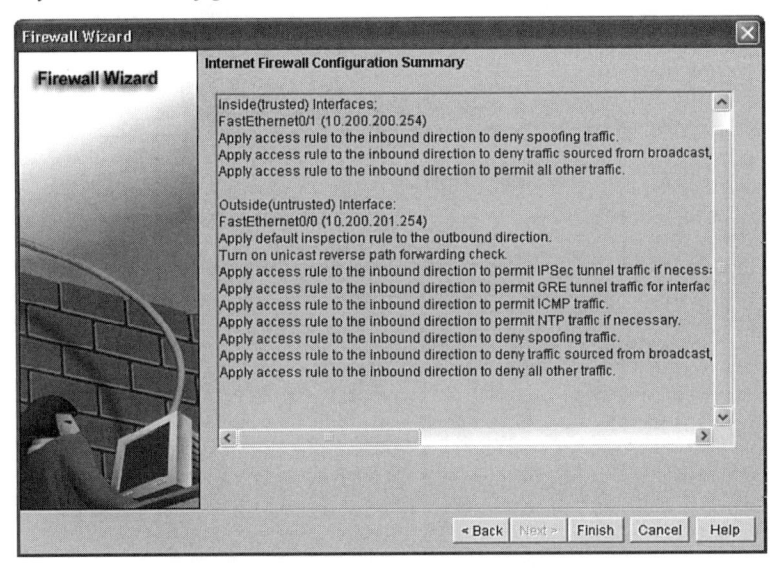

Figure 23-10 *Applying Configuration to the Router*

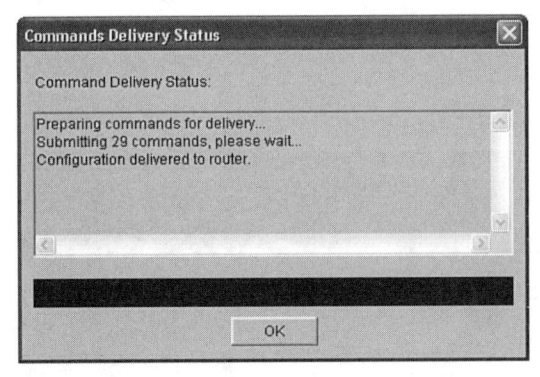

Firewall configuration changes may also be pushed to the router using the Edit Firewall/ACL tab Apply Changes button, as shown in the next section in Figure 23-12.

> **NOTE** If the Preview commands before delivering to router option was checked in the User Preferences window, the Deliver configuration to router window appears. In this window, you can view the command-line interface (CLI) commands that are delivering to the router.

Advanced Firewall Configuration Options

During the advanced firewall configuration setup, the wizard guides the user through the configuration steps for the firewall inspection rules and the DMZ service configuration options. The DMZ service configuration includes the IP address range and service type selection. The Inspection Rule

Editor window shows the default inspection rule and allows for the addition and modification of new rules. Figure 23-11 shows the creation of a new inspection rule set and the options available.

Figure 23-11 *Adding New Inspection Rule Set*

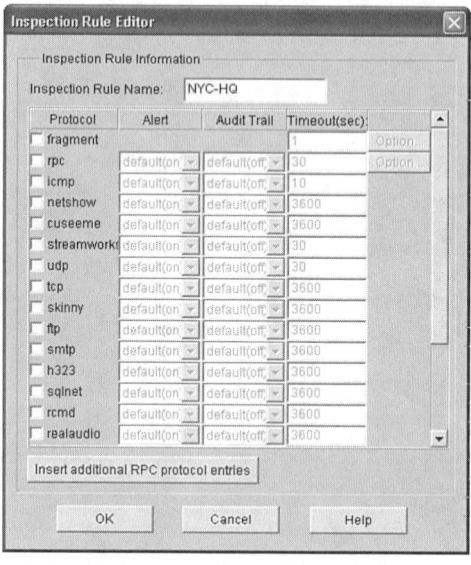

You can modify ACL rules under the Edit Firewall Policy/ACL configuration tab, as shown in Figure 23-12.

Figure 23-12 *Edit Firewall Policy/ACL Tab*

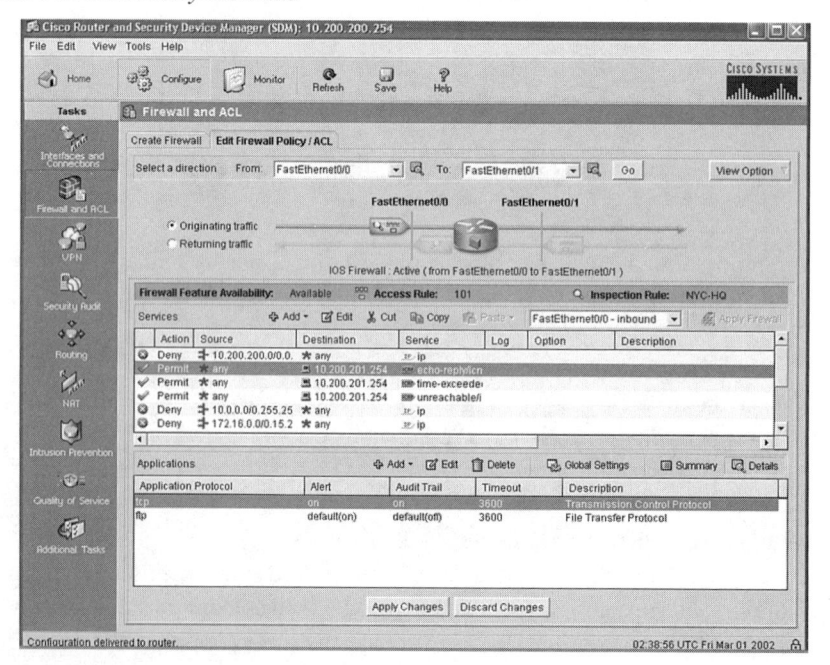

To create a new rule, choose **Add New** from the Services subwindow and enter the rule criteria. Figure 23-13 shows the rule creation window.

Figure 23-13 *Rule Creation Window*

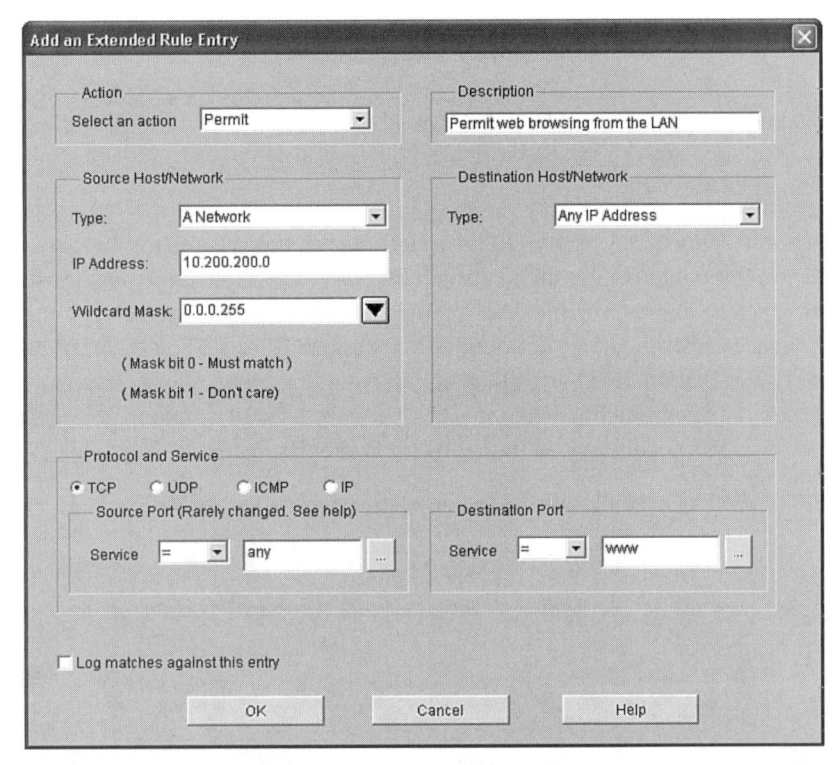

[Screenshot of "Add an Extended Rule Entry" dialog showing Action: Permit; Description: Permit web browsing from the LAN; Source Host/Network Type: A Network, IP Address: 10.200.200.0, Wildcard Mask: 0.0.0.255; Destination Host/Network Type: Any IP Address; Protocol and Service: TCP selected; Source Port Service: = any; Destination Port Service: = www; with OK, Cancel, Help buttons]

The rules are applied to the interface specified under the Interface drop-down box.

Activity on the firewall is monitored via the creation of log entries. For more information on logging, see the section "Using SDM Monitor Mode" later in this chapter.

Using SDM to Configure a VPN

SDM provides VPN configuration wizards for site-to-site VPN, Easy VPN Remote, Easy VPN Server, and Dynamic Multipoint VPN (DMVPN). The following sections discuss the configuration tasks for each one of these VPN types.

Site-to-Site VPN

Site-to-site VPNs protect traffic that travels between two sites over an unsecure transport, such as the Internet. This is a widely used method for sites with more than one location to connect their various branch offices to the headquarters location. SDM wizards guide the configuration of the end routers to encrypt all traffic between sites. The encryption protocols supported by SDM are Data Encryption Standard (DES), 3DES, AES-128, AES-192, and AES-256. Authentication algorithms that may be used are SHA-1 and Message Digest 5 (MD5).

There are two types of configuration for site-to-site VPNs:

- **Basic site-to-site VPN**—This option is used to configure a VPN tunnel between two routers via a preshared key or digital certificate. In case of preshared keys, the two sites must have matching keys.

- **Site-to-site VPN using secure GRE tunnel**—This option is used to configure a protected GRE to tunnel IPsec-encrypted traffic between two sites. GRE tunneling may prove useful when routing protocols are exchanged between the two end routers. Because routing protocols, such as Enhanced Interior Gateway Routing Protocol (EIGRP), require the neighbor to be on a directly connected subnet; GRE tunnels can be set up to provide the impression that the other end is directly connected. GRE tunnels may also be used in situations where the LAN protocol is a different protocol, such as Internetwork Packet Exchange (IPX).

Figure 23-14 shows the configuration screen to create a site-to-site VPN.

Figure 23-14 *Site-to-Site VPN Configuration Screen*

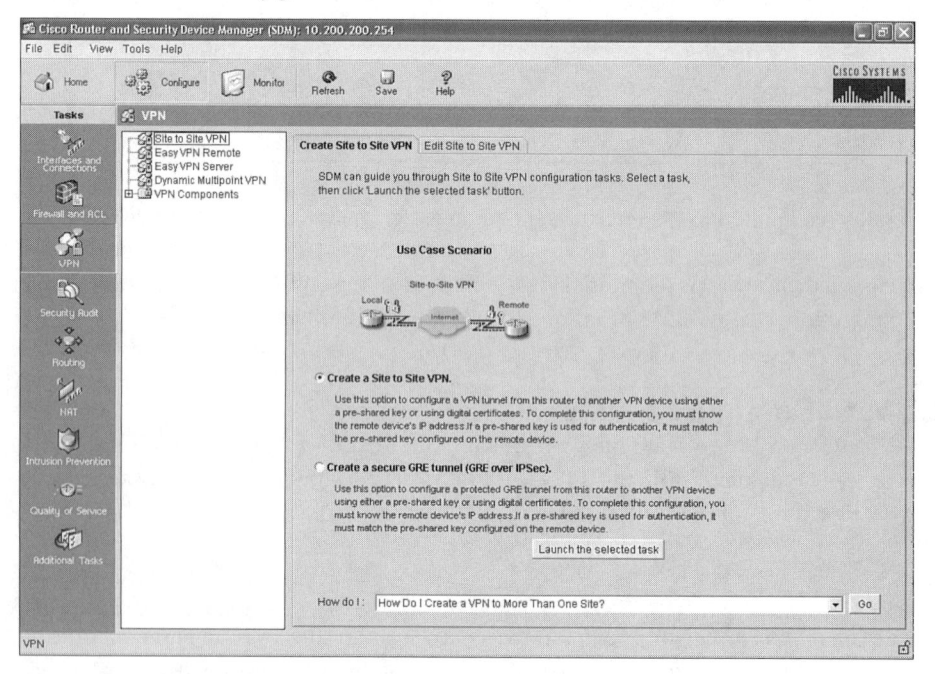

SDM Wizards 523

The Site-to-Site VPN Wizard has two options for stepping through the configuration:

- **Quick Setup**—If this option is selected, SDM will automatically provide a default transform set to control the encryption of the data, a default Internet Key Exchange (IKE) policy for authentication, and a default IPsec rule that will encrypt all traffic between the two routers.

- **Step-by-Step Wizard**—To configure a site-to-site VPN using specific nondefault parameters, use this option. Custom configuration for the VPN and stronger encryption can be selected via this wizard window.

Figure 23-15 shows the setup screen.

Figure 23-15 *Site-to-Site VPN Wizard Setup Screen*

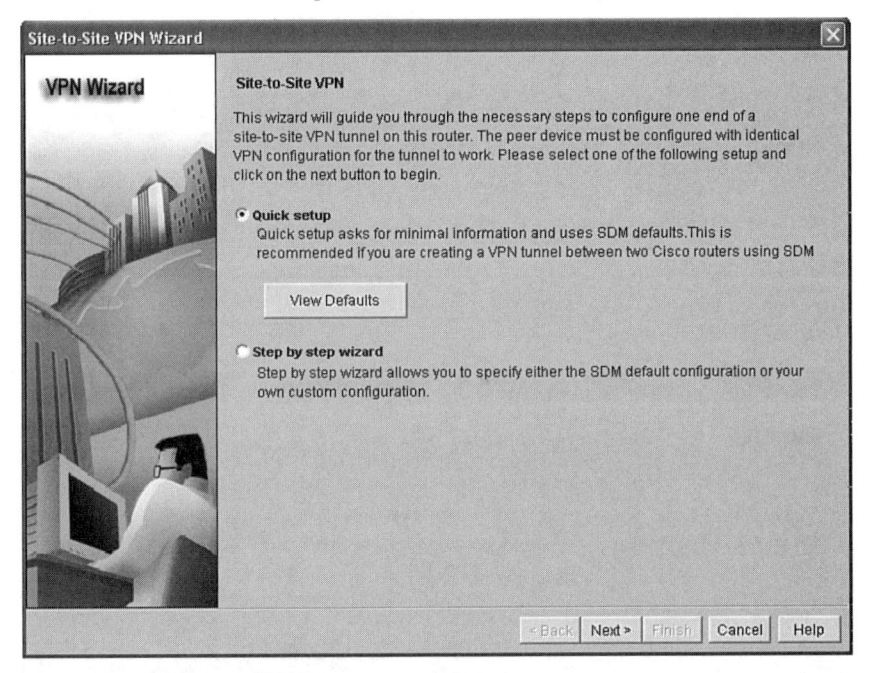

NOTE The Quick Setup option is not available if configuring the GRE tunnel option.

At a minimum, the user must specify the VPN connection interface, peer identity, and traffic to encrypt parameters. Figure 23-16 shows the setup screen for entering this information.

After the configuration steps have been completed, a summarized list displays for user review. Figure 23-17 shows this screen.

Figure 23-16 *Connection Information Screen*

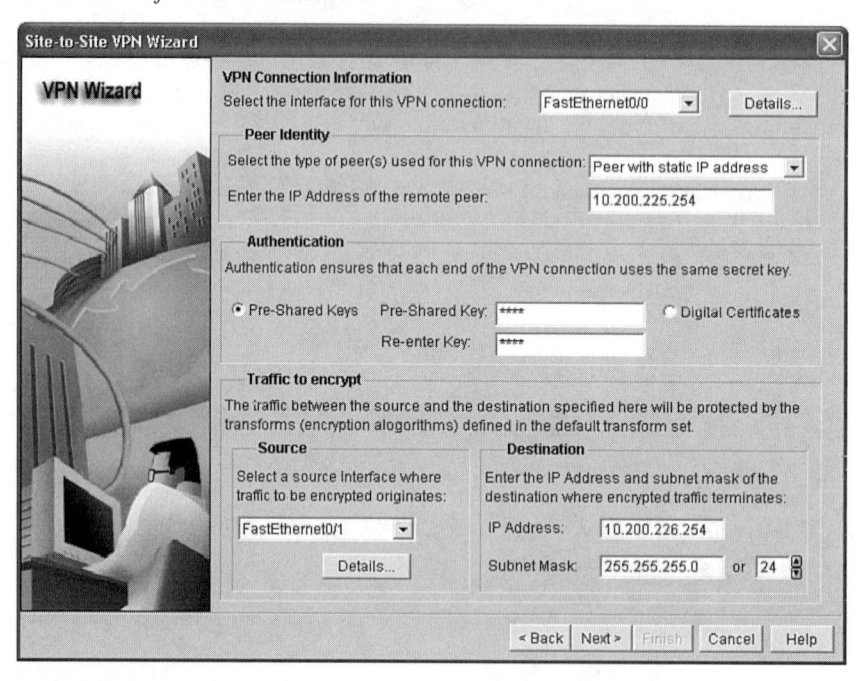

Figure 23-17 *Site-to-Site VPN Summary Screen*

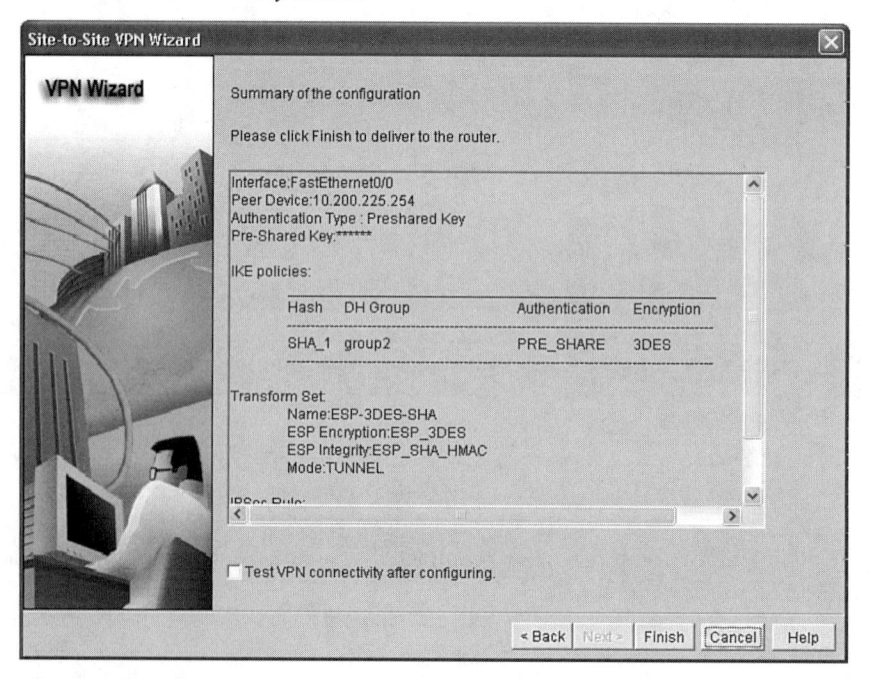

If the **Finish** button is clicked, the configuration is completed at that point and saved to the router's running configuration. Changes take effect immediately but require a manual save via the **write memory** command to store in the configuration in nonvolatile RAM (NVRAM) startup configuration.

Easy VPN Remote

SDM provides a wizard to guide the user through the configuration of a router as an Easy VPN client. The router must be running a Cisco IOS image that supports Easy VPN Phase 2. Figure 23-18 shows the Easy VPN configuration wizard window.

Figure 23-18 *Easy VPN Configuration Window*

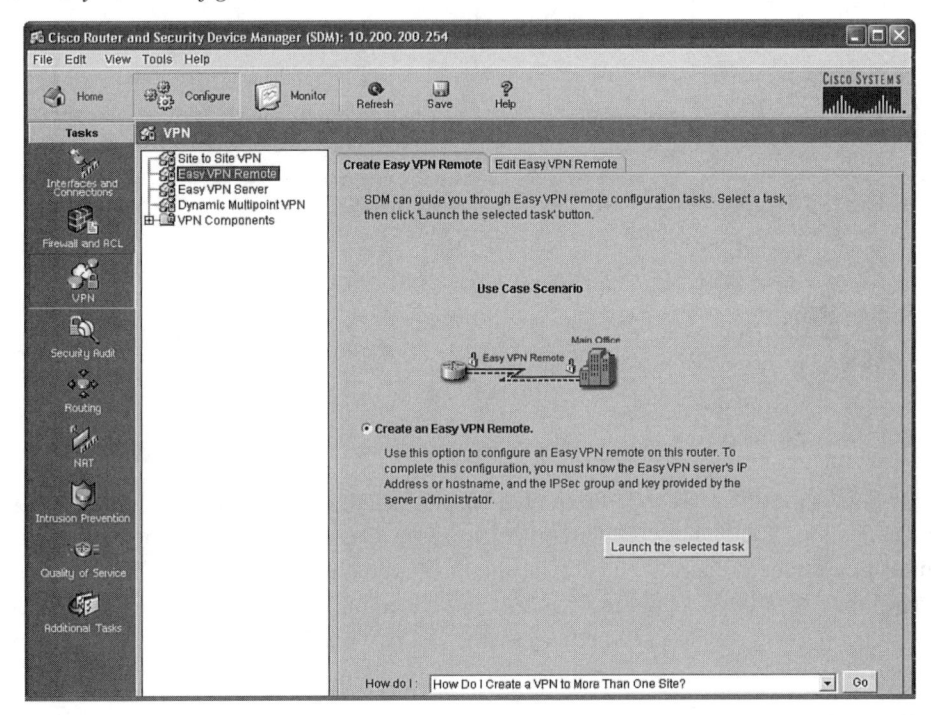

The information required to configure the remote client is the Easy VPN Server's IP address or host name, the key, and the IPsec group name. Figure 23-19 shows the configuration screen.

There are two modes of operation:

- **Client**—For use with PCs and other devices on the router's LAN that would create a private IP network. The IP addresses at the remote end will not be accessible by the server side. Network Address Translation (NAT) or Port Address Translation (PAT) is used to translate these addresses when accessing VPN server-side resources. An example of this mode would be a user working from a home office.

- **Network extension**—If this mode is used, the IP addresses on the remote client LAN will be visible and routable to the destination Easy VPN Server network. All the PCs and devices on each end will have direct access to each other.

Figure 23-19 *Easy VPN Remote Wizard Client Connection Setup*

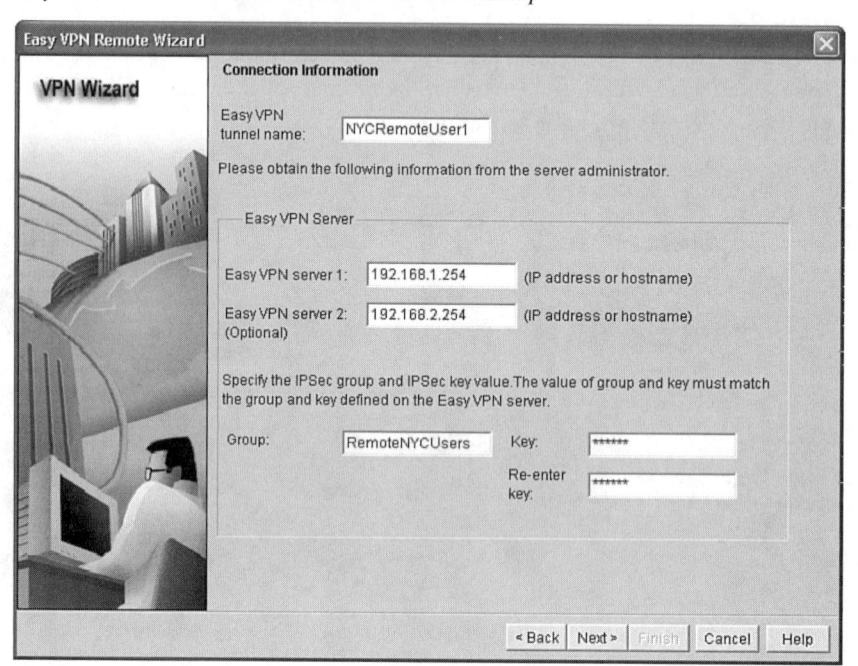

An option available in Easy VPN Remote Phase 3 is the User Authentication (Xauth) IKE extended authentication. Xauth allows all Cisco IOS Software AAA authentication methods to perform user authentication in a separate phase after the IKE authentication Phase 1 exchange. Xauth is an extension to IKE and does not replace IKE authentication. Although it is not recommended, SDM gives the option of saving the Xauth username and password in the configuration. Figure 23-20 shows this screen.

The inside and outside interface for the Easy VPN client must be specified to complete the configuration steps. Multiple inside interfaces (up to three in the Cisco 800 and 1700 series routers) may be configured if the interface is not being used as part of another Easy VPN configuration. An interface cannot be configured as both the inside and outside interface. Figure 23-21 shows the screen for this configuration.

The summary window provides a list of the Easy VPN configurations created and an option to save this configuration. Clicking the **Finish** button pushes the information to the router's running configuration. To change the Easy VPN configuration at a later time, use the Edit Easy VPN Remote window.

Figure 23-20 *Easy VPN Remote Setup Xauth Screen*

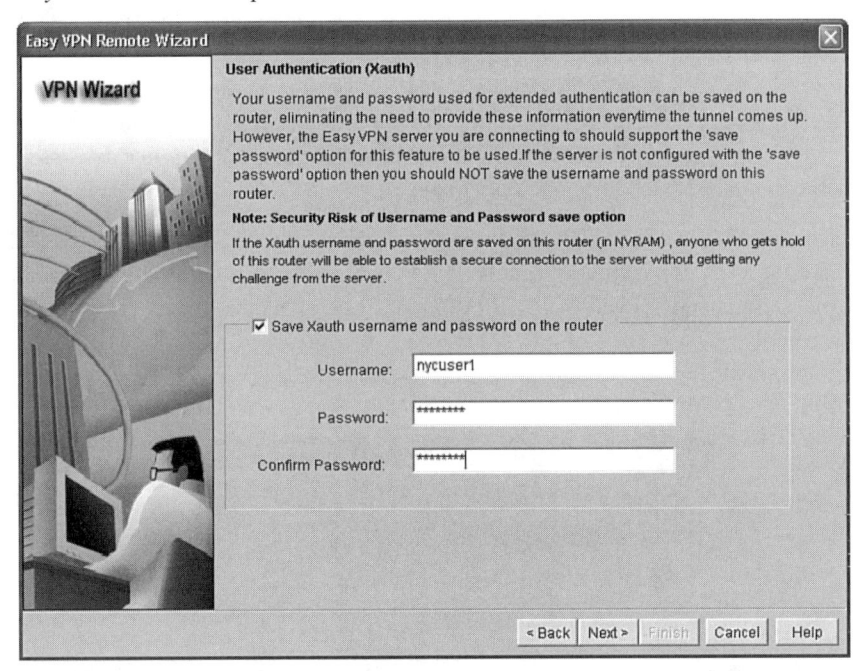

Figure 23-21 *Easy VPN Remote Interface Setup Screen*

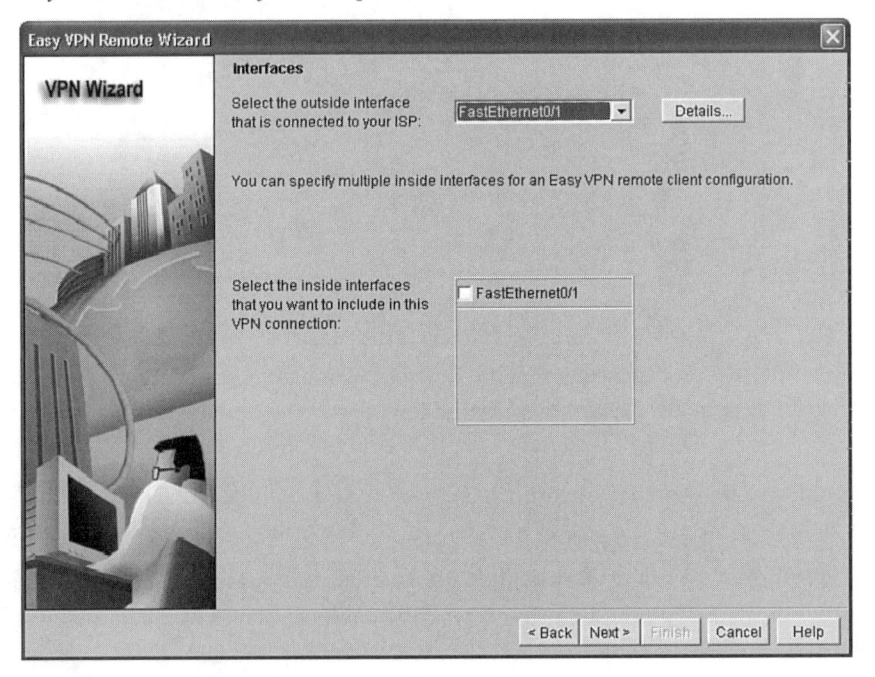

Easy VPN Server

The wizard will guide you through the necessary steps to configure the Easy VPN Server on a router. The server configuration steps are as follows:

1. Easy VPN Server requires AAA. The AAA configuration may be using a centralized server, such as asynchronous communications server (ACS) or a local router database. AAA may be enabled using SDM Additional Task menu.

2. Select the interface on which the client connections will be terminated. This is the interface where the VPN connections from the VPN clients will terminate. The interface may not be participating in GRE over IPsec, DMVPN, or Easy VPN client connection. Figure 23-22 shows this screen.

Figure 23-22 *Easy VPN Server Connections Setup Screen*

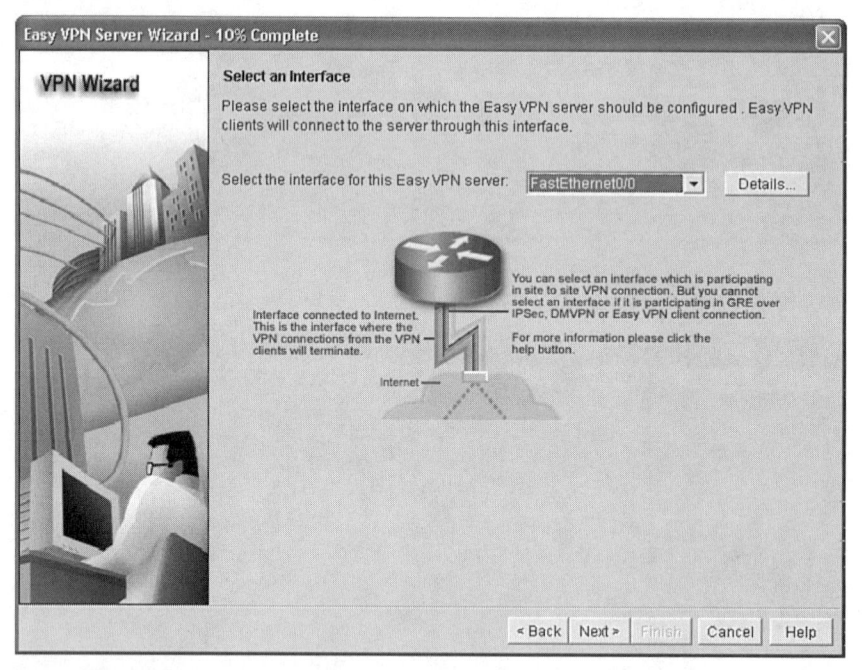

3. Configure IKE proposals that specifies the encryption algorithm, authentication algorithm, and key exchange method that is used by the router when negotiating a VPN connection with the remote device. Figure 23-23 shows the IKE configuration screen.

4. Specify the transform set that should be used for encryption and authentication algorithms to protect the data in a VPN tunnel. Figure 23-24 shows the transform set configuration screen.

5. Define the AAA authorization method list for group policy lookup or select an existing network method list. The options are local, RADIUS, and local only, TACACS+ and local only, or an existing AAA method list.

Figure 23-23 *Easy VPN Server IKE Proposal Screen*

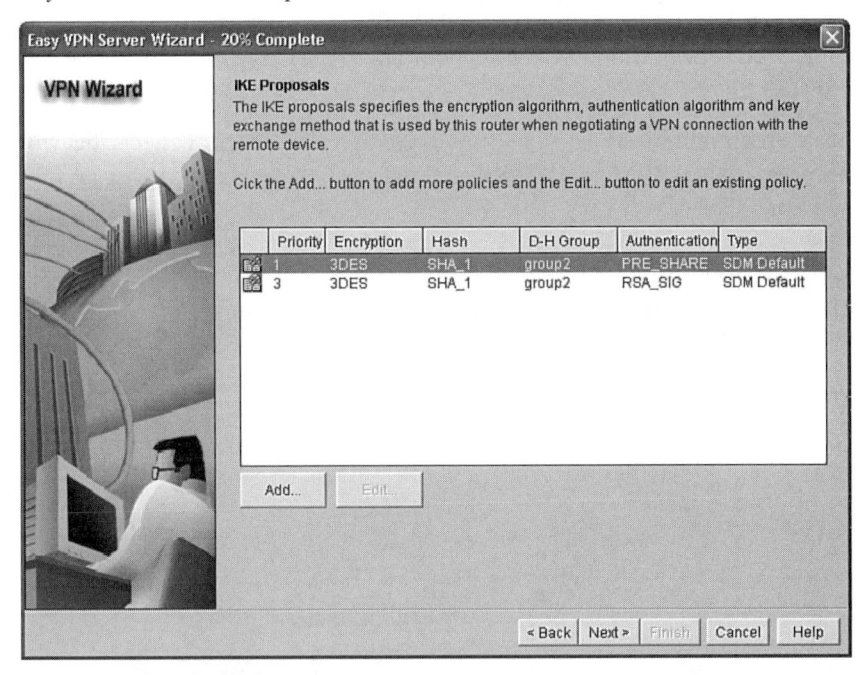

Figure 23-24 *Easy VPN Server Transform Set Screen*

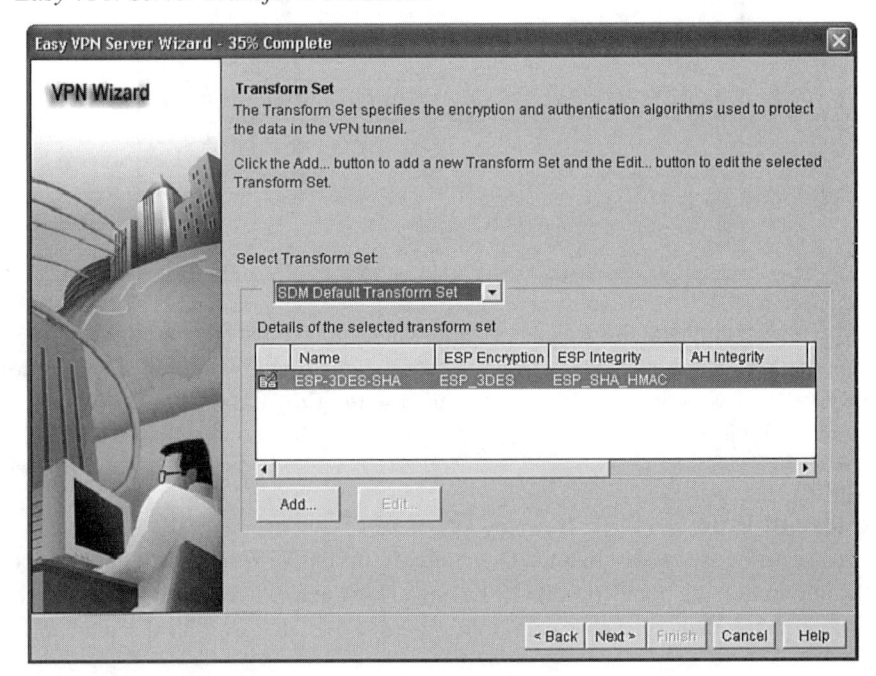

6. If you are supporting the Easy VPN Remote client Phase 3 option, Xauth must be configured. Xauth authenticates the user using the device after the device has been authenticated by the normal IKE authentication. Xauth supports any of the Cisco IOS Software AAA authentication methods.

7. Create group authorization and policies to group remote users or clients that will have common attributes such as DHCP pool, DNS, and Windows Internet Name Service (WINS). Figure 23-25 shows this screen.

Figure 23-25 *Easy VPN Server Add Group Policy Screen*

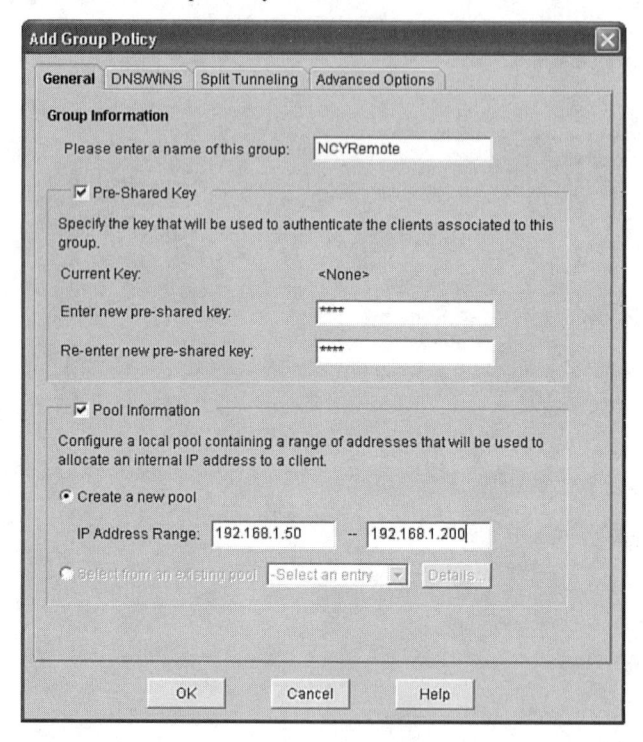

After you have completed these steps, the summary window lists the configurations specified. Clicking the **Finish** button pushes this information to the router's running configuration. To change the Easy VPN Server configuration at any time, use the **Edit Easy VPN Server** tab in the VPN configuration option.

Dynamic Multipoint VPN

This wizard assists you to configure a Dynamic Multipoint VPN (DMVPN) hub or spoke. A typical VPN configuration is a point-to-point IPsec tunnel between two routers. DMVPN creates a network with a central hub that connects to multiple spokes using GRE over an IPsec tunnel.

SDM Wizards

To configure a DMVPN hub, complete the following tasks in the wizard:

1. Specify the network topology. The network topology may be hub-and-spoke network or fully meshed. The hub-and-spoke option is used if the network has all spoke routers as point-to-point connections to a hub router. A fully meshed DMVPN configuration allows the spoke to establish direct IPsec tunnels to each other. A multipoint GRE tunnel is configured on the spoke to support this functionality. Support for a fully meshed VPN is available starting in Cisco IOS Software Release 12.3(T). Figure 23-26 shows the DMVPN Network Topology screen.

Figure 23-26 *DMVPN Network Topology Screen*

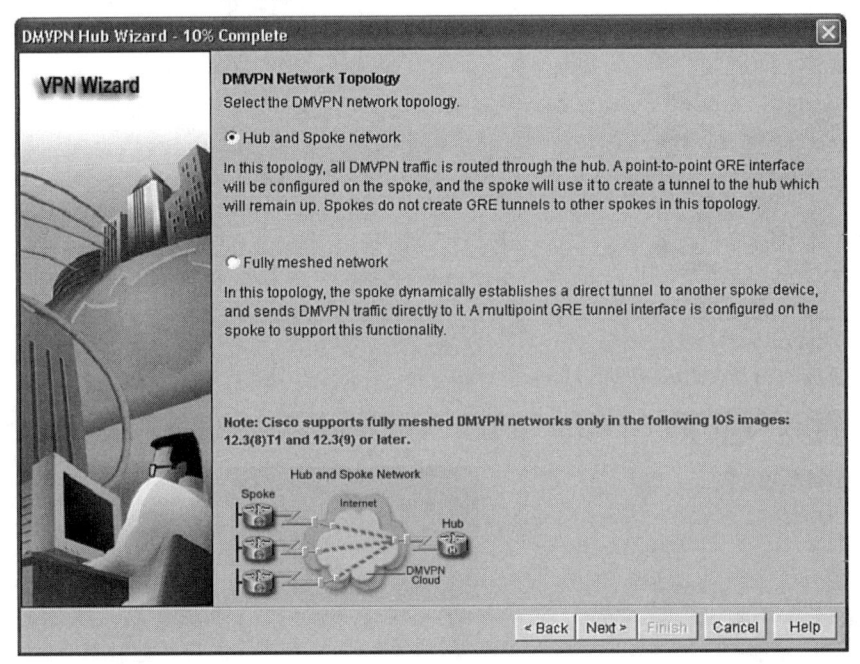

2. Specify the hub type. The hub type could be either the primary or the backup. Figure 23-27 shows the screen that displays.

3. Configure a multipoint GRE tunnel IP address that is on the same subnet as the other spokes and the hub. Figure 23-28 shows this screen.

4. Configure a preshared key or digital certificate to authenticate connections between DMVPN peers. If a digital certification is used, the router must have a valid certificate configured. If a preshared key is used, the key configured on the router must match the key configured on all the other routers in the DMVPN network. Figure 23-29 depicts this screen.

Figure 23-27 *DMVPN Hub Type Screen*

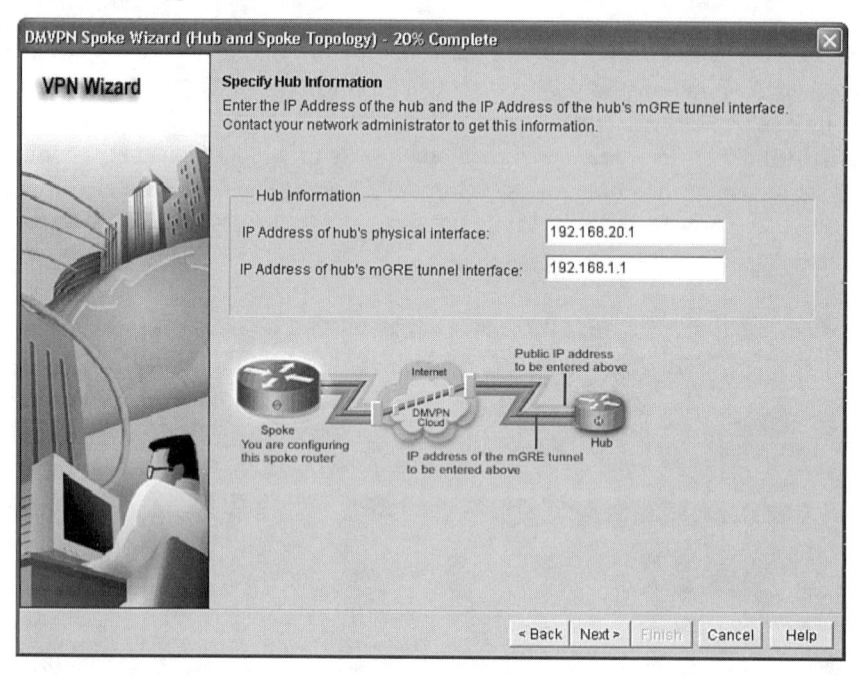

Figure 23-28 *DMVPN GRE Tunnel Interface Screen*

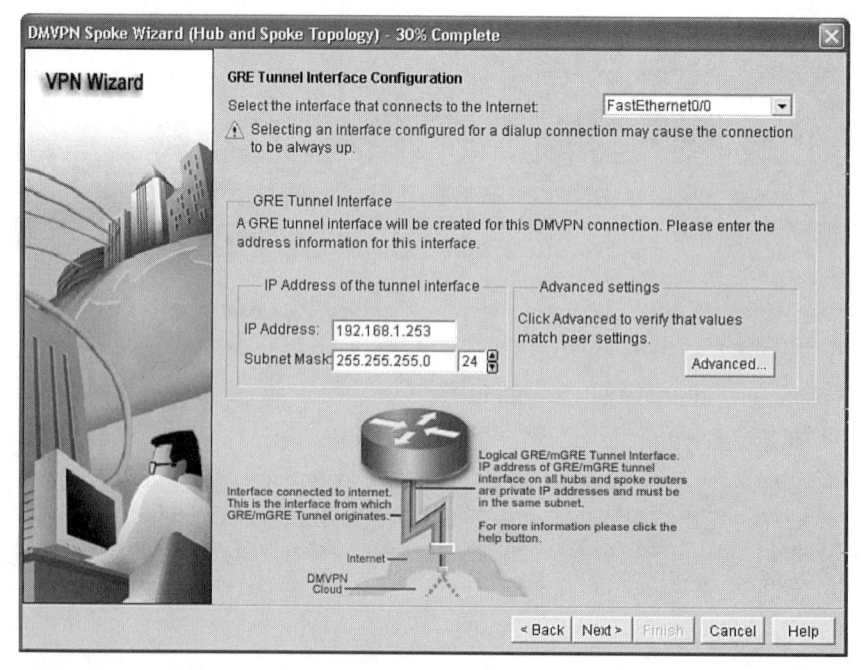

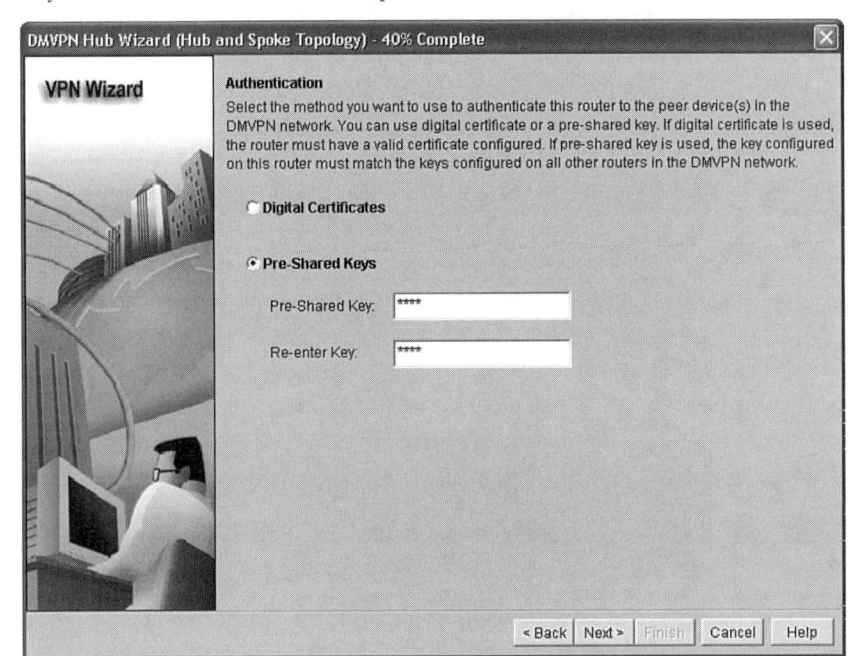

Figure 23-29 *Easy VPN Server Connections Setup Screen*

[Screenshot: DMVPN Hub Wizard (Hub and Spoke Topology) - 40% Complete. Authentication screen with Digital Certificates and Pre-Shared Keys options; Pre-Shared Keys selected with Pre-Shared Key and Re-enter Key fields.]

5. Configure IKE policies to specify the encryption algorithm, authentication algorithm, and key-exchange methods used by the router when attempting to establish a VPN connection.

6. Configure an IPsec transform set to specify the encryption and authentication algorithm used to protect the data in the VPN tunnel.

7. Configure a dynamic routing protocol to advertise the private networks behind the router to other routers in the DMVPN network. The available routing protocols are Open Shortest Path First (OSPF) and EIGRP. If no routing protocol is running, the routing process and network information must be entered, too.

To configure a DMVPN spoke, complete the following tasks in the DMVPN Wizard:

1. Specify the DMVPN network topology as a hub-and-spoke or fully meshed network.

2. Provide the hub physical and tunnel IP address.

3. Configure a GRE tunnel interface for the DMVPN connection.

4. Configure a preshared key or digital certificate.

5. Configure IKE policies to specify the encryption algorithm, authentication algorithm, and key exchange methods used by the router when attempting to establish a VPN connection.

6. Configure an IPsec transform set to specify the encryption and authentication algorithm used to protect the data in the VPN tunnel.

7. Configure a dynamic routing protocol to advertise the private networks behind the router to other routers in the DMVPN network. The available routing protocols are OSPF and EIGRP. If no routing protocol is running, the routing process and network information must be entered, too. Figure 23-30 displays the Select Routing Protocol screen.

Figure 23-30 *DMVPN Routing Protocol Selection Screen*

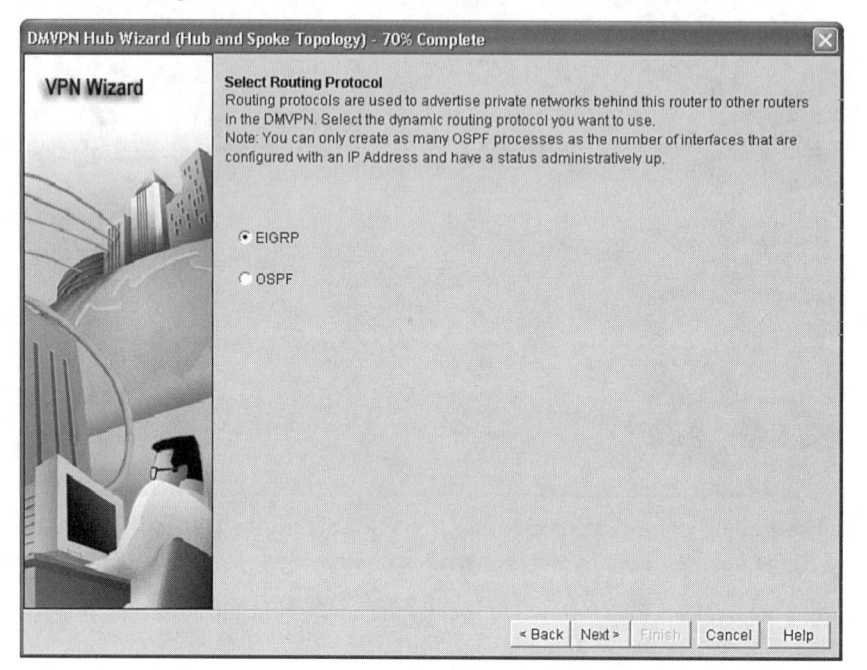

When you complete these steps, the summary window lists the configuration specified. Clicking the **Finish** button pushes this configuration to the router's running configuration.

Using SDM to Perform Security Audits

The security audit feature is based on the Cisco IOS AutoSecure feature, described in Chapter 5, "Secure Router Administration," which examines the existing configurations on the router. It then reconfigures the router with the best practices security recommendation to provide enhanced protection to the router and the network on which it resides. Figure 23-31 shows the main Security Audit screen.

The security audit feature has two modes of operation: Security Audit Wizard or one-step lockdown. The following sections describe each of these modes.

Security Audit Wizard

Security Audit Wizard examines the existing router configuration and provides a list of recommended configuration changes to the user. The user then selects which potential security-related configuration changes to implement on the router.

SDM Wizards   535

Figure 23-31 *Security Audit Main Screen*

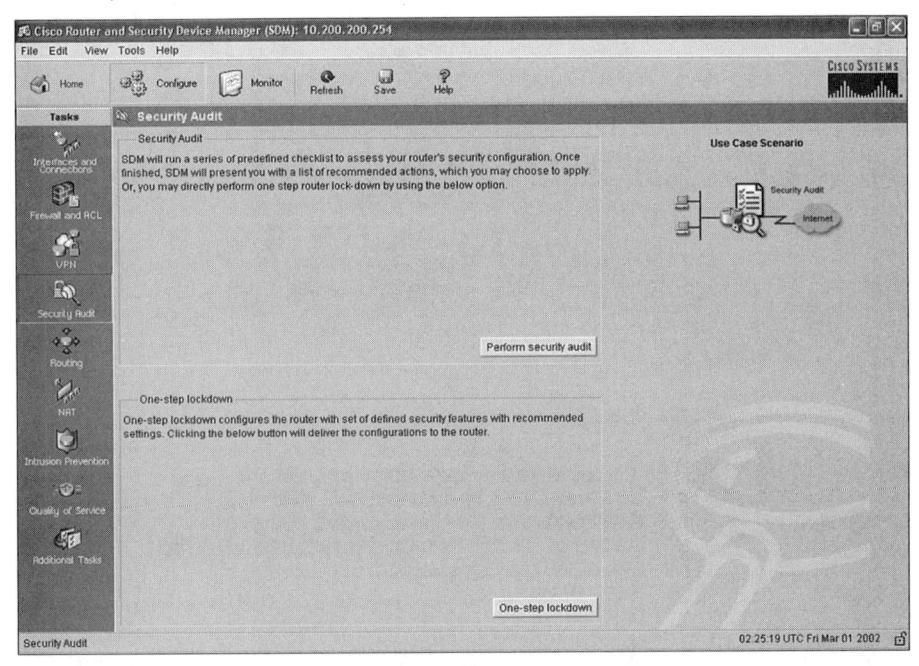

The user is prompted to identify the inside and outside interfaces. The wizard checks the router's configuration and provides a list of identified security problems. Figure 23-32 shows a sample list compiled by Security Audit Wizard.

Figure 23-32 *Security Audit Wizard Audit Check Screen*

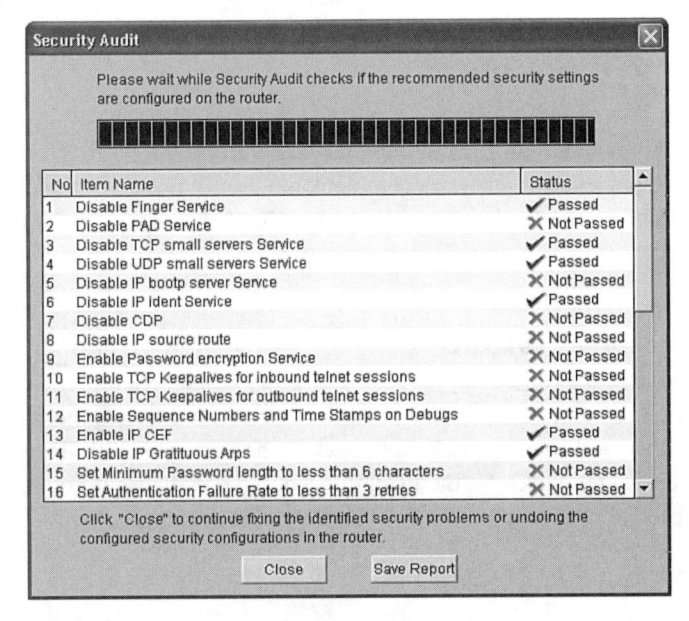

The user must select the problems that should be fixed by checking the Fix it box. The summary page provides the list of checked items to fix. Figure 23-33 shows the list of security problems found.

Figure 23-33 *Security Audit Wizard Security Problem List Screen*

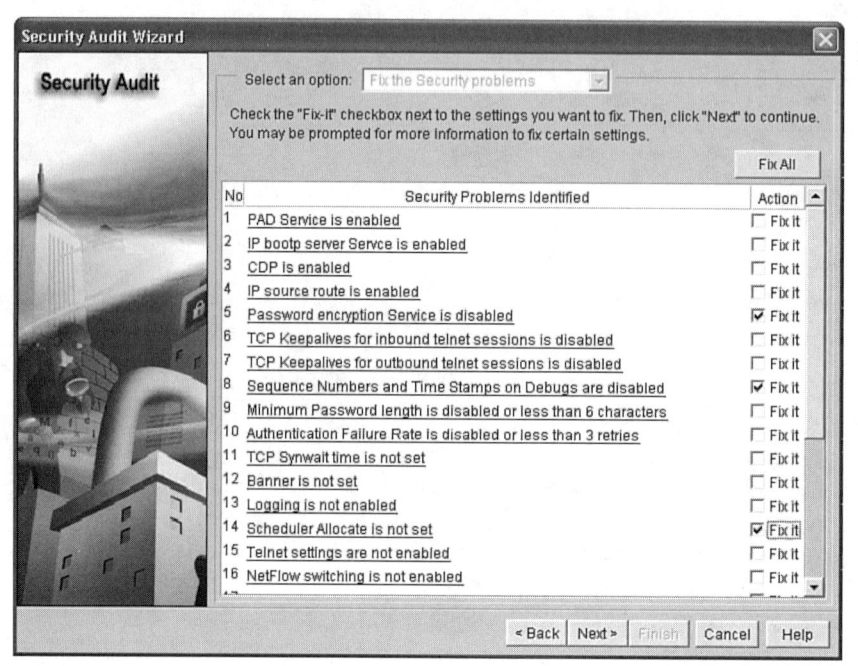

[Screenshot of Security Audit Wizard showing a list of 16 Security Problems Identified with Fix it checkboxes, including PAD Service is enabled, IP bootp server Servce is enabled, CDP is enabled, IP source route is enabled, Password encryption Service is disabled, TCP Keepalives for inbound telnet sessions is disabled, TCP Keepalives for outbound telnet sessions is disabled, Sequence Numbers and Time Stamps on Debugs are disabled, Minimum Password length is disabled or less than 6 characters, Authentication Failure Rate is disabled or less than 3 retries, TCP Synwait time is not set, Banner is not set, Logging is not enabled, Scheduler Allocate is not set, Telnet settings are not enabled, NetFlow switching is not enabled.]

Clicking the **Finish** button applies the fixes to the router's running configuration.

One-Step Lockdown

One-step lockdown makes all the recommended security configuration changes automatically. When the one-step lockdown check is complete, the user must accept the download of the fixes to the router by clicking the **Deliver** button. Figure 23-34 shows the screen associated with this configuration.

Using the Factory Reset Wizard

The configuration of the router may be reset to factory default. Before the router is set to factory default, the latest running configuration on the router is saved for recovery purposes. When the router is reset, the LAN IP address changes back to the factory default IP address of 10.10.10.1. To get back into the router via SDM, the PC will need to be on the same subnet as 10.10.10.0, or the CLI must be used to configure the LAN IP address of the router.

SDM Wizards 537

Figure 23-34 *Security Audit Wizard One-Step Lockdown Screen*

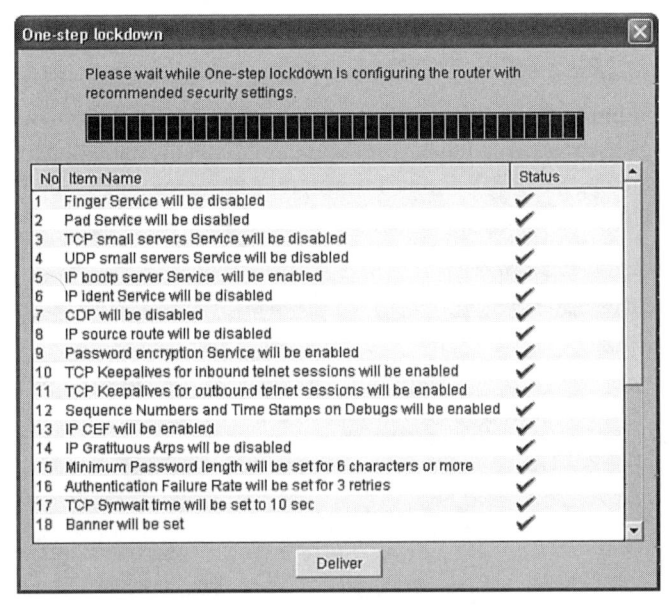

The reset to factory defaults feature is not supported on Cisco 3600 and 7000 series routers. This feature is also not supported when SDM is installed and running from a PC rather than the router itself. In this case, the screen shown in Figure 23-35 displays.

Figure 23-35 *Factory Reset Wizard Not Supported Screen*

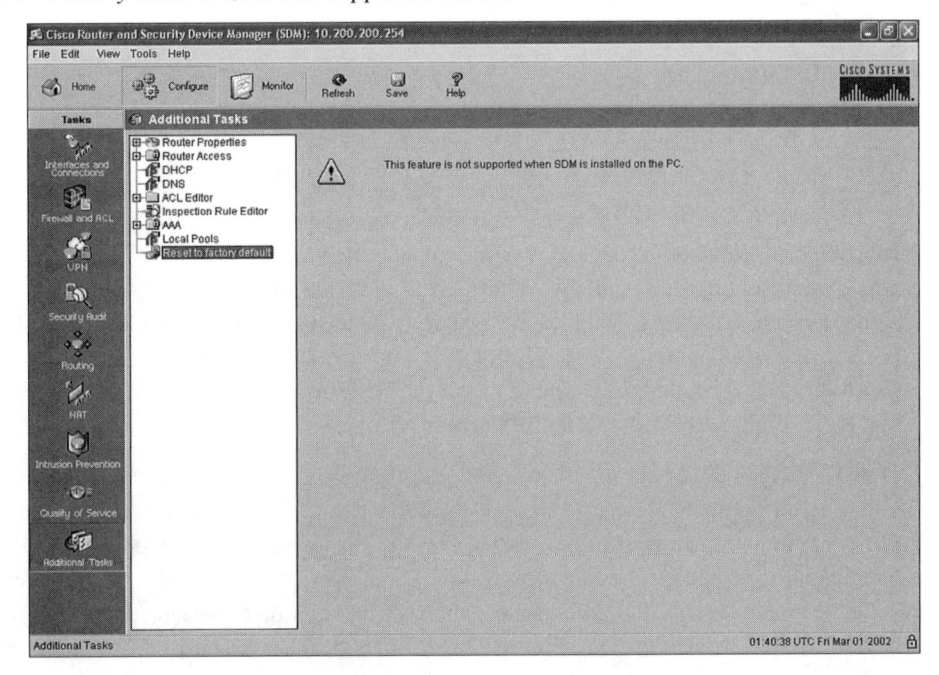

Using SDM Advanced Options

Additional configuration options are available in SDM. The following is a list and description of these options:

- **Routing configuration**—The routing window provides the option to view, add, modify, and delete routes in the routing table. A list of configured static and dynamic routes via routing protocols such as EIGRP, Routing Information Protocol (RIP), and OSPF displays in this window. Figure 23-36 shows the routing configuration screen.

Figure 23-36 *Routing Configuration Screen*

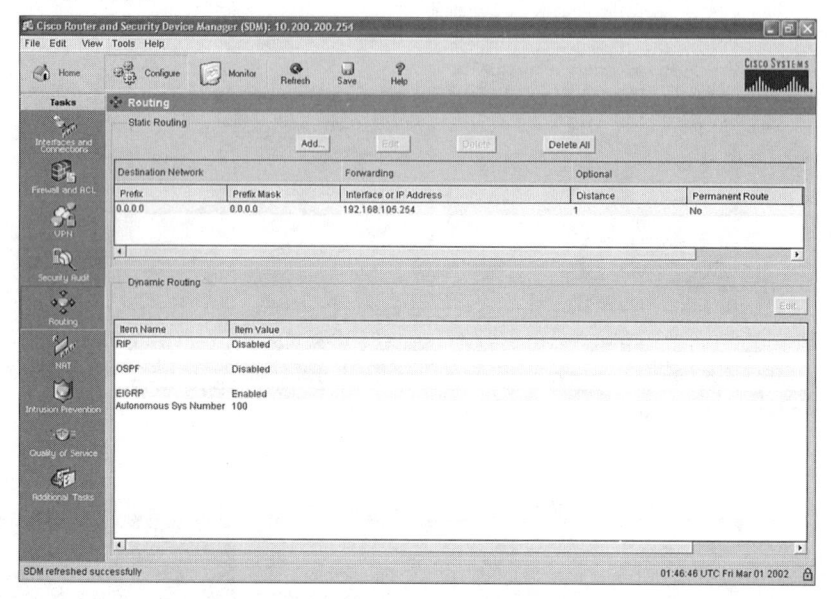

- **NAT**—The NAT window allows the designation of the inside and outside interfaces as well as the option to view NAT rules and address pools. It also allows the setting of translation timeouts. Figure 23-37 shows the NAT screen.

- **Intrusion prevention**—SDM allows the management of an IDS network module that has already been configured from the router CLI. It also provides a mechanism for configuring signatures and enabling Cisco IOS IPS Software on a Cisco router. The Cisco IOS IPS concept is discussed in Chapter 13, "Cisco IOS Intrusion Prevention System." However, configuring Cisco IOS IPS is beyond the scope of this book, so refer to Cisco.com for more information on this topic. Figure 23-38 shows the IPS screen.

- **QoS Wizard**—SDM enables you to configure basic quality-of-service (QoS) policies for outgoing traffic on WAN interfaces and IPsec tunnels. It creates a low-latency queuing (LLQ) service policy with its associated classes. This QoS policy allocates proportional bandwidth to handle real-time and business-critical traffic. Voice over IP (VoIP) and video teleconferencing could be categorized as real-time traffic. Database, network management, and routing data could be considered business critical. Figure 23-39 shows the QoS Wizard.

Using SDM Advanced Options 539

Figure 23-37 *NAT Screen*

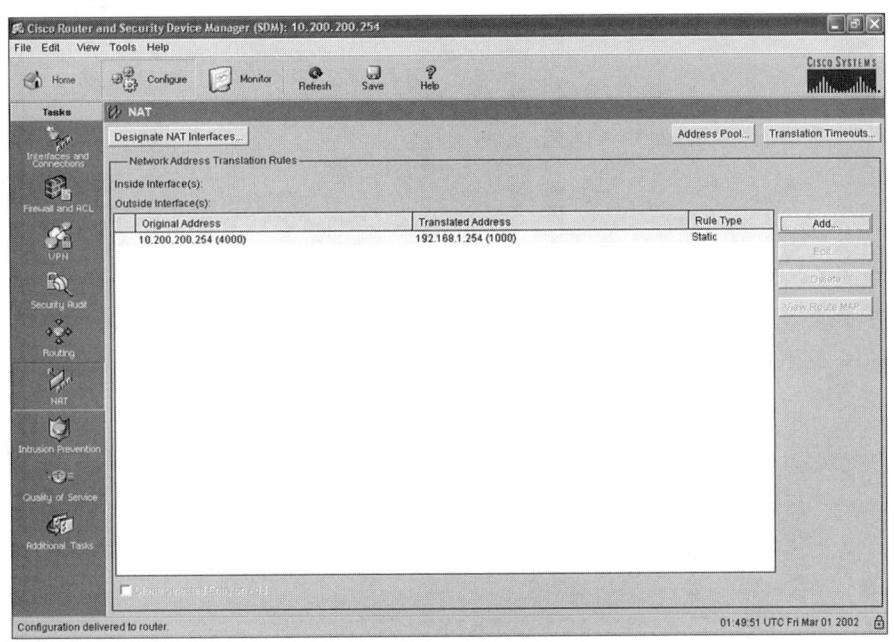

Figure 23-38 *IPS Screen*

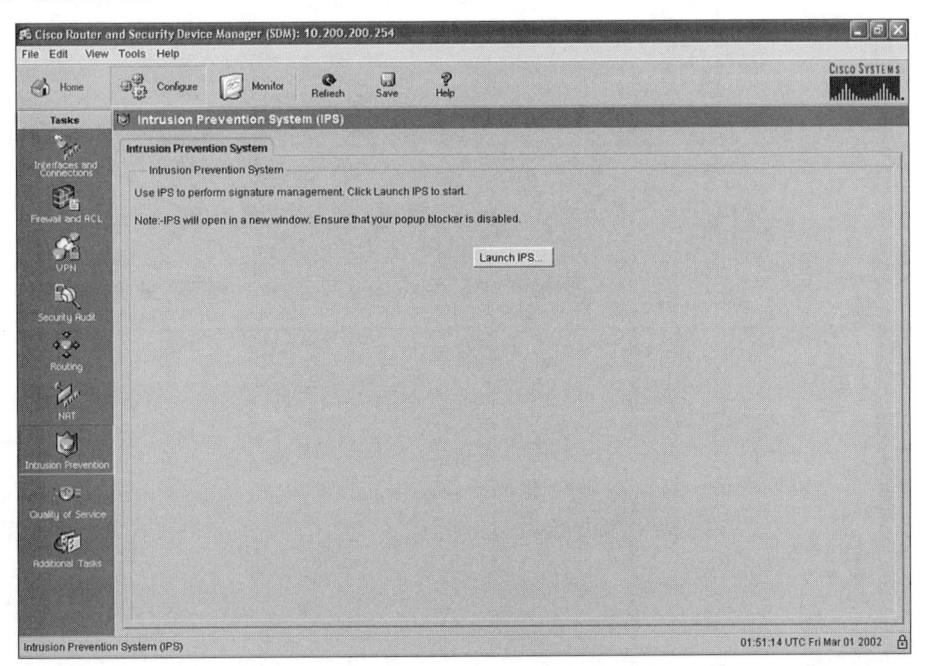

Figure 23-39 *QoS Wizard Screen*

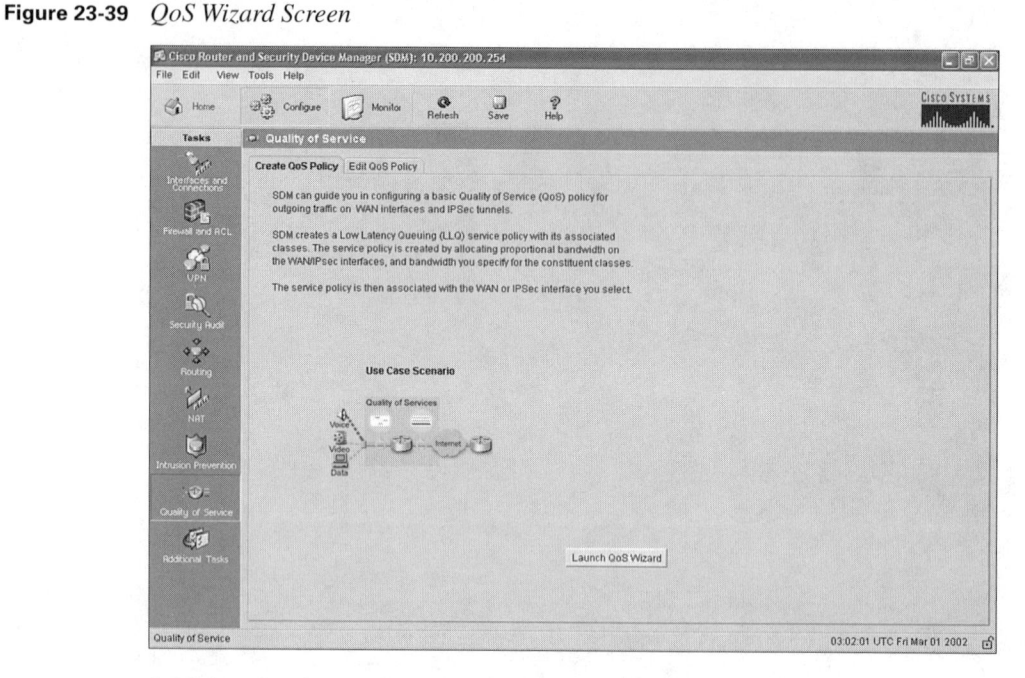

- **Additional tasks**—In this window, the router's basic configuration, such as associated access lists, router access parameters, DHCP/DNS, AAA, logging, and Network Time Protocol (NTP) information may be viewed or modified. Figure 23-40 shows the Additional Tasks screen.

Figure 23-40 *Additional Tasks Screen*

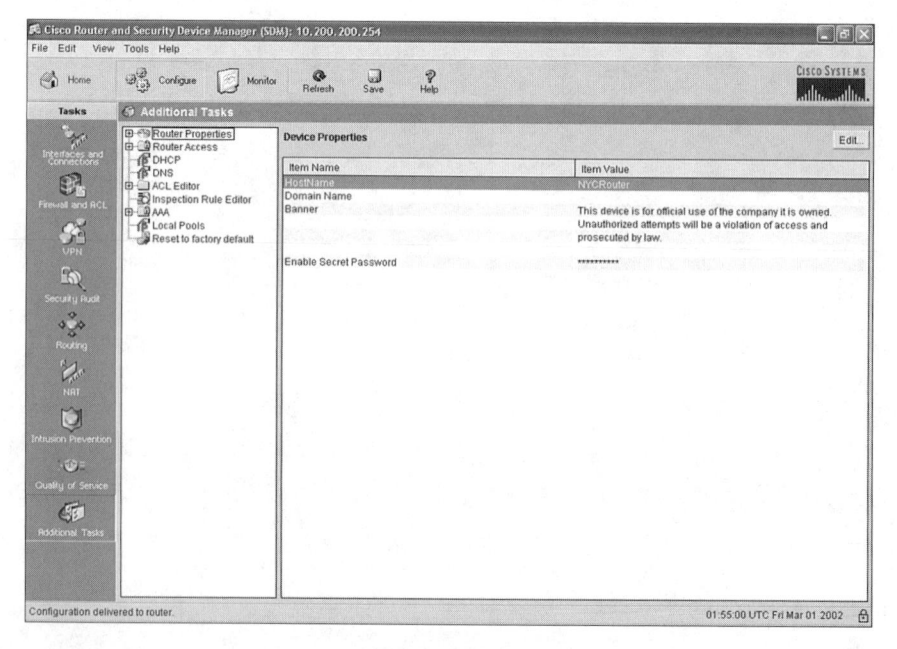

Using SDM Monitor Mode

The SDM monitor mode provides a snapshot of information about the router, the router interfaces, the firewall, and any active VPN connections. It also provides an option to view the router's event log. There are six screens under the monitor mode:

- **Monitor Overview**—Displays an overview of the router activity and statistics. The summary of the information in the other monitor modes is also included in this screen. Figure 23-41 shows the SDM Monitor Overview screen.

Figure 23-41 *Monitor Overview Screen*

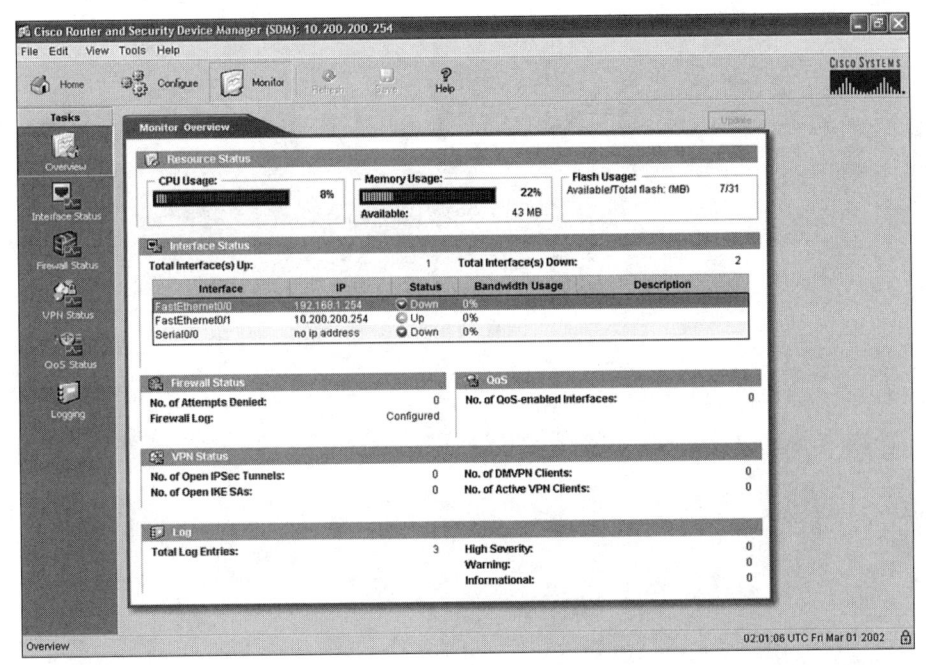

- **Interface Status**—Shows basic information about the interfaces installed on the router and their status. The information includes interface IP, status, description, and bandwidth usage. Figure 23-42 shows the Interface Status screen.

- **Firewall Status**—Shows the basic information about the firewall configuration on the router, including the number of attempts denied and the firewall log. Figure 23-43 shows the Firewall Status screen.

- **QoS Status**—Shows the number of interfaces associated with a router QoS policy. Figure 23-44 shows the QoS Status screen.

- **VPN Status**—Shows the basic information related to the VPN configured options on the router. This includes the number of open IKE security associations (SAs), number of open IPsec tunnels, and number of DMVPN clients. Figure 23-45 shows the VPN Status screen.

Figure 23-42 *Interface Status Screen*

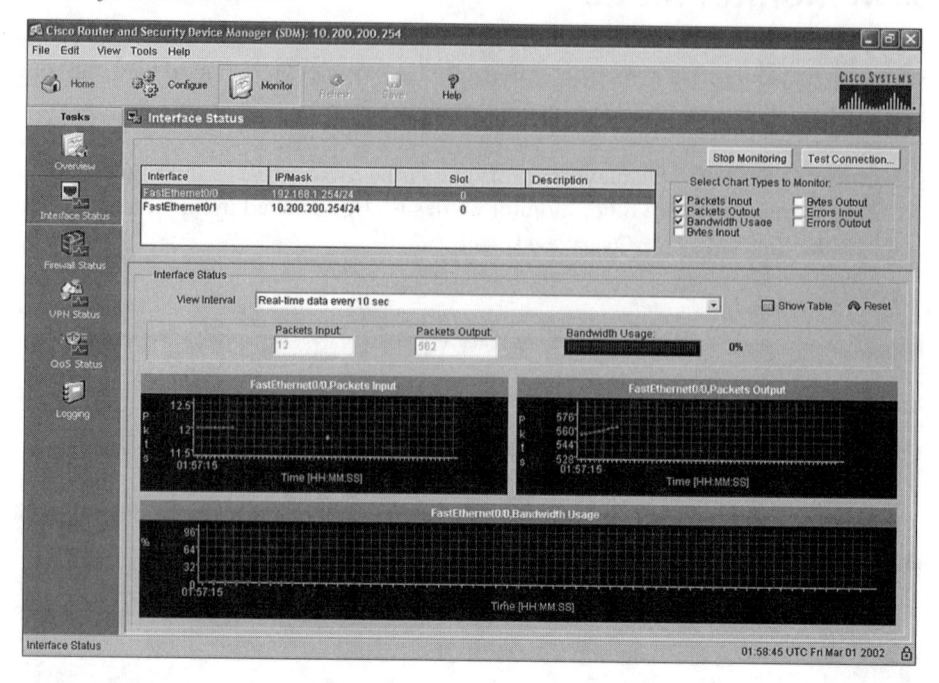

Figure 23-43 *Firewall Status Screen*

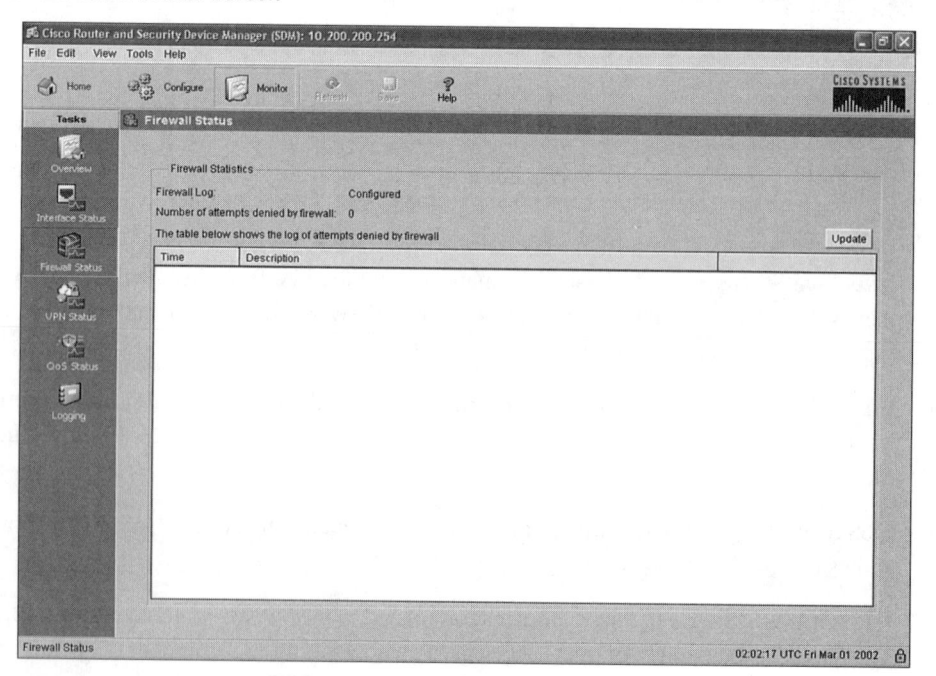

Using SDM Monitor Mode   **543**

Figure 23-44 *QoS Status Screen*

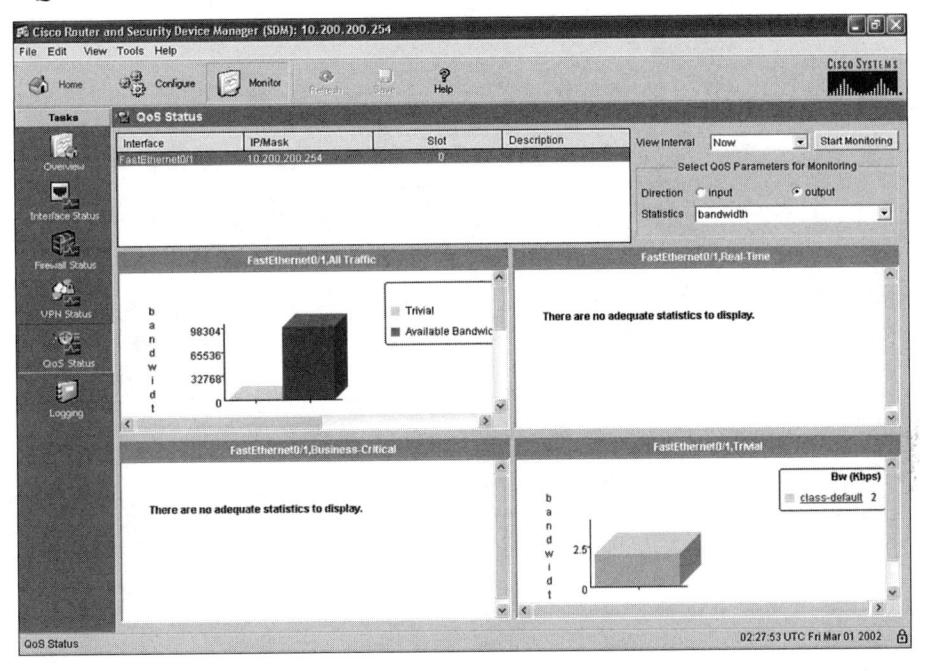

Figure 23-45 *VPN Status Screen*

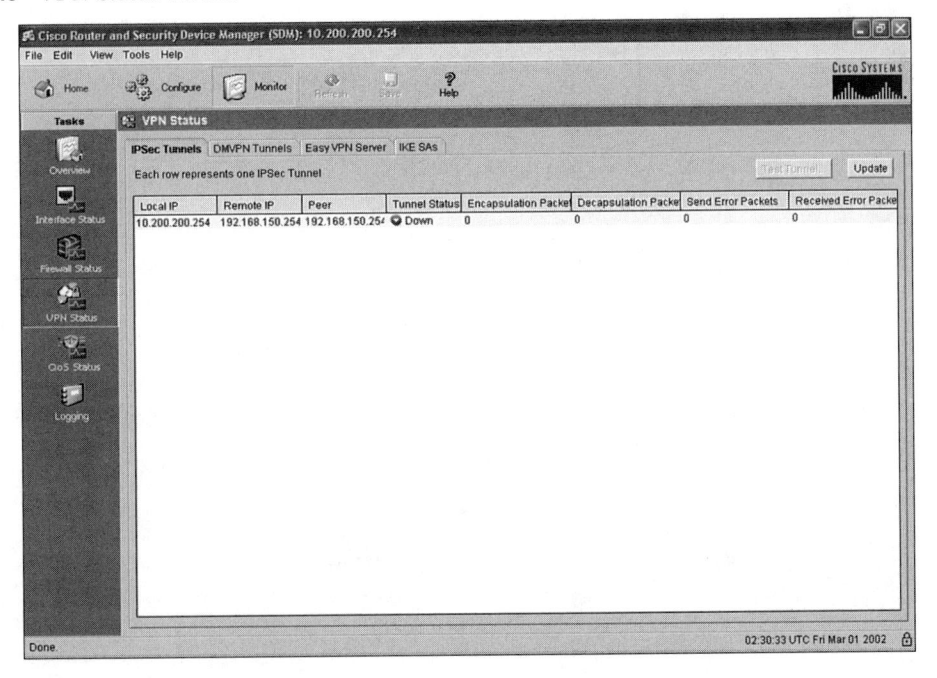

Chapter 23: Security Device Manager

- **Logging**—The Logging screen shows the basic information about the router resources, including total log entries and high-severity, warning, and information log entries. Figure 23-46 shows the Logging screen.

Figure 23-46 *Logging Screen*

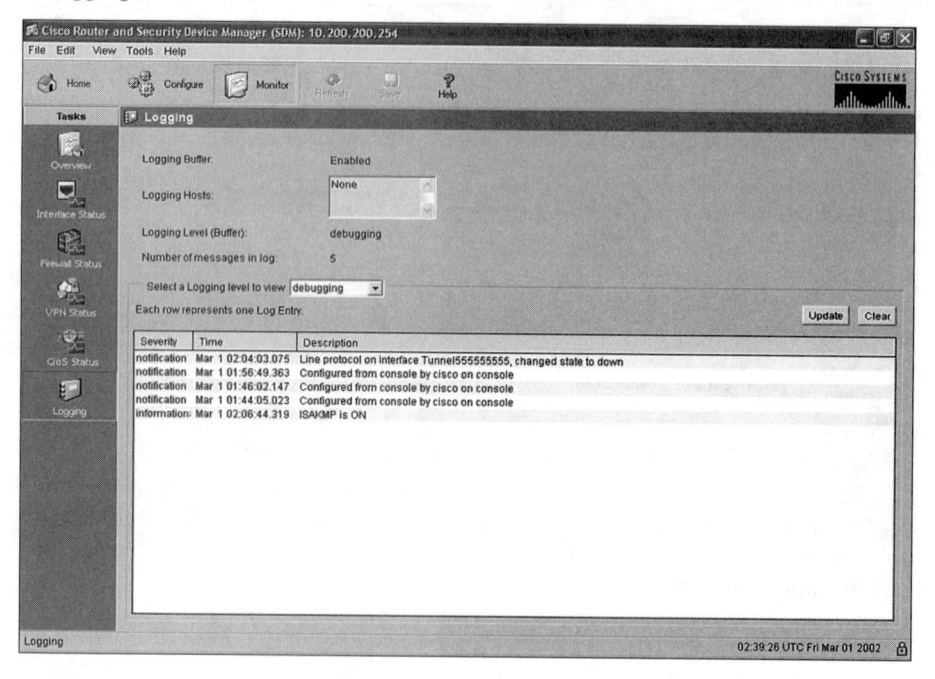

Foundation Summary

The "Foundation Summary" section of each chapter lists the most important facts from the chapter. Although this section does not list every fact from the chapter that will be on your SNRS exam, a well-prepared candidate should at a minimum know all the details in each "Foundation Summary" section before going to take the exam.

Cisco SDM provides a Java-based web management tool for easy deployment and monitoring of Cisco IOS routers.

The features of SDM include smart wizards, security audit, logical workflow.

The computer running SDM must be a Pentium III or later series processor. If the router is running the SDM files, it must have a minimum of 6 MB of free Flash memory on the router to store the SDM files. A minimum of 2 MB of free Flash memory is required on the router for Cisco SDM Express, and wireless management files requires 1.7 MB.

SDM is designed to run on a personal computer with any of the following operating systems: Windows XP Professional, Windows 2003 Server (Standard Edition), Windows 2000 Professional, Windows NT 4.0 Workstation (Service Pack 4), Windows Me, Windows XP Professional, Windows 2000 Professional. There's also language OS support for Japanese, Simplified Chinese, French, German, Spanish, and Italian for Microsoft XP Professional and Microsoft Windows 2000 Support.

The Cisco SDM must have a compatible browser installed. Cisco Secure ACS 3.3 has been tested with English language versions of the following browsers on Microsoft Windows operating systems: Microsoft Internet Explorer 5.5 or later, Netscape Navigator 7.1 and 7.2, Firefox 1.0.2. Java software is required either by using the built-in browser JVM, or Java plug-in (Java Runtime Environment version 1.4.2_05 or later).

SDM software may be configured to run from a PC local hard drive or from the router's Flash memory.

The Cisco IOS Software running on a router must support SDM, which is generally supported in Cisco IOS Software Release 12.3 or higher.

The SDM files are compressed in a zip file. A decompression utility must be installed on the PC to decompress these files.

For the router to support SDM, HTTP or HTTPS services must be enabled on the router.

To install SDM software, decompress the SDM software on the designated PC. In the extracted directory, locate the setup.exe file, and then run it.

The SDM wizards provide a step-by-step guide to configure different features on a router. SDM version 2.1.1 provides six wizards:

- **LAN Wizard**—Provides step-by-step configuration for a LAN interface on a router
- **WAN Wizard**—Provides step-by-step configuration for a WAN interface on a router
- **Firewall Wizard**—Provides step-by-step configuration for the Firewall feature set on a router
- **VPN Wizard**—Provides step-by-step configuration for the VPN feature set on a router
- **Security Audit Wizard**—Provides step-by-step examination of the router configuration and update of router's security posture
- **Reset Wizard**—Resets the configuration of the router to factory default

Additional configuration options are available in SDM, including routing configuration, NAT, IPS, a QoS Wizard, and additional tasks.

The SDM monitor mode provides a snapshot of information about the router, the router interfaces, the firewall, and any active VPN connections. It also provides an option to view the router's event log. There are six screens under the monitor mode: Monitor Overview, Interface Status, Firewall Status, QoS Status, VPN Status, and Logging.

Q&A

As mentioned in the section "How to Use This Book" in the Introduction, you have two choices for review questions: the Q&A questions here or the exam simulation questions on the CD-ROM. The questions that follow present a bigger challenge than the exam itself because they use an open-ended question format. By using this more difficult format, you can exercise your memory better and prove your conceptual and factual knowledge of this chapter. You can find the answers to these questions in the appendix.

1. SDM is designed to run on which personal computer operating systems?
2. Which browser versions are supported by SDM?
3. The SDM User Interface file menu offers what options?
4. What six wizards does SDM version 2.1.1 provide?
5. What are the main tasks required to configure the Cisco IOS Firewall feature set?
6. What does the Logging screen under the monitor mode provide?
7. What is the function of the NAT screen in SDM?
8. What is the SDM security audit feature based on?
9. What is a prerequisite to installing Easy VPN Server on a router?
10. What is user authentication Xauth?

Part VII: Scenarios

Chapter 24   Final Scenarios

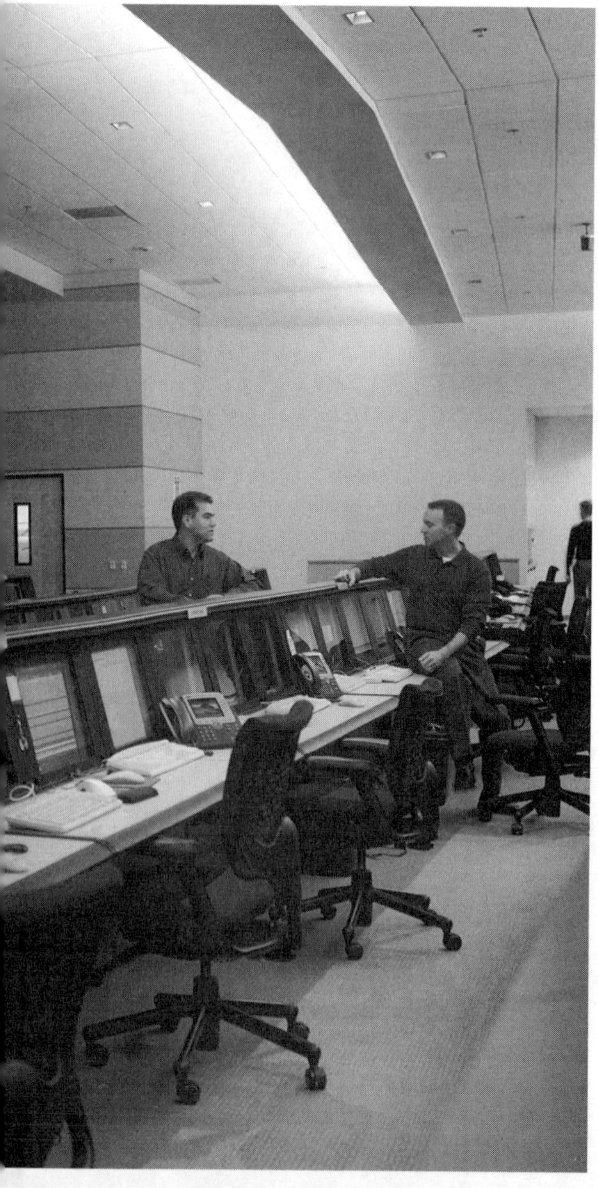

CHAPTER 24

Final Scenarios

The DHC group is continuing to expand operations in North America and you have been assigned the implementation of the networks supporting the facility in Miami, Florida. Figure 24-1 depicts the current configuration of the DHC enterprise network. The offices in New York, Chicago, and San Francisco are established; and Miami is your configuration responsibility.

Figure 24-1 *DHC Enterprise Network*

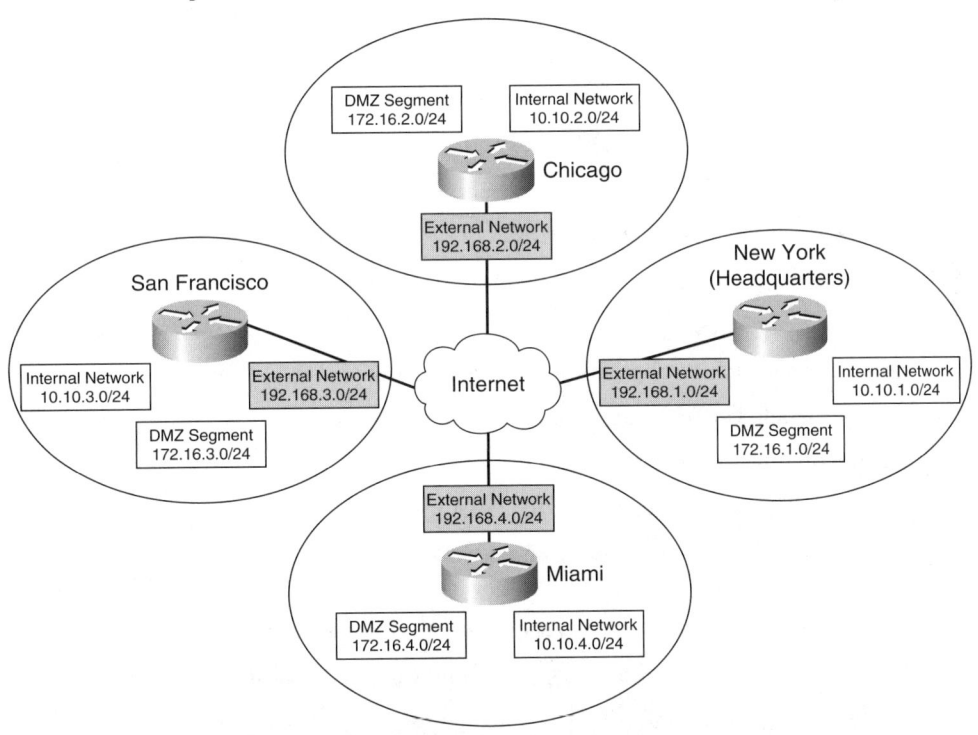

This is your first assignment as team leader, and you want to be sure that you have completed all the implementation steps and verified the results. You have several key tasks to complete; however, the overall result is that the facility must provide Internet connectivity to the users, secure connectivity with the New York headquarters, and remote connectivity for several users that occupy a branch office in Daytona Beach, Florida. You have studied DHC's security policies and the current configuration of the network and have identified the following tasks that must be completed to add Miami to the DHC enterprise network:

Step 1 Configure Cisco Secure Access Control Server (Cisco Secure ACS) to provide AAA services.

   a. Install Cisco Secure ACS for Microsoft Windows Server.
   b. Configure Cisco Secure ACS for Windows Server database for authentication.
   c. Configure the router to authenticate to the Cisco Secure ACS for Windows Server database.

Step 2 Configure and secure the perimeter router.

   a. Change all administrative access to the Miami routers.
   b. Configure local database authentication using authentication, authorization, and accounting (AAA).
   c. Configure a secure method for remote access of the routers.

Step 3 Configure 802.1x port-based authentication on a Catalyst 2950 switch.

   a. Enable 802.1x authentication.
   b. Configure the switch-to-RADIUS server communication.
   c. Enable periodic re-authentication.
   d. Manually re-authenticate a client connected to a port.
   e. Change the quiet period.
   f. Change the switch-to-client retransmission period.
   g. Set the switch-to-client frame-retransmission number.
   h. Enable multiple hosts.
   i. Reset the 802.1x configuration to the default values.
   j. Display the 802.1x statistics and status.

Step 4 Configure network switches and routers to mitigate Layer 2 attacks.

   a. Mitigate the content-addressable memory (CAM) table overflow attack.
   b. Mitigate virtual LAN (VLAN) hopping attacks.
   c. Prevent Spanning Tree Protocol manipulation.
   d. Mitigate Media Access Control (MAC) spoofing attacks.
   e. Defend private VLANs.
   f. Mitigate Dynamic Host Configuration Protocol (DHCP) starvation attacks.

Step 5 Configure Protected Extensible Authentication Protocol (PEAP) with Cisco Secure ACS.

   a. Obtain a certificate for the Cisco Secure ACS.
   b. Identify additional certification authorities (CAs) that the Cisco Secure ACS should trust.
   c. Configure the PEAP settings.
   d. Specify the access device.
   e. Configure the external user database.
   f. Restart the service.

Step 6 Prepare the network for IPsec using preshared keys.

   a. Establish a common convention for connectivity between locations.
   b. Configure initial setup of the router and verify connectivity.
   c. Prepare for Internet Key Exchange (IKE) and IPsec.
   d. Define the preshared keys.

Step 7 Configure IKE using preshared keys.

   a. Enable IKE.
   b. Create the IKE policy.
   c. Configure the preshared key.
   d. Verify the IKE configuration.

Step 8 Configure IPsec using preshared keys.

   a. Configure transform sets and security association (SA) parameters.
   b. Configure crypto access control lists (ACLs).
   c. Configure crypto maps.
   d. Apply the crypto map to an interface.

Step 9 Configure IKE and IPsec on a Cisco router.

   a. Enable IKE/ISAKMP.
   b. Create an IKE policy to use Rivest, Shamir, and Adelman (RSA) signatures.
   c. Configure transform sets and SA parameters.
   d. Configure crypto ACLs.
   e. Configure crypto maps.
   f. Apply the crypto map to an interface.

Step 10 Prepare the network for IPsec using digital certificates.

   a. Configure initial setup of the router and verify connectivity.
   b. Prepare for IKE and IPsec.
   c. Configure CA support.

Step 11 Test and verify IPsec CA configuration.

   a. Display IKE policies.
   b. Display transform Sets.
   c. Display configured crypto maps.
   d. Display the current state of IPsec SAs.
   e. Clear any existing SAs.
   f. Enable debug output for IPsec events.
   g. Enable debug output for ISAKMP events.
   h. Observe the IKE and IPsec debug outputs.
   i. Verify IKE and IPsec SAs.
   j. Ensure encryption is working.

Step 12 Configure authentication proxy on the Miami router.

   a. Verify initial router configuration.
   b. Configure Cisco Secure ACS.
   c. Configure AAA.
   d. Configure authentication proxy.
   e. Test and verify configuration.

Step 13 Configure Content-Based Access Control (CBAC) on the Miami router.

   a. Verify initial router configuration.
   b. Configure logging and audit trails.
   c. Define inspection rules and ACLs.
   d. Apply inspection rules and ACLs.
   e. Test and verify CBAC.

Step 14 Configure the Miami router with the Cisco IOS Intrusion Protection System (IPS).

   a. Verify initial router configuration.
   b. Initialize IPS on the router.
   c. Disable and exclude signatures.
   d. Create and apply audit rules.
   e. Verify the IPS router's configuration.
   f. Generate a test message.

Step 15 Verify and monitor the Miami router with IPS using SDM.

   a. Enable IPS SDEE.
   b. Configure SDF locations
   c. Show IPS SDEE status.

Step 16 Configure Easy VPN Server.

   a. Verify initial router configuration.
   b. Prepare a perimeter router for the Easy VPN Server.
   c. Enable policy lookup via AAA.
   d. Create an ISAKMP policy for remote client access.
   e. Define group policy information for a mode configuration push.
   f. Create a transform set.
   g. Create a dynamic crypto map.
   h. Apply mode configuration to the dynamic crypto map.
   i. Apply a dynamic crypto map to the router interface.

Step 17 Configure Easy VPN Remote.

   a. Install the Cisco VPN client 3.x.
   b. Create a new connection entry.
   c. Launch the Cisco VPN client.
   d. Test the remote-access connection.
   e. Configure extended authentication.
   f. Test extended authentication.

Task 1—Configure Cisco Secure ACS for AAA on Miami Network Devices

> **NOTE** The usernames and passwords used in the scenarios are neither strong nor complex. These usernames and passwords are for demonstration purposes and relate to the location that they are easy to identify when looking at the scenario configurations. The correct generation and use of passwords is discussed in Chapter 1, "Network Security Essentials."

Step 1 The first task is to log in to the Cisco Secure ACS using a supported web browser with an administrator account. Click the **Network Configuration** button, as shown in Figure 24-2.

Figure 24-2 *Cisco Secure ACS Network Configuration Screen*

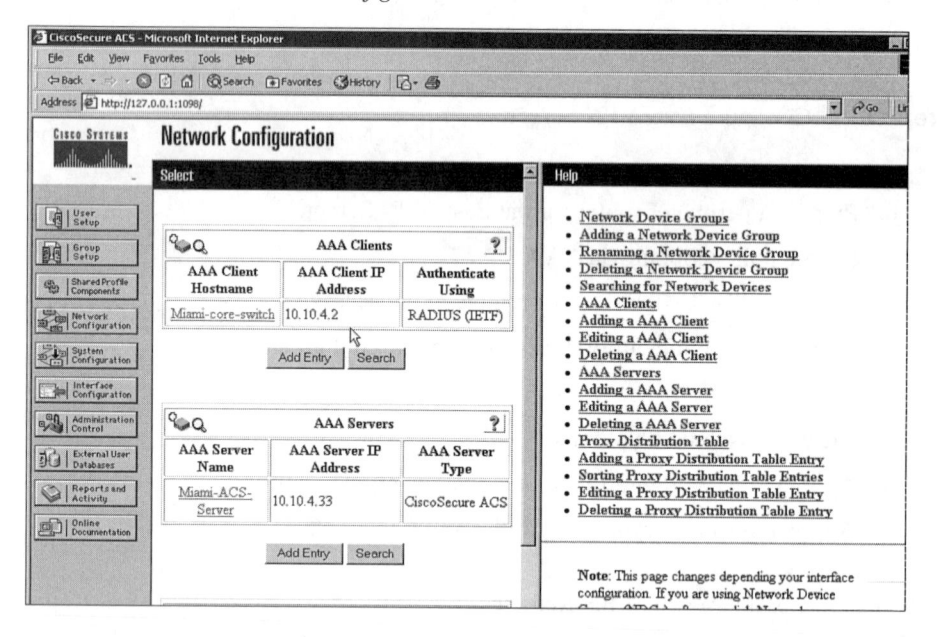

Step 2 Under the AAA client table, click the **Add Entry** button to add a new client, as shown in Figure 24-3.

Step 3 Add the name, IP address, authentication key, and authentication method to this screen.

Step 4 After the clients have been entered, click the **System Configuration** button, as shown in Figure 24-4.

Task 1—Configure Cisco Secure ACS for AAA on Miami Network Devices   557

Figure 24-3 *Cisco Secure ACS Add AAA Client Screen*

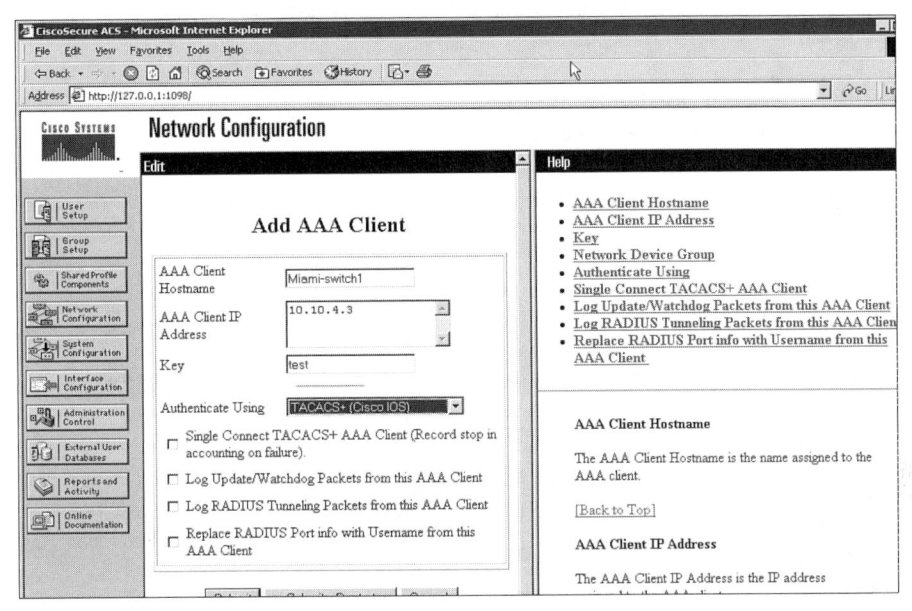

Figure 24-4 *Cisco Secure ACS System Configuration Screen*

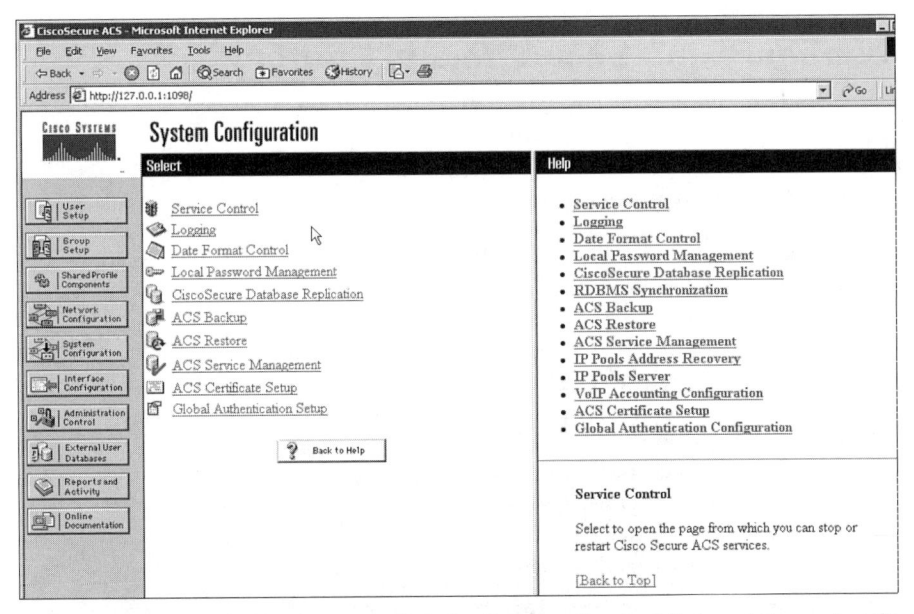

Step 5 To commit the changes, click the **Service Control** link and then select **Stop**. Click the **Restart** button to shut down and start the Cisco Secure ACS service, as shown in Figure 24-5.

Figure 24-5 *Cisco Secure ACS Service Control Screen*

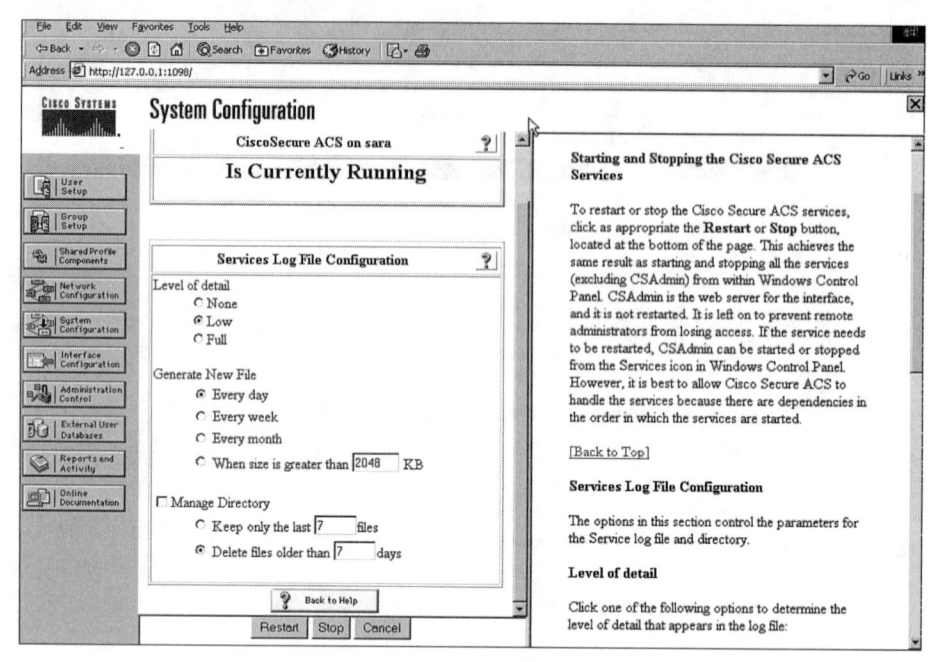

Task 2—Configure and Secure Miami Router

In this task, the Miami router is configured and locked down. The tasks involve the following subtasks:

   a. **Configure Router Interfaces**

The first task is to configure the interfaces on the Miami router. Three interfaces must be configured: LAN, WAN, and demilitarized zone (DMZ):

```
Miami-Router#configure terminal
Miami-Router(config)#interface fastethernet 0/0
Miami-Router(config-if)#description Connection to Miami WAN
Miami-Router(config-if)#ip address 192.168.4.1 255.255.255.0
Miami-Router(config-if)#exit
Miami-Router(config)#
Miami-Router(config)#interface fastethernet 0/1
Miami-Router(config-if)#description Connection to Miami LAN
Miami-Router(config-if)#ip address 10.10.4.1 255.255.255.0
Miami-Router(config-if)#exit
Miami-Router(config)#
Miami-Router(config)#interface Ethernet 1/0
Miami-Router(config-if)#description Connection to Miami DMZ
Miami-Router(config-if)#ip address 172.16.4.1 255.255.255.0
```

b. Change All Administrative Access to Miami Routers

This task is securing the Miami router. As part of this task, you replace the weak administrative access password on the site router with a strong password:

Step 1 Reconfigure the console port user-level password.

- vty password access to Miami:

```
Miami-Router(config)#line console 0
Miami-Router(config)#login
Miami-Router(config)#password Mi@conaccess
```

Step 2 Reconfigure the enable secret password on the Miami router.

- enable secret password of Mienable on the New York router:

```
Miami-Router(config)#enable secret Mienable
```

Step 3 Reconfigure the vty user-level password On the Miami router.

- vty password access to Miami:

```
Miami-Router(config)#line vty 0 15
Miami-Router(config)#login
Miami-Router(config)#password Mi$vtyaccess
```

Step 4 Reconfigure the auxiliary port user-level password.

- AUX password access to Miami:

```
Miami-Router(config)#line aux 0
Miami-Router(config)#login
Miami-Router(config)#password Mi!auxaccess
```

Step 5 After you change all the administrative access passwords to the router, encrypt all clear-text passwords using the **service** password-encryption command:

```
Miami-Router(config)#service password-encryption
```

c. Configure Local Database Authentication Using AAA

By requiring two tokens, a username and a password, instead of just a password, you can make the router more secure. You can do this by configuring a central RADIUS server such as the Cisco Secure ACS and local username and password on the router. AAA authentication must be configured to first check the RADIUS server. If RADIUS server is not available, the router uses local authentication:

```
Miami-Router(config)#aaa new-model
Miami-Router(config)#aaa authentication login default group radius local
Miami-Router(config)#username Miadmin password rtrxss4Mi
Miami-Router(config)#line con 0
Miami-Router(Config-line)#login authentication default
Miami-Router(config)#line vty 0 15
Miami-Router(config-line)#login authentication default
Miami-Router(config-line)#exit
Miami-Router(config)#radius-server host 10.10.4.33 auth-port 1812 key Miamikey
```

d. Configure a Secure Method for Remote Access of the Routers

The current use of Telnet to remotely access the routers is not a secure method of access. Configure Secure Shell (SSH) and disable Telnet. Six steps are required to enable SSH support on the routers:

Step 1   Verify that the router has a host name.

Step 2   Verify that the router has a DNS domain.

Step 3   Generate the SSH key to be used.

Step 4   Configure the SSH timeout value.

Step 5   Configure the allowed number of retries.

Step 6   Enable SSH transport support for the vty connections.

The host name on the router has been configured.

Step 1   Verify the host name. You can verify the router host name by looking at the command-line interface (CLI) and checking the prompt. In this case, each router is named after the office location (city):

```
Miami-Router#
```

Step 2   Verify that the router has a DNS domain:

```
Miami-Router(config)#ip domain-name Miami.Mi.com
```

Step 3   Generate the SSH key to be used:

```
Miami-Router(config)#crypto key generate rsa
The name for the keys will be: Miami.Mi.com
Choose the size of the key modulus in the range of 360 to 2048 for your
  Signature Keys. Choosing a key modulus greater than 512 may take
  a few minutes.

How many bits in the modulus [512]: 1024
Choose the size of the key modulus in the range of 360 to 2048 for your
  Encryption Keys. Choosing a key modulus greater than 512 may take
  a few minutes.

How many bits in the modulus [512]: 1024
% Generating 1024 bit RSA keys ...[OK]
% Generating 1024 bit RSA keys ...[OK]
```

Step 4   Configure the SSH timeout value. (In this case, use 60 seconds.)

```
Miami-Router(config)#ip ssh time-out 60
```

Step 5   Configure the allowed number of retries (3):

```
Miami-Router(config)#ip ssh authentication-retries 3
```

Step 6   Enable SSH transport support for the vty connections:

```
Miami-Router(config)#line vty 0 4
Miami-Router(config-line)#transport input SSH
```

Task 3—Configure 802.1x on Miami User Switches

In this task, a Miami user switch is configured with 802.1x. This configuration is duplicated to all the user switches on the LAN:

Step 1 Configure the Miami user switches for 802.1x authentication using the Miami Cisco Secure ACS server as the RADIUS server:

```
Miami-Switch#configure terminal
Miami-Switch(config)#aaa new-model
Miami-Switch(config)#aaa authentication dot1x default group radius
```

Step 2 Specify the IP address, port, and key for RADIUS authentication:

```
Miami-Switch(config)#radius-server host 10.10.4.33 auth-port 1812 key Miamikey
```

Step 3 Enable 802.1x global authentication:

```
Miami-Switch(config)#dot1x system-auth-control
```

Step 4 Configure the 2950 switch to authenticate ports fa0/1 through fa0/48 to the Miami Cisco Secure ACS server:

```
Miami-Switch(config)#interface-range fastethernet0/1 - 48
Miami-Switch(config-if)#switchport mode access
Miami-Switch(config-if)#dot1x port-control auto
Miami-Switch(config-if)#end
```

Task 4—Configure Miami User Switches and Router to Mitigate Layer 2 Attacks

The steps taken to secure the Miami's router and switches are as follows:

Step 1 Configure port security.

Configure the switch ports on the user switches as access ports and set dynamic port security with the maximum number of addresses learned to 1. The violation mode is to shut down the interface, sticky learning is enabled, and no static MAC addresses are configured:

```
Switch#configure terminal
Miami-Switch1(config)#interface range fastethernet0/1 - 48
Miami-Switch1(config-if)#switchport mode access
Miami-Switch1(config-if)#switchport access VLAN 100
Miami-Switch1(config-if)#switchport port-security
Miami-Switch1(config-if)#switchport port-security maximum 1
Miami-Switch1(config-if)#switchport port-security mac-address sticky
```

Step 2 Explicitly configure trunk ports.

For backbone switch-to-switch connections on the Gigabit Ethernet ports, explicitly configure trunking. Allow user VLANs 100 through 104 to pass via the trunk. Use the dedicated VLAN ID of 200 for the native VLAN of the trunk port:

```
Miami-Switch1(config)#interface GigabitEthernet0/1
Miami-Switch1(config-if)#switchport mode trunk
Miami-Switch1(config-if)#switchport trunk allowed vlan 100-104
Miami-Switch1(config-if)#switchport trunk native vlan 200
Miami-Switch1(config-if)#end
```

Step 3 Configure STP parameters.

Enable bridge protocol data unit (BPDU) guard to disable ports using portfast upon detection of a BPDU message and disable ports that would become the root bridge based on their BPDU advertisement:

```
Miami-Switch1#configure terminal
Miami-Switch1(config)#spanning-tree portfast bpduguard default
Miami-Switch1(config)#interface GigabitEthernet0/1
Miami-Switch1(config-if)#spanning-tree guard root
```

Step 4 Configure DHCP snooping.

Enable DHCP snooping for VLANs 100 through 104. Then configure a rate limit of 70 packets per second on all user ports:

```
Miami-Switch1#configure terminal
Miami-Switch1(config)#ip dhcp snooping
Miami-Switch1(config)#ip dhcp snooping vlan 100-104
Miami-Switch1(config)#ip dhcp snooping information option
Miami-Switch1(config)#interface range fastethernet 0/1 - 48
Miami-Switch1(config-if)#ip dhcp snooping limit rate 70
```

Step 5 Disable CDP on user ports.

Disable Cisco Discovery Protocol (CDP) on user ports with the following command:

```
Miami-Switch1#configure terminal
Miami-Switch1(config)#interface range fastethernet 0/1 - 48
Miami-Switch1(config-if)#no cdp enable
```

Step 6 Set the VTP password.

Configure a Virtual Terminal Protocol (VTP) domain password for all Miami LAN switches. All these switches must share the same password. Switches without a password or with the wrong password reject VTP advertisements:

```
Miami-Switch1#configure terminal
Miami-Switch1(config)#vtp password MiamiVTPPass
```

Task 5—Configure PEAP with Cisco Secure ACS

This task uses the following components:

- Cisco Secure ACS for Microsoft Windows version 3.3

- Microsoft Certificate Services (installed as enterprise root certificate authority)

- Cisco Catalyst switch 2950 with Cisco IOS Software Release 12.1(22)EA3
- PC running Microsoft Windows XP Professional with Service Pack 1

This task is to authenticate users via PEAP to Miami's Cisco Secure ACS server using its Microsoft Windows 2000 Server local database:

Step 1 Request and install a server certificate in the Cisco Secure ACS.

Prior to installing a server certificate, you need to request a certificate from a CA. Open a web browser window on the Cisco Secure ACS server and browse to the CA server by entering **http://certificate server address/certsrv** in the address bar. Log in to the CA server. The page shown in Figure 24-6 will display.

Figure 24-6 *CA Certificate Request*

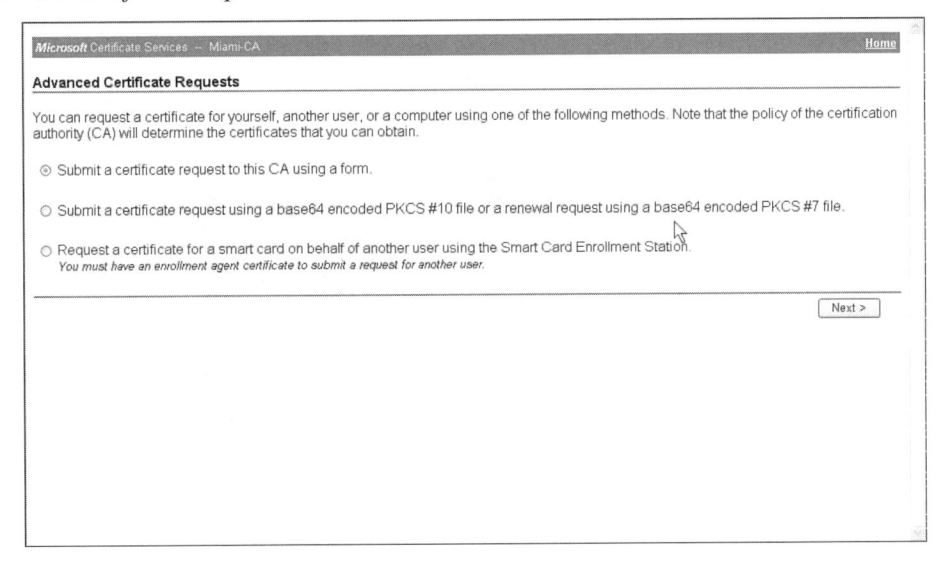

Step 2 Click the **Request a certificate** button, and then click **Next**. Figure 24-7 shows the next window displayed.

Step 3 Click the **Advanced request** button, and then click **Next**. In the next window, select **Submit a certificate request to this CA using a form**, and then click **Next**. Figure 24-8 shows the next window displayed.

564 Chapter 24: Final Scenarios

Figure 24-7 *Advanced Certificate Options*

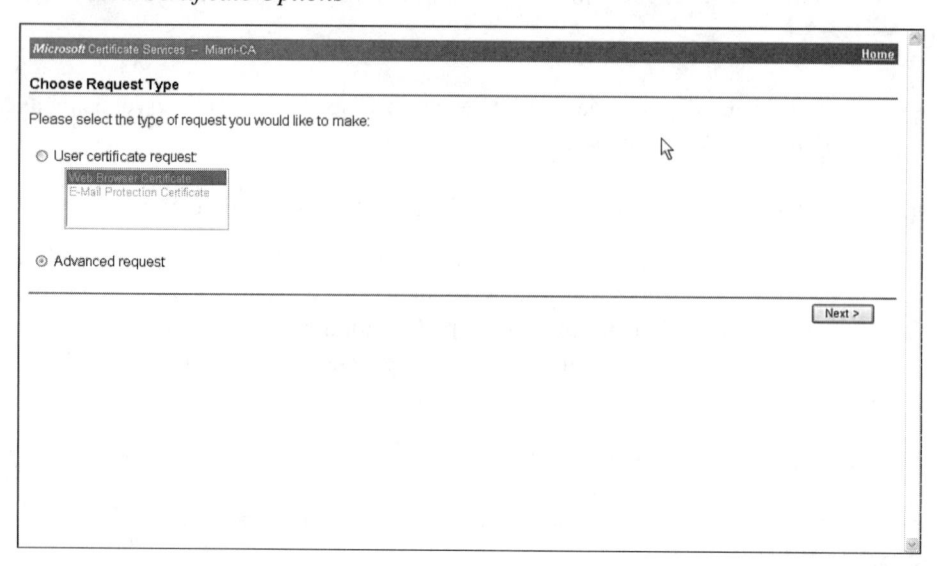

Figure 24-8 *Submit CA Form*

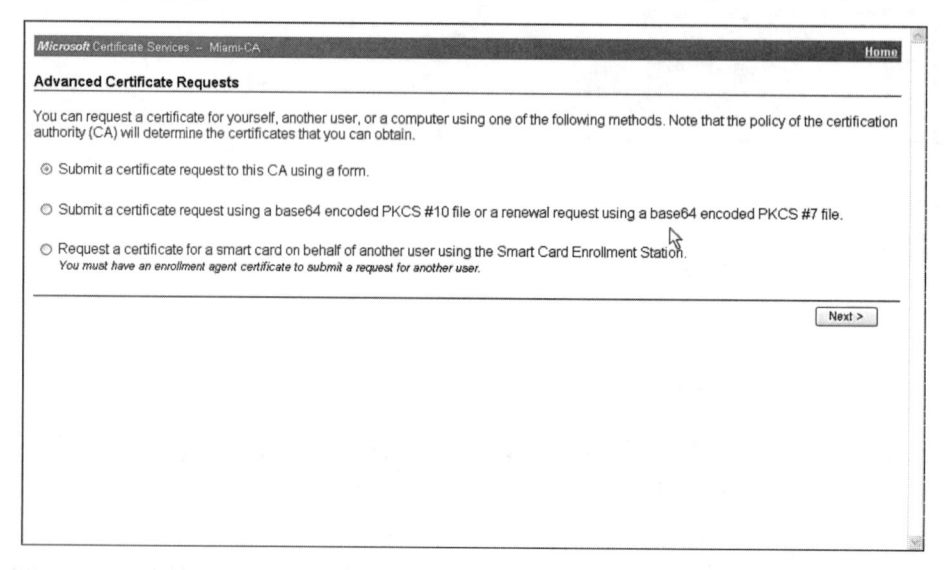

Step 4 Select **Web Server** as the template and enter the name of the Cisco Secure ACS server. Then set the key size to **1024**. Select the options for **Mark keys as exportable**, as depicted in the following figure. Figure 24-9 shows the next window displayed.

Figure 24-9 *Configure CA Options*

Step 5 Click **Install this certificate** in the next window. Upon successful installation, the confirmation message shown in Figure 24-10 displays.

Figure 24-10 *Successful Certificate Installation*

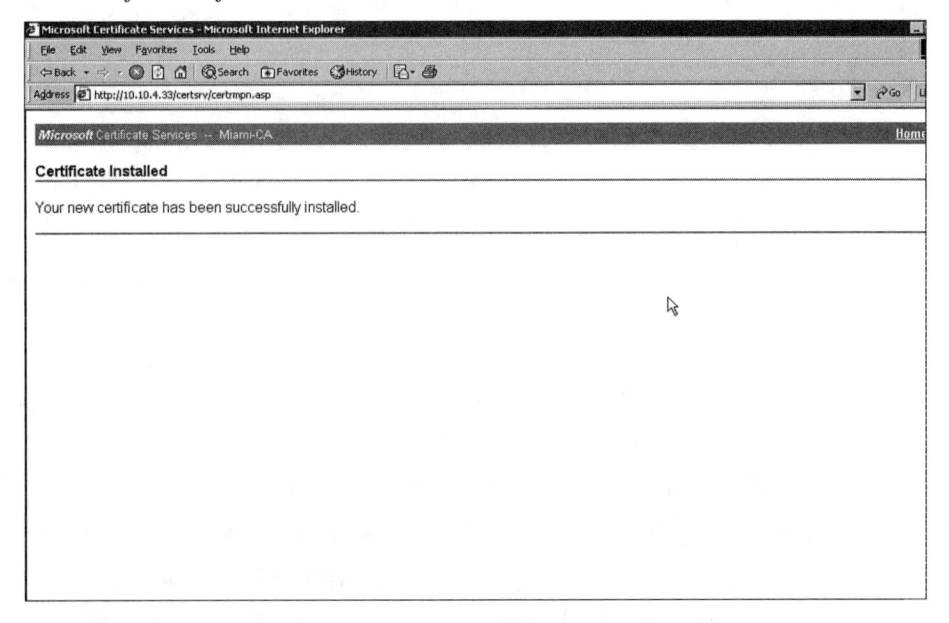

566 Chapter 24: Final Scenarios

The next step is to install the certificate that was created in the previous steps. To do so, follow these steps:

Step 1 In the main Cisco Secure ACS navigation bar, click **System Configuration**.

Step 2 Click the **ACS Certificate Setup** link from the following window, as shown in Figure 24-11.

Figure 24-11 *System Configuration Window*

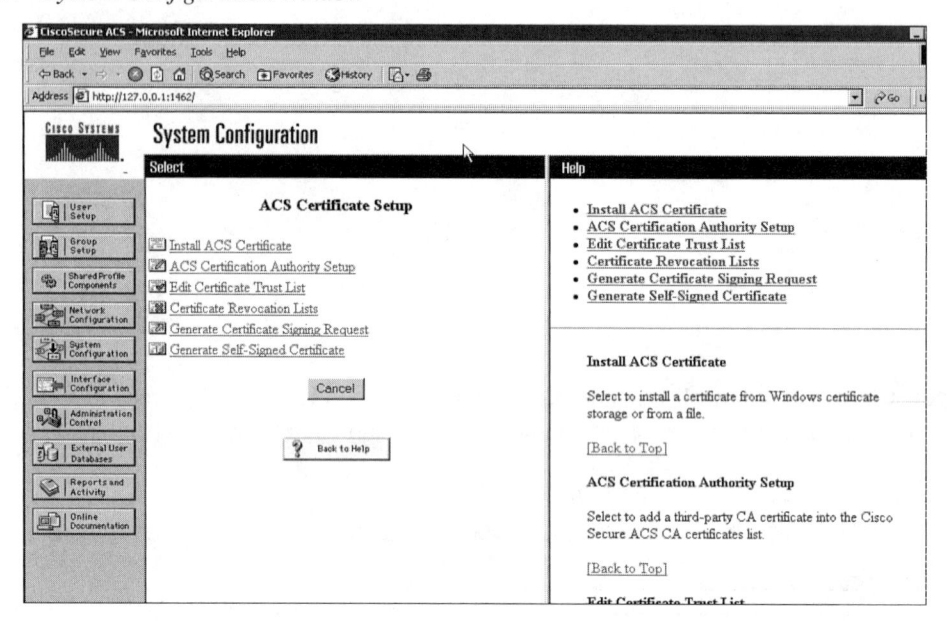

Step 3 Click the **Install ACS Certificate** link. Select the **Use certificate from storage** option, and then type the certificate created earlier. Then click the **Submit** button. CSACS displays the Installed Certificate Information table. After the certificate has been successfully installed, you must stop and restart the Cisco Secure ACS services for the System Configuration option to adopt the new settings. The screen shown in Figure 24-12 displays.

Step 4 From the ACS Certificate Setup, click **Edit Certificate Trust List**. Check that the Miami-CA was installed so that the Cisco Secure ACS will trust, and uncheck all the others that the Cisco Secure ACS should not trust.

   a. **Enable PEAP on the Global Authentication Setup Page**

Step 1 In System Configuration window, **click Global Authentication Setup**.

Step 2 Under the PEAP option, select **Allow EAP-MSCHAPv2**. Under the MS-CHAP configuration, select the **Allow MS-CHAP Version 1 Authentication** and **Version 2 Authentication** options. Figure 24-13 shows this window.

Task 5—Configure PEAP with Cisco Secure ACS

Figure 24-12 *Install Cisco Secure ACS Certificate*

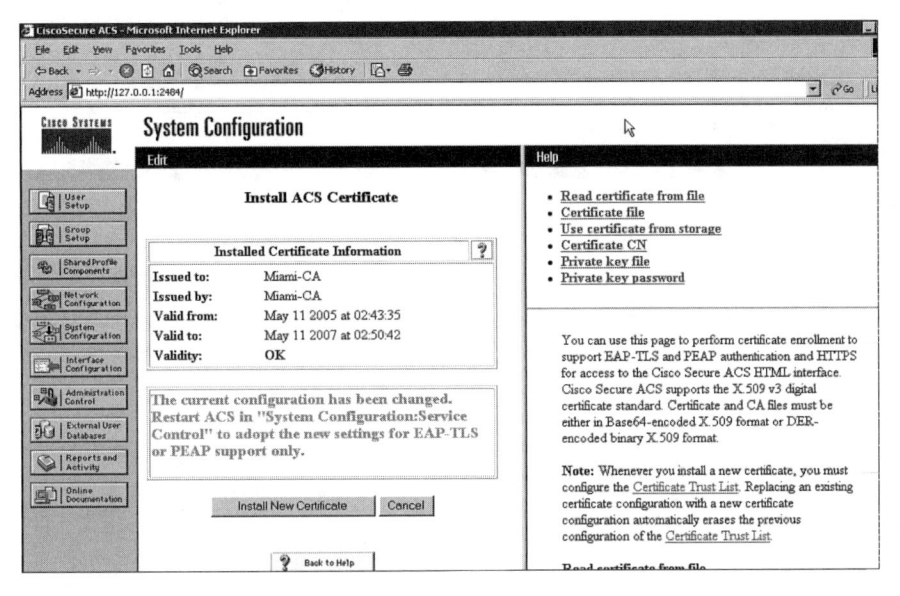

Figure 24-13 *Enabling PEAP in Global Authentication*

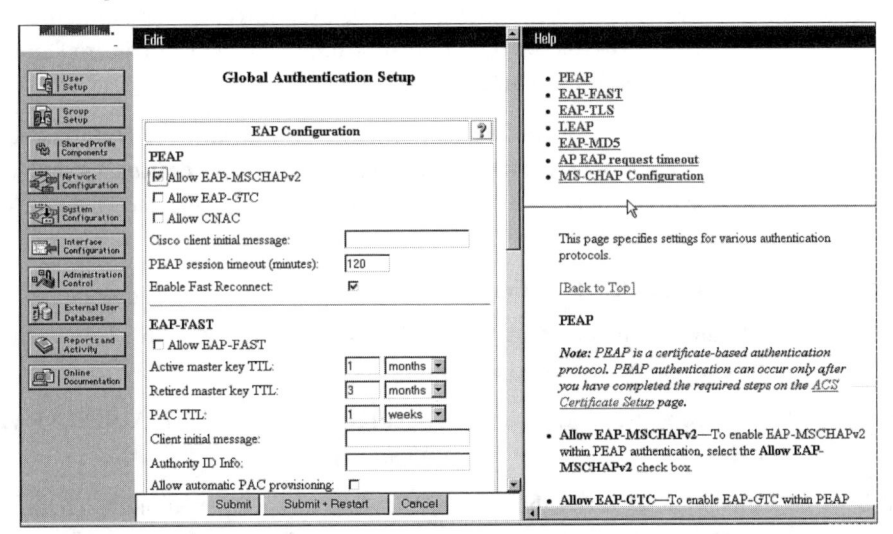

Then select **Submit + Restart** for the changes to take effect immediately.

b. Configure a User Database

Follow these steps to configure the external user databases:

Step 1   Click **External User Databases**, and then click **Database Configuration**, as depicted in Figure 24-14.

Figure 24-14 *Configure Windows Database*

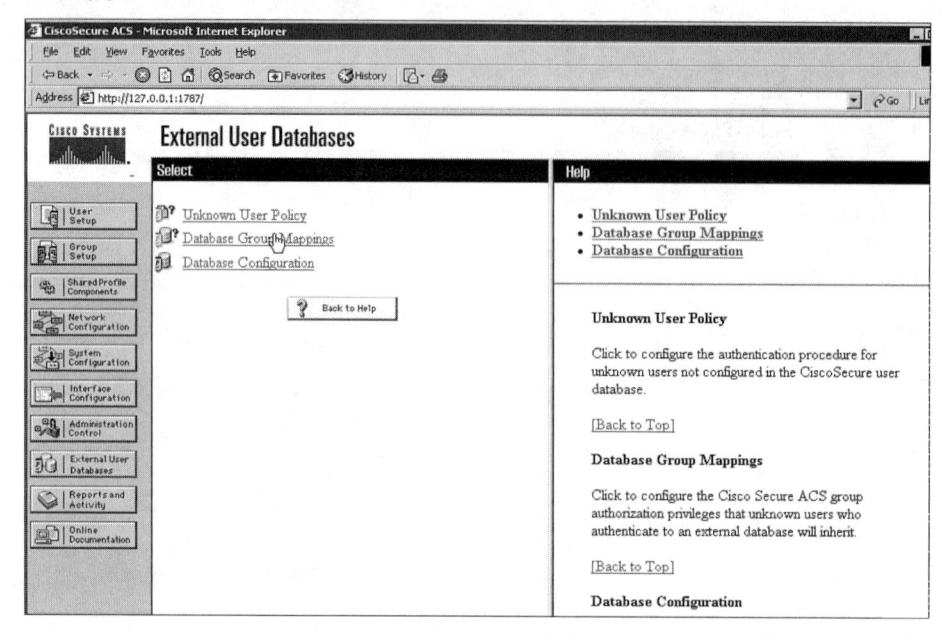

Step 2	Click **Windows Database**, and then click **Configure**. Under Configure Domain List, move the MIAMI domain from Available Domains to Domain List. Click the **Submit** button to apply the changes.
Step 3	Select **Unknown User Policy**, and then click **External User Databases**. Select the option for **Check the following external user databases**, and then move Windows Database from External Databases to Selected Databases. Then click **Submit**. Figure 24-15 shows this screen.

c. Configure PC for PEAP Authentication

Step 1	Configure PEAP authentication on the client on the LAN Area Connection properties Authentication tab. Figure 24-16 shows this screen.
Step 2	Select **Properties** and verify that **Secured password (EAP-MSCHAP v2)** is selected from the Select Authentication Method.
Step 3	At this point, to verify that the user has been authenticated on the client, go to **Control Panel, Network Connections**. On the menu bar, go to **View,** then select **Tiles**. The LAN connection should display the message "Authentication succeeded."

Figure 24-15 *Unknown User Policy Database*

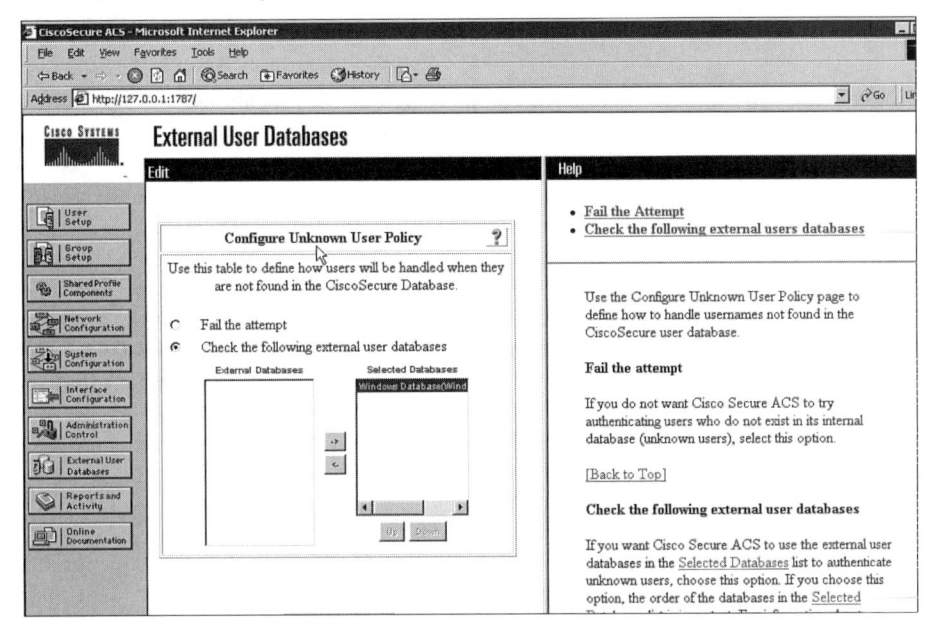

Figure 24-16 *Client Authentication Screen*

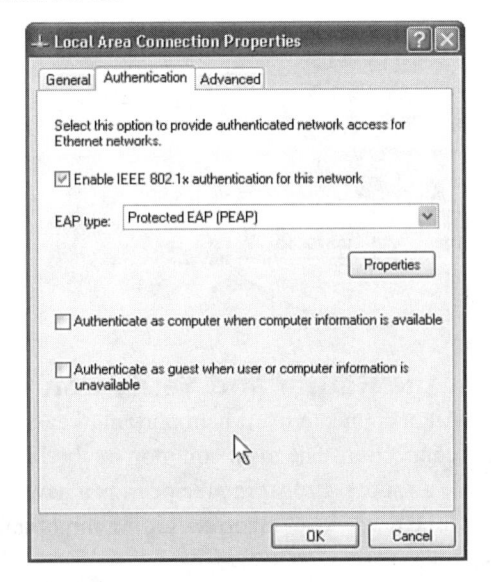

Task 6—Prepare the Network for IPsec Using Preshared Keys

This task consists of four subtasks:

- Establish a common convention for connectivity between locations.
- Configure initial setup of the router and verify connectivity.
- Prepare for IKE and IPsec.
- Define the preshared keys.

Establish a Common Convention for Connectivity Between Locations

When creating any connection between two or more locations, it is important that you have coordinated with both sides of the connection. Figure 24-17 depicts the network information for the connection between the Miami office and New York headquarters.

Figure 24-17 *Miami and New York Addressing*

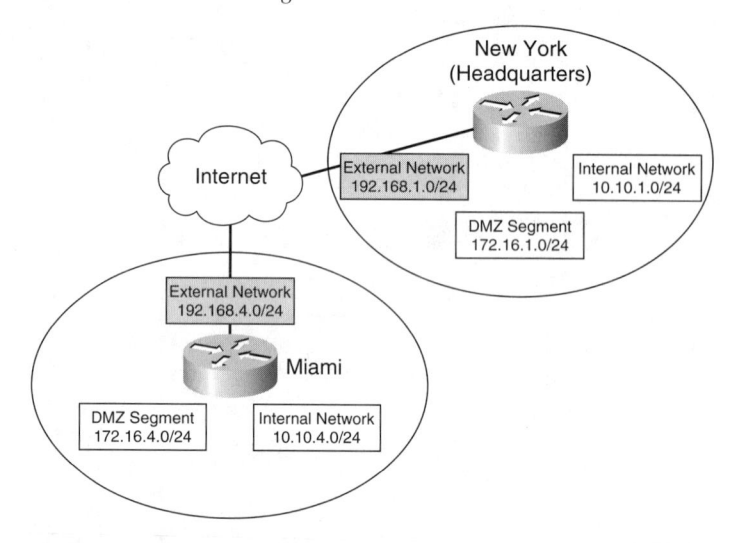

Configure Initial Setup of the Router and Verify Connectivity

Before you configure any VPN connectivity, it is important to ensure that you can successfully complete a nonencrypted connection. The most common method for verifying connectivity is to ping the peer router. If you can successfully ping the peer, you have connectivity. Next, you want to verify the router date and time because an incorrect setting might impact communication with the CA server and/or the negotiation with the peer:

Step 1   Ping the peer (New York) to verify connectivity:

```
Miami#
Miami#ping 192.168.1.1

Type escape sequence to abort.
```

Task 6—Prepare the Network for IPsec Using Preshared Keys

```
Sending 5, 100-byte ICMP Echos to 192.168.1.1, timeout is 2 seconds:
!!!!!
Success rate is 100 percent (5/5), round-trip min/avg/max = 1/1/4 ms
Miami#
```

Step 2 Set/verify the router date and time. Some platforms have a hardware clock and a software clock. In Cisco IOS Software, the hardware clock is commonly referred to as the *calendar* and the software clock is referred to as the *clock*. The clock is what is referenced when negotiating communications, but it is never a good idea to allow the clock and calendar to be different. Several commands enable you to configure the time/date in these devices:

- **clock set**—This command is executed in the privileged EXEC mode and is used to set the software clock.

- **calendar set**—This command is executed in the privileged EXEC mode and is used to set the hardware clock.

- **clock timezone**—This command is executed in the global configuration mode and is used to set a time zone (offset) on the clock. To terminate this command and return to coordinated universal time (UTC), use the **no clock timezone** command.

- **clock read-calendar**—This command is executed in the privileged EXEC mode and is used to set the software clock by pulling the configuration of the hardware clock.

- **clock update-calendar**—This command is executed in the privileged EXEC mode and is used to push the settings of the software clock to the hardware clock.

- **show clock**—This command is executed in the privileged EXEC mode and is used view the current software clock setting.

- **show calendar**—This command is executed in the privileged EXEC mode and is used view the current hardware software clock setting:

```
Miami#show clock
17:06:41.015 UTC Thu May 5 2005
Miami#
```

Prepare for IKE and IPsec

DHC Enterprises has established a configuration standard for VPN connectivity between locations. When configuring the router in Miami for the connection with New York, use the following protocols:

- Encryption peers: Miami (192.168.4.1); New York (192.168.1.1)

- Key exchange method: IKE

- Authentication method: Preshared key

- Encryption algorithm: AES-256

- Hash algorithm: SHA-1
- Diffie-Hellman (D-H) group: 5
- IKE SA lifetime: 86,400 seconds
- IPsec transforms: esp-aes-256, esp-sha-hmac
- IPsec SA lifetime: 1800 seconds
- Preshared key: abc123

Define the Preshared Key

Preshared keys are used to authenticate the peers to each other as the first step of the IKE negotiation. The preshared key selected for the scenarios does not meet the standard for complexity for preshared keys. We have selected simple keys so that they are easy to identify in the configuration examples.

Task 7—Configure IKE Using Preshared Keys

In this task, you use the configuration settings that were defined in the previous task. This task includes four steps:

- Enable IKE.
- Create the IKE policy.
- Configure the preshared key.
- Verify the IKE configuration.

Enable IKE

IKE is enabled by default, but it is a good practice to ensure that it is still enabled before you start to configure the router for a VPN. IKE is enabled using a single command in the global configuration mode:

```
Miami#configure terminal
Miami(config)#crypto isakmp enable
```

Create the IKE Policy

The IKE policy defines several aspects of the communication between the ISAKMP peers. Both peers must have matching values in an IKE policy to negotiate successfully the IKS SA. The IKE SA must be established to begin negotiation of the IPsec SA and create the encrypted connection. The IKE policy defined the following items:

- Authentication method
- Encryption algorithm
- Hash algorithm

- D-H group

- IKE SA lifetime

Configuring the IKE policy requires a single command followed by several subcommands:

```
Miami#configure terminal
Miami(config)#crypto isakmp policy 100
Miami(config-isakmp)#authentication pre-share
Miami(config-isakmp)#encryption aes 256
Miami(config-isakmp)#hash sha
Miami(config-isakmp)#group 5
Miami(config-isakmp)#lifetime 86400
```

Configure the Preshared Key

The preshared key is the method used by each peer to authenticate each other. The preshared keys must match exactly on both peers. The configuration command includes the key and the address/identity of the peer:

```
Miami#configure terminal
Miami(config)#crypto isakmp key abc123 address 192.168.1.1 255.255.255.255
```

Verify the IKE Configuration

The best way to troubleshoot a connection is to ensure that your configuration settings are correct in the first place. Because of the complexity required for VPN connections and the requirement that both peer configurations must match it is always a good idea to verify the configuration. If possible, it is best to compare the configurations from both peers at the same time. To verify the IKE policy configuration, use the **show crypto isakmp policy** command:

```
Miami#show crypto isakmp

Global IKE policy
Protection suite of priority 100
   encryption algorithm:    AES - Advanced Encryption Standard (256 bit keys).        hash
algorithm:            Secure Hash Standard
        authentication method:  Pre-Shared Key
        Diffie-Hellman group:   #5 (1536 bit)
        lifetime:               86400 seconds, no volume limit
Default protection suite
        encryption algorithm:   DES - Data Encryption Standard (56 bit keys).
        hash algorithm:         Secure Hash Standard
```

Task 8—Configure IPsec Using Preshared Keys

Now that you have completed the configuration of the IKE parameters, you must complete the configuration of the IPsec parameters. In this task, use the configuration settings that were defined in the previous task. This task includes five steps:

- Configure transform sets and SA parameters.

- Configure IPsec SA lifetimes.

- Configure crypto ACL.
- Configure crypto maps.
- Apply the crypto map to the interface.

Configure Transform Sets and SA Parameters

The transform sets define the encryption and hash algorithms to be used for the IPsec SA. The transform set is defined within a single command:

```
Miami#configure terminal
Miami(config)#crypto ipsec transform-set 20 esp-aes 256 esp-sha-hmac
```

Configure IPsec SA Lifetimes

Just as the IKS SA has a defined lifetime, so does the IPsec SA. A common weakness of any cryptography is that given a sufficient sample of the traffic and enough time, any encryption can be broken. The use of IKE and IPsec lifetimes forces the peers to rekey the connection, changing the parameters of the connection and preventing a potential hacker from gathering a sufficient sample of traffic:

```
Miami#configure terminal
Miami(config)#crypto ipsec security-association lifetime 1800 seconds
```

Configure Crypto ACLs

The crypto ACL defines for the router what traffic must be encrypted. If the router receives traffic on the internal interface that is destined for the VPN network segment, it will encrypt that traffic before forwarding it on to the peer. If the router receives traffic that is from the peer, and that traffic is not encrypted, the router will drop the traffic:

```
Miami#configure terminal
Miami(config)#access-list 105 permit ip 10.10.4.0 0.0.0.255 10.10.1.0 0.0.0.255
```

Configure Crypto Maps

The crypto map matches the IKE/IPsec configuration to the crypto ACL and puts into motion the other aspects of the VPN negotiation. The crypto map defines the following items:

- Which crypto ACL addresses the connection
- The IKE and IPsec peer addresses
- Sets Perfect Forward Secrecy
- Defines the applicable transform set

- Sets the IPsec SA lifetime

```
Miami#configure terminal
Miami(config)#crypto map NewYork 120 ipsec-isakmp
Miami(config-crypto-map)#match address 105
Miami(config-crypto-map)#set peer 192.168.1.1
Miami(config-crypto-map)#set pfs group5
Miami(config-crypto-map)#set transform-set 20
Miami(config-crypto-map)#set security-association lifetime seconds 1800
```

Apply the Crypto Map to the Interface

The crypto map is applied to the interface, telling the router which traffic to look for:

```
Miami#configure terminal
Miami(config)#interface fa0/1
Miami(config-if)#crypto map NewYork
```

Task 9—Configure IKE and IPsec on a Cisco Router

This task is similar to the previous two tasks combined, with the exception that you now use RSA signatures rather than preshared keys. This task requires the following seven steps:

- Enable IKE.

- Create an IKE policy using RSA signatures.

- Configure transform sets and SA parameters.

- Configure IPsec SA lifetimes.

- Configure crypto ACL.

- Configure crypto maps.

- Apply the crypto map to the interface.

Enable IKE

IKE is enabled by default, but it is a good practice to ensure that it is still enabled before you start to configure the router for a VPN. IKE is enabled using a single command in the global configuration mode:

```
Miami#configure terminal
Miami(config)#crypto isakmp enable
```

Create an IKE Policy Using RSA Signatures

The IKE policy defines several aspects of the communication between the ISAKMP peers. Both peers must have matching values in an IKE policy to negotiate successfully the IKS SA. The IKE

SA must be established to begin negotiation of the IPsec SA and create the encrypted connection. The IKE policy defines the following items:

- Authentication method
- Encryption algorithm
- Hash algorithm
- D-H group
- IKE SA lifetime

Configuring the IKE policy requires a single command followed by several subcommands:

```
Miami#configure terminal
Miami(config)#crypto isakmp policy 100
Miami(config-isakmp)#authentication rsa-sig
Miami(config-isakmp)#encryption aes 256
Miami(config-isakmp)#hash sha
Miami(config-isakmp)#group 5
Miami(config-isakmp)#lifetime 86400
```

Configure Transform Sets and SA Parameters

The transform sets define the encryption and hash algorithms to be used for the IPsec SA. The transform set is defined within a single command:

```
Miami#configure terminal
Miami(config)#crypto ipsec transform-set 20 esp-aes 256 esp-sha-hmac
```

Configure IPsec SA Lifetimes

Just as the ISAKMP SA has a defined lifetime, so does the IPsec SA. A common weakness of any cryptography is that given a sufficient sample of the traffic and enough time, any encryption can be broken. The use of IKE and IPsec lifetimes forces the peers to rekey the connection, changing the parameters of the connection and preventing a potential hacker from gathering a sufficient sample of traffic:

```
Miami#configure terminal
Miami(config)#crypto ipsec security-association lifetime 1800 seconds
```

Configure Crypto ACLs

The crypto ACL defines for the router what traffic must be encrypted. If the router receives traffic on the internal interface that is destined for the VPN network segment, it will encrypt that traffic before forwarding it on to the peer. If the router receives traffic that is from the peer, and that traffic is not encrypted, the router will drop the traffic:

```
Miami#configure terminal
Miami(config)#access-list 105 permit ip 10.10.4.0 0.0.0.255 10.10.1.0 0.0.0.255
```

Configure Crypto Maps

The crypto map matches the IKE/IPsec configuration to the crypto ACL and puts into motion the other aspects of the VPN negotiation. The crypto map defines the following items:

- Which crypto ACL addresses the connection

- The IKE and IPsec Peer addresses

- Sets Perfect Forward Secrecy

- Defines the applicable transform set

- Sets the IPsec SA lifetime

```
Miami#configure terminal
Miami(config)#crypto map NewYork 120 ipsec-isakmp
Miami(config-crypto-map)#match address 105
Miami(config-crypto-map)#set peer 192.168.1.1
Miami(config-crypto-map)#set pfs group5
Miami(config-crypto-map)#set transform-set 20
Miami(config-crypto-map)#set security-association lifetime seconds 1800
```

Apply the Crypto Map to the Interface

The crypto map is applied to the interface, telling the router which traffic to look for:

```
Miami#configure terminal
Miami(config)#interface fa0/1
Miami(config-if)#crypto map NewYork
```

Task 10—Prepare the Network for IPsec Using Digital Certificates

This task consists of three subtasks:

- Configure initial setup of the router and verify connectivity.

- Prepare for IKE and IPsec.

- Configure CA support.

Configure Initial Setup of the Router and Verify Connectivity

Verify the connectivity as you did for Task 6 and ensure that the clock is correct:

```
Miami#show clock
17:06:41.015 UTC Thu May 5 2005
Miami#
```

Prepare for IKE and IPsec

DHC Enterprises has established a configuration standard for VPN connectivity between locations. When configuring the router in Miami for the connection with New York, use the following protocols:

- Key exchange method: IKE

- Authentication method: RSA signatures
- Encryption algorithm: AES-256
- Hash algorithm: SHA-1
- D-H group: 5
- CA server: 192.168.242.42

Configure CA Support

Working with the CA server administrator, configure CA support on the router by defining the router's domain name, defining the CA server's static host name-to-IP address mapping, generating RSA usage keys, and configuring the CA server trustpoint:

Step 1 Configure the router host name and domain name:

```
router#configure terminal
router(config)#hostname Miami
Miami(config)#ip domain-name Mi.com
```

Step 2 Generate the RSA key pair. The RSA usage keys are generated using the selected modulus. In this case, we are using 1024 bits:

```
Miami(config)#crypto key generate rsa usage-keys
The name for the keys will be: Miami.Mi.com
Choose the size of the key modulus in the range of 360 to 2048 for your
   Signature Keys. Choosing a key modulus greater than 512 may take
   a few minutes.

How many bits in the modulus [512]: 1024
Choose the size of the key modulus in the range of 360 to 2048 for your
   Encryption Keys. Choosing a key modulus greater than 512 may take
   a few minutes.

How many bits in the modulus [512]: 1024
% Generating 1024 bit RSA keys ...[OK]
% Generating 1024 bit RSA keys ...[OK]

Miami(config)#
```

Step 3 Define the CA server and verify connectivity. Depending on the location of your CA server, you might need to define specific routes to get to it. DHC hosts a CA server publicly (192.168.242.42) so that the perimeter routers default route to it. Again, the easiest way to verify connectivity is to initiate a ping.

Define the CA server:

```
Miami#configure terminal
Miami(config)#ip host CA-Server 192.168.242.42
```

Verify connectivity with the CA server:

```
Miami#
Miami#ping 192.168.242.42
```

```
            Type escape sequence to abort.
            Sending 5, 100-byte ICMP Echos to 192.168.1.1, timeout is 2 seconds:
            !!!!!
            Success rate is 100 percent (5/5), round-trip min/avg/max = 1/1/4 ms
            Miami#
```

Step 4 Configure the CA trustpoint. Now that you have verified connectivity with the CA server, you simply need to designate the CA server using the **crypto ca trustpoint** command, and then define the enrollment URL and mode:

```
NewYork(config)#crypto ca trustpoint CA-Server
NewYork(ca-trustpoint)#enrollment url http://CA-Server/certserv/mscep/mscep.dll
NewYork(ca-trustpoint)#enrollment mode ra
```

Task 11—Test and Verify IPsec CA Configuration

After completing the configuration of the IPsec VPN, you want to test the connection and verify that it is functioning correctly before placing that connection into production. There are several methods for testing the connection. Of course, the most direct method is to pass traffic through the VPN tunnel and verify that it reaches the final destination. In this case, take a more thorough approach and look at each individual component of the configuration, verifying each item individually. This task consists of 10 individual subtasks:

- Display IKE policies.
- Display transform sets.
- Display your configured crypto maps.
- Display the current state of IPsec SAs.
- Clear any existing SAs.
- Enable debug output for IPsec events.
- Enable debug output for ISAKMP events.
- Observe the IKE and IPsec debug outputs.
- Verify IKE and IPsec SAs.
- Ensure encryption is working.

Display IKE Policies

You can display the IKE policy by using the **show crypto isakmp policy** command:

```
Miami#sh crypto isakmp policy

Global IKE policy
Protection suite of priority 100
```

encryption algorithm: AES - Advanced Encryption Standard (256 bit keys).
                hash algorithm: Secure Hash Standard
                authentication method: Rivest-Shamir-Adleman Encryption
                Diffie-Hellman group: #5 (1536 bit)
                lifetime: 86400 seconds, no volume limit
        Default protection suite
                encryption algorithm: DES - Data Encryption Standard (56 bit keys).
                hash algorithm: Secure Hash Standard
                authentication method: Rivest-Shamir-Adleman Signature
                Diffie-Hellman group: #1 (768 bit)
                lifetime: 86400 seconds, no volume limit
        Miami#
```

## Display Transform Sets

You can display the transform sets by using the **show ipsec transform-set** command:

```
Miami#sh crypto ipsec transform-set
Transform set 20: { esp-256-aes esp-sha-hmac  }
   will negotiate = { Tunnel,  },

Miami#
```

Display Configured crypto maps

You can display the crypto map sets by using the **show crypto map** command:

```
Miami#sh crypto map
Crypto Map "NewYork" 120 ipsec-isakmp
        Peer = 192.168.1.1
        Extended IP access list 105
            access-list 105 permit ip 10.10.4.0 0.0.0.255 10.10.1.0 0.0.0.255
        Current peer: 192.168.1.1
        Security association lifetime: 4608000 kilobytes/1800 seconds
        PFS (Y/N): Y
        DH group:  group2
        Transform sets={ 20,  }
        Interfaces using crypto map NewYork:
                FastEthernet0/1

Miami#
```

Display the Current State of IPsec SAs

You can display the IPsec SAs by using the **show crypto ipsec sa** command:

```
Miami#sh crypto ipsec sa

interface: FastEthernet0/1
    Crypto map tag: Miami, local addr. 192.168.4.1

   protected vrf:
   local  ident (addr/mask/prot/port): (10.10.4.0/255.255.255.0/0/0)
   remote ident (addr/mask/prot/port): (10.10.1.0/255.255.255.0/0/0)
   current_peer: 192.168.1.1:500
     PERMIT, flags={origin_is_acl,}
    #pkts encaps: 321, #pkts encrypt: 321, #pkts digest 321
    #pkts decaps: 321, #pkts decrypt: 321, #pkts verify 321

```
            #pkts compressed: 0, #pkts decompressed: 0
            #pkts not compressed: 0, #pkts compr. failed: 0
            #pkts not decompressed: 0, #pkts decompress failed: 0
            #send errors 4, #recv errors 0

             local crypto endpt.: 192.168.4.1, remote crypto endpt.: 192.168.1.1
             path mtu 1500, ip mtu 1500, ip mtu idb FastEthernet0/1
             current outbound spi: AE399F1B

             inbound esp sas:
              spi: 0xFE4F2F8(266662648)
                transform: esp-256-aes esp-sha-hmac ,
                in use settings ={Tunnel, }
                slot: 0, conn id: 2000, flow_id: 1, crypto map: NewYork
                sa timing: remaining key lifetime (k/sec): (4570617/1455)
                IV size: 16 bytes
                replay detection support: Y

             inbound ah sas:

             inbound pcp sas:

             outbound esp sas:
              spi: 0xAE399F1B(2923011867)
                transform: esp-256-aes esp-sha-hmac ,
                in use settings ={Tunnel, }
                slot: 0, conn id: 2001, flow_id: 2, crypto map: newYork
                sa timing: remaining key lifetime (k/sec): (4570617/1453)
                IV size: 16 bytes
                replay detection support: Y

             outbound ah sas:

             outbound pcp sas:
      Miami#
```

Clear Any Existing SAs

The command for clearing established SAs is **clear crypto sa**.

Enable Debug Output for IPsec Events

The command for debugging IPsec events is **debug crypto ipsec**.

Enable Debug Output for ISAKMP Events

The command for debugging IKE events is **debug crypto isakmp**.

Observe the IKE and IPsec Debug Outputs

The following is the output from the different **debug** commands on the New York router:

```
NewYork#debug crypto isakmp
09:19:27: ISAKMP (8): beginning Main Mode exchange
09:19:27: ISAKMP (8): processing SA payload. message ID = 0
09:19:27: ISAKMP (8): Checking ISAKMP transform 1 against priority 100 policy
09:19:27: ISAKMP: encryption AES-256
09:19:27: ISAKMP: hash SHA
```

Chapter 24: Final Scenarios

```
09:19:27: ISAKMP: default group 5
09:19:27: ISAKMP: auth pre-share
09:19:27: ISAKMP (8): atts are acceptable. Next payload is 0
09:19:27: ISAKMP (8): SA is doing pre-shared key authentication
09:19:28: ISAKMP (8): processing KE payload. message ID = 0
09:19:28: ISAKMP (8): processing NONCE payload. message ID = 0
09:19:28: ISAKMP (8): SKEYID state generated
09:19:28: ISAKMP (8): processing ID payload. message ID = 0
09:19:28: ISAKMP (8): processing HASH payload. message ID = 0
09:19:28: ISAKMP (8): SA has been authenticated
09:19:28: ISAKMP (8): beginning Quick Mode exchange, M-ID of 192036475
09:19:28: ISAKMP (8): processing SA payload. message ID = 192036475
09:19:28: ISAKMP (8): Checking IPsec proposal 1
09:19:28: ISAKMP: transform 1, AES-256
09:19:28: ISAKMP: attributes in transform:
09:19:28: ISAKMP: encaps is 1
09:19:28: ISAKMP: SA life type in seconds
09:19:28: ISAKMP: SA life duration (basic) of 1800


NewYork#debug crypto key-exchange
CRYPTO-KE: Sent 4 bytes.
CRYPTO-KE: Sent 2 bytes.
CRYPTO-KE: Sent 2 bytes.
CRYPTO-KE: Sent 2 bytes.
CRYPTO-KE: Sent 64 bytes.
NewYork# debug crypto key-exchange
CRYPTO-KE: Received 4 bytes.
CRYPTO-KE: Received 2 bytes.
CRYPTO-KE: Received 2 bytes.
CRYPTO-KE: Received 2 bytes.
CRYPTO-KE: Received 49 bytes.
CRYPTO-KE: Received 15 bytes.
NewYork# debug crypto ipsec
09.22.55: IPsec(sa_request):            <<<< IPsec requests and negotiations begin
(key eng. msg.) src= 192.168.1.1, dest= 192.168.3.1,
src_proxy= 192.168.1.1/255.255.255.255/0/0 (type=1),
dest_proxy= 192.168.3.1/255.255.255.255/0/0 (type=1),
protocol= ESP, transform= esp-256-aes  esp-sha-hmac ,
lifedur= 1800s and 4608000kb,
spi= 0x0(0), conn_id= 0, keysize= 0, flags= 0x4
09.22.55: IPsec(sa_request): ,
(key eng. msg.) src= 192.168.1.1, dest= 192.168.3.1,
src_proxy= 192.168.1.1/255.255.255.255/0/0 (type=1),
dest_proxy= 192.168.3.1/255.255.255.255/0/0 (type=1).,
protocol= AH, transform= ah-sha-hmac ,
lifedur= 1800s and 4608000kb,
spi= 0x0(0), conn_id= 0, keysize= 0, flags= 0x0.    <<<< IKE initiates SPI Exchange
00:24:34: IPsec(key_engine): got a queue event...
00:24:34: IPsec(spi_response): getting spi 302974012ld for SA
from 192.168.3.1 to 192.168.1.1 for prot 3
00:24:34: IPsec(spi_response): getting spi 525075940ld for SA
from 192.168.3.1 to 192.168.1.1 for prot 2
00:24:34: IPsec(validate_proposal_request): proposal part #1,   <<<< IKE Initiates SA
    proposal
(key eng. msg.) dest= 192.168.3.1, src= 192.168.1.1,
dest_proxy= 192.168.3.1/255.255.255.255/0/0 (type=1),
src_proxy= 192.168.1.1/255.255.255.255/0/0 (type=1),
protocol= ESP, transform= esp-256-aes esp-sha-hmac ,
lifedur= 0s and 0kb,
spi= 0x0(0), conn_id= 0, keysize= 0, flags= 0x4
09.23:00: IPsec(initialize_sas): ,        <<<< IKE Negotiation complete INBOUND SA
    established
(key eng. msg.) dest= 192.168.1.1, src= 192.168.3.1,
dest_proxy= 192.168.1.1/255.255.255.255/0/0 (type=1),
src_proxy= 192.168.3.1/255.255.255.255/0/0 (type=1),
```

```
protocol= ESP, transform= esp-256-aes esp-sha-hmac ,
lifedur= 120s and 4608000 kb,
spi= 0x120F043C(302974012), conn_id= 29, keysize= 0, flags= 0x4
09.23:00: IPsec(initialize_sas): ,
(key eng. msg.) src= 192.168.1.1, dest= 192.168.3.1,
src_proxy= 192.168.1.1/255.255.255.255/0/0 (type=1),
dest_proxy= 192.168.3.1/255.255.255.255/0/0 (type=1),
protocol= ESP, transform= esp-256-aes esp-sha-hmac ,
lifedur= 120s and 4608000kb,
spi= 0x38914A4(59315364), conn_id= 30, keysize= 0, flags= 0x4
09.23:00: IPsec(create_sa): sa created,
(sa) sa_dest= 192.168.1.1, sa_prot= 50,
sa_spi= 0x120F043C(302974012),
sa_trans= esp-256-aes esp-sha-hmac , sa_conn_id= 29
09.23:00: IPsec(create_sa): sa created,
(sa) sa_dest= 192.168.3.1, sa_prot= 50,
sa_spi= 0x38914A4(59315364),
sa_trans= esp-256-aes esp-sha-hmac , sa_conn_id= 30
```

Verify IKE and IPsec SAs

The following is the output from the different **show** commands on the New York router:

```
NewYork#show crypto ipsec sa

interface: FastEthernet0/1
    Crypto map tag: SanFran, local addr. 192.168.1.1

  protected vrf:
  local   ident (addr/mask/prot/port): (10.10.1.0/255.255.255.0/0/0)
  remote ident (addr/mask/prot/port): (10.10.3.0/255.255.255.0/0/0)
  current_peer: 192.168.3.1:500
    PERMIT, flags={origin_is_acl,}
   #pkts encaps: 321, #pkts encrypt: 321, #pkts digest 321
   #pkts decaps: 321, #pkts decrypt: 321, #pkts verify 321
   #pkts compressed: 0, #pkts decompressed: 0
   #pkts not compressed: 0, #pkts compr. failed: 0
   #pkts not decompressed: 0, #pkts decompress failed: 0
   #send errors 4, #recv errors 0

    local crypto endpt.: 192.168.1.1, remote crypto endpt.: 192.168.3.1
    path mtu 1500, ip mtu 1500, ip mtu idb FastEthernet0/1
    current outbound spi: AE399F1B

    inbound esp sas:
     spi: 0xFE4F2F8(266662648)
       transform: esp-256-aes esp-sha-hmac ,
       in use settings ={Tunnel, }
       slot: 0, conn id: 2000, flow_id: 1, crypto map: SanFran
       sa timing: remaining key lifetime (k/sec): (4570617/1455)
       IV size: 16 bytes
       replay detection support: Y

    inbound ah sas:

    inbound pcp sas:

    outbound esp sas:
     spi: 0xAE399F1B(2923011867)
       transform: esp-256-aes esp-sha-hmac ,
       in use settings ={Tunnel, }
       slot: 0, conn id: 2001, flow_id: 2, crypto map: SanFran
       sa timing: remaining key lifetime (k/sec): (4570617/1453)
       IV size: 16 bytes
```

```
        replay detection support: Y

    outbound ah sas:

    outbound pcp sas:
NewYork#
```

Task 12—Configure Authentication Proxy on the Miami Router

This task consists of four subtasks:

- Configure AAA.

- Configure the HTTP server.

- Configure authentication proxy.

- Test and verify the authentication proxy configuration.

Figure 24-18 depicts the addressing of the network segments in Miami and the location of the TACACS server.

Figure 24-18 *Miami Network Addressing*

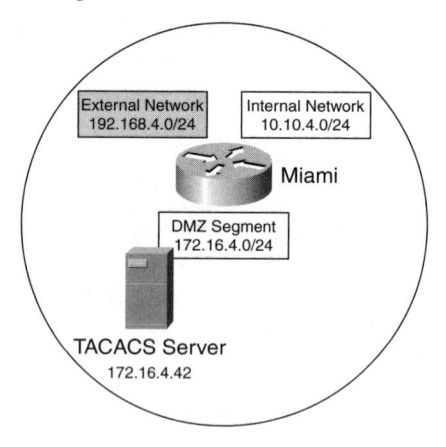

Configure AAA

Several commands are required to configure AAA on the router.

```
Miami#configure terminal
Miami(config)#aaa new-model
Miami(config)#aaa authentication login default tacacs+
Miami(config)#aaa authorization auth-proxy default group tacacs+
Miami(config)#aaa accounting auth-proxy default start-stop group tacacs+
Miami(config)#tacacs server host 172.16.4.42
Miami(config)#tacacs server key abc123
Miami(config)#access-list 110 permit tcp host 172.16.4.42 eq tacacs 172.17.4.254
Miami(config)#end
Miami#
```

Configure the HTTP Server

The Miami router must now be configured as an HTTP server to facilitate the communication between the requestor, the router, and the TACACS server:

```
Miami#configure terminal
Miami(config)#ip http server
Miami(config)#ip http authentication aaa
Miami(config)#access-list 10 deny any any
Miami(config)#ip http access-class 10
Miami(connfig)#end
Miami#
```

Configure Authentication Proxy

The router must now be configured to facilitate the authentication proxy:

```
Miami#configure terminal
Miami(config)#ip auth-proxy auth-cache time 30
Miami(config)#ip auth-proxy auth-proxy-banner
Miami(config)#ip auth-proxy name allowed-inbound http
Miami(config)#interface fa0/1
Miami(connfig-if)#ip auth-proxy allowed-inbound
Miami(config-if)#end
Miami#
```

Test and Verify the Authentication Proxy Configuration

A simple way to verity the authentication proxy is to try the connection. You can also verify the authentication proxy by capturing the connection with the **debug auth-proxy** command:

```
Miami#debug auth-proxy

16:23:33: AUTH-PROXY creates info:
cliaddr - 192.168.122.5, cliport - 36583
seraddr - 192.168.4.1, serport - 80
ip-srcaddr 192.168.122.5
pak-srcaddr 0.0.0.0
16:23:47:AUTH_PROXY OBJ_CREATE:acl item 61AD60CC
16:23:47:AUTH-PROXY OBJ_CREATE:create acl wrapper 6151C7C8 -- acl item 61AD60CC
16:23:47:AUTH-PROXY Src 172.16.4.254 Port [0]
16:23:47:AUTH-PROXY Dst 172.16.4.22 Port [22]
16:23:47:AUTH-PROXY OBJ_CREATE:acl item 6151C908
16:23:47:AUTH-PROXY OBJ_CREATE:create acl wrapper 6187A060 -- acl item 6151C908
16:23:47:AUTH-PROXY Src 172.16.4.254 Port [0]
16:23:47:AUTH-PROXY Dst 172.16.4.22 Port [20]
16:23:47:AUTH-PROXY OBJ_CREATE:acl item 61A40B88
16:23:47:AUTH-PROXY OBJ_CREATE:create acl wrapper 6187A0D4 -- acl item 61A40B88
16:23:47:AUTH-PROXY Src 172.16.4.254 Port [0]
16:23:47:AUTH-PROXY Dst 172.16.4.22 Port [21]
16:23:47:AUTH-PROXY OBJ_CREATE:acl item 61879550
16:23:47:AUTH-PROXY OBJ_CREATE:create acl wrapper 61879644 -- acl item 61879550
16:23:47:AUTH-PROXY Src 172.16.4.254 Port [0]
16:23:47:AUTH-PROXY Dst 172.16.4.22 Port [23]
```

Task 13—Configure CBAC on the Miami Router

For this task, CBAC is being configured to inspect inbound traffic to the Miami LAN. The interface FastEthernet 0/1 is the protected network, and interface FastEthernet 0/0 is the unprotected network.

The security policy for the protected site uses ACLs to restrict inbound traffic on the unprotected interface to specific Internet Control Message Protocol (ICMP) traffic, denying inbound access for TCP and UDP traffic. Inbound access for specific protocol traffic is provided through dynamic ACLs, which are generated according to CBAC inspection rules:

Step 1 Create an access list 110. ACL 110 denies TCP and UDP traffic from any source or destination while permitting specific ICMP traffic. The final **deny** statement is not required but is included. The final entry in any ACL is an implicit denial of all IP traffic:

```
Miami-Router(config)#access-list 110 deny tcp any any
Miami-Router(config)#access-list 110 deny udp any any
Miami-Router(config)#access-list 110 permit icmp any any echo-reply
Miami-Router(config)#access-list 110 permit icmp any any time-exceeded
Miami-Router(config)#access-list 110 permit icmp any any traceroute
Miami-Router(config)#access-list 110 permit icmp any any unreachable
Miami-Router(config)#access-list 110 deny ip any any
```

Step 2 Apply ACL 110 to interface FastEthernet 0/0 to block all access from the unprotected network to the protected network:

```
Miami-Router(config)#interface FastEthernet 0/0
Miami-Router(config-if)#ip access-group 110 in
```

Step 3 Change the size of the session hash table to 2048 buckets to improve CBAC performance:

```
ip inspect hashtable 2048
```

Step 4 Create an inspection rule for intranet users:

```
Miami-Router(config)#ip inspect name intranetusers ftp
Miami-Router(config)#ip inspect name intranetusers http
Miami-Router(config)#ip inspect name intranetusers rcmd
Miami-Router(config)#ip inspect name intranetusers realaudio
Miami-Router(config)#ip inspect name intranetusers smtp timeout 3600
Miami-Router(config)#ip inspect name intranetusers tftp timeout 30
Miami-Router(config)#ip inspect name intranetusers udp timeout 15
Miami-Router(config)#ip inspect name intranetusers tcp timeout 3600
```

Step 5 Apply the inspection rule inbound at interface FastEthernet0/1 to inspect traffic from users on the protected network:

```
Miami-Router(config)#interface FastEthernet0/1
Miami-Router(config-if)#ip inspect intranetusers in
```

Task 14—Configure Miami Router with IPS Using SDM

This task will be configured using Cisco SDM. The wizard guides you through the necessary steps to configure IPS on the Miami router. The configuration steps are as follows:

Step 1 After you launch IPS from the SDM Configure Intrusion Prevention window, the main IPS home page is loaded, as shown in Figure 24-19.

Figure 24-19 *IPS Home Page Screen*

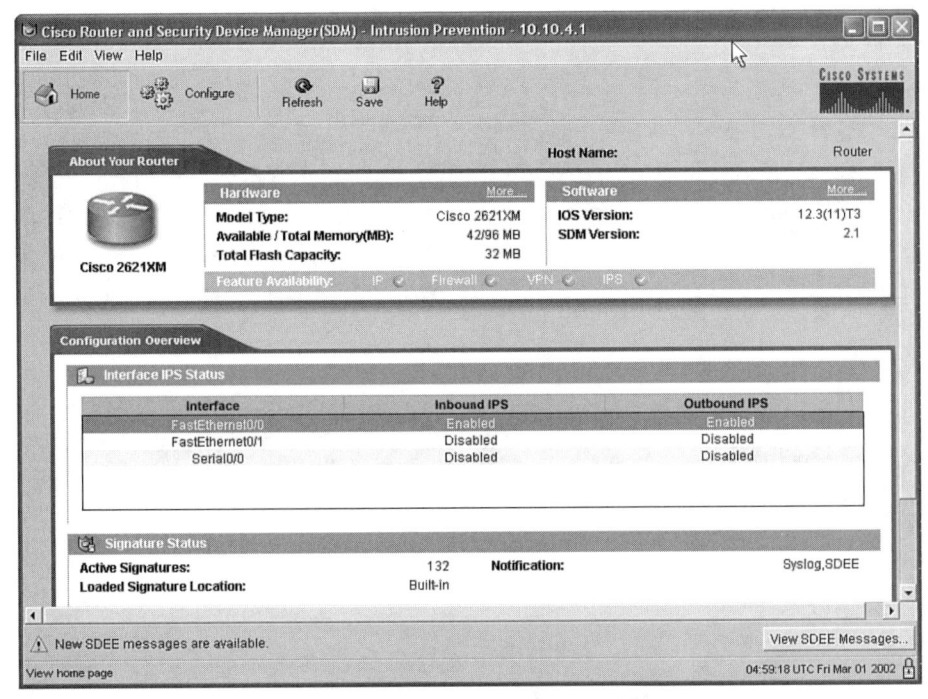

Step 2 In the Configure window, select **Global Settings** and add the Configured SDF Locations for the attack-drop.sdf, as shown in Figure 24-20.

Figure 24-20 *Add a Signature Location Screen*

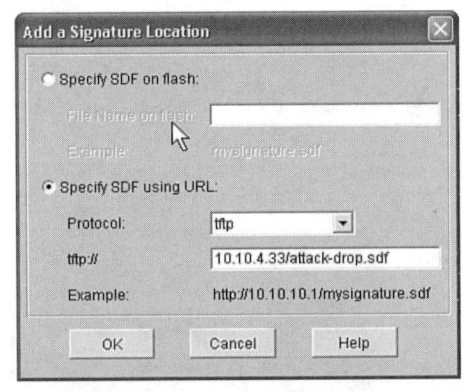

Step 3 In the Configure window, select **Rules** mode, as shown in Figure 24-21.

Step 4 Click the **Enable** button and the WAN **FastEthernet0/0** interface. To complete this task, you must configure an inbound filter. Click the **Inbound** button and use **Internet-access** as the name of the inbound filter, as shown in Figure 24-22.

588 Chapter 24: Final Scenarios

Figure 24-21 *IPS Rules Screen*

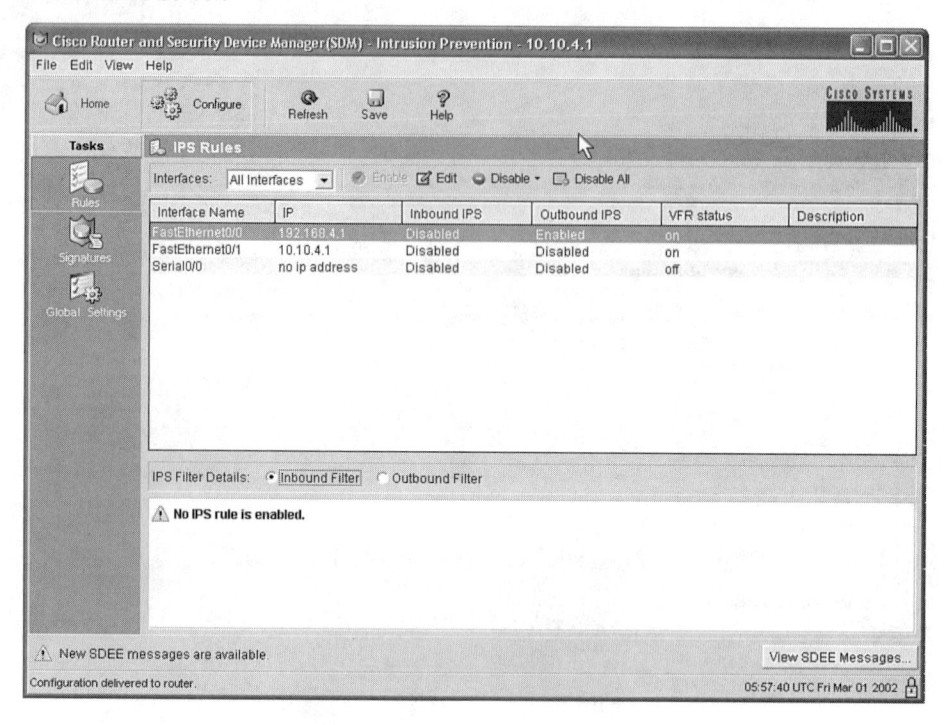

Figure 24-22 *Edit IPS on an Interface Screen*

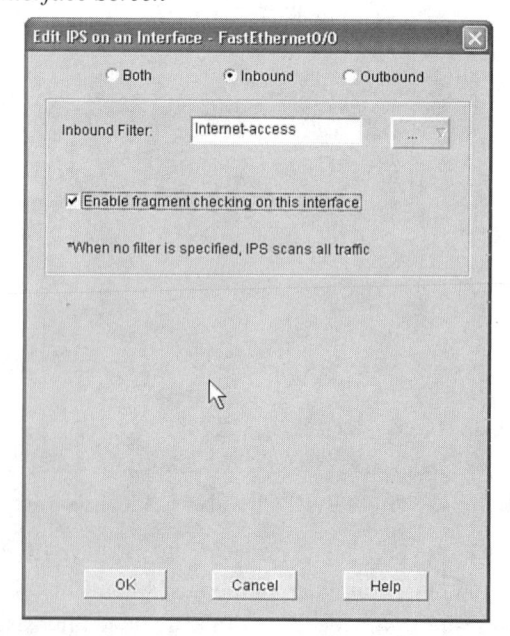

Task 14—Configure Miami Router with IPS Using SDM

Step 5 Create a new access list with the Miami LAN as the destination subnet, as shown in Figure 24-23.

Figure 24-23 *Add an Extended Rule Entry Screen*

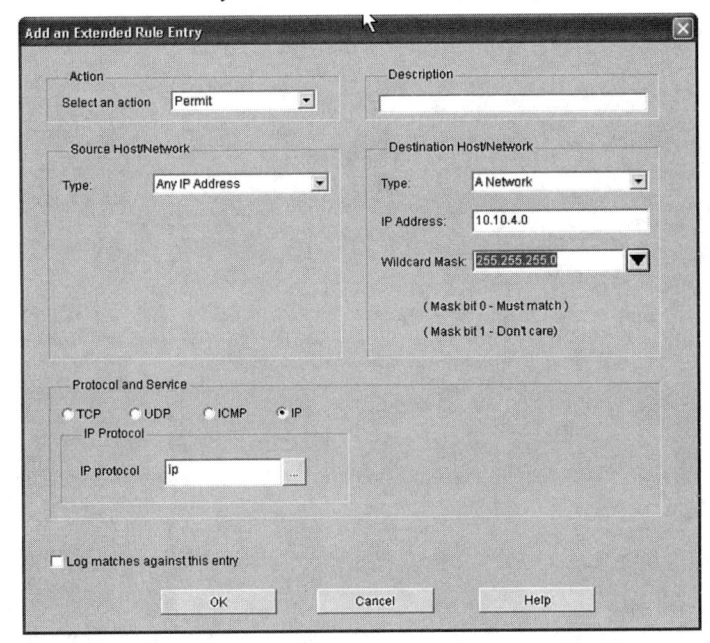

Step 6 Click **OK** twice to complete the entry and load the signatures on to the router. Figure 24-24 shows the completed Rule Entry screen.

Figure 24-24 *Completed Rule Entry Screen*

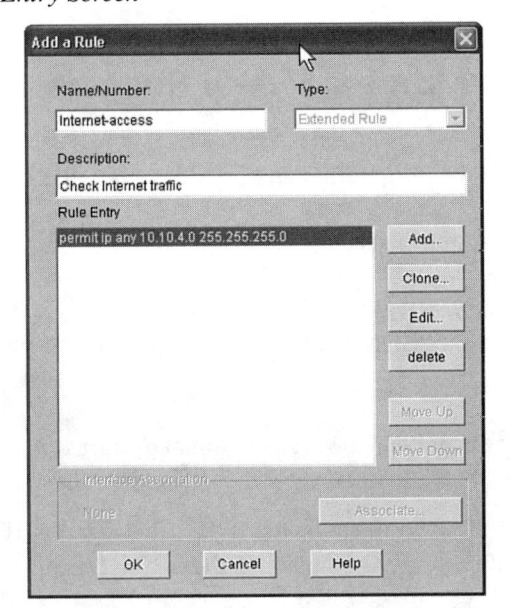

590 Chapter 24: Final Scenarios

Step 7 In the Configure window, click the **Signatures** button. Figure 24-25 shows the IPS Signatures screen.

Figure 24-25 *IPS Signatures Screen*

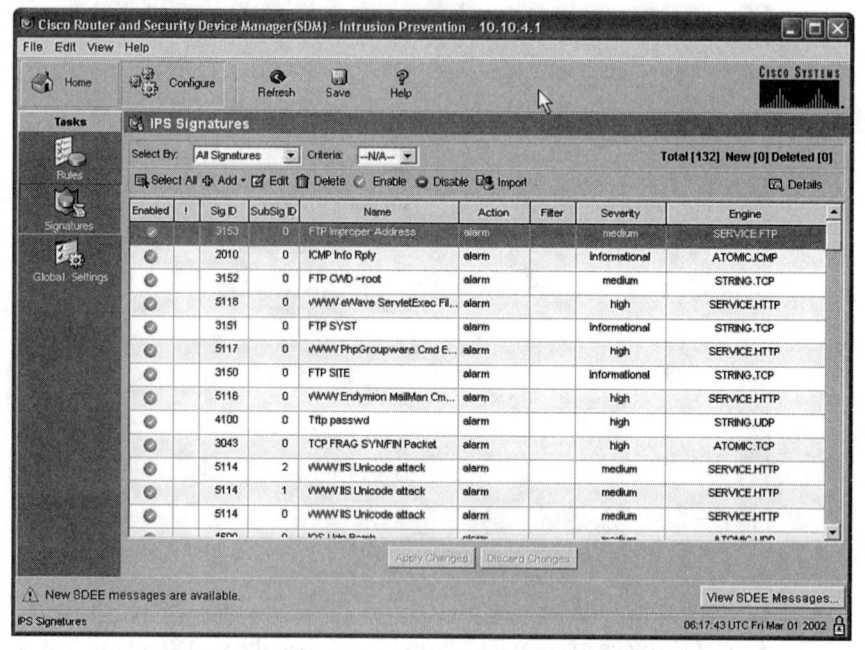

At this point, the signatures are loaded on the router's running configuration. Save this configuration by clicking the **Save** button in the IPS window to push the configuration to Miami router's startup configuration file.

Task 15—Verify and Monitor Miami Router with IPS Using SDM

This task verifies the IPS configuration and provides examples of tools available to monitor it.

To enable IPS SDEE events, you can use the CLI as follows:

```
Miami-Router(config)#ip ips notify sdee
```

Alternatively, you can use the SDM Global Settings task to enable the SDEE notification, as shown in Figure 24-26:

Step 1 In the Configure window, **select Global Settings** and then add the Configured SDF Locations for the attack-drop.sdf.

Step 2 View the SDEE notifications, and then click **View SDEE Messages** at the bottom of the screen, as shown in Figure 24-27.

Figure 24-26 *Global Settings Screen*

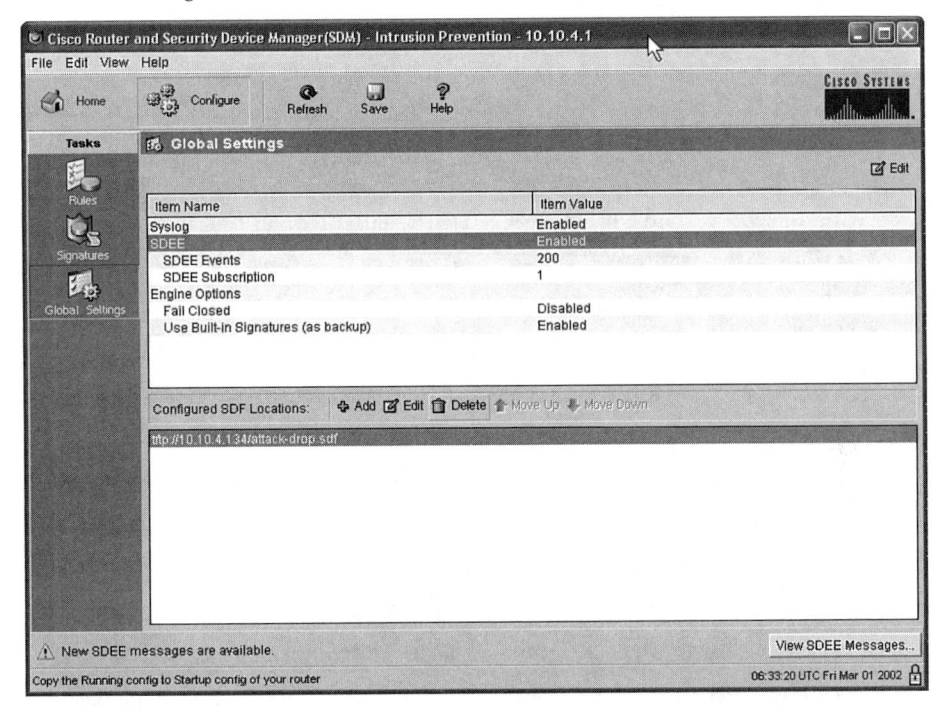

Figure 24-27 *SDEE Messages Screen*

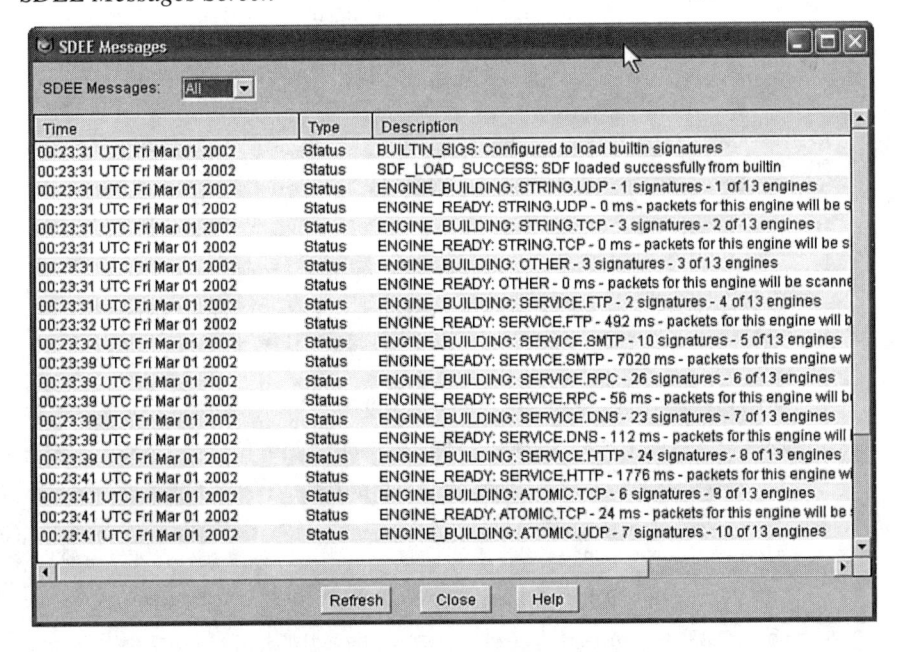

Chapter 24: Final Scenarios

You can also perform this task from the CLI with the **show ip sdee status** command:

```
Miami-Router#sho ip sdee stat
Event storage: 200 events maximum, using 131200 bytes of memory

                               SDEE Status Messages
          Time                 Message                                          Description
    1: 000 00:23:31 UTC Mar 1 2002    BUILTIN_SIGS: Configured to load builtin signatures
    2: 000 00:23:31 UTC Mar 1 2002    SDF_LOAD_SUCCESS: SDF loaded successfully from builtin
    3: 000 00:23:31 UTC Mar 1 2002    ENGINE_BUILDING: STRING.UDP - 1 signatures - 1 of 13 engines
    4: 000 00:23:31 UTC Mar 1 2002    ENGINE_READY: STRING.UDP - 0 ms - packets for this engine will be scanned
    5: 000 00:23:31 UTC Mar 1 2002    ENGINE_BUILDING: STRING.TCP - 3 signatures - 2 of 13 engines
    6: 000 00:23:31 UTC Mar 1 2002    ENGINE_READY: STRING.TCP - 0 ms - packets for this engine will be scanned
    7: 000 00:23:31 UTC Mar 1 2002    ENGINE_BUILDING: OTHER - 3 signatures - 3 of 13 engines
    8: 000 00:23:31 UTC Mar 1 2002    ENGINE_READY: OTHER - 0 ms - packets for this engine will be scanned
    9: 000 00:23:31 UTC Mar 1 2002    ENGINE_BUILDING: SERVICE.FTP - 2 signatures - 4 of 13 engines
   10: 000 00:23:32 UTC Mar 1 2002    ENGINE_READY: SERVICE.FTP - 492 ms - packets for this engine will be scanned
   11: 000 00:23:32 UTC Mar 1 2002    ENGINE_BUILDING: SERVICE.SMTP - 10 signatures - 5 of 13 engines
   12: 000 00:23:39 UTC Mar 1 2002    ENGINE_READY: SERVICE.SMTP - 6912 ms - packets for this engine will be scanned
   13: 000 00:23:39 UTC Mar 1 2002    ENGINE_BUILDING: SERVICE.RPC - 26 signatures - 6 of 13 engines
   14: 000 00:23:39 UTC Mar 1 2002    ENGINE_READY: SERVICE.RPC - 52 ms - packets for this engine will be scanned
   15: 000 00:23:39 UTC Mar 1 2002    ENGINE_BUILDING: SERVICE.DNS - 23 signatures - 7 of 13 engines
   16: 000 00:23:39 UTC Mar 1 2002    ENGINE_READY: SERVICE.DNS - 112 ms - packets for this engine will be scanned
   17: 000 00:23:39 UTC Mar 1 2002    ENGINE_BUILDING: SERVICE.HTTP - 24 signatures - 8 of 13 engines
   18: 000 00:23:41 UTC Mar 1 2002    ENGINE_READY: SERVICE.HTTP - 1768 ms - packets for this engine will be scanned
   19: 000 00:23:41 UTC Mar 1 2002    ENGINE_BUILDING: ATOMIC.TCP - 6 signatures - 9 of 13 engines
   20: 000 00:23:41 UTC Mar 1 2002    ENGINE_READY: ATOMIC.TCP - 24 ms - packets for this engine will be scanned
   21: 000 00:23:41 UTC Mar 1 2002    ENGINE_BUILDING: ATOMIC.UDP - 7 signatures - 10 of 13 engines
   22: 000 00:23:41 UTC Mar 1 2002    ENGINE_READY: ATOMIC.UDP - 8 ms - packets for this engine will be scanned
   23: 000 00:23:41 UTC Mar 1 2002    ENGINE_BUILDING: ATOMIC.ICMP - 14 signatures - 11 of 13 engines
   24: 000 00:23:41 UTC Mar 1 2002    ENGINE_READY: ATOMIC.ICMP - 12 ms - packets for this engine will be scanned
   25: 000 00:23:41 UTC Mar 1 2002    ENGINE_BUILDING: ATOMIC.IPOPTIONS - 7 signatures - 12 of 13 engines
   26: 000 00:23:41 UTC Mar 1 2002    ENGINE_READY: ATOMIC.IPOPTIONS - 4 ms - packets for this engine will be scanned
   27: 000 00:23:41 UTC Mar 1 2002    ENGINE_BUILDING: ATOMIC.L3.IP - 6 signatures - 13 of 13 engines
   28: 000 00:23:41 UTC Mar 1 2002    ENGINE_READY: ATOMIC.L3.IP - 4 ms - packets for this engine will be scanned
   29: 000 06:03:44 UTC Mar 1 2002    BUILTIN_SIGS: Configured to load builtin signatures
   30: 000 06:03:45 UTC Mar 1 2002    SDF_LOAD_SUCCESS: SDF loaded successfully from builtin
   31: 000 06:03:45 UTC Mar 1 2002    ENGINE_BUILDING: STRING.UDP - 1 signatures - 1 of 13 engines
```

```
    32: 000 06:03:45 UTC Mar 1 2002    ENGINE_READY: STRING.UDP - 0 ms - packets for this
engine will be scanned
    33: 000 06:03:45 UTC Mar 1 2002    ENGINE_BUILDING: STRING.TCP - 3 signatures - 2 of
13 engines
    34: 000 06:03:45 UTC Mar 1 2002    ENGINE_READY: STRING.TCP - 0 ms - packets for this
engine will be scanned
    35: 000 06:03:45 UTC Mar 1 2002    ENGINE_BUILDING: OTHER - 3 signatures - 3 of 13
engines
    36: 000 06:03:45 UTC Mar 1 2002    ENGINE_READY: OTHER - 0 ms - packets for this engine
will be scanned
    37: 000 06:03:45 UTC Mar 1 2002    ENGINE_BUILDING: SERVICE.FTP - 2 signatures - 4 of
13 engines
    38: 000 06:03:45 UTC Mar 1 2002    ENGINE_READY: SERVICE.FTP - 180 ms - packets for
this engine will be scanned
    39: 000 06:03:45 UTC Mar 1 2002    ENGINE_BUILDING: SERVICE.SMTP - 10 signatures - 5
of 13 engines
    40: 000 06:03:48 UTC Mar 1 2002    ENGINE_READY: SERVICE.SMTP - 3208 ms - packets for
this engine will be scanned
    41: 000 06:03:48 UTC Mar 1 2002    ENGINE_BUILDING: SERVICE.RPC - 26 signatures - 6
of 13 engines
    42: 000 06:03:48 UTC Mar 1 2002    ENGINE_READY: SERVICE.RPC - 64 ms - packets for
this engine will be scanned
    43: 000 06:03:48 UTC Mar 1 2002    ENGINE_BUILDING: SERVICE.DNS - 23 signatures - 7
of 13 engines
    44: 000 06:03:48 UTC Mar 1 2002    ENGINE_READY: SERVICE.DNS - 112 ms - packets for
this engine will be scanned
    45: 000 06:03:48 UTC Mar 1 2002    ENGINE_BUILDING: SERVICE.HTTP - 24 signatures - 8
of 13 engines
    46: 000 06:03:50 UTC Mar 1 2002    ENGINE_READY: SERVICE.HTTP - 1764 ms - packets for
this engine will be scanned
    47: 000 06:03:50 UTC Mar 1 2002    ENGINE_BUILDING: ATOMIC.TCP - 6 signatures - 9 of
13 engines
    48: 000 06:03:50 UTC Mar 1 2002    ENGINE_READY: ATOMIC.TCP - 24 ms - packets for this
engine will be scanned
    49: 000 06:03:50 UTC Mar 1 2002    ENGINE_BUILDING: ATOMIC.UDP - 7 signatures - 10 of
13 engines
    50: 000 06:03:50 UTC Mar 1 2002    ENGINE_READY: ATOMIC.UDP - 8 ms - packets for this
engine will be scanned
    51: 000 06:03:50 UTC Mar 1 2002    ENGINE_BUILDING: ATOMIC.ICMP - 14 signatures - 11
of 13 engines
    52: 000 06:03:50 UTC Mar 1 2002    ENGINE_READY: ATOMIC.ICMP - 12 ms - packets for
this engine will be scanned
    53: 000 06:03:50 UTC Mar 1 2002    ENGINE_BUILDING: ATOMIC.IPOPTIONS - 7 signatures -
12 of 13 engines
    54: 000 06:03:50 UTC Mar 1 2002    ENGINE_READY: ATOMIC.IPOPTIONS - 4 ms - packets for
this engine will be scanned
    55: 000 06:03:50 UTC Mar 1 2002    ENGINE_BUILDING: ATOMIC.L3.IP - 6 signatures - 13
of 13 engines
    56: 000 06:03:50 UTC Mar 1 2002    ENGINE_READY: ATOMIC.L3.IP - 8 ms - packets for
this engine will be scanned
Router#
```

Task 16—Configure Easy VPN Server Using SDM

Easy VPN Server requires AAA. The AAA configuration of the Miami Cisco Secure ACS server may be used in this task.

The wizard guides the user through the necessary steps to configure the Easy VPN Server on a router. The server configuration steps are as follows:

Step 1 After launching the Create Easy VPN Server task, select the interface on which the client connections will terminate. This is the interface where the VPN connections from the VPN clients will be terminated. The interface may not be participating in GRE over IPsec, DMVPN, or an Easy VPN client connection. Figure 24-28 shows this screen.

Figure 24-28 *Select an Interface Screen*

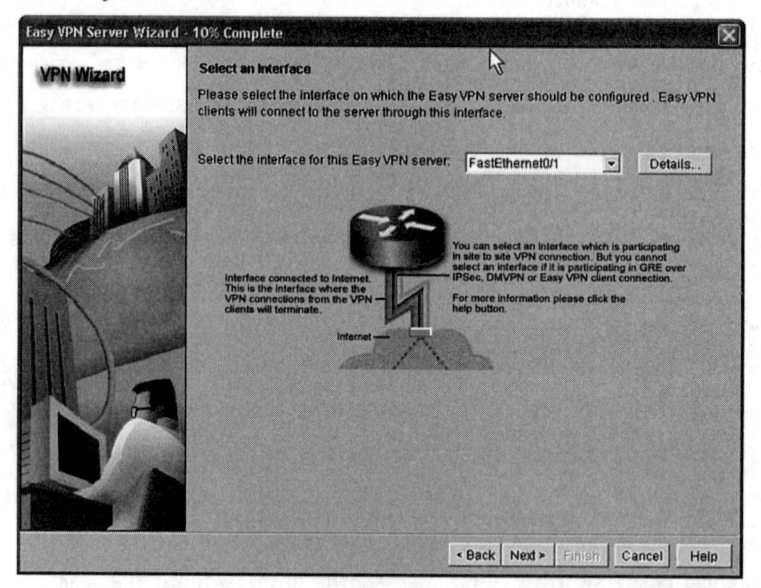

Step 2 Configure IKE proposals that specifies the encryption algorithm, authentication algorithm, and key-exchange method used by the router when negotiating a VPN connection with the remote device. Figure 24-29 shows the IKE configuration screen.

Figure 24-29 *IKE Proposal Screen*

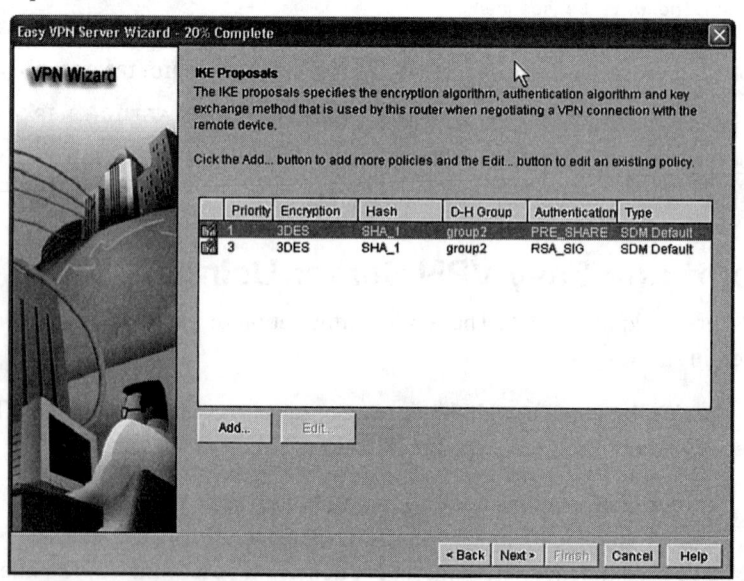

Step 3 Configure an IPsec transform set that specifies the encryption and authentication algorithms used to protect the data in the VPN tunnel. You may use the default transform sets provided in SDM, as shown in Figure 24-30.

Figure 24-30 *Transform Set Screen*

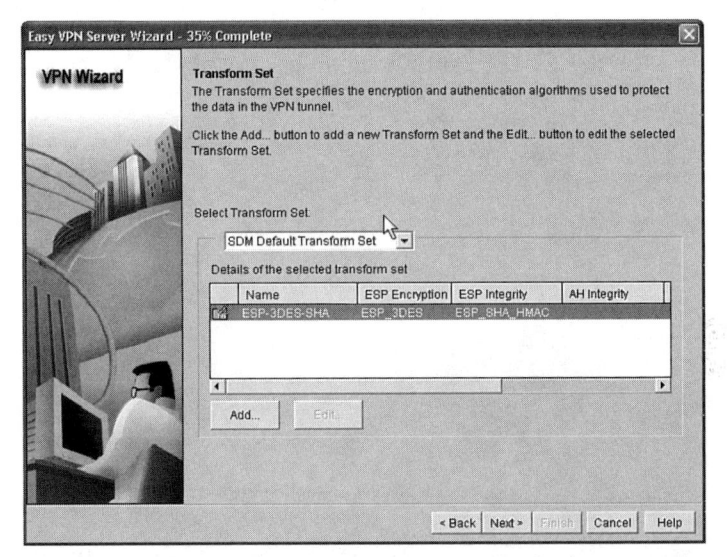

Step 4 Configure the group policy lookup method as **RADIUS and Local Only**. Specify the Miami Cisco Secure ACS server as the RADIUS server. Figure 24-31 shows the Group Authorization/Group Policy Lookup screen.

Figure 24-31 *Group Authorization/Group Policy Lookup Screen*

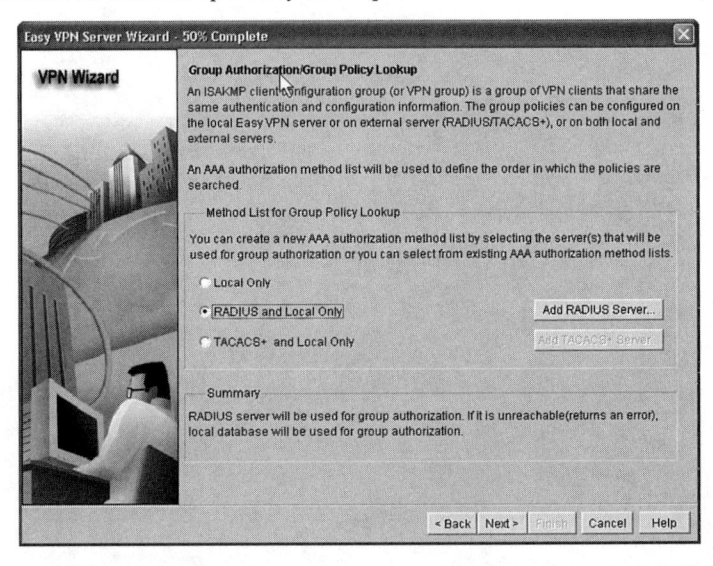

Step 5 Configure user authentication as the **Existing AAA method list**.

Chapter 24: Final Scenarios

Step 6 Configure group policies on the local router by adding the Miami Remote VPN Users group with a preshared key and the pool of IP addresses that would be assigned to clients, as shown in Figure 24-32.

Figure 24-32 *Add Group Policy Screen*

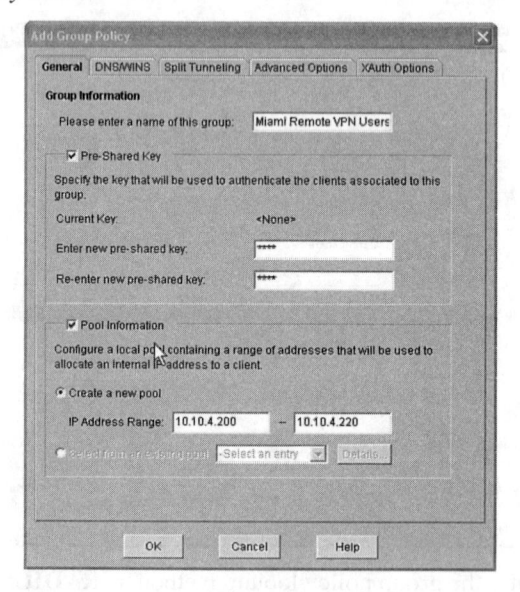

After the previously mentioned steps are completed, the summary window lists the configurations specified, as shown in Figure 24-33.

Figure 24-33 *Configuration Summary Screen*

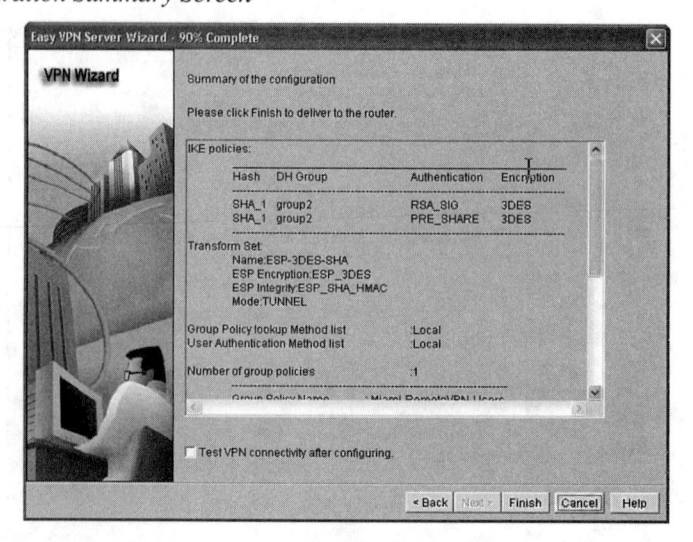

Clicking the **Finish** button pushes this information to the Miami router's running configuration.

Task 17—Configure Easy VPN Remote Using SDM

The wizard guides the user through the necessary steps to configure the Easy VPN Remote on a router as a client. The router must be running a Cisco IOS image that supports Easy VPN Phase 2. The client configuration steps are as follows:

Step 1 After launching the Create Easy VPN Remote task, select the interface on which the client connections will terminate. This is the interface where the VPN connections from the VPN clients will be terminated. The interface may not be participating in GRE over IPsec, DMVPN, or an Easy VPN client connection.

Step 2 The information that will be required to configure the remote client is the Miami router Easy VPN Server's IP address or host name, the key, and the IPsec group name. Figure 24-34 shows the Connection Information screen.

Figure 24-34 *Connection Information Screen*

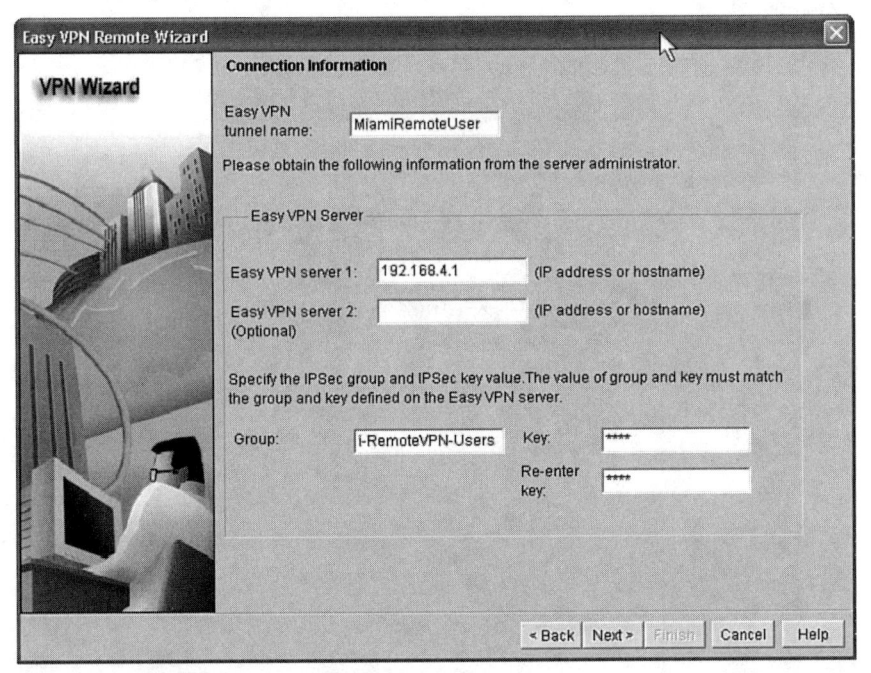

Step 3 Select **Client** for the mode of operation. Select the IPsec tunnels to be initiated automatically, as shown in Figure 24-35.

Step 4 Enter the Xauth username and password for the router. Figure 24-36 shows the User Authentication (Xauth) screen.

Figure 24-35 *Connection Characteristics Screen*

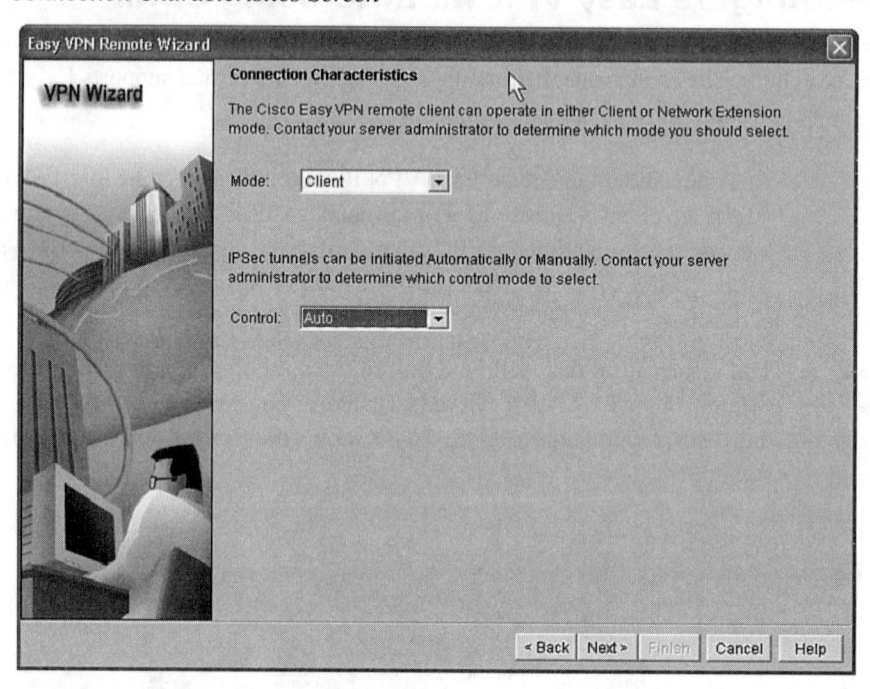

Figure 24-36 *User Authentication (Xauth) Screen*

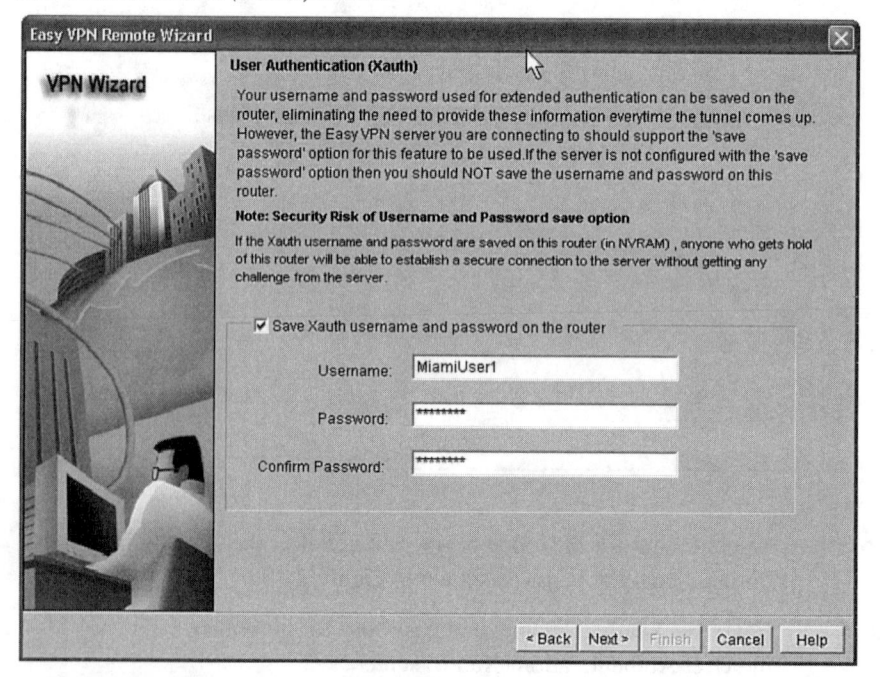

Step 5 Select the inside and outside interface for the Easy VPN client to complete the configuration steps, as shown in Figure 24-37.

Figure 24-37 *Interfaces Screen*

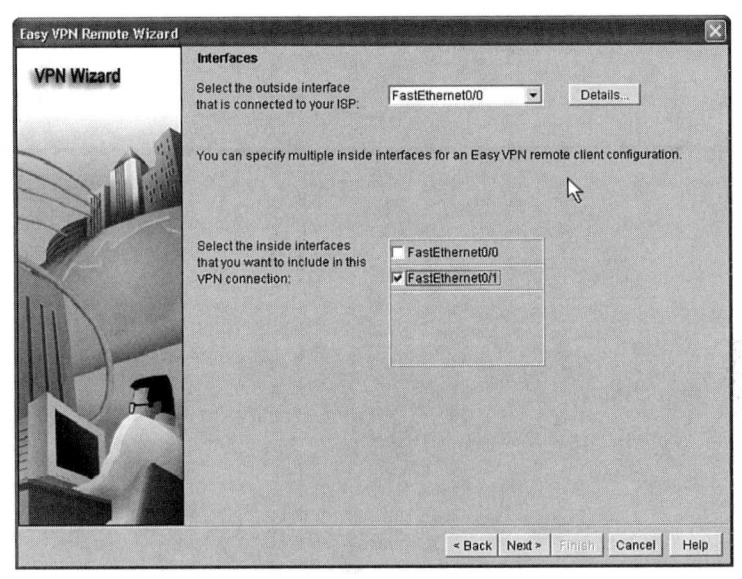

After you have completed these steps, the summary window, as shown in Figure 24-38, lists the configurations specified.

Figure 24-38 *Summary Window*

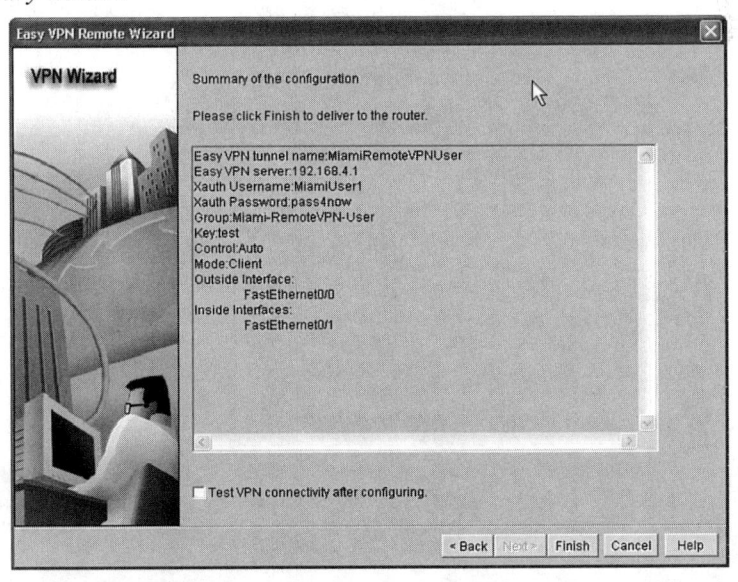

Clicking the **Finish** button pushes this information to the Miami remote client's running configuration.

Part VIII: Appendix

Appendix A Answers to the "Do I Know This Already?" Quizzes and Q&A Sections

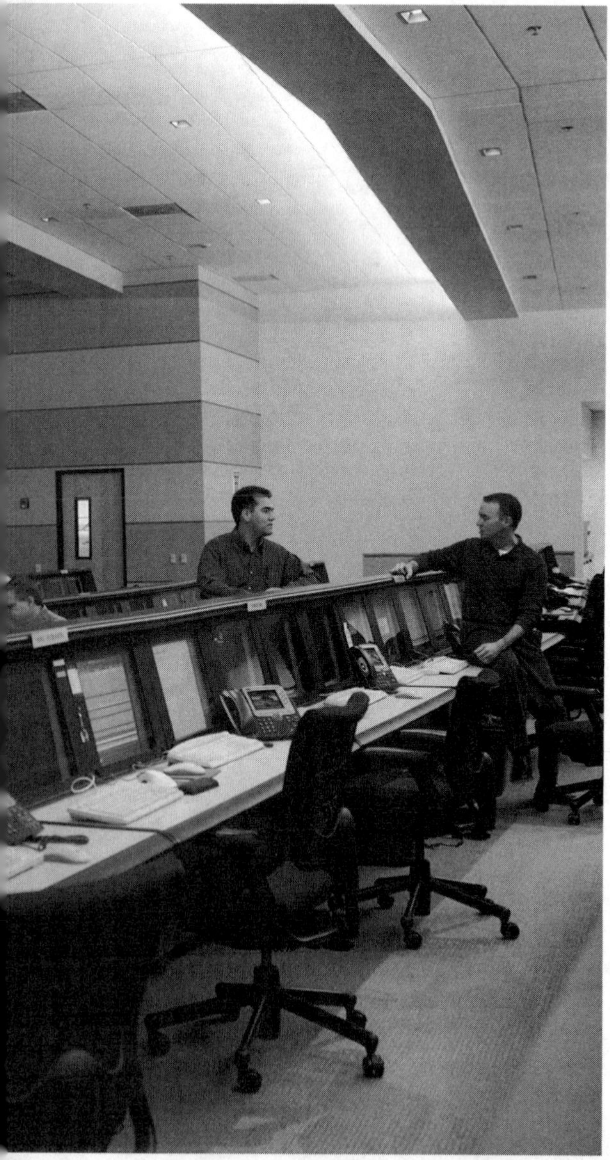

APPENDIX A

Answers to the "Do I Know This Already?" Quizzes and Q&A Sections

Chapter 1

"Do I Know This Already?"

1. f
2. b
3. e
4. c
5. f
6. b
7. a
8. c
9. d
10. a

Q&A

1. *Network security is the fine art of balancing security versus _____.*

   Answer: Business needs

2. *The policy that details information about software and hardware standards utilized within an organization and how they are installed is called _____.*

   Answer: Server and workstation configuration policy

3. *A new vulnerability has come out, and the equipment that is being managed is vulnerable. What item will help to guide your actions to resolve this vulnerability?*

   Answer: Patch management policy

4. *What is the biggest benefit and business driver for network security from a manager's perspective?*

Answer: Cost savings realization

5. *One of the goals of a security policy is to define consequences for _____. This can be in the form of management or human resources discipline, demotion, or termination.*

Answer: Violating the security policy

6. *The security policy must have support from _____ because lack of such may impede various business functions or processes.*

Answer: Management

7. *Security polices should define user access and control not by their position, but by what factor?*

Answer: Job function/role

8. *The security policy needs to be specific to define all the requirements, but it must also be _____ to respond to the industry.*

Answer: Flexible

9. *So that an organization can respond quickly and efficiently to an exposed vulnerability or a suspected compromise of a network resource, the network security policy should also contain an _____.*

Answer: Incident-response plan

10. *Network security is considered a _____ because organizations are expected to ensure that personal information is kept secure from public access or leakage. Any leakage of this information may result in legal actions.*

Answer: Legal issue

Chapter 2

"Do I Know This Already?"

1. a
2. b
3. f
4. b

5. f
6. b
7. a
8. b

Q&A

1. An operating system application weakness that allows a user to alter software or gain access to a system or network is called a _____.

   Answer: vulnerability

2. What are the three reasons that attacks can occur and damage networks?

   Answer: Lack of an effective policy, configuration weaknesses, and technology weaknesses.

3. If an organization has a high turnover rate for its system administrators, this is considered a security policy weakness because of _____.

   Answer: lack of continuity

4. Because smaller organizations have a limited IT staff, there is a greater chance that this organization would be susceptible to which kinds of configuration issues?

   Answer: Misconfigured equipment, weak or exposed passwords, misconfigured Internet services, and using default settings

5. What is one of the most common default settings overlooked by most network and security administrators?

   Answer: Passwords

6. Describe an operating system weakness?

   Answer: An operating system weakness is an inherent flaw in the operating software of a computer, often discovered by random chance or through detailed analysis. This flaw can be exploited to attack or steal data from the system.

7. What is a structured threat and why does the Federal Bureau of Investigation (FBI) consider it important?

   Answer: A structured threat is a targeted attack against a specific organization or system. The FBI looks at these attacks carefully because they are implemented typically for corporate espionage or for criminal activity.

8. *What are the five core reasons for intruding on a system or network?*

    Answer: Curiosity, fun and pride, revenge, profit, and political purposes

9. *The types of reconnaissance attacks are DNS queries, ping sweep, vertical scans, horizontal scans, and _____.*

    Answer: block scans

10. *What is a DoS attack and why is it considered so destructive?*

    Answer: A DoS attack takes away resources from valid users and denies them access. It can be destructive by causing outages and disruptions, which can be financially devastating.

Chapter 3

"Do I Know This Already?"

1. e
2. d
3. f
4. e
5. b
6. e
7. d
8. c
9. d
10. d

Q&A

1. *Setting up a layered security model for an organization helps to _____.*

    Answer: better protect network resources from complex attacks

2. *A DoS or DDoS attack against a router or firewall is used to _____.*

    Answer: force redirection of traffic or deny access to the network

3. *An _____ is the centerpiece of any organizations security implementation.*

    Answer: effective security policy

4. In addition to providing controlled access to resources, AAA implementations provide detailed _____ and _____ to help expose and mitigate risks.

   Answer: audit, accounting logs

5. The use of RFC 1918 addressing on internal networks prevents attacks that originate from the Internet unless _____.

   Answer: those segments are statically one to one NAT'd at the network perimeter

6. What is the fundamental premise behind an anomaly-based IDS/IPS?

   Answer: It establishes a baseline of valid and approved activities. Any activity that goes against this norm is considered an attack.

7. Why is monitoring important?

   Answer: To determine the state of the network, detect anomalies, and detect attacks quickly and effectively and allow for a quick response

8. Performing a network assessment would be considered part of which step of the effective security process for an organization?

   Answer: Testing

9. From where do an estimated 70 percent of the network attacks originate?

   Answer: Internal sources

10. Where should network IDS/IPS be implemented?

    Answer: At critical points on the network

Chapter 4

"Do I Know This Already?"

1. c
2. b
3. e
4. c
5. f
6. b

Appendix A: Answers to the "Do I Know This Already?" Quizzes and Q&A Sections

7. c
8. c
9. e
10. d

Q&A

1. To access the ROM monitor mode, you have to press the _____ key during the first 60 seconds of the router boot sequence.

   Answer: Break

2. From which command mode can a user access the global configuration mode and which command is utilized?

   Answer: Privileged EXEC and **configure terminal**

3. Can abbreviated commands be utilized in Cisco IOS for all commands?

   Answer: Yes, because system recognizes command based on first few letters. Abbreviated commands can be limited by the number of letters used if several command begin with the first few letters.

4. What mode does the prompt **router(config)#** command depict?

   Answer: This command shows that the device named router is in global configuration mode.

5. What is considered the most secure method for administering a router and why?

   Answer: Console connection, because it requires physical access to the system.

6. What additional steps must you take to configure the router to accept SSH connections?

   Answer: Enable the SSH server and configure the SSH parameters.

7. If you want to configure the console port with a password of @bc123, which commands would you type at the RouterA(config-line)# prompt?

   Answer: **login**
   password @bc123
   end

8. Which type of access list allows specific traffic to maintain state?

   Answer: Reflexive access list

9. *Which type of code does the firewall feature set block?*

    Answer: Java

10. *What is PAM and why is it a useful feature of the Firewall feature set?*

    Answer: Port-to-application mapping enables administrators access to specific ports for applications. This allows the use of nonstandard ports. Security is enhanced.

Chapter 5

"Do I Know This Already?"

1. b
2. b
3. b and d
4. b
5. b and c
6. a
7. d
8. a and d
9. b
10. c
11. c
12. c
13. e
14. b

Q&A

1. *Which steps can be taken to secure administrative access to a network device?*

    Answer: Password protection and enforcing privilege levels

2. *What is level 15 access called on the Cisco command-line interface?*

    Answer: Privileged EXEC mode

3. *Access to the Cisco device console can be secured using which two methods?*

    Answer: A password, local authentication database, or remote authentication database

610 Appendix A: Answers to the "Do I Know This Already?" Quizzes and Q&A Sections

4. Can the same password be used for the enable password and enable secret password?

   Answer: Yes, but you will get a warning indicating that it's not recommended.

5. If you set the privilege level of the **show ip route** command, what other effects need to be considered?

   Answer: All **show** and **show ip** commands are set to the same privilege level.

6. What are the two modes of interactive access to lines on a Cisco router, and what is a good rule of thumb for these connections?

   Answer: Vty and tty are the two modes of interactive access to lines. A good rule of thumb is to us an authentication mechanism, especially with lines connected to modems or untrusted access points.

7. Which command enables you to implement secured vty access using only the **telnet** and **ssh** commands?

   Answer: **transport input telnet ssh** command

8. Describe the two methods for configuring SSH on a Cisco router or switch.

   Answer: A user can configure SSH by first using a host name and domain name in the configuration and then generating keys. The second is by using RSA key pairs and by generating encryption keys.

9. If an incorrect MAC address is detected on a switch that has port security enabled, the port can be configured to go to _____.

   Answer: shutdown, protect, or restrictive mode

10. What does the command **set port security 2/4 enable 00-B0-34-03-2C-01** do on a Cisco switch?

    Answer: Enables port security on port 2/4, and allows 00-B0-34-03-2C-01 as the secure MAC address

Chapter 6

"Do I Know This Already?"

1. a and c
2. a and c
3. e

4. a
5. c
6. c
7. a
8. d
9. a
10. a

Q&A

1. *Which port is reserved for TACACS+ use?*

   Answer: TCP port 49

2. *Which versions of the TACACS protocol in Cisco IOS Software have officially reached end-of-maintenance?*

   Answer: TACACS and XTACACS

3. *In the RADIUS security architecture, what is the network access server?*

   Answer: Authenticator

4. *Which method of strong authentication for client/server applications utilizes secret-key cryptography.*

   Answer: Kerberos

5. *Who developed and designed the Kerberos authentication protocol?*

   Answer: MIT

6. *Which two popular authentication methods does PPP support?*

   Answer: PAP and CHAP

7. *Why is PAP considered insecure compared to other authentication protocols, such CHAP and MS-CHAP?*

   Answer: In PAP, the authenticated username and password are sent across the link in clear text.

8. *Which type of encryption algorithm does CHAP uses during the three-way handshake?*

   Answer: One-way hash algorithm (typically MD5)

612 Appendix A: Answers to the "Do I Know This Already?" Quizzes and Q&A Sections

9. *Give one difference between CHAP and MS-CHAP?*

Answer: MS-CHAP is enabled by negotiating CHAP algorithm 0x80 in LCP option 3, Authentication Protocol.

The MS-CHAP response packet is in a format designed to be compatible with Windows. This format does not require the authenticator to store a clear or reversibly encrypted password.

MS-CHAP provides an authenticator-controlled authentication retry mechanism.

MS-CHAP provides an authenticator-controlled change-password mechanism.

MS-CHAP defines a set of reason-for-failure codes returned in the failure packet's Message field.

10. *Where is EAP used in Cisco devices?*

Answer: Wireless LANs, which use EAP to authenticate wireless clients to a centralized authentication server, such as RADIUS
IEEE 802.1x, which uses EAP for port-based network access control of clients to a centralized authentication server
Remote access, which uses EAP to authenticate remote-access users using PPP

Chapter 7

"Do I Know This Already?"

1. c
2. d
3. b
4. c
5. d
6. a and b
7. b
8. e
9. b
10. c

Q&A

1. *Which command enables AAA on a router/NAS?*

   Answer: **aaa new-model** command

2. *Which of the AAA services can be used for billing and auditing?*

   Answer: Accounting

3. *What is the difference between console and network AAA authorization supported on the Cisco IOS Software?*

   Answer:

   network—Applies to network connections. This can include a PPP, SLIP, or ARAP connection.

   console—Applies to the authorization of console line. This command must be used in conjunction with the authorization command on the console port.

4. *Which AAA command would you use to configure authentication for login to an access server?*

   Answer: **auth-proxy** command

5. *Where is authorization information stored for each user?*

   Answer: Locally or on the ACS

6. *Which command enables you to troubleshoot a AAA accounting problem?*

   Answer: **debug aaa accounting** command

7. *What types of auditing methods can be specified in a AAA configuration?*

   Answer:

   network—Provides information for all PPP, SLIP, or ARAP sessions, including packet and byte counts.

   exec—Provides information about user EXEC terminal sessions of the NAS.

   commands—Provides information about the EXEC mode commands that a user issues. Command accounting generates accounting records for all EXEC mode commands, including global configuration commands, associated with a specific privilege level.

connection—Provides information about all outbound connections made from the NAS, such as Telnet.

system—Provides information about system-level events.

resource—Provides start and stop records for calls that have passed user authentication, and provides stop records for calls that fail to authenticate.

8. *What is the difference between a FAIL response and an ERROR response in a AAA configuration?*

    Answer: FAIL means authentication or authorization is denied, ERROR means the authentication or authorization did not complete.

9. *How do you display all the detailed accounting records for actively accounted functions?*

    Answer: **debug aaa accounting** command

10. *What command disables AAA functionality on your access server?*

    Answer: **no aaa new-model** command

Chapter 8

"Do I Know This Already?"

1. a
2. b
3. b
4. a
5. a
6. a, c, and d
7. c
8. b
9. a and d
10. c

Q&A

1. *What command specifies a TACACS server?*

   Answer: **tacacs-server host** command

2. *What is the purpose of the **radius-server retry method reorder** command?*

   Answer: During periods of high load or when server failure occurs, the RADIUS Server Reorder on Failure feature provides for failover to another server in the server group, which speeds up authentication.

3. *What purpose does the **tacacs-server key** command serve?*

   Answer: The **tacacs-server key** command specifies the encryption key that will be used.

4. *What is the purpose of the keyword **local** in the following configuration line?*

   aaa authentication ppp test1 tacacs local

   Answer: Keyword **local** indicates that the authentication is attempted using the local database on the router if the TACACS server returns an error.

5. *Is it possible to change the default port used by RADIUS authentication?*

   Answer: Yes. The **radius-server host** command can change the default port.

6. *What command deletes a RADIUS server configuration?*

   Answer: **no-radius server host** command

7. *What command enables network-level authorization to use a TACACS+ server?*

   Answer: **aaa authorization network list group tacacs** command

8. *Which testing and verifying command used for TACACS+ produces a substantial amount of output?*

   Answer: **debug tacacs events** command

9. *What default port is reserved for TACACS?*

   Answer: TCP 49

10. *Is it possible to have both RADIUS and TACACS configured on a single router/NAS?*

    Answer: A single router can have both a RADIUS and TACACS server configured.

Chapter 9

"Do I Know This Already?"

1. a and b
2. a
3. a, b, and c
4. d
5. a
6. a, c, and d
7. d
8. c
9. b
10. b and e
11. a, c, and d
12. b

Q&A

1. *Where does the Cisco Secure ACS write its accounting records?*

    Answer: Cisco Secure ACS writes accounting records to a CSV log file or ODBC database.

2. *Give one example of the user repository that Cisco Secure ACS supports?*

    Answer: Windows AD, generic LDAP, Novell NDS, CRYPTOCard token server, SafeWord token server, PassGo token server, RSA SecureID token server, AXENT, LEAP proxy agent, Secure Computing Safeword, ActivCard token server, and Vasco token server

3. *Give one advantage of using Cisco Secure ACS?*

    Answer: Centralized access control and accounting

4. *Give two examples of the password protocols that Cisco Secure ACS supports?*

    Answer: Cisco Secure ACS supports many common password protocols, including EAP-CHAP, EAP-TLS, LEAP, ARAP, ASCII/PAP, CHAP, and MS-CHAP.

5. *Roger is a network administrator at an engineering firm. He would like to restrict access to consultants during the weekend. Can Cisco Secure ACS help him?*

    Answer: The Cisco Secure ACS access-restrictions feature enables Roger to deny logins based on day of the week.

6. *What are the core services of Cisco Secure ACS 3.3?*

    Answer: CSAdmin, CSAuth, CSDBSync, CSLog, and CSMon

7. *What is the function of CSAdmin?*

    Answer: CSAdmin is the service that provides the HTML interfaces for Cisco Secure ACS.

8. *What are the benefits of database replication?*

    Answer: Replicate parts of the primary Cisco Secure ACS configuration; create schedules and timing for the replication process; export selected configuration from the primary Cisco Secure ACS; transport selected configuration data from primary to secondary Cisco Secure ACS servers securely

9. *Which core service of the Cisco Secure ACS for Windows provides synchronization with external RDBMS applications?*

    Answer: The CSDBSync service provides synchronization of the Cisco Secure user database with an RDBMS application.

10. *Name two types of accounting logs generated by Cisco Secure ACS?*

    Answer: Administrative accounting and RADIUS accounting

Chapter 10

"Do I Know This Already?"

1. a, b, and c
2. b
3. a, b, c, and d
4. a and b
5. c
6. c
7. c
8. c
9. d
10. b

Q&A

1. *Which factors should you consider when deploying Cisco Secure ACS?*

   Answer:

   - Number of users
   - Network topology
   - Access policy
   - Network latency and reliability
   - Remote-access policy
   - Administrative-access policy

2. *What are the minimum hardware requirements to install Cisco Secure ACS?*

   Answer:

   - Pentium III processor, 550 MHz or faster.
   - 256 MB of RAM.
   - At least 250 MB of free disk space. If you are running your database on the same machine, more disk space is required.
   - Minimum graphics resolution of 256 colors at 800 by 600 lines.

3. *How does Cisco Secure ACS provide control for remote-access policies?*

   Answer: Cisco Secure ACS provides control by using a central authentication and authorization of a remote user.

4. *Where would be a good place to start to troubleshoot Cisco Secure ACS related AAA problems?*

   Answer: A good place to start troubleshooting Cisco Secure ACS–related AAA problems is checking the Failed Attempts Report under Reports and Activity.

5. *Does a browser using a proxy server have any effect in the administration of a Cisco Secure ACS remotely?*

   Answer: Yes, it does. If the browser used for an administrative session is configured to use a proxy server, Cisco Secure ACS sees the administrative session originating from the IP address of the proxy server rather than from the actual address of the computer and will not work.

Chapter 11

6. *What should you check when there are authentication failures in the failed attempts report against a Windows 2000 user database?*

   Answer: Verify whether Cisco Secure ACS is configured to authenticate to the Windows 2000 user database. Verify whether the correct username and password is being used. Confirm the existence of the username. Check whether the user account has User Must Change Password at Next Login selected. If this option is selected, deselect it. Confirm that Cisco Secure ACS is configured to reference the Grant Dial-In permission to User.

   Verify whether the retry interval is too brief.

7. *Which protocol and port number does UCP use for the web application?*

   Answer: TCP port 2000

Chapter 11

"Do I Know This Already?"

1. b
2. a, b, d, and e
3. a
4. c
5. b
6. a and c
7. d
8. b and d
9. b
10. b

Q&A

1. *Describe one of the functions of an ACL.*

   Answer: To control the transmission of packets coming in to or out from an interface

   To control virtual terminal line access

   To restrict contents of routing updates

   To define interesting traffic

2. **What is the difference between a standard and extended ACL?**

    Answer: Standard ACLs use source addresses for matching operations, whereas enhanced ACLs use source and destination addresses for matching operations and optional protocol type information for finer granularity of control.

3. **Which types of ACLs do the corresponding number identifiers of 150, 750, and 1400 implement?**

    Answer: Extended IP, Ethernet address, and standard IP

4. **What command sets an ACL that will deny inbound traffic from 192.168.10.0/24 from interface E0/0?**

    Answer:

    ```
    Firewall(config)#access-list 1 deny 192.168 10.0 0.0.0.255
    Firewall(config)#interface Ethernet 0/0
    Firewall(config-if)#ip address 192.168.10.0 255.255.255.0
    Firewall(config-if)#ip access-group 1 in
    ```

5. **Where is the traffic filtered when an ACL is configured with the "out" command?**

    Answer: When traffic has already been through the router and is on its way out

6. **What command enables you to set the ACL time to limit Telnet access to Sunday evenings between 8 p.m. and 10 p.m.?**

    Answer:

    ```
    Firewall(config-time-range)#periodic Sunday 20:00 to 22:00
    ```

7. **For security reasons, why is SNMPv1 ill suited for configuring devices that are directly connected to the Internet?**

    Answer: Because it uses clear-text strings, and most implementations send messages repeatedly

8. **Why would the following command be issued?**

    ```
    Firewall(config)#ip http secure-server
    ```

    Answer: This command enables HTTPS so that users administering the system via the web browser are secured via SSL.

9. **Why is it a good practice to limit directed broadcasts on the Cisco devices?**

    Answer: Other devices may allow crafted packets that target the broadcast range (for example, a Smurf attack).

10. *Can routing protocols utilize ACLs? If so, what is the logic behind it?*

    Answer: Yes. Primarily to protect the routing tables from spoof attacks or rogue device updates that can corrupt the normal routing operation of the device

Chapter 12

"Do I Know This Already?"

1. f
2. f
3. d
4. f
5. a, c, and d
6. e
7. c
8. a
9. f
10. e

Q&A

1. *The LLQ is used to provide QoS for what type of encrypted traffic type?*

    Answer: Voice and video

2. *The Cisco IOS Firewall provides URL filtering support, Context-Based Access Control, and complete vulnerability detection and prevention. True or false?*

    Answer: False. The IOS Firewall does not provide complete vulnerability detection and prevention, just limited DoS.

3. *The authentication proxy dynamically opens connections on the inbound ACL of the input interface where the proxy is enabled, as well as on the outbound ACL of the output interface where the packet exits. True or false?*

    Answer: True

4. *What command must you use to enable the audit trail of firewall messages?*

    Answer: **ip inspect audit-trail** command

5. Can system-defined PAM entries be modified to allow Telnet through port 80?

   Answer: Yes, but only for host-specific PAM.

6. Mapping port 9000 to run Telnet services via the PAM table can be done using which kind of mapping?

   Answer: User-defined mapping

7. Which command must you use to configure PAM?

   Answer: **ip port-map** command

8. What does the list option in the **port mapping** command give?

   Answer: Maps ports to ACL

9. The Cisco IOS Firewall feature set provides DoS protection. True or False?

   Answer: True

10. The Cisco IOS Firewall was designed for which type of network?

    Answer: Smaller networks (to provide cost savings)

Chapter 13

"Do I Know This Already?"

1. d
2. c
3. c
4. d
5. b
6. c
7. b
8. a
9. d
10. b
11. b
12. b and d

Q&A

1. *What three actions are performed by the Cisco IOS IPS when malicious traffic is discovered?*

   Answer: Notify, drop, or reset

2. *Why would you want to disable a signature?*

   Answer: To prevent alarms based on authorized traffic

3. *What is SDEE?*

   Answer: Security Device Event Exchange (SDEE) is an ICSA standardized IPS communications protocol and message format.

4. *Which type of signatures create greater load on the router performance?*

   Answer: Compound signatures

5. *How do you exclude a signature?*

   Answer: Using the **ip ips signature** command and an ACL

6. *What is contained in SDF?*

   Answer: The SDF contains the signature definitions and configurable actions for each signature. This file is in XML format.

7. *What notification types can you configure up to Cisco IOS Software Release 12.3(14)T?*

   Answer: Cisco IOS logging, POP, and SDEE

8. *What is the benefit of the new Cisco IOS IPS?*

   Answer: Expanded signature capability, parallel signature scanning, and extended ACL support

9. *What default action does Cisco IOS take on incoming packets while compiling new signatures for a particular engine?*

   Answer: Cisco IOS passes the packets without scanning for that particular engine.

10. *What tools can you use to configure the Cisco IOS IPS?*

    Answer: CLI, SDM, and CiscoWorks VMS IDS

Chapter 14

"Do I Know This Already?"

1. b
2. b
3. d
4. c
5. b
6. a
7. a and b
8. b
9. b
10. b
11. a, c, and e

Q&A

1. *What are the most common types of Layer 2 attacks?*

    Answer: CAM table overflow, VLAN hopping, STP manipulation, MAC address spoofing, attacks on private VLANs, and DHCP "starvation"

2. *Describe the CAM table overflow attack.*

    Answer: In a CAM table overflow attack, an attacker sends thousands of bogus MAC addresses from one port, which looks like valid hosts' communication to the switch. The goal is to flood the switch with traffic by filling the CAM table with false entries. Once flooded, the switch broadcasts traffic without a CAM entry on its local VLAN, thus allowing the attacker to see other VLAN traffic that would not display otherwise.

3. *Explain the three categories of action that can be taken when a port security violation occurs.*

    Answer:

    - **Protect**—If the number of secure MAC addresses reaches the limit allowed on the port, packets with unknown source addresses are dropped until a number of MAC addresses are removed or the number of allowable addresses is increased. You receive no notification of the security violation in this type of instance.

- **Restrict**—If the number of secure MAC addresses reaches the limit allowed on the port, packets with unknown source addresses are dropped until some number of secure MAC addresses are removed or the maximum allowable addresses are increased. In this mode, a security notification is sent to the SNMP server (if configured), and a syslog message is logged. The violation counter is also incremented.

- **Shutdown**—If a port security violation occurs, the interface changes to error-disabled and the port LED will be turned off. It sends an SNMP trap, logs to a syslog message, and increments the violation counter.

4. *When a secure port is in the error-disabled state, how can it be brought out of this state?*

   Answer: When a secure port is in the error-disabled state, you can bring it out of this state by entering the **errdisable recovery cause psecure-violation** global configuration command, or you can manually reenable it by entering the **shutdown** and **no shutdown** interface configuration commands.

5. *How can you mitigate VLAN hopping attacks?*

   Answer: Mitigating VLAN hopping attacks requires the following configuration modifications:

   - Always use dedicated VLAN IDs for all trunk ports.
   - Disable all unused ports and place them on an unused VLAN.
   - Set all user ports to nontrunking mode by disabling DTP. Use the **switchport mode access** command in the interface configuration mode.
   - For backbone switch-to-switch connections, explicitly configure trunking.
   - Do not use the user native VLAN as the trunk port native VLAN.
   - Do not use VLAN 1.

6. *What is involved in an STP attack?*

   Answer: An STP attack involves an attacker spoofing the root bridge in the topology. The attacker broadcasts the STP configuration/topology change BPDU in an attempt to force an STP recalculation. The BPDU sent out announces that the attacker's system has a lower bridge priority. The attacker can then see a variety of frames forwarded from other switches to it. STP recalculation may also cause a DoS condition on the network by causing an interruption of 30 to 45 seconds each time the root bridge changes.

7. *How does MAC spoofing–man-in-the-middle attacks work?*

   Answer: MAC spoofing involves the use of a known MAC address of another host authorized to access the network. The attacker attempts to make the target switch forward frames destined for the actual host to the attacker device instead.

Appendix A: Answers to the "Do I Know This Already?" Quizzes and Q&A Sections

8. *How can you mitigate MAC spoofing attacks?*

    Answer: Use the **port-security** command to specify MAC addresses connected to particular ports. Use DHCP snooping to filter untrusted DHCP messages. Use DAI to intercept and verify IP-to-MAC address bindings and discard invalid ARP packets.

9. *Describe how a proxy attack bypasses access restrictions of private VLANs.*

    Answer: In a proxy attack, frames are forwarded to a host on the network connected to a promiscuous port, such as a router. The network attacker sends a packet with its source IP and MAC address and a destination IP address of the target system but the destination MAC address of the router, instead of the target system.

    The switch forwards the frame to the router. The router routes the traffic, rewrites the destination MAC address as that of the target, and sends the packet out. Because the router is authorized to communicate with the private VLANs, the packet is forwarded to the target system.

10. *Explain how a DHCP starvation attack is performed.*

    Answer: A DHCP starvation attack works by broadcasting DHCP requests with spoofed MAC addresses. This scenario is achieved with attack tools such as gobbler. Gobbler looks at the entire DHCP scope and tries to lease all the DHCP addresses available in the DHCP scope. This is a simple resource starvation attack, similar to a SYN flood attack.

Chapter 15

"Do I Know This Already?"

1. c
2. a
3. b
4. d
5. a and d
6. a, b, and c
7. b, c, and d
8. b
9. a
10. a
11. a
12. a, b, and c

Q&A

1. *What are the steps in the CBAC configuration process?*

   Answer: Pick an interface, configure an IP access list at the interface, configure global timeouts and thresholds, define an inspection rule, and apply the inspection rule to an interface.

2. *Are inspection rules a requirement for CBAC configuration?*

   Answer: Yes. Inspecting rules is a mandatory requirement for CBAC configuration.

3. *What three categories of **debug** commands are commonly used to debug CBAC configuration?*

   Answer: The three categories for debugging CBAC configurations are generic, transport level, and application level.

4. *Can CBAC be configured to inspect all TCP, UDP, and ICMP packets?*

   Answer: No. CBAC is available only for TCP, UDP, and common ICMP-type protocol traffic.

5. *Which command enables you to show a complete CBAC inspection configured on the Cisco IOS Firewall?*

   Answer: **show ip inspect config** command

6. *Which command do you use to turn on audit trail messages?*

   Answer: **ip inspect audit trail** command

7. *What indicators in half-open sessions does CBAC measure before it takes steps to prevent a DoS attack?*

   Answer: CBAC measures both the total number of existing half-open sessions and the rate of session establishment attempts.

8. *Does CBAC block malicious Java applets that are in JAR format?*

   Answer: No. CBAC cannot block any Java applet that is wrapped in a ZIP or JAR format.

9. *Name two features of the CBAC.*

   Answer: Some of CBAC features include securing per-application access control, filtering on generic TCP UDP packets, detecting and preventing DoS attacks, and generating real-time alerts and audit trails.

10. *Name one restriction with using CBAC.*

    Answer:

    Some of the restrictions when using CBAC include the following:

    - Packets with the firewall as the source or destination address are not inspected by CBAC.

- If you reconfigure your ACLs when you configure CBAC, be aware that if your ACLs block TFTP traffic into an interface, you will not be able to netboot over that interface. (This limitation is not a CBAC specific, but is part of existing ACL functionality.)

- CBAC is available only for IP and common ICMP-type protocol traffic. Only TCP and UDP packets are inspected. Other IP traffic, such as ICMP, cannot be inspected with CBAC and should be filtered with extended ACLs instead.

- H.323v2 and RTSP inspection supports only the following multimedia client/server applications: Cisco IP/TV, RealNetworks RealAudio G2 Player, and Apple QuickTime 4.

11. *What are half-open sessions, and how does CBAC mitigate to many half-open sessions?*

    Answer: A half-open session is a session open in only one direction. For TCP traffic, this might mean that the session did not complete the three-way handshake. For UDP traffic, this means that return traffic was not detected.

    CBAC can measure the rate and the active number of half-open sessions several times per minute. If the number of sessions or the rate of the new connection attempts increases above the threshold, the software deletes the half-open sessions. This deletion process continues until the rate drops below the threshold.

    You can set the thresholds for the number and rate of half-open sessions by using the **max-incomplete high**, **max-incomplete low**, **one-minute high**, and **one-minute low** commands.

12. *RTSP may use many different data transport modes. What transport modes does CBAC support?*

    Answer:

    Real Data Transport (RDT)—RDT is a proprietary protocol developed by RealNetworks for data transport. It uses RTSP for communications control and RDT for data connection and retransmission of lost packets.

    Interleaved (tunnel mode)—RTSP uses the control channel to tunnel RTP or RDT traffic.

    Synchronized Multimedia Integration Language (SMIL)—SMIL is a simple HTML-like language that enables creation of interactive multimedia presentations consisting of multiple elements of music, voice, images, text, video, and graphics. It is a proposed specification of the W3C.

    Real-Time Transport Protocol (RTP)—RTP is the transport protocol for RTSP. RTP uses UDP packet format for transmitting real-time data. You can use it in conjunction with RTCP to get feedback on quality of data transmission and information about participants in a streaming session.

13. *What performance improvements does the new release of Cisco IOS Software CBAC provide?*

    Answer:

    Layer 4 processing—The code path for connection initiation and teardown was rewritten, and enables quicker creation of connections per second and reduces CPU utilization per connection.

    Hash table function—Dynamic configuration of the hash table size, from 1K to 8K, and ensures better distribution of the hash function.

    Application module tuning—By improving the connections per seconds for a session, it enhances the performance of the associated application.

Chapter 16

"Do I Know This Already?"

1. a, b, c, and d
2. e
3. c
4. f
5. a, c, and d
6. a
7. a, d, e, and f
8. b
9. d
10. a

Q&A

1. *What is the job of the authentication proxy?*

    Answer: Enables administrators the ability to restrict access to resources on a per-user basis and tailor privileges of each individual user.

2. *After the authentication is completed, the Cisco IOS Firewall receives authorization from the AAA server in what form?*

    Answer: Dynamic access list

630   Appendix A: Answers to the "Do I Know This Already?" Quizzes and Q&A Sections

3. What keyword must be added to the following AAA authentication command to enable the authentication proxy?

   aaa authorization xxxxxxxxxx default

   Answer: auth-proxy

4. What does the following command do?

   router1(config)#**aaa accounting auth-proxy default start-stop group tacacs+**

   Answer: Activates the authentication proxy accounting

5. What command enables you to create an authentication proxy name called testauth?

   Answer: **ip auth-proxy name testauth http** command

6. Through the user of the Cisco Secure ACS, TACACS+ and RADIUS authentication are supported for authentication proxy. True or false?

   Answer: True

7. Does the authentication proxy support concurrent user access to the same host?

   Answer: No, access is granted to the first user accessing it.

8. What is the command for enabling authentication proxy using HTTPS?

   Answer: **ip http secure-server** command

9. Authentication proxy is configured as a new service in the Cisco Secure ACS in which window?

   Answer: Interface configuration window, TACACS Services window

10. Can authentication proxy be trigger by connecting to the Cisco IOS Firewall?

    Answer: No, the connection must be made to the destination.

Chapter 17

"Do I Know This Already?"

1. c
2. a, c, and d
3. a, c, and f
4. a and e

5. d

6. b

Type	Methods
2. Challenge-response	A. EAP-MD5, LEAP, and EAP-MSCHAPv2
1. Cryptographic	B. EAP-TLS
4. Tunneling-based	C. PEAP, EAP-TTLS, and EAP-FAST
3. Other	D. EAP-GTC

7. c

8. a

9. b

10. a and c

11. c

Q&A

1. *What benefits does IBNS provide?*

Answer: Improved user flexibility and mobility, increased network and resource connectivity, reduced operating cost and user productivity, and timely access to resources.

2. *What enhancements to the IEEE 802.1x standard are offered by Cisco?*

Answer: VLAN assignments, 802.1x guest VLAN, port security, voice VLAN ID, high availability, and access control list (ACL) assignment.

3. *Explain the parts of the EAP frame format.*

Answer: The Code field is 1 octet and identifies the type of EAP packet. EAP only defines codes 1 through 4. Any EAP packets with other codes must be silently discarded. The Identifier field is 1 octet and aids in matching Responses with Requests. The Length field is 2 octets and indicates the length, in octets, of the EAP packet including the Code, Identifier, Length, and Data fields. The Data field is 0 or more octets. The Code field determines the format of the Data field.

4. *Describe the method EAP-MD5 uses for challenge-response authentication.*

Answer: Using this method, the client identity is transmitted over the network but the password is not sent. The server generates a random string and sends it to the user as a challenge. The client MD5 hashes the challenge using their password as the key. The server then authenticates the subscriber by verifying the user's MD5 hash password.

5. **What benefits may be realized as a result of combining IEEE 802.1x and ACS?**

    Answer: User accounting and auditing with the ability to track and monitor users on the LAN; strong authentication using a variety of authentication methods such as smart cards, tokens, PKI, and external user databases; flexible policy assignments, such as per-user VLANs; and granular control of supplicants on the LAN via centralized means.

6. **What is the purpose of the guest VLAN?**

    Answer: This is an important feature in environments where there is a mix of 802.1x clients and clients that do not support 802.1x. Using this feature, the non-802.1x-compatible users or devices gain access to network resources via the guest VLAN.

7. **Describe the concept of mutual authentication used by Cisco LEAP.**

    Answer: Cisco LEAP uses the concept of mutual authentication to validate a user. Mutual authentication relies on a shared secret and the user's password, which is known by the client and the network. The authentication server sends a challenge to the client. The client uses a one-way hash of the of the user password to send a response to the challenge. The server creates its own response based on the user database information and compares it to the response received from the client. When the server authenticates the client, the same process is repeated in reverse so the client could authenticate the server. When this process is completed, an EAP success message is sent to the client and both the client and the authentication server derive the dynamic WEP key.

    This type of mutual authentication reduces the risk of access point vulnerability to man-in-the-middle attacks; however, as with MD5 authentication, the user password remains vulnerable to attackers.

Chapter 18

"Do I Know This Already?"

1. d
2. e
3. e
4. a, b, and d
5. c
6. b
7. a
8. d
9. e
10. a and b

Q&A

1. *The 802.1x protocol support is available on which types of ports?*

   Answer: Layer 2 static access, voice VLAN, and MVAP

2. *During 802.1x port authentication, the client only receives EAPOL, CDP, and _____ traffic through the port.*

   Answer: STP

3. *To enable 802.1x port-based AAA, use the _____ command in global configuration mode.*

   Answer: **aaa new-model**

4. *The three parameters of the **dot1x port-control** command are _____, **force-authentication**, and **force-unauthorized**.*

   Answer: auto

5. *The combination of which two parameters creates a unique identifier for RADIUS servers providing redundancy and availability?*

   Answer: Name or IP address and specific port number

6. *The **radius-server timeout**, **radius-server retransmit**, and **radius-server key** commands set what keys for the RADIUS servers specified on the switch?*

   Answer: Timeout, retransmission, and encryption

7. *What is meant by the quiet period when dealing with authentication on the Catalyst switch?*

   Answer: The idle time between failed client authentications

8. *If one of the attached clients fails authentication when multiple clients are connecting on an 8021.x-enabled port, then _____.*

   Answer: All attached clients are denied access to the network

9. *Because the switch keeps a history of _____ packets for clients that failed to authenticate, only non-802.1x-capable clients can join a guest VLAN.*

   Answer: EAPOL

10. *If a client has not been authenticated, or is in the process of being authenticated when the **show dot1x all** command is run, the port status would be _____.*

    Answer: UNAUTHORIZED

Chapter 19

"Do I Know This Already?"

1. a
2. d
3. e
4. f
5. d
6. a, b, c, and f
7. e
8. d
9. d
10. a and c

Q&A

1. When an organization wants to connect two or more offices to one another over VPNs, this is called _____.

   Answer: A site-to-site VPN

2. What kind of authentication method is used if the endpoints agree on the key of "VPNK3y" on their routers?

   Answer: Preshared key

3. To access information on the far end of a VPN that is destined behind the VPN endpoint, it is best to use what IPSec mode?

   Answer: Tunnel

4. What is the primary function of the AH (authentication header)?

   Answer: Provides only origin authentication and verification via IP

5. What command enables you to display the current ISAKMP policy configured on the router?

   Answer: **show crypto isakmp policy** command

6. Why is it important to verify that the ACLs on the perimeter router are correct?

   Answer: They need to allow IPSec traffic through and must allow IKE UDP port 500 traffic as well as ESP and AH protocol traffic (protocol numbers 50 and 51).

7. *What values need to be placed in the following IKE Phase 1 policy?*

   - Key distribution:ISAKMP
   - Authentication:Preshared key
   - Encryption algorithm:?
   - Hash algorithm:?
   - Diffie-Hellman group:2
   - IKE SA lifetime:86400

   Answer: AES and MD5

8. *What is the following command used for when established an IPSec policy?*

   router(config)#**crypto isakmp key abc123 address 200.10.41.254 255.255.255.255**

   Answer: This command sets the preshared key for use with the peer device with the IP address of 200.10.41.254.

9. *Why is the IPSec SA lifetime so important to an IPSec policy?*

   Answer: The IPSec SA lifetime identifies interval at which devices will renegotiate the connection. This ensures that keys between these two devices are more difficult to guess.

10. *Is it possible to establish multiple transform sets on a router? If so, why is this beneficial?*

    Answer: Yes. Allows multiple tunnel types to be established because it will search through until it finds one that works. The most secure transforms should be listed first.

11. *For a crypto map to take effect, it must be applied to _____.*

    Answer: The interface facing the peer

12. *What are the tasks required to configure IPSec using RSA-encrypted nonces?*

    Answer:

    1. Select the IKE and IPSec parameters.

    2. Configure the RSA keys.

    3. Configure IKE.

    4. Configure IPSec.

    5. Test and verify the IPSec configuration.

13. *After you receive your peer's RSA public key, what command do you use to enter this key into the crypto map?*

    Answer: **crypto key pubkey-chain rsa** command

Chapter 20

"Do I Know This Already?"

1. e
2. e
3. b
4. d and f
5. a, b, c, and f
6. f
7. e
8. a
9. a, b, d, and f
10. a, b, d, and e

Q&A

1. *What is a digital signature?*

    Answer: A Public Key Infrastructure component use to digitally identify and authenticate a device or user

2. *What keys are exchanged between peers that utilize a certificate authority?*

    Answer: Public keys

3. *The CA maintains a list of certificates that are no longer valid, which is called _____.*

    Answer: CRL

4. *The _____ facilitates the transfer of digital certificates between certificate authorities on Cisco devices.*

    Answer: SCEP

5. *What is the X.509v3 certificate?*

    Answer: Allows peers to exchange digital certificates to authenticate their identity

6. *Why is the host name, domain name, router date, time, and time zone so important when configuring the Cisco device for CA support?*

    Answer: The host name and domain name are written into the key pair that is generated. The router date, time, and time zone must be accurate to enroll with the CA server.

7. *What command enables you to authenticate the CA server Chicago_CA for the router?*

    Answer: **crypto ca authenticate Chicago_CA** command

8. *What does the **no crypto ca enroll** command do when implemented on a router?*

    Answer: This deletes the current enrollment from a CA server.

9. *The option on the router that enables you to delete keys and certificates from the router is called what?*

    Answer: Key management

10. *What commands enable you to debug and validate the CA configuration on the router?*

    Answer: **crypto ca identity** command
    debug crypto pki command
    show crypto ca certificates command

Chapter 21

"Do I Know This Already?"

1. d
2. d
3. a
4. a, c, d, and f
5. d
6. e
7. c
8. b
9. a, b, d, and e
10. e

Q&A

1. *When you configure any type of encrypted communication between devices, it is critical that the configuration of both devices _____.*

    Answer: Match

2. *If you want to debug IPsec, what are the various **debug** commands you can utilize?*

Answer: **debug crypto isakmp** command

```
debug crypto engine command
debug crypto ipsec command
debug crypto pki messages command
debug crypto pki transactions command
```

3. *What does the **show crypto ca certificates** command do on the router?*

Answer: Displays information about the certificate, the CA, and the registration authorities

4. *What does the **peer** option on the **show crypto ipsec sa** command show?*

Answer: Identifies all virtual routing and forwarding SAs based on the address or vrf name

5. *To show the SA lifetime value, which command do you use?*

Answer: **show crypto ipsec security-association lifetime** command

6. *Which command generates the following output?*

```
Codes: M - Manually Configured, C - Extracted from certificate
Code Usage IP-address Name
M Signature 192.168.3.1 FanFrancisco.SF.com
M Encryption 192.168.3.1 FanFrancisco.SF.com
```

Answer: **show crypto key pubkey-chain rsa** command

7. *Which **debug** command enables you to display all IKE events?*

Answer: **debug crypto isakmp** command

8. *The **debug crypto key-exchange** command displays the exchange of keys between the CA and peers. True or False?*

Answer: False

9. *Which command enables you to display information about the communication between the router and the CA?*

Answer: **debug crypto pki messages** command

10. *Which command enables you to clear the SA counters?*

Answer: **clear crypto sa counters** command

Chapter 22

"Do I Know This Already?"

1. b

2. c

3. a
4. b
5. b
6. d
7. a
8. d
9. b, d, and e
10. d

Q&A

1. *How does the Easy VPN Server control VPN policies for remote clients?*

   Answer: The Easy VPN Server manages all IPSec policies centrally and pushes the policy out to the client.

2. *What is DPD?*

   Answer: Dead peer detection incorporates a series of "keepalive" messages between the IPSec peers when there is no traffic passing through the VPN tunnel.

3. *How does the **aaa new model** command prepare the router for Easy VPN Server?*

   Answer: The first task is to enable AAA on the router.

4. *What must you do before selecting your IKE parameters for remote VPN clients?*

   Answer: You must ensure the ISAKMP is enabled on the router.

5. *What servers should you designate when defining the group policy for mode configuration push?*

   Answer: DNS and WINS servers (if applicable)

6. *What must you do to make a dynamic crypto map function?*

   Answer: Apply the dynamic crypto map to the interface.

7. *What is the difference between **crypto isakmp keepalive seconds** and **retries**?*

   Answer: Keepalive seconds is the time router waits before sending a keepalive. Keepalive retries is the time the router waits before sending another keepalive after not receiving a response from a previous keepalive.

8. *What is Xauth?*

   Answer: Extended authentication (Xauth) is a process for using AAA authentication for VPN connections. Produces a pop-up username and password box.

9. *How many different remote Phase 2 modes does Easy VPN Server support?*

   Answer: Three. client mode, network extension mode, and network extension plus.

10. *Which remote Phase 2 modes do not support NAT or PAT?*

    Answer: Network extension mode and network extension plus

Chapter 23

"Do I Know This Already?"

1. a, b, and c
2. b
3. a, b, and e
4. b and e
5. a and c
6. a and e
7. a
8. a and d
9. b and d
10. d
11. a, b, and d

Q&A

1. *SDM is designed to run on which personal computer operating systems?*

   Answer: Windows 2003 Server (Standard Edition), Windows NT 4.0 Workstation (Service Pack 4), Windows Me, Windows XP Professional, Windows 2000 Professional, and language OS support in Microsoft XP Professional or Microsoft Windows 2000 Professional for Japanese, Simplified Chinese, French, German, Spanish and Italian

2. *Which browser versions are supported by SDM?*

   Answer: Microsoft Internet Explorer 5.5 or later, Netscape Navigator 7.1 and 7.2, Firefox 1.0.2, JVM built-in browsers required, Java plug-in (Java Runtime Environment version 1.4.2_05 or later)

3. *The SDM User Interface file menu offers what options?*

    Answer: Save running configuration to PC, SDM deliver to router, write to startup config, reset to factory defaults, and exit

4. *What six wizards does SDM version 2.1.1 provide?*

    Answer: LAN Wizard, WAN Wizard, Firewall Wizard, VPN Wizard, Security Audit Wizard, and Reset Wizard

5. *What are the main tasks required to configure the Cisco IOS Firewall feature set?*

    Answer: Select the type of firewall setup, configure the firewall interface, apply the firewall configuration to the router, and advanced firewall configuration options.

6. *What does the Logging screen under the monitor mode provide?*

    Answer: This screen shows the basic information about the router resources, including total log entries and high-severity, warning, and information log entries.

7. *What is the function of the NAT screen in SDM?*

    Answer: The NAT window allows the designation of the inside and outside interfaces, and enables you to view NAT rules and address pools. It also allows the setting of translation timeouts.

8. *What is the SDM security audit feature based on?*

    Answer: The Cisco AutoSecure feature.

9. *What is a prerequisite to installing Easy VPN Server on a router?*

    Answer: Easy VPN Server requires AAA.

10. *What is user authentication Xauth?*

    Answer: Xauth allows all Cisco IOS Software AAA authentication methods to perform user authentication in a separate phase after the IKE authentication Phase 1 exchange. Xauth is an extension to IKE and does not replace IKE authentication.

Index

Numerics

3DES (Triple Data Encryption Standard), 406
802.1x standard
    authentication, 359, 377, 561
    Cisco Secure ACS, 369, 377
    client negotiation, 377–378
    components, 357–359
    configuration, verifying, 387–389
    EAP, 359
    operation, 359–362
    options
        defaults, resetting, 386
        frame-retransmission number, 384–385
        guest VLANs, 385–386
        multiple hosts, 385
        quiet period, 383
        re-authentication, 382–383
        retransmission times, 384
    port-based authentication, 378–382
    port state, 362
    servers, authentication, 359
    supplicants, 359
    switch ports, 381
    unauthorized state, 362

A

AAA (authentication, authorization, and accounting)
    802.1x protocol, 377
    access control, 125
    accounting
        Cisco IOS Software, 130, 140
        Cisco Secure ACS, 181–182
        configuring, 140
        RADIUS, 164–165
        TACACS+, 157
    attributes, IETF, 115
    authentication
        Cisco Secure ACS, 178–179
        configuring, 129–136
        proxy, 72, 75, 129–130, 338, 584, See also authentication proxy
        RADIUS, 162–163
        TACACS+, 155–156
    authorization
        Cisco IOS Software, 130, 136–137
        Cisco Secure ACS, 180–181, 210
        configuring, 136–140
        RADIUS, 163–164
        TACACS+, 156
    Cisco Secure ACS, 173, 178–182, 556–557
    clients, 205
    configuring
        accounting, 140
        authentication, 131–136
        authorization, 136–140
        overview, 130, 146, 338
        RADIUS, 160–162, 169
        TACACS+, 153–154, 169
    databases
        backup, 186
        replication, 183–186
    defense in depth, 50, 55
    Easy VPN Remote, 597–599
    Easy VPN Server, 593–596
    ERROR responses, 130
    FAIL responses, 130
    IETF attributes, 115
    RADIUS
        accounting, 164–165
        authorization, 163–164
        configuring, 160–162, 169
        overview, 122

testing/troubleshooting, 166–167
RDBMS synchronization, 186
responses, 130
server support, 72, 75, 116
switch ports, 381
TACACS+
  accounting+, 157
  authentication, 155–156
  authorization, 156
  configuring, 153–154, 169
  overview, 122
  testing/troubleshooting, 157
troubleshooting, 142
abbreviated commands, 67
ACCEPT responses, RADIUS, 115
acceptable use policies, 10
access
  attacks, 37–39, 42
  control
    accounting, 130
    authentication, 129–130
    authorization, 130
    CBAC configuration, 585–586
    interactive, 91
    overview, 125, 129
access control lists
  See ACLs
Access VPNs, 402
accounting
  AAA, 130
  Cisco IOS Software, 130, 140
  Cisco Secure ACS, 181–182
  RADIUS, 164–165
  TACACS+, 157
ACEs (access control list entries), 221
ACLs (access control lists), 71, 75, 221, 237, 247
  802.1x protocol, 358
  ACEs, 221
  bypassing feature, 313

CBAC, 222, 309–318
Cisco Secure ACS, 180
configuring, 222, 228–230
crypto ACLs, 417, 574–576
dynamic ACLs, 71, 75
functions, 221
IP
  certificate-based, 228
  extended, 226
  overview, 222–226
  reflexive, 226–227
  time-based, 227–228
packets, 228–230
permit statements, 222
VACLs, 296
Active Directory, Windows, 192
addresses
  address pools, 490–491
  addressing schemes, 10
  dynamic secure, 286
  lookup tables, 279
  MAC addresses, 283
  spoofing, 291–294
  static/sticky secure, 286
Advanced Security feature set
  firewall engine, 245–246
  routing support, 248
AH (Authentication Header), 409
alerts, 72, 76, 310–311
Apple clients, 179
applets, 247
applications
  attack targets, 50, *See also attacks*
  weaknesses, 33
Arspoof tool, 292
atomic signatures, 260
Attack-drop.sdf, 260, 266
attacks
  access attacks, 37–42
  applications, 50

attacks

Arpspoof, 292
Attack-drop.sdf, 260, 266
block scans, 38
CAM table overflow attacks, 283–288
data, 50
DDoS attacks, 39, 49
DHCP
    snooping, 292–293, 562
    starvation, 296
dictionary attacks, 92
DNS
    queries, 37
    spoofing, 93
DoS attacks
    CBAC, 310
    DDoS attacks, 39, 49
    firewalls, 248
    overview, 37–42, 75, 92, 234, 260
EAP attacks, 296–298
exploits, 260
fabrication, 39
firewalls, 49
FTP sweeps, 38
gobbler tool, 296
horizontal scans, 38
hosts, 50
interception, 38
IP
    source routing, 93
    spoofing, 93
Layer 2 attacks, 283–300, 561–562
MAC address spoofing, 291–294
Macof, 285
management components, 50
man-in-the-middle attacks, 93, 291–294
misuse, 260
mitigating
    CAM table overflow attacks, 286–288
    DHCP starvation attacks, 296
    EAP attacks, 297–300
    MAC address spoofing, 292–294
    man-in-the-middle attacks, 292–294
    on private VLANs, 295
    STP manipulation attacks, 290–291
    VLAN hopping attacks, 289
modification, 39
motivations for, 30–36, 41
networks, 50
physical security, 54
ping sweeps, 38
reconnaissance attacks, 37–42, 260
routers, 49
scans, 38
session-hijacking attacks, 93
signatures
    atomic, 260
    attack signatures, 259
    unsupported signatures, 261
Smurf attacks, 232
SNMP traps, 92
spoofing
    DNS, 93
    IP, 93
    MAC addresses, 292–294
STP manipulation attacks, 289–291
switches, 49, 279
SYN floods, 71, 310
targets, 49
unstructured threats, 34
Trojan horses, 39
vertical scans, 38
viruses, 39
VLANs
    hopping attacks, 288–289
    vulnerabilities, 294–295
vulnerabilities, 29, 40–41
worms, 39
attributes field, RADIUS, 115
auditing
    audit trails, 247–249
    CBAC, 310–311
    overview, 71–75
authentication
    802.1x
        client negotiation, 377–378
        defaults, resetting 386
        configuration, verifying, 387–389
        frame-retransmission number, 384–385
        guest VLANs, 385–386
        multiple hosts, 385
        quiet period, 383
        re-authentication, 382–383
        retransmission time, 384
        port-based, 378–382
        overview, 377, 561
    Active Directory, 192
    CAs, 438–440
    Cisco Secure ACS, 178–179, 189–191, 210
    configuring

authorization   645

line password authentication,
  111–112
login authentication, 132–133
overview, 131–132
PPP authentication, 134–136
usernames, 112
EAP
  EAP-FAST, 366–367
  EAP-MD5, 363
  EAP-TLS, 364–365
  LEAP, 363
  network requirements, 368
  PEAP, 365–366
Easy VPN, 488
firewalls, 72
frame-retransmission number, 384–385
hosts, multiple 385
IKE, 405, 439
IPsec, 439
line password authentication, 111–112
login authentication, 132–136
MD5, 234, 363
method list, 130
neighbor routers, 72, 76
NFS, 33
OTPs, 111
passwords, 111, 133–134
peer routers, 247
PPP, 134–136
protocols
  CHAP, 117–119
  EAP, 120–121
  MS-CHAP, 119–120
  PAP, 118
  routing, 233–234
proxy, 72, 75, 129–130, 338, 584
quiet periods, 383
RADIUS, 162–163
re-authentication, 382–383
remote security servers
  CHAP, 118–119
  EAP, 120–121
  Kerberos, 117
  MS-CHAP, 119–120
  PAP, 118
  RADIUS, 114
  TACACS, 112–114
remote users, 118–122
retransmission time, 384
routers

CA support, 438–439
neighbor routers, 72, 76
peer routers, 247
routing protocols, 233–234
RSA keys, 439
SNMPv3, 231
switch ports, 381
TACACS+, 155–156
token cards, 111
usernames, 111–112
users, remote 117
VLANs, 385–386
X.509 standard, 439
Authentication Header (AH), 409
authentication proxy
AAA, 72, 75, 129–130, 338, 584
CBAC, 336
Cisco IOS Firewall, 248, 333, 336
Cisco Secure ACS, 344–349
configuration
  types, 337
  overview, 337–339
  AAA, 584
  HTTP servers, 585
  inbound/outbound authentication,
    341–342
  routers, 585
  testing/verifying 584–585
features, 336
firewalls, 336
HTTP
  connections, 335
  servers, 339
IPsec encryption, 336
limitations, 349
RADIUS, 347–349
TACACS+, 344–347
VPNs, 336
**authentication, authorization,
  and accounting**
  See AAA
authenticator field, RADIUS, 114
authorization
AAA, 130
Cisco Secure ACS, 180–181, 210
configuring, 136–140
RADIUS, 163–164
overview, 72, 76
TACACS+, 156

AutoSecure feature, 99
auxiliary connections, 68, 74
availability, 129, 358

B

backup databases, 186
banners, 90–91
block scan attacks, 38
border routers, 222
BPDU guard, STP, 290
broadcasts, 232–233
bypassing feature, ACL, 313

C

CAM (content-addressable memory)
    entries, 98
    lookup tables, 279
    overflow attacks, 283–288
CAs (certificate authorities)
    configurations, verifying, 450
    digital signatures, 437
    interoperability, 433
    IPsec, 440–450, 578–579
    peer authentication, 440
    routers, 441
    SCEP, 439–440
    show commands, 460
    X.509 standard, 439
CBAC (Context-Based Access Control)
    ACLs, 222, 309–318
    alerts/audit trails, 309–311
    authentication proxy, 336
    Cisco IOS Firewall, 585–586
    configuring, 318, 324–325, 585–586
    CPU/performance impact, 317–318
    debug commands, 323–324
    DoS attacks, 310
    firewalls, 585–586
    H.323 inspection, 316
    ICMP types supported, 310–311
    inspection rules, 320–323
    interfaces, 318
    Java applets, 322
    operation, 311–314
    overview, 71–75
    PAM, 319–320, 251
    protocols, 314–316
    restrictions, 314
    RTSP, 315–316
    sessions, 311–314
    show commands, 323–324
    timeouts/thresholds, 319
    utilization, improving, 317
    verifying/debugging, 323–324
CDP (Cisco Discovery Protocol), 235, 562
certificates
    certificate-based IP ACLs, 228
    certificate revocation lists (CRLs), 440
    Cisco Secure ACS, 566
    Easy VPN, 488
    IPsec, 577–579
    overview, 437, 566
    X.509 standard, 439
Challenge Handshake Authentication Protocol (CHAP), 118–122, 179
    See CHAP
CHALLENGE responses, RADIUS, 115
change management, 10
CHANGE PASSWORD responses, RADIUS, 115
CHAP (Challenge Handshake Authentication Protocol), 118–122, 179
Chargen port, 234
Cisco Discovery Protocol (CDP), 235
Cisco IOS Firewall
    advanced firewall engine, 246
    authentication proxy, 248, 333, 336
    CBAC
        alerts/audit trails, 309–311
        configuring, 318, 585–586
        CPU/performance impact, 317–318
        DoS attacks, 310
        H.323 inspection, 316
        inspection rules, 320–322
        operation, 311–314
        PAM, 319–320
        protocols, 314–315
        restrictions, 314
        RTSP, 315–316
    DoS attacks, 248
    feature set, 361, 241–246 333
    features/functions, 71–74 245
    IDS, 336, 255
    IPS, 71, 75, 246, 255
    overview, 61, 241, 305
    transparency, 247
    URLs, 251
    VPNs, 246
    See also firewalls

commands 647

Cisco IOS Intrusion Detection System (IDS), 336, 255
Cisco IOS Intrusion Prevention System (IPS)
    CLI, 262
    configuration
        process, 262–266
        verifying 270–273
    deployment strategies, 273–274
    false positive/negative, 267
    features, 259–260
    firewalls, 71, 75, 246, 255
    functions, 260
    malicious traffic, 262
    notification types, 263
    rules, 268–269
    SDF, 260, 266
    signatures, 260–261
    SME, 260
Cisco IOS Software
    accounting, 130, 140
    Advanced Security feature set, 248
    authorization, 130, 136–137
    configuring
        RADIUS, 160–162
        TACACS+, 153–154
Cisco Router and Security Device Manager
    See SDM
Cisco Secure ACS (Access Control Server)
    802.1x authentication, 369, 377
    AAA, 173, 178–182, 556–557
    ACLs, 180
    accounting, 181–182
    administration, 182–183
    Apple clients, 179
    attributes, vendor-specific, 181
    authentication
        overview, 178–179
        proxy, 344–349
        users, 189–191
    authorization, 180–181
    certificates, 566
    CHAP, 179
    databases
        backup/replication, 183–186
        external users, 178
        local, 191–192
        support, 189–191
    deploying, 203–209
    LDAP, 193
    maximum numbers
        AAA clients, 205
        users, 204
    PAP, 179
    passwords
        aging, 179
        protocols, 179, 190–191
    PEAP, 562–568
    RDBMS synchronization, 186
    services, 187–189
    token server, 193
    transactions per second, 205
    troubleshooting, 209–211
    UCP, 194
    vendor attributes, 181
    Windows
        installing for, 205, 208
        Windows NT/2000, 192–193
Cisco Secure Policy Manager (CSPM), 264
Cisco Security Device Manager
    See SDM
Cisco Unix Director, 264
CiscoSecure Integrated Software (CSIS), 71
CiscoWorks VPN/Security Management Solution IDS Management Center (VMS IDS), 263
clear commands
    clear crypto isakmp command, 473
    clear crypto sa command, 473
    clear crypto sa counters, 474
    connecttivity, 460, 472
    overview, 273
CLI (command-line interface) routers, 68–70, 74
client mode, 525
clock commands, 571
Code field, EAP, 360
code field, RADIUS, 114
commands
    ?, 67
    802.1x, enabling, 297, 381
    AAA commands
        accounting, 141
        authentication, 131
        authentication attempts login, 136
        authentication enable default, 133
        authentication login, 132
        authentication ppp, 134
        authorization, 138
        troubleshooting 142–143

audit trails, 249
abbreviated, 67
attacks
    private VLAN proxy attacks, mitigating 295
    router attacks, preenting 237
authentication
    802.1x port-based, 380
    proxy, 339
banner login commmand, 90
CA configurations, verifying 450
clear commands, 273, 460, 472–474
configuring
    AAA, 338
    authentication proxy, 339
    AutoSecure, 100
    console interface, 85
    DAI, 294
    DHCP snooping, 293
    dynamic port security, 288
    HTTP servers, 339
    interface inspection rules, 323
    notification types, 264
    queue size, 265
    Reverse Telnet, 96
    SCP, 94–96
    SSH
crypto ca trustpoint command, 444
debug commnds, 273
    802.1x authentication, 390
    aaa accounting, 145, 167
    aaa authentication, 144, 157
    aaa authorization, 144
    connectivity, 460, 467
    crypto engine, 468
    crypto ipsec, 468
    crypto isakmp, 467
    crypto key-exchange, 467
    crypto pki, 470
    crypto pki transactions, 471
    radius, 166
    tacacs, 158
    tacacs events, 158
enable commands
    password command, 85
    secret command, 86–87
encryption
    MD5, 88
    passwords, 112

EXEC mode, accessing, 85
exec-timeout command, 92
IPS configurations, veifying, 270–273
logging, 249
login, blocking 93
MD5 encryption, 88
passwords, encrypting 112
period re-authentication, 382
policies, attaching to signatures, 267
ppp authentication command, 135
private VLAN proxy attacks, mitigating 295
privilege command, 88
privilege level, 86
radius-server host command, 160
radius-server retry method command, 161
router attacks, preventing, 237
SDFs, 266
service password-encryption command, 88
show commands
    802.1x authentication, 387–389
    aaa servers, 142
    aaa user, 143
    accounting, 167
    overview, 89, 270
    VPN connectivity, 459–466
tacacs-server host command, 153
transport input, command 91
username secret command, 88
compound signatures, 260
confidentiality, 93, 129
configuration
    802.1x protocol, 387–389
    AAA, 131–140
    ACLs, 222, 228–230
    CAs, 450
    CBAC, 585–586
    IKE, 412–414, 572–573
    IPSs, 262–266
    IPsec, 449–450
    management, 10
    network, 31–32
    RADIUS servers, 160–162, 169
    routers modes
        global configuration mode, 66, 73
        interface configuration mode, 66
        line configuration mode, 66
        privileged EXEC mode, 65, 73
        ROM monitor mode, 65, 73
        user EXEC mode, 65, 73
    TACACS, 153–154, 169

console access, 84–85
    connections, 68, 74
Context-Based Access Control
    See CBAC
continuity, 30, 40
correlation, 53, 56
crackers, 34, 41
CRLs (certificate revocation lists), 440
crypto
    ACLs, 417, 574, 576
    maps, 418–419, 464, 494–495, 574–580
CSAdmin service, 187–188
CSAuth service, 187–188
CSDBSync service, 186–188
CSIS (CiscoSecure Integrated Software), 71
CSLog service, 187–188
CSMon service, 187, 189
CSPM (Cisco Secure Policy Manager), 264
CSRadius service, 187, 189
CSTacacs service, 187, 189
cyber-attacks
    See attacks
cyber-warfare, 36

D

DAI (Dynamic ARP inspection) feature, 293
Data Encryption Standard (DES), 406
Data field, EAP, 360
Database Replication Report, 186
Daytime port, 234
DDoS (distributed denial-of-service) attacks, 39, 49
dead peer detection (DPD), 487, 495
debug command, 273
    802.1x authentication, 390
    AAA, 142–145
    aaa accounting command, 145, 167
    aaa authentication command, 144, 157
    aaa authorization command, 144
    CBAC, 323–324
    connectivity, 460, 467
    crypto engine command, 468
    crypto ipsec command, 468
    crypto isakmp command, 467
    crypto key-exchange command, 467
    crypto pki command, 470
    crypto pki transactions command, 471
    IKE/IPSsec, 581
    radius command, 166

tacas command, 158
tacas events command, 158
defense in depth, 49–55
demilitarized zones (DMZs), 298
denial-of-service attacks
    See DoS attacks
DES (Data Encryption Standard), 406
DHCP (Dynamic Host Configuration Protocol)
    guest VLANs, 386
    snooping, 292–293, 562
    starvation attacks, 296
dictionary attacks, 92
Diffie-Hellman encryption, 407
digital certificates
    Easy VPN, 488
    IPsec, 57–579
    X.509 standard, 439
    See also certificates
directed broadcasts, disabling, 232–233
diaster recovery, 30, 40
Discard port, 234
distributed denia-of-service (DDoS) attacks, 39, 49
    See DoS attacks
distribution, security policies, 17
DMVPNs (Dynamic Multipoint VPNs), 530–534
DMZs (demilitarized zones), 298
DNS (Domain Name Server)
    domains, 560
    queries, 37
    spoofing, 93
domain names, 94
DoS (denial-of-service) attacks
    CBAC, 310
    DDoS attacks, 39, 49
    firewalls, 248
    overview, 37–42, 49, 75, 92, 234
    SYN floods, 71, 310
double tagging, 289
DPD (dead peer detection), 487, 495
Dynamic ARP inspection (DAI) feature, 293
Dynamic Multipoint VPNs, 530–534

E

EAP (Extensible Authentication Protocol)
    802.1x, 359
    attacks, 296–298
    EAP-FAST, 366–367

EAP-MD5, 363
EAPOL, 360
EAP-TLS, 364–365
frame format, 360–362
LEAP, 363
network requirements, 368
overview, 120–122
PEAP, 365–366
response frames, 384
EAP Flexible Authentication via Secure Tunneling (EAP-FAST), 366–367
EAP over LANS (EAPOL), 360
EAP Transport Layer Security (EAP-TLS), 364–365
EAP-FAST (EAP Flexible Authentication via Secure Tunneling), 366–367
EAPOL (EAP over LANs), 360
EAP-TLS (EAP Transport Layer Security), 364–365
Easy VPN
address pools, 490–491
authentication, 488
configuring, 489–496
crypto maps, 494–495
DPD, 495
group policies, 492–493
group policy lookup, 491
ISAKMP policies, 492
modes, 496
operation, 488–489
overview, 479
Remote, 484–485, 525–526, 597–599
routers, 491
SDM
Remote, 525–526, 597–599
Server, 528–530
Server
configuring, 489–496
functionality, 483–488
SDM, 528–530
transform sets, 493–494
Xauth, 496
Echo port, 234
e-mail, 10
enable password, 85–86
Encapsulating Security Payload (ESP), 409
encryption
Diffie-Hellman, 407
interception attacks, 38

IPsec, 94, 336, 404
MD5, 88, 407
NFS, 33
passwords, 87–88
public key cryptography, 437
RSA-encrypted nonces, 405, 423–427
SNMPv3, 231
SSH, 68, 94
endpoints, 401
ERROR responses, AAA, 130
escalation, privilege, 39
ESMTP (Extended Simple Mail Transport Protocol), 315
ESP (Encapsulating Security Payload), 409
Ethernet switches, 97–98
events
buffers, 265
Event Log, 186
logging, 72, 76, 247
security, 10
EXEC mode, 85
Execution Shell, 94
exploit attacks, 260
Extended Simple Mail Transport Protocol (ESMTP), 315
Extensible Authentication Protocol
See EAP
extranet VPNs, 402

F

fabrication attacks, 39
Factory Reset Wizard, SDM 536–537
FAIL responses, AAA, 130
Failed Attempts Report, 209
failover, 247
false negatives/positives, 267
feasibility, security policies, 15
feature sets
Advanced Security, 241
Firewall, 61, 241–246 333
IPS, 259
filtering
traffic, 71, 247
URLs, 247, 251
finger service, 235
fingerprinting, 235
Firewall feature set, 61, 241–246 333
firewalls
applets, blocking, 247

attack target, 49
audit trails, 247
authentication proxy, 248, 333, 336
CBAC
    alerts/audit trails, 309–311
    configuring, 318
    CPU/performance impact, 317–318
    defining inspection rules, 320–322
    DoS attacks, 310
    H.323 inspection, 316
    operation, 311–314
    PAM, 319–320
    protocols, 314–315
    restrictions, 314
    RSTP, 315–316
event logging, 247
features, 71–72
filtering
    traffic, 247
    URLs, 247, 251
managing, 247
mitigating DoS attacks, 247–248
NAT, 247
overview, 61, 79, 241, 245, 305
port mapping, 247
SDM, 515, 541
    advanced options, 519–521
    applying config to routers, 518–519
    configuring interfaces, 516–518
    selecting type, 515–516
traffic, filtering, 247
URLs, filtering 247, 251
flexibility, security policies, 16
frames, 360–362, 384
FTP sweeps, 38

G

GARP (Gratuitous ARP), 292
Generic Lightweight Directory Access Protocol (LDAP), 193
global
    configuration mode, 66, 73
    filtering, URLs, 251
    implementation, security policies, 15
goals
    network attacks, 37
    security policies, 12–13, 20
gobbler tool, 296
group policies, Easy VPN, 491–493

guidelines, security policies, 13–21

H

H.323 inspection, 316
hackers
    ethical hackers, 34
    hacktivism, 36
    Microsoft, 32
    motivations, 34–36, 41
half-open sessions, 314
handshakes, 118
hardware weaknesses, 33
hash, 234
high availability, 129, 358
hijacking, sessions, 93, 297
HMAC (Keyed-Hashing for Message Authentication), 406
hopping attacks, VLANs, 288–289
horizontal scan attacks, 38
hosts
    802.1x authentication, 385
    attack target, 50
    host-based defenses, 53
    hostile, 52
    IPSs, 56
    names, 94, 560
    port mapping, 320
HTTP
    authentication proxy, 585
    connections, 335
    SDM, 511
    servers, 339
HTTPS (Secure HTTP), 232, 511

I

IBNS (Identity-Based Networking Services), 353
ICMP types, CBAC, 310–311
Identifier field
    EAP, 360
    RADIUS, 114
Identity frames, 384
IDS (intrusion detection system), 255, 336
IDS Management Center (MC), 264
IEEE 802.1x standard
    See 802.1x standard
IETF (Internet Engineering Task Force) attributes, RADIUS, 115

IKE (Internet Key Exchange)
    configuring, 412–414, 572–573
    crypto maps, 580
    debut command output, 581
    DPD, 487, 495
    enabling, 575
    IPsec, preparing for, 571–572
    ISAKMP peers, 406
    keys, 405
    overview, 404, 439–441, 448
    Phase 1, 405–408
    policies, 579–580
    preshared keys, 405, 572–573
    RSA
        signatures, 405, 575–576
        encrypted nonces, 405
    SAs, 583
    transform sets, 580
    Xauth, 487
in traffic, 226
incident-response plans, 17–18
infrastructure, network, 10
inspection, 320–322
integrity, 93, 129
interactive access, 91
interception
    attacks, 38
    passwords, 93
interesting traffic, 417
interface configuration mode, 66
interleaved mode, 316
Internet
    access, 10
    services, 31, 41
Internet Control Message Protocol (ICMP)
    types, CBAC, 310–311
Internet Key Exchange
    See IKE
Internet Security Association and Key Management Protocol (ISAKMP), 461
internetworking, 49
interoperability, CAs, 433
intranet VPNs, 402
intruders, 34–36, 41
intrusion detection system
    See IDS
intrusion prevention system
    See IPS

IP
    ACLs
        certificate-based, 228
        extended, 226
        reflexive, 226–227
        standard, 222–226
        time-based, 227–228
    source routing, 93
    spoofing, 93
IPS (intrusion prevention system)
    Cisco, 586–592
    CLI, 262
    configuring, 262–266
    deployment strategies, 273–274
    false positive/negative, 267
    features, 259–260
    functions, 260
    malicious traffic, 262
    notification types, 263
    rules, 268–269
    SDF, 260, 266
    SDM, 586–592
    signatures, 260–261
    SME, 260
IPsec
    CAs, 433
    certificates, 577–579
    configuration
        ACLs, 412, 417–418
        connectivity, verifying, 411
        crypto maps, 417–418
        IKE, 404–407, 412–414
        transform sets, 415–416
        overview, 449–450
        policies, 408–411
        manual configuration, 423
        preshared keys, using, 573–575
        router configurations, verifying, 411
        SA lifetimes, 416–417
        testing/verifying, 419
    debugging, 581
    encryption, 94, 336
    IKE, 404, 412–414
    modes, 408
    overview, 403
    preshared keys, configuring with, 574–575
    policies, 408–411

RSA-encrypted nonces
    configuring host/domain name,
        423–424
    planning for, 424
    RSA keys, 424–427
SAs, 416–417, 462, 574–583
transforms, 410
transparent mode, 408
tunnel mode, 408
VPNs
    CAs, 437
    configuring, 440–450
    digital signatures, 437
    tunnels, 403–404
ISAKMP (Internet Security Association and Key Management Protocol)
debugging
    commands, 467
    events, 581
overview, 461
peers, 406
policies, 406, 492
See also IKE
ISAKMP/Oakley key exchange, 439

J–K

jargon, security policies, 15
Java
blocking, 322
code, 71, 75
inspection, 322
Kerberos, 72, 75, 117
Keyed-Hashing for Message Authentication (HMAC), 406
keys
IKE, 405
distribution, 405
preshared
    common convention, establishing, 570
    Easy VPN, 488
    IKE, 405, 572–573
    IPsec, 573–575
    overview, 413
    router setup, 570–571
RSA
    deleting, 443
    overview, 95, 439
    show commands, 465–466
    signatures, 448
    encrypted nonces, 423

SSH, 560

L

LAN Wizard, SDM, 513
Layer 2 attacks
CAM table overflow attacks, 283–288
DHCP starvation attacks, 296
IEEE 802.1x EAP attacks, 296–298
MAC address spoofing, 291–294
man-in-the-middle attacks, 291–294
mitigating
    CAM table overflow attacks, 286–288
    DHCP starvation attacks, 296
    EAP attacks, 297–300
    MAC address spoofing, 292–294
    man-in-the-middle attacks, 292–294
    private VLAN attacks, 295
    STP manipulation attacks, 290–291
    VLAN hopping attacks, 289
overview, 561–562
STP manipulation attacks, 289–291
VLAN hopping attacks, 288–289
Layer 3 switching, 49
LDAP (Generic Lightweight Directory Access Protocol), 193
LEAP (Lightweight Extensible Authentication Protocol), 363
least-privilege policy, 10
legal issues, security, 19, 90
Length field
EAP, 360
RADIUS, 114
line configuration mode, 66
link-down traps, 98
LLQ (low-latency queuing), 248
load balancing, 161, 169
Local Authentication Service, 358
lock-and-key traffic filtering, 71
lockdown, MAC address, 97
logging
AAA, 140
logs, 10
overview, 72, 76, 249
SDM, 544
logins
attempts, limiting, 136
authentication, 132–133, 335
features, 92
RADIUS, 115

lookup tables, 279
low-latency queuing (LLQ), 248

M

MAC addresses
   lockdown, 97
   modes, 97
   port security, configuring 99
   spoofing, 291–294
Macof tool, 285
man-in-the-middle attacks, 93, 291–297
manipulation attacks, STP, 289–291
maps, crypto, 418–419, 464, 494–495, 574–580
MC (Management Center), IDS, 264
MD5 (Message Digest 5)
   authentication, 234, 363
   encryption, 88, 407
message
   digests, 234, 406
   integrity, SNMPv3, 231
method lists, 130
Microsoft Challenge Handshake Authentication Protocol (MS-CHAP), 119–122
misuse attacks, 260
modes
   client, 525
   Easy VPN Server, 496
   EXEC, 85
   interleaved, 316
   IPsec, 408
   MAC addresses, 97
   monitor, 541, 544
   network extension, 526
   push configuration, 492–493
   restrictive, 97
   router configuration
      global configuration, 66, 73
      interface configuration, 66
      line configuration, 66
      privileged EXEC, 65, 73
      ROM monitor, 65, 73
      user EXEC, 65, 73
   shutdown, 97
modification attacks, 39
monitoring
   general, 10, 53
   monitor mode, 541, 544

MS-CHAP (Microsoft Challenge Handshake Authentication Protocol), 119–122
multicast/multiprotocol support, 248

N

naming conventions, 10
NAT (Network Address Translation), 72, 76, 247, 538
network extension mode, 526
Network File System (NFS), 33
Network Time Protocol (NTP), 235
networks
   addressing schemes, 10
   attack targets, 50
   configuration weaknesses, 31–32
   hostile networks, 52
   infrastructure policies, 10
   overview, 5
   security, *See* security
   segmentation, 52, 55
NFS (Network File System), 33
notification, legal 90, 263
NTP (Network Time Protocol), 235

O–P

one-step lockdown, SDM, 536
one-time passwords (OTPs), 111
out traffic, 226
out-processing procedures, 36
overflow attacks, 283–288
Packet Type field, EAPOL, 361
packets, 114, 228, 259
PAM (port to application mapping) feature
   CBAC, 319–320
   overview, 72, 75, 249
   port mapping
      host-specific, 251
      system-defined, 249
      user-defined, 250–251
PAP (Password Authentication Protocol), 118, 122, 179
passwords
   aging, 179
   authentication, 111
   case sensitivity, 111
   Cisco Secure ACS, 179, 190–191
   console access, securing, 84
   default, 32

enable password, 85–86
encryption, 87–86
interception, 93
privilege levels, 84, 89, 133–134
SSH, 92
Telnet, 92
weak/exposed, 31, 41
PAT (Port Address Translation), 72
patches
    managing, 30
    security policy, 10
PEAP (Protected EAP), 365–366, 562–568
Perfect Forward Secrecy, 411
perimeter security, 53–55
permit statements, 222
Phase 1, IKE, 405–407
Phase 2, IKE, 408
phreakers, 34, 41
physical terminal lines, 91
ping sweeps, 38
PKI debug commands, 470–471
policies, security, 10–20
POP (PostOffice Protocol), 263
Port Address Translation (PAT), 72
ports
    mapping
        host-specific, 251
        overview, 247–249, 320
        system-defined, 249
        user-defined, 250–251
        See also *PAM*
    security, 97–98
        802.1x security, 358
        Cisco Secure ACS, 208
        NFS, 33
        private VLANs, 294
        small server services, 234
    trunk, 98, 561
Post Office Protocol (POP), 263
preshared keys
    configuring/verifying initial router setup, 570–571
    Easy VPN, 488
    establishing common convention, 570
    IKE, 405, 572–573
    IPsec, 573–575
    See also keys
private VLANs, 294–295
privileges

escalation, 39
levels, 84
    authentication proxy, 333
    configuring multiple levels, 88–89
    enable password, 85–86
    passwords, 89
unauthorized escalation, 39
privileged EXEC mode, 65, 73, 84–85
process, security, 18–21, 54–56
protect
    actions, 286
    mode, 97
Protected EAP (PEAP), 365–366
    protocols
    See specific protocol
proxy, authentication
    AAA, 72, 75, 129–130, 338, 584
    CBAC, 336
    Cisco IOS Firewall, 248, 333, 336
    Cisco Secure ACS, 344–349
    configuration
        types, 337
        overview, 337–339
        AAA, 584
        HTTP servers, 585
        inbound/outbound authentication, 341–342
        routers, 585
        testing/verifying 584–585
    features, 336
    firewalls, 336
    HTTP
        connections, 335
        servers, 339
    IPsec encryption, 336
    limitations, 349
    RADIUS, 347–349
    TACACS+, 344–347
    VPNs, 336
public key cryptography, 437, 439

Q–R

QoS (quality of service)
    LLQ, 248
    SDM, 538, 541
question mark (?) command, 67
quiet period, 92, 383

RADIUS (Remote Authentication Dial-In User Service)
ACCEPT responses, 115
authentication
802.1x 381–382
overview, 162–163
proxy, 347–349
accounting, 164–165
authorization, 163–164
CHALLENGE responses, 115
CHANGE PASSWORD responses, 115
Cisco Secure ACS, 347–349
configuring, 160–162
firewalls, 72–75
load balancing, 161
login, 115
overview, 114, 122, 169
packet fields, 114
REJECT responses, 115
responses, 115
TACACS+, versus, 116
testing/troubleshooting, 166–167
VLAN assignment, 380
RDBMS (relational database management system) synchronization, 186
RDT (RealNetworks Real Data Transport) protocol, 316
Real-Time Streaming Protocol (RTSP), 315–316
Real-Time Transport Protocol (RTP), 316
re-authentication, 382–383
reconnaissance attacks, 37–38, 42, 260
redundancy, 247
reflexive
ACLs, 71, 75
IP ACLs, 222, 226–227
REJECT responses, RADIUS, 115
relational database management system synchronization (RDBMS), 186
remote access, Easy VPN, 479
Remote Authentication Dial-In User Service
See RADIUS
remote security servers
CHAP, 118–122, 179
EAP, *See* EAP
Kerberos, 72, 75, 117
MS-CHAP, 119–120
PAP, 118, 122, 179
RADIUS, *See* Radius
TACACS, 112–114, *See also* TACACS+

Remote Site IEEE 802.1x Local Authentication Service, 358
remote users
CHAP authentication, 118–119
EAP authentication, 120–121
MS-CHAP authentication, 119–122
PAP authentication, 118
remote-access VPNs, 402
replication, databases, 183–186
responses
AAA, 130
incident-response plans, 17–18
RADIUS, 115
security events, 11
responsibilities/roles, security policies, 16
restrict actions, 286
restrictive mode, 97
retransmission, 384–385
reverse route injection (RRI), 489
reverse Telnet, 96
risk, analyzing, 11
roles/responsibilities, security policies, 16
ROM monitor mode, 65, 73
root guard, STP, 290
routers
ACLs, 221–222
certificate-based IP, 228
configuring, 228–230
extended IP, 226
reflexive IP, 226–227
standard IP, 223–226
time-based IP, 227–228
advanced support, 248
attack target, 49
authentication
CA support, 438–439
neighbor routers, 72, 76
peer routers, 247
proxy, 585
routing protocols, 233–234
border routers, 222
browsers, controlling access via, 231
CAs, 438–441
CBAC, 585–586
CDP, 235
Cisco IOS
Firewall, 61, 245
router, 61
CLI, 68–70, 74
configuration
process, 558–560
modes, 65–68, 73

directed broadcasts, disabling, 232–233
Easy VPN, 491
finger service, 235
global configuration mode, 66, 73
in/out traffic, 226
interface configuration mode, 66
IPsec VPNs, 437, 440–450
line configuration mode, 66
memory, 265
modes
    global configuration, 66, 73
    interface configuration, 66
    line configuration, 66
    privileged EXEC, 65, 73
    ROM monitor, 65, 73
    user EXEC, 65, 73
NTP, 235
peer routers, authenticating, 247
preshared keys, 570–571
privileged EXEC mode, 65, 73
ROM monitor mode, 65, 73
routing protocol, authenticating, 233–234
securing, 558–560
small server services, 234
SNMP, 230–231
SSH, 94–96, 69, 560
support, advanced, 248
TCP services, 234
Telnet, 32, 68
UDP services, 234
user EXEC mode, 65, 73
routing tables, 233
RRI (reverse route injection), 489
RSA (Rivest, Shamir, and Adelman) keys
    deleting, 443
    encrypted nonces, 405, 423–427
    overview, 95, 439
    show commands, 465–466
    signatures, 405, 448, 575–576
RTP (Real-Time Transport Protocol), 316
RTSP (Real-Time Streaming Protocol), 315–316

S

SAFE, 54
SAs (security associations)
    clearing, 581
    counters, resetting, 474
    IKE, 583
    IPsec, 416–417, 462, 574–583
    transform sets, 576

scan attacks, 38
SCEP (Simple Certificate Enrollment Protocol), 439–440
SCP (Secure Copy Protocol), 94–96
script kiddies, 34, 42
SDEE (Security Device Event Exchange), 263
SDF (signature definition file), 260, 266
SDM (Security Device Manager)
    advanced options, 538–540
    audits, 534–536
    browsers, 510
    configuring
        DMVPN, 530–534
        Easy VPN Remote, 525–526
        Easy VPN Server, 528–530
        site-to-site, 521–525
    Easy VPN
        Remote, 525–526, 597–599
        Server, 528–530, 593–596
    firewalls, 515–521, 541
    hardware, 509
    intrusion prevention, 538, 586–592
    logging, 544
    monitor mode, 541, 544
    NAT, 538
    OS requirements, 509
    options, advanced 538–540
    overview, 263, 505, 509
    QoS, 538, 541
    routing configuration, 538
    security audits, 534–536
    smart wizards, 509
    software, 510–511
    user interface, 511–513
    wizards
        Firewall, 515–521
        Factory Reset, 536–537
        LAN, 513
        Security Audit, 534–536
        smart wizards, 509
Secure Copy Protocol (SCP), 94–96
Secure Hash Algorithm 1 (SHA-1), 406
Secure HTTP (HTTPS), 232
Secure Shell
    See SSH
Secure Sockets Layer (SSL), 232
Security
    associations, *See* SAs
    audits, 509, 534–536

events, 10
policies, 10–20
posture assessments (SPAs), 18
servers, remote, 112–120
wheel, 18, 54
zones, 298
See also attacks
Security Audit Wizard, SDM, 534–536
Security Device Event Exchange (SDEE), 263
Security Device Manager
See SDM
segmentation, 51, 55
session hijacking attacks, 93, 297
SHA-1 (Secure Hash Algorithm), 406
show commands
802.1x authentication, 387–389
AAA, troubleshooting, 142–145
CBAC, 323–324
overview, 270
privilege levels, 89
VPN connectivity
show crypto ca certificates, 460
show crypto ipsec sa, 462
show crypto ipsec security-assocation lifetime, 463
show crypto ipsec transform-set, 464
show crypto isakmp key, 464
show crypto isakmp policy, 461
show crypto key mypubkey rsa, 466
show crypto key pubkey-chain rsa, 465
show crypto map, 464
shutdown
actions, 286
mode, 97
signatures
atomic, 260
attack, 259
Attack-drop.sdf, 260
compound, 260
definition file, *See* SDF
IPSs, 53, 260–261
false positive/negative, 267
micro-engine, *See* SME
memory consumption, 261
merging files, 266
RSA keys, 405, 448, 575–576
SDF, 260, 266

unsupported, 261
VPNs, 437
Simple Certificate Enrollment Protocol (SCEP), 439–440
Site Security Handbook, 9, 19
site-to-site VPNs, 401, 521–525
smart wizards, SDM, 509
SME (signature micro-engine), 260
SMIL (Synchronized Multimedia Integration Language), 316
Smurf attacks, 232
SNMP (Simple Network Management Protocol)
routers, 230–231
traps, 92, 98
snooping, DHCP, 292–293, 562
soft tokens, 111
SPAs (security posture assessments), 18
SPAN (Switched Port Analyzer) ports, 98
Spanning Tree Protocol (STP)
BPDU guard, 290
manipulation attacks, 289–291
parameters, 562
root guard, 290
split tunneling, 486
spoofing, 93, 291–294
DNS spoofing, 93
IP spoofing, 93
man-in-the-middle attacks, 291–294
switches, 288
SSH (Secure Shell)
AutoSecure feature, 101
configuring
routers, 69
SSHv2 using host/domain name, 94
SSHv2 using RSA key pairs, 95–96
keys, 560
overview, 68, 87, 93–94
passwords, 92
reverse Telnet, 96
routers, 560
SCP, 94–96
setting up, 94
SSL (Secure Sockets Layer), 232
standard
ACLs, 71, 75
IP ACLs, 222–224, 226

starvation attacks, DHCP, 296
state, maintaining, 71
static/sticky secure MAC addresses, 286
STP (Spanning Tree Protocol)
    BPDU guard, 290
    manipulation attacks, 289–291
    parameters, 562
    root guard, 290
structured threats, 34
supplicants, 802.1x, 359
support, management, 14
sweep attacks, 38
Switched Port Analyzer (SPAN)
    ports, 98
switches
    attack target, 49
    attacks, 279
    ports, 98, 381
    security, 298
    spoofing, 288
    SSH, 94–96
SYN floods, 71, 310
Synchronized Multimedia Integration Language (SMIL), 316
syslog servers, 249
system
    auditing, 71, 75
    management, 10
    port mapping, 320

T

TACACS (Terminal Access Controller Access Control System), 112–114
TACACS+,
    accounting, 157
    authentication, 155–156
    authentication proxy, 344–347
    authorization, 156
    Cisco Secure ACS, 344–347
    configuring, 153–154
    firewalls, 72, 75
    overview, 113–114, 122, 169
    RADIUS, versus, 116
    testing/troubleshooting, 157–158
tagging, 289
targets, attacks, 49
TCP
    half-open sessions, 314
    inspection, CBAC, 322
    intercept, 71, 75
    services, 234
    SYN floods, 71, 310
    three-way handshake, 310
TCP/IP, 33
Telnet
    CLI, 68
    connections, 68, 74
    default, 32
    interactive access, 91
    passwords, 92, 112
    reverse Telnet, 96
    routers, 32, 68
Terminal Access Controller Access Control System
    See TACACS/TACACS+
terminal lines, 111–112
TFTP servers
    enable secret command, 87
    MD5 encryption, 88
threats, 34, 41
    See also attacks
three-way handshake, 310
thresholds, 319
throughput, 317
timeouts
    CBAC, 319
    vts, 92
TLS (Transport Layer Security), 232
token cards, 111
traffic, interesting 417
transform sets
    Easy VPN, 493–494
    IKE, 580
    IPsec, 415–416
    overview, 410, 464, 574–576
transparent mode, 408
Transport Layer Security (TLS), 232
transport protocols, 116
trending, 53, 56
Triple Data Encryption Standard (3DES), 406
Trojan horses, 39
trunk ports, 98, 561
ttys (physical terminal lines), 91
tunnels, 316, 389, 408

U

UCP (User Changeable Passwords), 194
UDP
    CBAC
        inspection, 322
        sessions, 313
    half-open sessions, 314
    services, 234
unauthorized state, 802.1x, 362
Unity Protocol, 485
unstructured threats, 34
    See also *attacks*
URLs, filtering, 247, 251
usage policy statements, 9
User Changeable Passwords (UCP) feature, 194
user EXEC mode, 65, 73, 84
usernames, 111–112
users
    account policies, 10
    group security, 298
    interface, SDM, 511, 513
    filtering, URLs, 251
    port mapping, 320

V

V3PNs (Voice and Video over VPNs), 248
VACLs (VLAN ACLs), 296
varsdb.MDB file, 192
vertical scan attacks, 38
violations, security policies, 17
Virtual Terminal Protocol (VTP), 562
virtual ttys
    See vtys
viruses, 39
VLAN ACLs (VACLs), 296
VLANs
    802.1x
        authentication, 385–386
        guest VLANs, 358
    hopping attacks, 288–289
    IBNS, 358
    ports, 294
    VACLs, 296
    voice ID, 358
    vulnerabilities, 294–295
VMS IDS (CiscoWorks VPN/Security Management Solution IDS Management Center), 263

Voice and Video over VPNs (V3PNs), 248
voice VLAN ID, 358
VoIP, 248
VSAs (vendor-specific attributes), 115
VTP (Virtual Terminal Protocol), 562
vtys (virtual ttys)
    passwords, 112
    securing access, 91–92
vulnerabilities, 29, 40–41
    See also attacks

W–X

warning banners, 90–91
weaknesses
    applications, 33
    hardware, 33
    network configuration, 31–32
    operating systems, 32
    passwords, 41
    protocol suites, 32
    security policies, 30–31
    TCP/IP, 33
    technology, 32–33
websites
    ACLs, certificate-based, 228
    authentication proxy, 337
    Kerberos, 117
    SAFE, 54
wheel, security, 18, 54
wizards
    Firewall, 515–521
    Factory Reset, 536–537
    LAN, 513
    Security Audit, 534–536
    smart wizards, 509
workstation configuration policy, 10
worms, 39
X.509 standard, 439
Xauth, 487, 496, 526, 530